WATERSHEDS: PROCESSES, ASSESSMENT, AND MANAGEMENT

WATERSHEDS: PROCESSES, ASSESSMENT, AND MANAGEMENT

PAUL A. DeBARRY

WILEY

JOHN WILEY & SONS, INC.

Library of Congress Cataloging-in-Publication Data:
DeBarry, Paul A.
    Watersheds : processes, assessment, and management / Paul
DeBarry.
        p.   cm.
    Includes bibliographical references and index.
    ISBN 0-471-26423-7 (cloth)
    1. Watersheds. I. Title.
    TC409D35 2004
    627—dc22   2003017778

Printed in the United States of America

10 9 8 7

CONTENTS

PREFACE

Management of our water resources on a comprehensive watershed basis is perhaps the most popular environmental topic today. Currently, there exists no comprehensive source to guide a user through a watershed assessment, planning, and management project. Many books are available on the individual scientific fields that are required for watershed assessment, such as hydrology, hydraulics, aquatic biology, fluvial geomorphology (FGM), aquatic biology, and groundwater. However, in researching the literature, I could not find a single book that provides a comprehensive approach to watershed management from process to planning and that includes most facets of the comprehensive watershed management approach: stormwater management, erosion and sediment pollution control, floodplain management, water quality management, groundwater management, stream restoration, fisheries resource management, and land use management. This book is intended to be that missing link. It presents a comprehensive approach for analyzing watersheds. That said, it would be very difficult, if not impossible, to include everything on the topic, as the wealth of data is overwhelming.

Comprehensive watershed management requires an interdisciplinary approach to solve complex problems. Complicating the effort, however, is the fact that the various disciplines frequently use different definitions to describe the same thing, or classify a physical feature or phenomenon to suit their particular needs. Therefore, one of the goals for this book is to provide many of the terms, definitions, and/or classifications in one reference, to allow readers to see the similarities and differences and, ideally, to encourage interdisciplinary consistency. The book also aims to aid in the process of coordinating and standardizing the various methods and classifications, and to define the terms used by each of the scientific disciplines that utilize watershed management principles, in order to provide for future planning consistency. That said, it's important to point out that, although correlations between the various classifications are highlighted, there are always exceptions to the rule; hence, each situation should be evaluated in the field.

I was fortunate to know at a young age that I wanted to work in watershed management. I was also fortunate to be able to take many courses related to watershed management in pursuit of my B.S. degree from Penn State, including biology, chemistry, agronomy, hydrology, hydrogeology, geology, and engineering. The knowledge I gained from these courses provided me with the many facets to piece together a comprehensive understanding of the holistic watershed management approach. My love for the science gave me the impetus to continue in civil engineering, where I specialized in hydraulics, hydrology, and computer modeling; further, my experience over the past 20 years, working in many aspects of watershed management, has allowed me to collaborate with many individuals with specialized expertise. For their transfer of knowledge, I am truly grateful.

I have found, for example, that engineers don't always fully understand the biological issues facing them; conversely, biologists don't always fully appreciate engineering issues. Moreover, professionals don't always have a reliable reference source in which to find the answers to questions they encounter outside of their fields. This book will serve as a reliable source for engineers, conservationists, biologists, planners, and organizations alike, which they can use while developing a tailored comprehensive watershed plan.

I have also found, in developing watershed plans with the aid of students and young professionals, that through their coursework they have obtained bits and pieces of the various components required to put together a watershed plan. However, they lack the understanding of how the processes work together versus how they should work together. For instance, they may have taken several courses in geology, but are lacking study in soils; or they have studied fluvial geomorphology without a background in hydrology and hydraulics. This book is intended as a capstone to the majority of the disciplines required for watershed planning and management.

The greater purpose of this book, then, is to go beyond theory, to apply planning, management, and hydrologic engineering principles to practical applications in comprehensive watershed management. This text promotes the integrated approach to watershed management—that is, integrating the various disciplines to develop comprehensive assessment and management of water resources in our watershed—therefore, theoretically, it could have been written as one long comprehensive chapter. But for the ease of reading, subjects are divided logically into three parts and 23 chapters. However, readers are encouraged to think holistically as they proceed, about how each subject interrelates.

- Part A, "Watershed Processes," focuses on the scientific relationships among the physical features of the watershed.

- Part B, "Watershed Assessment," details the assessment process, featuring discussions on assembling the watershed team, Geographic Information Systems (GIS), available GIS data, and data collection, including sampling.
- Part C, "Watershed Management," concentrates on the solutions. It addresses the regulatory framework, tools available to aid in watershed analysis such as GIS and models, and development of a watershed plan, with a strong emphasis on GIS analysis for water resources management. The development of a holistic watershed plan process is also discussed. This part also covers various avenues for implementation of a watershed plan, various structural and nonstructural techniques for management in a particular watershed, and implementing best management practices (BMPs) for water quality control.

The goals and objectives process explained in this book will aid the practitioner in identifying what is important in the watershed being studied. Readers must keep in mind that, although there are usually one or two driving issues in a watershed, usually caused by some human-induced problem, the interaction among the various natural and man-made functions warrants a comprehensive look at long-term, holistic, watershed management to achieve sustainable water and land resources.

Watersheds: Processes, Assessment, and Management is unique in that it explores certain watershed management concepts not yet presented on an international level, but that are much needed and widely applicable at that level. It compares various management alternatives in order to evaluate the takings issue. It provides a concise reference manual for engineers, planners, administrators, and citizens' groups, to aid in evaluation of proper watershed analysis, planning, and management procedures. And because it takes a comprehensive approach to covering the topics required for watershed assessment, the book does not attempt to explain the topics in exhaustive detail, but rather to provide readers with enough information to enable them to understand the process and to point them in the right direction so that they can obtain more detailed information if required.

The intended audience for *Watersheds* includes students, professors, and practitioners with environmental science, natural resources engineering, and planning backgrounds. Many universities, natural resources agencies, nongovernmental organizations (NGOs), and professionals will also find the book useful in their work. The book incorporates the holistic approach to watershed management with a focus on state-of-the-art technology. Special features include the latest computer models and 3-D analysis with GIS. Readers will benefit from the ability to take a watershed assessment and management project from start to finish utilizing the most up-to-date modeling and GIS tools. For this reason, readers will note that many of the references cited in this text are "official" Web pages; the Internet is the only medium that can keep pace with the rapidly advancing technology in this field.

As I wrote the book, I realized that attempting to incorporate all my ideas and concepts was even more challenging than I anticipated. It could have been, like the hydrologic cycle, a never-ending process, as one idea flowed into another. Sciences and concepts to improve watershed management are continually being explored, refined, and developed, and I take off my hat to all those scientists out there with a vision to refine science to conserve our natural resources. Even as I was writing, methodologies, procedure protocols, and storage retrieval Web sites were being refined, improved, and updated. Probably, this will necessitate a very early second edition.

Finally, I would like to point out that much of my work has been on the East Coast, and that many of the principles mentioned herein represent a similar geographical experience. I feel that this is an advantage, because development pressures and population demands in this part of the country have forced proactive planning, lessons of which can be utilized elsewhere as growth continues around the world. Thus, I would be very interested in hearing from readers from other parts of the world, so that I may incorporate their experiences into the next edition. I welcome comments and suggestions for the second edition, which can be submitted to *watersheds@hotmail.com.* Your interest in this book and in watershed management is appreciated.

Additional material from this book is available on the publisher's website at www.wiley.com.

ACKNOWLEDGMENTS

I would like to thank, first of all, my father and mother, who instilled in me the appreciation for conservation of the land and water for use by our children's children, and for always supporting me in everything I've done.

I would also like to thank my immediate family, wife, and children for their support and understanding through the long hours of writing this book. I'm also indebted to Dr. David Kibler, Virginia Tech; Dr. Gert Aron, Professor Emeritus at the Pennsylvania State University; and Dr. Shaw Yu, University of Virginia, for transferring their knowledge and for always being supportive. I also thank all the engineers, planners, biologists, geologists, and citizens who opened different perspectives to me and from whom I learned a lot. I would especially like to acknowledge the staff at Wiley (who were a joy to work with), and the technical and editorial reviewers of this book, who took their personal time to aid in the transfer of this knowledge. They are:

Reviewer	Affiliation	Subject
Dr. Gert Aron, P.E.	Professor Emeritis, Penn State University	Hydrology, Stormwater
Chris Capelli	ESRI	GIS
Dr. Robert Cook, P.G.	Keystone College	Geology, Hydrogeology
Christina Cianfrani, PhD Candidate	University of Vermont	Ecology
Dr. Jeri Jewett-Smith	East Stroudsburg University	Ecology

Gerry Longenecker, P.E.	Skelly-Loy	FGM
Orianna Roth Richards	Monroe County Conservation District	Editorial
Ronald Thompson	U.S. Geological Survey	Hydrology, Stream Data
Dr. Robert Traver, P.E.	Villanova University	Water Quality, Stormwater
Edward White, CPSS, ARCPACS	State Soil Scientist, Natural Resources Conservation Service	Soils/Digital Data

CHAPTER 1

INTRODUCTION: COMPREHENSIVE WATERSHED ASSESSMENT AND MANAGEMENT

1.0 INTRODUCTION

A *watershed* is an area of land that captures water in any form, such as rain, snow, or dew, and drains it to a common water body, i.e., stream, river, or lake. All land is part of the watershed of some creek, stream, river, or lake. The watershed boundary is defined by the higher elevations or ridges that define which direction the rainwater will flow, as shown in Figures 1.1 and 1.2. It is analogous to a bathtub, where water that falls on the inside brim of the bathtub, or watershed, will flow to the drain, or outlet. Water that falls outside the brim, or watershed, divides, and ends up on the floor—in the case of a bathtub—or in another watershed. The entire continental landmass is made up of watersheds.

The downstream-most point of a watershed is defined by the required point of analysis, or where flows, samples, or design criteria might be required. These are usually stream *confluences* (where two streams merge), bridges, problem areas, dams, or some other type of outlet where the analysis ends. The downstream-most location is referred to as the *point of interest* for analysis purposes.

Delineating the watershed begins by identifying the point of interest, then drawing a line perpendicular to the contours, picking the high points on a topographic map, and continuing until returning back to the point of interest. A typical watershed, as defined on a United States Geological Survey (USGS) 7½-minute topographic quadrangle, is shown in Figure 1.3.

A *watershed assessment* is a detailed evaluation of the specific processes, influences, and problems in a watershed so that a plan of action to preserve the watershed can be developed. The watershed management plan should be

Figure 1.1   A watershed boundary is defined by the higher elevations or ridges that direct the flow in one direction or another, as shown from the ground.

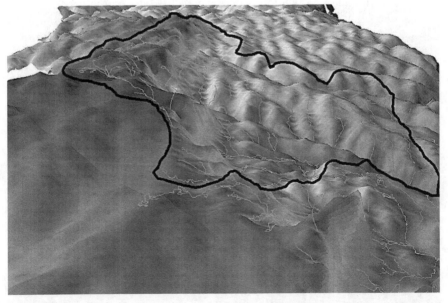

Figure 1.2   A watershed boundary as shown in a GIS digital elevation model (DEM).

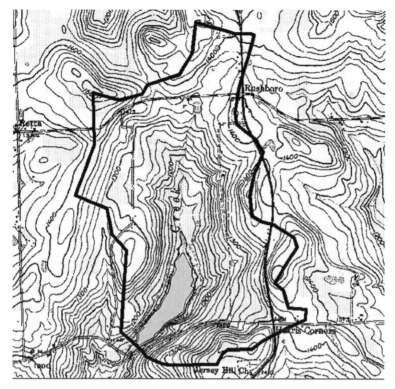

Figure 1.3   Typical watershed defined by high points or ridges, draining to the point-of-interest, in this case the dam of a lake.

a systematic approach to preserve or restore the watershed and its hydrologic regime, or to correct problems based on the comprehensive assessment. A detailed implementation program should be a component of the management plan.

1.1 SUSTAINABILITY

Sustainability can be defined as "creating, to pass along to our children and to theirs, a natural resource base whose yields in economic prosperity, social improvement, environmental quality, and natural beauty will go on and on—tomorrow and forever—because of the political choices we are willing to make today" (Bucks County Planning Commission, 2002). As seen throughout this text, a watershed in hydrologic equilibrium should be the goal for sustainability of water resources and for the health, safety, and welfare of humans in general. Lives depend upon it, thus water resource sustainability is paramount for the survival of human beings. Sustainability can be achieved through "good science and smart planning" (Schaffhausen, 2002).

1.2   WATERSHED ASSESSMENT AND MANAGEMENT

Each watershed has a unique personality that needs to be explored to develop a truly personalized management plan. For instance, the Rio Santa Catarina watershed evaluated in Nueva Leon, near Monterrey, Mexico, has an elevation relief from 3,100 to 1,420 meters or 1,680 vertical meters within 6,500 meters horizontal, for an average slope of twenty-five (25) percent and hard desert soils with very little infiltration capability. The result: flash flooding with loss of life during Gulf Coast storm events. The urban Darby Creek watershed near Philadelphia, Pennsylvania, suffered flash flooding, reduced base flow, and water quality problems due to urbanization and the increase in impervious area. In contrast, the Wysox Creek watershed in rural Bradford County, Pennsylvania, had stream bank erosion and agricultural nonpoint source pollution problems due to stream banks comprised of *unconsolidated glacial deposits* and mismanaged cattle grazing. The Solomon Creek in Wilkes-Barre, Pennsylvania, watershed had a severe acid mine drainage (AMD) problem due to past mining activities. As a recommendation of the assessment, funds were procured to design and install an AMD treatment facility. The analogy of a doctor seeing a patient for the first time can be used, whereby the doctor first evaluates the individual's entire body before concentrating on the symptoms and then recommending a remedy, as shown in Figure 1.4.

A watershed is like an interdisciplinary puzzle; that is, the watershed assessment collects the biological, physiographic, hydrologic, hydraulic, political and social pieces of the puzzle and the management plan puts all the pieces of the puzzle together. The puzzle can fall apart, however, unless it is laminated and preserved in a frame for long-term enjoyment, hence the implementation phase. Perhaps the most important phase of a project is the implementation phase, thus it must be in the preparer's mind throughout the project, for if the plan cannot be implemented, it will become a typical report collecting dust on the shelf. The best watershed management plans are those whose covers are the most worn.

1.3   COMPREHENSIVE WATERSHED MANAGEMENT CONCEPTS

Why manage our land and water resources on a watershed-wide basis? Because watersheds are formed by natural land masses and water flows into a common waterbody. In other words, watersheds are defined by natural hydrology. Streams and rivers do not follow political boundaries, and the flow of water, pollution, problems, etc. does not stop at political boundaries. In addition, managing the whole is better than managing or correcting the sum of its parts.

The Federal Emergency Management Agency (FEMA) used to conduct Flood Insurance on a municipality-by-municipality basis. Unobstructed 100-year flood flows were reported less in downstream communities than those

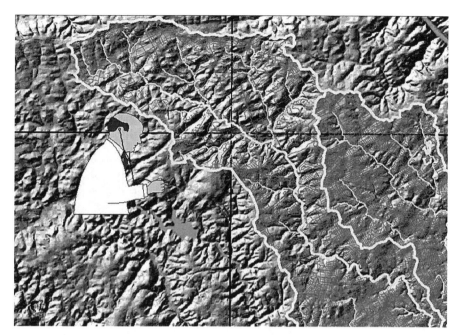

Figure 1.4 A watershed assessment involves closely examining the physical features of the watershed, determining what ails it, and prescribing a prescription plan to improve its health.

upstream, a concept we know not to be practical. This was caused by the studies being performed at separate points in time and by different consultants. It is good to see that FEMA is now conducting flood insurance studies on a watershed basis. Studies performed on a watershed basis allow efficiency in data collection, monitoring, report writing, grant appropriation, and eventual permitting. Managing on a watershed basis provides consistency amongst municipalities within a watershed and takes into account how changes in one portion of the watershed may affect it in another area.

The watershed approach is a coordinating framework for environmental management that focuses on public- and private-sector efforts to address the highest-priority problems within hydrologically defined geographic areas, taking into consideration both ground and surface water.

Watershed management can be undertaken in two ways: the proactive approach and the reactive approach. Humans typically react to floods, water quality problems, and stream bank erosion. Once an event or series of events creates a problem, then people react to fix the problem they created. People build in the floodplain, then build levees to protect their buildings. A flood greater than the levee capacity occurs, and the government pays for the cleanup and to make the levees higher. However, as we are slowly learning, taking a proactive approach—performing a watershed assessment and putting a watershed management plan in place to strive to maintain the natural hydrologic regime—prevents flooding, maintains groundwater quantity and quality, maintains stream flow and quality, prevents stream bank erosion, preserves environmentally sensitive areas, and so on.

Natural, undisturbed watersheds are in hydrologic, hydraulic, and geologic equilibrium. As humankind alters the land surface, the landform, hydrologic budget, and stream processes are ultimately affected. Improper use of our natural resources causes a number of problems, including flooding, erosion and sedimentation, stream bank erosion, water quality problems, and reduction of groundwater and base flow augmentation. The goal of any watershed management plan should therefore be to maintain the hydrologic budget. In order to properly manage a watershed, the comprehensive picture or holistic approach must be followed. This book summarizes the physical features that must be analyzed in order to accomplish true comprehensive watershed and ultimately stormwater management.

Ensuring sustainable water resources requires comprehensive management of the many facets of water, including water supply (i.e., groundwater and surface water), stormwater management, flood control, nonpoint pollution control, and wastewater treatment and reuse. Water resources management begins with understanding the various sources, paths and uses of surface and groundwater, stormwater, floodwaters, recreational waters, drinking water, irrigation water, and so on. Generally, surface water includes rivers, streams, reservoirs, lakes, and ponds. Groundwater can be classified as shallow and deep. Water uses include municipal, industrial, and commercial uses, and residential wells and springs. Human water uses include irrigation, manufacturing, recreation, and consumption. An adequate supply of clean water is essential for maintaining the quality and health of natural ecosystems such as fisheries, forests, wetlands, and aquatic habitats.

Water resources management is a broad and wide-ranging effort that encompasses activities such as identifying and delineating source water protection areas, minimizing discharges, and managing stormwater. Zoning and land use regulations and growth management techniques are effective mechanisms for directing development to areas that can best support it. Using a watershed-based approach further ensures that down gradient areas are not adversely impacted.

In order to develop a comprehensive watershed management plan, a number of steps must be followed. Most important, a comprehensive analysis of all of the physical features of the watershed should be conducted. These features include geology, soils, topography or slopes, stream channel sections, floodplains, and wetlands. In addition to the physical feature parameters, socioeconomic and political parameters should also be considered. Once these physical feature, socioeconomic, and political parameters are analyzed, first individually, then as a connected whole, the goals and objectives of each individual watershed management plan can be developed. For instance, in southeastern Pennsylvania, long-term groundwater supply is a crucial physical, socioeconomic, and political issue. Development of a plan for southeastern Pennsylvania may concentrate on replenishing the groundwater by recharging stormwater. The following checklist itemizes factors to consider when developing a comprehensive watershed management plan. The list is not exhaustive; other items can be added to address the physical characteristics, issues, and goals within the study area.

Stormwater management	Urban sprawl
Stormwater-related problems	Riparian buffers
Floodplain management	Lumbering activities
Flood control	Agriculture activities
Hydrologic regime	Bridge capacities/obstructions
Wellhead protection areas	Existing flood control facilities
Sinkholes	Agricultural concerns
Regional facilities	Citizen concerns
Wildlife habitat management	Water quality
Wetlands preservation	Nonpoint source pollution
Invasive species	Total maximum daily loads (TMDLs)
Fluvial geomorphology assessment	Geology
Natural channel restoration	Limestone
Stream bank erosion protection	Water supply areas
Erosion/sedimentation problems	Base flow augmentation
Sustainable development	Steep slopes
Conservation planning	Gravel bars
Public health, safety, and welfare	Cost/benefits
Infill/redevelopment	Other
Urbanization	

The items in this list, which have been incorporated into seven major classifications in the following sections, may dictate the goals and objectives of the watershed plan. Determining the primary goals and objectives of a watershed management plan is one of the first steps in developing a watershed assessment and plan. Geographic Information Systems (GIS) is an efficient tool to aid in developing a comprehensive watershed management plan.

1.3.1   Stormwater Management: Maintain the Hydrologic Regime

As development occurs in a watershed, the amount of infiltration/recharge decreases while the runoff generated from increased impervious areas increases the peak and volume of stormwater runoff. Since stormwater runoff is a major component of the hydrologic cycle, increasing the runoff has a major impact on the hydrology, biology, chemistry, and other physical features of the watershed. Therefore, the stormwater management plan must be an integral part of any watershed management plan. Developing one is complex and involves detailed hydrologic modeling and development of standards and criteria for new development, as will be discussed in Chapters 12, 18, and 19.

The ultimate stormwater management scenario to achieve zero runoff would be to maintain the existing hydrologic regime, recharging that which originally infiltrated to maintain both the existing peak *and* volume. Although in the majority of the situations this may not be practically feasible, keeping this goal in mind will aid in the prevention of stormwater problems. Another goal of comprehensive stormwater management should be to coordinate stormwater management, erosion and sedimentation control, and water quality control for best management practices (BMPs).

1.3.2 Groundwater Recharge

As mentioned in the introduction, if base flow augmentation is a priority, the plan should encourage groundwater recharge by providing relevant standards and criteria. This could include such measures as infiltration and recharge structures, porous pavement, and impervious surface reduction. However, caution should be exercised in limestone areas so as not to create sinkholes or provide a direct conduit to groundwater reserves.

Maintaining groundwater *quality* is another factor that must be considered when developing a plan. Evaluating the pollution vulnerability of specific sites would be imperative in proposing recharge structures. GIS can be utilized in a number of applications, including development of pollution vulnerability mapping from physical feature overlays (Reese and DeBarry, 1994). The features that would affect the vulnerability of groundwater to potential contamination—geology (fractures, limestone, etc.), soils (permeability), land slope, and streams (inflow or outflow)—can be overlaid to determine pollution vulnerability. A pollution vulnerability map can be utilized not only to aid in evaluating the pollution potential of recharge structures but also for new well siting, developing emergency spill response plans, and transportation planning (by overlaying major commercial highways).

Another facet that should be considered in development of watershed stormwater management plans in critical water supply areas are individual well recharge areas or zones of influence. These land areas could be mapped into the GIS from MODFLOW or related groundwater programs, and specific recharge evaluation criteria developed for these areas.

1.3.3 Surface Water Quality

Maintaining or improving water quality should always be a goal of any storm-watershed management plan because of long-term implications. With an increased emphasis on managing the quality of stormwater runoff and controlling nonpoint source pollution through the use of BMPs, the use of GIS can again provide watershed managers with a significant tool to analyze pollutant loads and target priority management areas. As early as 1991, GIS was used to compute nonpoint source pollution loads on a subwatershed basis utilizing the unit aerial loading approach based upon National Urban Research Project (NURP) and local area land use/loading data (DeBarry, 1991). Those subwatersheds contributing the most pollution for existing and proposed conditions could then be displayed graphically in the GIS to develop priority management areas or require BMPs in those areas. A similar approach using grids to determine pollutant loads by multiplying event mean concentrations (EMCs) by the expected runoff also utilizes GIS (Quenzer and Maidment, 1998). Water quality models have also been programmed into ArcView (Quenzer and Maidment, 1998). The Environmental Protection Agency's (EPA) total maximum daily load (TMDL) program, established under Section 303(d) of the Clean Water Act, is a written, quantitative assessment of water quality problems and contributing pollutant sources, and should be coordinated with any watershed management plan.

1.3.4 Flood Control

The objective of most watershed plans is to prevent flooding, whether at a local drainage level or on a regional stream level. The past history of flooding on a local and regional level should be determined, and the existing and potential development in a floodplain should be identified. Where there will be major damage in a floodplain during a specific flooding event, a watershed plan should manage that storm to reduce or prevent further exacerbating flooding problems. In typical floodplains, this may include placing design criteria for stormwater detention on new development for the larger storms— that is, the 50- and 100-year—as well as the smaller storms. However, where flood damage is not a problem due to lack of development in floodplains, the emphasis should be on prevention. In this case, it may be more prudent to concentrate on the smaller design storms—that is, the 2-, 5-, and 10-year storms—than to detain the 100-year storm. This objective can be further accomplished by placing tighter restrictions on developing in the floodplains, even preventing development in floodplains. Regional stormwater management facilities should also be considered to maintain the hydrologic regime of the watershed.

1.3.5 Stream Bank Erosion

In one particular study, through input from concerned citizens, it was found that a stream was causing property damage because it had very erodible stream banks. Studies have shown that stream banks begin to erode when the velocities in the stream reach the "critical velocity" (DeBarry and Stolinas, 1994). This velocity has been found to correspond to approximately the bank full flow or the one- and one-half-year storm. Theoretically, preventing the flows in the stream from ever reaching this threshold would prevent the stream banks from eroding. Naturally, this is impossible to achieve, but if the watershed management plan incorporates reducing the peak flows from development to these smaller (1.5-year) storms through the use of a subregional detention facility, for instance, the frequency of stream bank erosion would be reduced. Detention of the larger storms would typically be in the overbank areas and would not significantly contribute to the erosion in the stream. However, in implementing such a scheme, care should be taken to ensure that one is not increasing the duration of flows above the critical value for the larger storms.

1.3.6   Conservation Planning (Nonstructural Stormwater Management) Master Planning: Opportunity and Constraints

One of the most effective means of stormwater management is through open space conservation planning. By minimizing impervious areas and concentrating development in those areas most suitable for development, runoff can be minimized.

Large- or small-scale master planning can be developed utilizing the GIS. By identifying areas that are suitable for development and, conversely, those areas that are most sensitive, a development approach that best conserves the environment can be undertaken. Mapping development constraints (wetlands, floodplains, steep slopes, historic structures, critical habitat, sinkholes, erodible soils, etc.), and then overlaying each of these within the GIS, will conversely display those areas most suitable for development. By preserving the conservation areas and developing only in those areas most suitable for development, the "cluster" method of development occurs. Thus, many objectives for sound stormwater management can be achieved, such as floodplain preservation, impervious surface reduction, determining the best location for stormwater management measures, and preservation of natural drainage patterns.

1.3.7   Habitat Identification/Preservation

Habitat preservation should be a prime concern in any watershed management program. Identifying stream buffers, wetlands, and even prime trout habitat through geomorphologic techniques should be performed. The use of GIS can aid in this evaluation. Once identified through the process outlined in the previous section on conservation planning, the preservation/management of each area can be accomplished. As can be seen by this example, this process should be coordinated with the conservation planning and stream bank erosion strategies mentioned previously.

1.4   POLITICAL VERSUS NATURAL RESOURCE MANAGEMENT

Management of our land and water resources in the past has been based mostly on areas defined by political boundaries, and proper water resource management can be accomplished only by evaluating the comprehensive picture. However, as will be explained in this text, our land and water resources are not separated by political boundaries. Land and water resources are integrated and are divided by drainage areas, and ground- and surface waters are interconnected. A watershed is a natural resource management unit; therefore, for a sustainable future, land and water resources must be managed on a watershed basis, which includes an understanding and coordination of surface and groundwater systems, reservoirs and aquifers, point and nonpoint source pollution, wetlands and uplands, wastewater and drinking water, lakes and streams, and physical, biological, and chemical characteristics of water. Physical characteristics would include parameters such as temperature, flow, mixing, and habitat. Biological characteristics would include the health and integrity of biotic communities; chemical characteristics would include ambient conditions as well as pollutants.

Water resources are increasingly being addressed at the watershed level instead of only at the political boundary level. When watersheds cross political boundaries, land use regulations need to be consistent across borders to ensure that upstream land and water uses in one jurisdiction do not conflict or adversely impact water quality and quantity in downstream jurisdictions.

Regulations also tend to follow various disciplines. For instance, there are individual regulations relating to water for flood control, stormwater management, erosion and sediment pollution control, point source discharges, nonpoint source discharges, groundwater withdrawal, and drinking water supply. As we will see, water is water; it is all connected through the hydrologic cycle. So why, then, are different aspects of water resources regulated differently? Our lack of integrated water resources management is illustrated by state regulations that require *waste water disposal drain fields* to be a minimum of 100 feet from existing wells, but don't require that wells be drilled at least 100 feet from *waste water disposal drain fields.* Theoretically, a well could be drilled right next to a drain field. Likewise, in the past, we have seen channelization or levee projects constructed for flood control that have so drastically altered the flow regime that they totally undermine the stream's capability to support a viable aquatic habitat, and biota was lost.

Water resources management requires cooperation between state, county, and local officials, and involves proper planning, engineering, construction, operation, and maintenance. This involves educating the public and local officials, program development, financing, revising policy, and development of workable criteria and adoption of ordinances. The goal of a watershed management plan should be to enable future development to occur within the watershed, while using both structural and nonstructural measures to properly manage water resources.

Regulations in the past have tended to be *reactive,* those passed due to an observed problem such as water pollution or flooding. In the future, hopefully, now that we better understand the sciences relating to watersheds and the integration between them, future regulations and policy can be developed to be *proactive,* putting into place measures to prevent problems from occurring.

One option for the protection of water resources is to incorporate the development of a standardized "water resources protection plan" for each new or increased land development or water withdrawal. Such a plan would incorporate all the existing water-related requirements such as stormwater management and floodplain management items, and it would include additional computations to balance land use with water budget. These plans would comprehensively describe specific performance requirements that, when implemented, would strive to ensure that the land development proposal does not adversely affect water resources.

By necessity, all policies, standards, and recommendations included in the watershed plan should be consistent with sound environmental planning and engineering practices and applicable laws, regulations, policies, and procedures in effect at the federal, regional, state, and municipal levels. Examples include best management practices for stormwater management, stream water quality standards, riparian protection areas, and wetland buffer standards.

Although a comprehensive watershed approach had been utilized in a variety of projects in early years, the Watershed Protection Approach (WPA) Framework was not formalized until 1991 by the U.S. EPA to meet the nation's water quality goals. It is encouraging to note that things are changing. Pennsylvania, for example, is drafting a new "comprehensive water management" policy and regulations that tie together many aspects of water issues to be regulated. The U.S. Army Corps of Engineers, which had been criticized in the past for destruction of aquatic ecosystems, is now employing biologists and fluvial geomorphologists and incorporating natural channel design and water quality concerns into its projects. Societies are now opening their doors to other disciplines, forming new "institutes" for multidisciplinary interaction through cosponsored conferences and workshops. For instance, the American Society of Civil Engineers has formed the Environmental and Water Resources Institute for just this reason.

The Federal Clean Water Act's National Pollutant Discharge Elimination System (NPDES) regulations utilize the WPA, and many states are now revising or developing new regulations to manage water resources on a watershed-wide basis, as opposed to political boundaries. Some states are also providing funding for watershed assessment and management plans. And, although plans are now being accomplished on a watershed basis, many programs still artificially separate groundwater from surface water, wastewater from "clean water," nonpoint sources of pollution from point sources of pollution, water supply from stormwater runoff, even though "water resources" is really one integrated, connected, continuous system.

Funding and grant programs are also set up accordingly, whereas certain grants, for instance, may have to address only nonpoint source pollution; and even though this pollution may be indirectly affecting a groundwater supply by recharging within the wellhead delineation zone, funding from a drinking water source program would not allow for the nonpoint source pollution assessment. Or a state stormwater management fund would not allow a nonpoint source pollution assessment because water quality was not specifically mentioned in the legislative document providing the funding. These are two real-world examples encountered when applying for funds for a comprehensive water resource approach. Ideally, state and federal funds relating to water should be combined to allow funding of comprehensive watershed assessment and management plan implementation. Legislatures, whose members typically do not have a water resources background, should be educated as to the economic and scientific advantages of performing the comprehensive water resources management approach.

The plan should be comprehensive, with the intent to present all information that may be required in order to implement the plan. It should cover legal, scientific, and municipal government topics, which, when combined, form the basis for development of model ordinance language that will be considered for adoption by each municipality. Sample stormwater management and wellhead protection ordinances should be incorporated into the plan.

Plan implementation will require working within existing regulatory programs and applying available regulatory, planning, and management tools to implement the plan at the municipal level. The intent of the plan should not be to limit growth but to provide a scientific approach for analysis of the water resources and to apply sound planning principles to implement the recommendations of the plan for the benefit of future generations.

EPA (1997) has developed a publication entitled "Top 10 Watershed Lessons Learned." In summary, they are:

- The best plans have clear visions, goals, and action items.
- Good leaders are committed and empower others.
- Having a coordinator at the watershed level is desirable.
- Environmental, economic, and social values are compatible.
- Plans only succeed if implemented.
- Partnerships equal power.
- Good tools are available.
- Measure, communicate, and account for progress.
- Education and involvement drive action.
- Build on small successes.

1.5 SUMMARY

The watershed is the framework from which the web of life is structured, from the microbe to the mouth of the watershed, with the common element being water, its lifeblood (Figure 1.5). Affecting even a small component, such as microorganisms or soil particles, causes a chain of events that could affect other components in the watershed. This text therefore reviews even

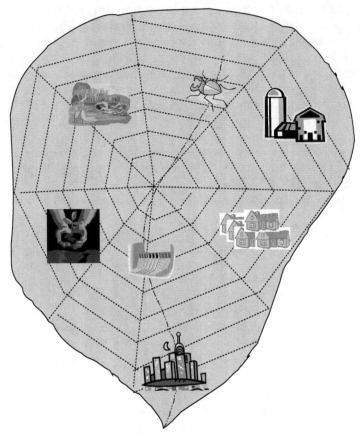

Figure 1.5 The watershed is the framework from which the web of life is supported.

the smallest component of the watershed, and the reader should keep in mind when reading those sections how these relate to the processes in one. The text concerns things as small as soil particles to things as large as the Mississippi River watershed.

Except in rare instances, for instance where a water company owns all the watershed land to their reservoir, a watershed cannot be managed by a single entity. However, one can manage the various parts that comprise the watershed: land use, lakes, stream banks, water withdrawals, and so on. Of these, land use change has the greatest impact on the watershed.

In order to develop a truly comprehensive watershed management plan, all facets of the watershed, including physical features and socioeconomic and political factors, must be considered. All of these factors should be analyzed individually and then combined to determine goals and objectives unique to the particular watershed being studied. This will allow watershed managers to better and more efficiently manage watersheds to address their particular concerns, whether it is stormwater management, floodplain management, water quality control, or conservation planning. Each factor will be mentioned in more detail in the following chapters.

The watershed should be divided into "management districts" based upon similar biological, chemical, hydrologic, hydraulic, land use, geologic, soils, and political and regulatory characteristics. Therefore, evaluating the ecosystem, physiographic provinces and other classification systems aids in developing these districts. GIS provides an efficient tool to aid in watershed planning and management. This book will expand upon the items described in this chapter and will discuss in detail the requirements for a successful, comprehensive watershed management plan.

REFERENCES

Bucks County Planning Commission. "Pennridge Area Water Resources Plan," Bucks County, PA, August 2002.

DeBarry, Paul A., and Raymond J. Stolinas Jr. "The Geography of a Rural Watershed: A Citizen/Consultant Approach to Solving Accelerated Stream Bank Erosion through Act 167," *Storm Water Runoff and Quality Management* (University Park, PA: Penn State University), 1994.

DeBarry, Paul A. "Comprehensive Stormwater Management: Checklist for Success." Proceedings from the Pennsylvania Stormwater Management Symposium, Villanova University, Villanova, PA, 1998.

———. "GIS Applications in Storm Water Management and Non-point Source Pollution," Proceedings of the ASCE National Conference on Hydraulic Engineering, Nashville, TN, 1991.

Quenzer, Ann M., and David R. Maidment. "Constituent Loads and Water Quality in the Corpus Christi Bay System." Proceedings of the ASCE Water Resources Engineering '98 Conference, Volume I, Edited by Steven R. Abt, Jayne Young-Pezeshk & Chester C. Watson, ASCE, Memphis, TN, pp 790–795, 1998.

Reese, Geoffery, and Paul A. DeBarry. "The GIS Landscape for Wellhead Protection." Proceedings of the ASCE National Hydraulics Conference, San Francisco, CA, 1993.

Schaffhausen, Eric. Direct quote from a conversation with the author, December 21st, 2001.

PART A:

WATERSHED PROCESSES

CHAPTER 2

PHYSIOGRAPHY

2.0 INTRODUCTION

In order to properly analyze a watershed's characteristics, one must be first familiar with the social, political, and physical environment within a watershed. The social environment includes the agendas of stakeholders, citizen's (watershed) associations, and educational opportunities. The political environment includes political jurisdiction boundaries and their political agendas, and regulations. These will be discussed later in the book. The physical environment includes climate, precipitation, geology, soils, hydrology, stream characteristics, topography, and slope. This chapter addresses the physical characteristics of the watershed.

2.1 PHYSIOGRAPHY AND PHYSIOGRAPHIC PROVINCES

Physiography refers to the physical features and changes in the earth. The land is divided into regions known as *physiographic provinces,* which are similar in topography, geology, and climate, three characteristics that affect the type of soils, flora, fauna, stream patterns, micro-climate, and even land use. Each makes its own contribution to the characteristics of the watershed, for example by releasing or storing nutrients and sediments. The physiographic provinces of the world can be seen in Figure 2.1.

2.2 GEOLOGY

In order to better appreciate the hydrologic and water quality characteristics of streams in a watershed, it is important to understand the physical makeup

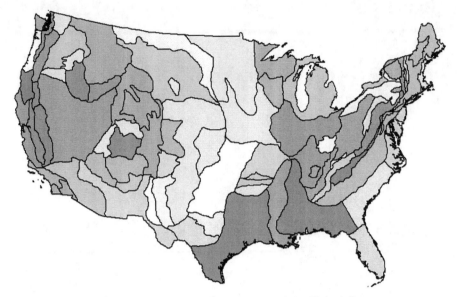

Figure 2.1 Physiographic provinces of the United States.

of the watershed starting from its origins, or bedrock geology. *Geology* is the science of the solid earth. *Bedrock* is the solid unweathered rock at the surface of the earth or just below superficial deposits of gravel, sand, or soil. The *lithosphere* is the solid mineral realm of our planet. The earth's crust is the thin outermost layer of this realm and averages about 10 miles thick. This is the source of soil and other sediment vital to life. The 11 most abundant elements with approximate percent composition of the earth's crust are shown in Table 2.1 (Strahler and Strahler, 1976). These elements are important due to their reaction to the environment and ability to influence the quality of

TABLE 2.1 Most Abundant Elements in Earth's Crust (*Source:* Strahler, 1976.)

Element	Percentage
Oxygen	47%
Silicon	28%
Aluminum	8.1%
Iron	5.0%
Calcium	3.6%
Sodium	2.8%
Potassium	2.6%
Magnesium	2.1%
Titanium	0.44%

water that comes into contact with them. For instance, stream flow in a limestone (calcium) area will have a higher pH than stream flow through a bedrock that contains a high amount of iron. *Lithology* is the description of rocks on the basis of such physical characteristics as manner of origin, composition, and texture. The fundamental map unit is the *formation. Stratigraphy* is the arrangement of rocks as classified by geographic position and chronological order.

The earth's crust is comprised of the inorganic mineral matter that contains many of the nutrients required for life processes. The composition of crystal rocks is the primary control of the composition of groundwater and surface water chemistry. Bedrock exposed at the earth's surface is the primary source of these nutrients. Bedrock is usually composed of two or more minerals. *Minerals* are naturally occurring inorganic solids possessing a definite chemical composition and characteristic atomic structure. The most common minerals are silicates such as quartz (silicon dioxide) and feldspars that contain potassium (K), aluminum (Al), calcium (Ca), and sodium (Na), in addition to silicate (Si) and oxygen (O_2).

Bedrock geology plays an important role in the hydrologic processes of watersheds. Soils are developed from the bedrock material. Therefore, the bedrock has a major influence on the soil properties and the hydrology of the watershed. The bedrock also plays an important role in the quality of groundwater and the transfer and flow of groundwater, and, in turn, stream base flow conditions. The various geologic factors and their impacts on watershed characteristics are described in the following sections.

2.2.1 Sedimentary Rocks

In order to fully understand the role of sedimentary rocks in watershed processes, particularly surface-groundwater interaction and the hydrologic balance, one must consider how they have formed, in geologic time. As streams erode their banks, meander, shift, and carry sediment, the particles are continually washed downstream into lakes, estuaries, bays, or oceans. When water velocities slow, the larger particles settle first. As the velocity continues to decrease due to a flatter slope (decreased hydraulic gradient), increased dimensions (depth, width, cross-sectional area), and position of the channel, smaller and smaller particles will continue to settle, forming stratified sediments. Sedimentary rocks are formed of compacted and cemented layers called *beds* of sediment (under extreme pressure) derived from the disintegration and decomposition of older rocks (such as igneous rocks) that have been weathered by water, wind, ice, or organisms. For this reason, sedimentary rocks are often layered, or stratified, into layers called *strata*. The arrangement of these overlapping beds is called *bedding*. Where individual beds meet each other, they form distinct planar surfaces between them called *bedding planes,* which affect groundwater movement.

Originally, beds were horizontal, but *tectonic* (uplifting) forces tilt, break, bend, and move these planes. The various layers, with their various grain sizes, will exhibit different porosity and permeability. *Folding* or *faulting* of the rock strata may fracture the bedding planes in the vertical or horizontal direction. This angle of displacement is extremely important in groundwater movement and often exerts control on drainage patterns. These fractures allow rapid groundwater movement, as shown in Figure 2.2. Sedimentary rocks cover approximately 75 percent of the earth's surface. These fracture planes can sometimes be almost vertical, as shown in Figure 2.3.

It is important in a watershed assessment to know which type of sedimentary rocks are prevalent. *Conglomerate rocks* consist of cemented sand, pebbles, and cobbles. Also important is the material from which the sedimentary rocks are formed. The most common sedimentary rocks are sandstone, siltstone, shale, and limestone. In geologic time, sand-sized sediments become *lithified* (turned to rock) into *sandstone,* silt into *siltstone,* clay into *mudstone* and *shale,* calcareous sediments into *limestone.* The chemical composition of the various layers may lead to variances in water quality. Sandstone, mudstone, siltstone, shale, and limestone constitute approximately 99 percent of the sedimentary rocks (Strahler and Strahler, 1976).

Sandstone is formed from sand-grade sediment cemented into a solid rock by calcium carbonate or silicon dioxide. Siltstone and shale are rocks formed from cemented silt and clay, respectively. The chief constituent of *limestone* is the mineral *calcite,* which is comprised primarily of *calcium carbonate* ($CaCO_3$). Limestone is formed from the sedimentation and compaction of the

Figure 2.2   Fractured shale bedrock with near horizontal planes.

Figure 2.3   Near vertical fracture planes in fractured shale bedrock.

calcite skeletal structure of dead marine organisms such as corals, clams, and algae. *Dolostone* has very similar characteristics to limestone but is composed of the mineral *dolomite* (calcium-magnesium carbonate). Soils formed from limestone or dolostone typically have neutral pH, which aids in buffering the soils and prevents acid streams.

The presence of carbonate bedrock is extremely important in watershed processes. Because carbonate rocks are very soluble relative to silicate rocks, water chemistry and their dissolution can lead to the formation of sinkholes. Limestone is often found close to the surface, as shown in Figure 2.4. Although sinkholes can form in marble, they are rare. Therefore the following discussion will relate more to limestone and dolomite.

Carbonic acid dissolves limestone and forms sinkholes, a process termed *dissolution*. Atmospheric carbon dioxide dissolves in precipitation to produce naturally acidic rain. Water combines with carbon dioxide to form the carbonic acid, as shown in equation 2.1:

$$H_2O + CO_2 \rightarrow H_2CO_3 \tag{2.1}$$

Acid rain aids in this dissolution process. As the water enters the natural breaks, or fractures, of the limestone or dolomite, the acidic water dissolves the surrounding exposed bedrock, widening the fractures. Uneven dissolution of the limestone typically occurs due to differential resistance of the bedrock and the amount of noncarbonate minerals present. The enlarged fractures act like natural pipes, and could channel surface water directly to the groundwater

Figure 2.4   Limestone exposed on the surface.

system, often with little or no treatment of pollutants. Decomposition and respiration by soil organisms also produce carbon dioxide gas, which contributes to acidic conditions.

As the carbonate bedrock geology continues to dissolve, voids are often created beneath the overlying soil. As a void becomes larger, it can succumb to the weight of the overburden, forming a surface *sinkhole*. Sometimes the collapse is not as quick, and the soil is slowly carried into the sinkhole, a process called *piping*. Sinkholes are typically circular and often form directly after heavy precipitation periods or after periods of prolonged drought when the water table lowers and the absence of hydrostatic pressure no longer can hold the walls of the void intact. An area where sinkholes are prevalent is termed *karst topography* and has characteristics of a "pocked" landscape with sinkholes, depressions, and caves. These features should be evaluated both using published geologic and topographic maps and through field reconnaissance when performing a watershed assessment.

Carbonate rocks typically have high-yielding wells due to the availability of water in the fractures. However, the quality of the water may be in question due to the absence of filter material. Mine and quarry pumping or overpumping from domestic, public, or industrial wells may lower the water table, also causing subsidence.

Stormwater runoff in karst topography is a major concern. Sinkholes exacerbate the direct surface water/groundwater connection problem. When stormwater is concentrated and collected into stormwater detention basins, a sinkhole situation is often created at the bottom of the detention basin, as shown in Figure 2.5, providing a direct conduit of stormwater pollutants into

Figure 2.5 Sinkhole created by and at the bottom of a stormwater detention basin, Lehigh Valley, Pennsylvania. (Photo supplied by Dave Jostenski, PA DEP.)

the groundwater system. Water main breaks can also be a menace in karst terrain. A leaking pipe may go undetected and continue to dissolve the limestone until the ground collapses into a sinkhole.

Often, streams in carbonate geology will flow for awhile, then disappear into the fractures, only to reappear somewhere downstream. These are termed *disappearing streams* and can be clearly seen on USGS topographic maps. A sample of disappearing streams can be seen in Figure 2.6.

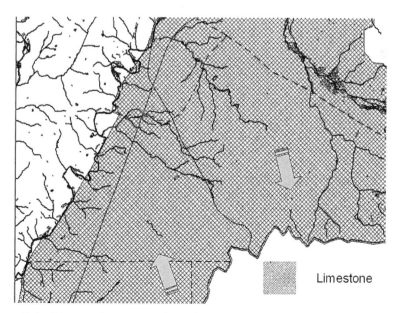

Limestone

Figure 2.6 Disappearing streams in carbonate geology. Notice the completely different drainage pattern in the karst topography.

Clastic sediment, such as sandbars in a riverbed, is derived from breakage of the parent material (igneous, sedimentary, or metamorphic rock) and consists of mineral particles. *Chemically precipitated sediment* such as rock salt is the accumulation of dissolved inorganic material that has settled. *Organic sediment* such as peat consists of the accumulation of organic tissues of plants or animals. The type of sediment and, in turn, the particle size will affect transportability, soil erodibility, and water quality. Naturally, the finer the sediment, the more easily it stays in suspension and the farther it will travel. The coarser particles settle out first, a process termed *sorting*. If, for example, one were to place a soil sample in a glass of clean water, stir it, and then set it aside to allow the particles to settle, one would notice a stratification of the particles, with the larger particles settling to the bottom of the glass first and the finer particles staying in suspension longest, in accordance with Stokes Law.

Stokes Law can be summarized as:

$$\text{Force due to Acceleration} = \text{Gravity Force} - \text{Buoyancy Force}$$
$$- \text{Drag Force} \tag{2.2}$$

Or more specifically:

$$v = (2gr^2)(d1 - d2)/9\mu \tag{2.3}$$

Where:

v = velocity of fall (cm sec^{-1})
g = acceleration of gravity (cm sec^{-2})
r = "equivalent" radius of particle (cm)
$d1$ = density of particle (g cm^{-3})
$d2$ = density of medium (g cm^{-3})
μ = viscosity of medium (dyne sec cm^{-2})

In this equation, the assumption is that the soil particles have an equivalent radius (r), where in fact, sediment particles are seldom spheroidal. The settling rate (velocity) of a particle is proportional to the square of the particle's "equivalent" diameter. This is an extremely important concept and explains the *aggradation/degradation* process of streams, settling of sediment in detention and sedimentation basins, and prevention of nonpoint source pollution from suspended solids.

2.2.2 Igneous Rocks

Igneous rocks form under high temperatures and harden from molten material that originates beneath the earth's surface. The majority of igneous rocks consist of silicate minerals, which are compounds containing silicon atoms and oxygen atoms, and may have one or more metallic elements (aluminum, iron, calcium, sodium, potassium, and magnesium) attached. Igneous rocks vary widely in composition. *Magma,* which hardens beneath the earth's surface, forms *intrusive igneous rocks,* whereas magma that reaches the earth's surface is referred to as *lava,* and rocks formed from hardened lava are called *extrusive igneous rocks.* Intrusive and extrusive rocks are texturally and geometrically distinct.

2.2.3 Metamorphic Rocks

Both sedimentary and igneous rocks can be transformed into metamorphic rocks by changing the temperature, pressure, and chemical environment in which they initially formed. The most common metamorphic rocks are slate, phyllite, schist, amphibolite, gneiss, marble, quartzite, hornfels and anthracite (coal) (Strahler and Strahler, 1976). The chemical composition of the metamorphic rocks, their textures, and the types and orientations of geologic structures are important in the watershed. For instance, as the original organic matter that would eventually make up the coal undergoes heat and pressure, volatiles carry away with them hydrogen and oxygen, leaving the carbon. Bituminous coal has approximately 80 percent carbon and 10 percent oxygen, whereas the anthracite is comprised of approximately 95 percent carbon and 2 percent oxygen. Other impurities such as sulfur, silica (SiO_2), and aluminum oxide contribute to the formation of acid mine drainage when exposed to water and air.

2.2.4 Weathering

Weathering is the process whereby solid rock is transformed by wind, temperature extremes, rain, glaciers, waves, chemical reaction, and streams into a layer of *disaggregated rock* (soft mineral matter) called *regolith.* This regolith is the decayed rock that has disintegrated into sand, silt, and clay and cannot support plant life due to the absence of organics and nutrients. However, regolith is further broken down though *physical* (*mechanical*), *chemical,* and *biological* forces to form soil. All processes by which the landform is weathered and the weathering products are eventually transported to the sea via streams and rivers are called *denudation.* The net effect of denudation is a lowering of the elevations of the landforms and sediment deposition in *waterways.* The transport of rock, regolith, and soil through the action of gravity is termed *mass wasting.* This is a very important concept as it sets the foundation for soils, land cover, and the *fluvialgeomorphic* (*FGM*) processes, which will be discussed in Chapter 5. Frost and other weathering action on the bedrock on mountain slopes detaches the weathered rock fragments, which then accumulate at the base of the slope as *talus,* as shown in Figure 2.7. Talus, due to its unconsolidated nature, is generally very porous.

Figure 2.7 Talus slope, Mt. Rainier, Washington. (*Source:* American Memory Library of Congress, 2003, http://memory.loc.gov/ammem/award97/icuhtml/ aepSubjects26.html.)

Physical or *mechanical change* is the physical breakdown of rocks and minerals primarily from expansion and contraction of rock materials and water in rock *fissures* through freezing and thawing and heat change processes. The rocks are broken into smaller fragments exposing even more surface area to chemical and biological forces. Abrasion from gravity, wind, ice, and water transport also breaks down the rocks, and the resulting weathered material ranges in size from boulders to silt.

Chemical change, also known as *mineral alteration,* is due primarily to exposure of rocks to oxygen (*oxidation*), carbon dioxide (*carbonation*), organic acids, and water (*hydrolysis*). The end result is minerals broken down into a more stable mix of new minerals and dissolved constituents. When the dissolved oxygen in water comes in contact with the mineral surface, it combines with the metallic elements such as calcium, iron, and magnesium, a process called oxidation. Carbon dioxide, in contrast, forms carbonic acid when in solution and causes a *carbonic acid reaction,* dissolving certain minerals, especially calcium carbonate. Water combines with silicate mineral compounds in a true chemical change of the material, a process known as hydrolysis. When decaying organic matter is present, organic acids are formed, also increasing the chemical reaction process. End products of silicate mineral alteration often include clays. Clay-rich soils play an important part in the hydrologic processes of watersheds, including runoff potential, infiltration, and erosion potential.

2.2.5 Biological Change

Geologic material is changed through interaction with the *biosphere,* which includes all living organisms and the environment in which organisms live. The exchange of matter and energy within the environment is the science of *ecology,* and the whole ecological system in which organisms live is known as the *ecosystem.* The energy exchanged between the food chain is what aids in the transformation of the bedrock to soils. For instance, root masses grow and break apart rocks, animals' burrows aid in transport, and fungi and microorganisms physically and chemically alter the rock structure on which they grow. Recycling of *macronutrients* (hydrogen, carbon, and oxygen, otherwise known as carbohydrates, nitrogen, phosphorus, calcium, potassium, silicon, magnesium, sulfur, aluminum, and sodium) and *micronutrients* (iron, copper, zinc, boron, molybdenum, manganese, and chlorine) also occurs in the biosphere. Alteration of the ecosystem affects long- and short-term equilibrium in the watershed. Conversely, the geologic makeup of the regolith, along with other factors such as climate, exposure, and drainage, dictates the type of ecosystem that will occur in any particular location.

2.2.6 Transport

The downslope movement of *detrital* material by gravity is termed mass wasting. Detrital material is the organic debris formed by the decomposition of plants and animals. Through the process of mineral alteration, solid rock is softened and fragmented, yielding particles of many sizes, which can then be transported in a fluid medium (air, ice, or water). The material that is then deposited is called *sediment,* and can include both inorganic and organic matter. Dissolved mineral constituents can also be transported through these media. Due to decreasing velocities and stream dimensions, the sediment, as it's transported, is sorted by size and density. The regolith and soils that are transported through water action and eventually deposited in the stream and river valley floors are referred to as *alluvium.* Alluvial deposits are highly variable in grain size, composition, and thickness. Alluvial deposits consisting mainly of sand and gravel derived from glaciation are termed *outwash,* and the flat lowlands where it is deposited are called *outwash plains.* Outwash plains are some of the most productive *aquifers* in the world. (Aquifers will be discussed in Chapter 3.) The sediment that accumulates, primarily on seabeds, and undergoes physical and chemical changes under pressure, becomes compacted and hardened to form the *sedimentary* rocks. Dissolved ion constituents may precipitate and aid to cement the sediment. Human activities often lead to sediment being transported quickly through stormwater runoff or wind erosion processes to cause problems in aquatic systems.

2.3 SOILS

Soils have many physical properties that are important in a watershed, a list too long to reproduce here; therefore, only some of the most important will be discussed. It is important to note that soil properties change from the surface to the bedrock. These changes in the "column" or "profile" of the soil type can be classified into soil *horizons*. A soil horizon is a layer of similar soil *color, texture, structure,* or *porosity,* oriented approximately horizontally to the earth's surface. Most soils, from the surface down, have an *O,* or *organic* horizon; an *A, B,* and *C* mineral horizon; and an *R,* or *consolidated* bedrock horizon. The *A* horizon, often referred to as *topsoil,* has organic matter mixed with the mineral matter. This is the horizon that is typically scraped and sold for lawn topsoil or stockpiled for reapplication on construction sites. Organic and mineral matter from the *A* horizon are carried downward by percolating water into the *B* horizon, often referred to as the *subsoil,* which is the horizon of accumulation of these materials. This accumulation often results in a higher-density soil with clay accumulation and pronounced soil structure. The *C* horizon has characteristics of the *unconsolidated,* weathered *parent material.*

Each of the soil horizons may be further classified into smaller subsets, as shown for the Woodson soil series in Table 2.2. The description of each soil horizon consists of its ranges in distance from the land surface; its color, texture, structure, consistency, course fragments, and chemical characteristics if any; and a description of its boundaries. The color description comes from the Munsell Color Chart, which is a book of seven plates containing 175 color chips arranged by hue (dominant wavelength or color of light), usually an expression of yellows and reds (YR), value (brilliance or quantity of light), and chroma (dominant wavelength of light). Number values from 1 to 10 are assigned to the hue and value. The combination provides a description of the color, such as dark gray in Table 2.2. Therefore the color of 10YR 3/1 in the Al horizon below would have a hue of 10YR, a value of 3, and a chroma of 1, forming a dark gray soil.

A *soil profile* showing typical *A, B,* and *C* horizons for the same Woodson series is shown in Figure 2.8. The physical and structural constitution of the soil horizons is referred to as soil *morphology.*

As can be seen from Table 2.2, the soil horizons have various physical and chemical differences. When performing a watershed analysis, it is important to keep in mind that these differences affect soil porosity, permeability, erodibility, and so on. For instance, if one were concerned with erodibility, the upper or 'A' horizon may be the horizon of concern. Or, if one were constructing an infiltration device for stormwater management purposes, the prop-

TABLE 2.2   **Soil Description for Each Horizon of the Woodson Series***

Taxonomic Class	Fine, smectitic, thermic Abruptic Argiaquolls
Typical Pedon	Woodson silt loam, on a nearly level area in native grass. (Colors are for moist soil unless otherwise stated.)
Al	0 to 8 inches; very dark gray (10YR 3/1) silt loam, gray (10YR 5/1) dry; weak fine granular structure; slightly hard, friable; many roots; medium acid; abrupt smooth boundary (6 to 14 inches thick).
B21t	8 to 19 inches; very dark gray (10YR 3/1) silty clay, dark gray (10YR 4/1) dry; few fine faint dark brown (10YR 3/3) mottles; moderate fine blocky structure; extremely hard, very firm; common roots; medium acid; gradual smooth boundary.
B22t	19 to 31 inches; very dark gray (10YR 3/1) silty clay, gray (10YR 5/1) dry; many medium distinct strong brown (7.5YR 5/6) and a few medium distinct olive brown (2.5Y 4/4) mottles; moderate fine and medium blocky structure; extremely hard, very firm; fine and medium concretions; medium acid; gradual smooth boundary. (Combined thickness of the B2 horizon is 16 to 35 inches.)
B3	31 to 38 inches; gray (10YR 5/1) silty clay, gray (10YR 6/1) dry; few medium distinct dark reddish-brown (5YR 3/4) mottles; weak medium blocky structure; extremely hard, very firm; few roots; few fine black concretions; common fine gypsum particles; medium acid; diffuse wavy boundary (6 to 20 inches thick).
C	38 to 60 inches; gray (10YR 5/1) silty clay, gray (10YR 6/1) dry; many fine distinct olive (5Y 4/4) mottles; moderate fine and very fine blocky structure; extremely hard, very firm; very few roots; common fine black concretions; common black stains; few fine gypsum particles; medium acid.
Type Location	Allen County, Kansas; 4 miles south of LaHarpe, Kansas; 100 feet south and 1,420 feet east of the northwest corner of sec. 23, T. 25 S., R. 19 E.
Range in Characteristics	The thickness of the solum ranges from 30 to 60 inches.

*The Woodson series consists of deep, somewhat poorly drained, very slowly permeable soils that formed in silty and clayey sediments. These soils are on uplands.

erties of the horizon at the depth of the proposed structure (or the horizon with the most restrictive permeability) would be of concern, most likely the *B* or *C* horizon. In looking at the spatial extent of the soils as with GIS mapping, one should be aware of which horizon's attributes are being displayed. These soil characteristics are all important in the hydrologic budget and land use suitability and all influence the nature and extent of anthropogenic changes in the watershed landscape.

Figure 2.8 Typical soil profile showing A, B, and C horizons. Notice the well-defined horizon differentiation on the left and the not-so-well-defined distinction on the right. Horizon differentiation is typically not well defined. (*Source:* UDA—NRCS, http:// soils.usda.gov/gallery/profiles/img0086.htm. Photo by Jim R. Fortner, USDA, NRCS.)

2.3.1 Soil Mapping Terminology

Soils have been mapped in the United States by the USDA Natural Resources Conservation Service (NRCS) (formerly the Soil Conservation Service, or SCS). Soils may be classified based upon their morphology or form based upon the nature of their profile, termed the *taxonomic classification*. Soils may also be displayed based upon their supportive use, or the *interpretive classification* (Hausenbuiller, 1985). Soils are portrayed spatially as "mapping units," which may vary as to the map type and the amount of detail required based upon the scale of mapping. The mapping unit is defined in terms of a mappable pattern of geographically associated taxonomic units defined by the properties of the units and their pattern (USDA, 1951). There is often much confusion as to the types of soil mapping units required for watershed assessment and in soil map unit terminology. USDA NRCS has standardized, yet is constantly improving on, the soil classification terminology.

Mapping units are areas that can be shown on the maps. In detailed maps, they may consist of *consociations* (one soil type dominates), *complexes* (patterns of two or more soils types that cannot be separated at the scale of mapping), or *associations* (two or more soils where the use and management is similar and there was no need to separate them in mapping).

Detailed mapping "units" are soil types, or *series,* with a defined set of physical properties (color, texture, structure, horizons), such as the Woodson series described above or a Hagerstown silt loam, and are used for detailed analyses on a specific site or small watershed level. A *component* is a phase of a soil series. For example, Hagerstown silt loam, 3 to 8 percent slopes, is a gently sloping silt loam surface texture of the Hagerstown series. A grouping of soils series such as Hagerstown-Conestoga-Opequon is called an *association* and is usually identified at the county scale. Soil associations are groups of soils that exhibit a regularly repeating pattern based on common occurrence within a type of landscape. Soils may be even further grouped into *complexes.* A *soil complex* is a mapping unit wherein two or more units are so intermixed that it is impractical to separate them, based upon the scale of the application, and are shown together due to the necessity for clear cartographic presentation. (These classifications will be discussed in further detail in Chapter 10, "Geographic Information Systems.")

The smallest volume of soil that displays similar properties or the one that provides the most detail is the *pedon.* A *polypedon* is a grouping of two or more pedons sufficiently alike to be classified as one soil type. A typical polypedon would be the soil series. Soil series typically are named for nearby locations where they were first mapped, along with a surface texture class and sometimes a phrase name, for example, take slope classification such as WmB, where Wm stands for Wellsboro and the B is the slope classification. The soil series typically are comprised of several pedons called *inclusions.*

Slope classifications as subscripts to the series name are generally as shown in Table 2.3. As can be seen from the table, slope classes overlap due to localized soil and topographic conditions. These slopes are important for runoff and erosion potential in the watershed.

Drainage in a soil description refers to the rapidity and extent of the removal of water from the soil.

Samples of *interpretive* soil classifications prior to 2003 would come from NRCS's Map Unit Interpretations Record (MUIR) tables and after 2003 from the National Soil Information System (NASIS) database, and might include soil erodibility, permeability, and so on. Two types of data are contained in NASIS: properties and interpretations.

TABLE 2.3   Typical Slope Classes

Class	Slope Groupings
A or no slope class	0–3%, 0–5%
B	0–8%, 3–8%
C	8–15%, 8–20%, 8–25%
D	15–25%, 15–30%,
E	25–70%
F	45–65%

Another important interpretive classification NRCS has originally developed for conservation planning, which is useful in watershed assessment, land use planning, opportunities and constraints analysis, and to aid in buffer restrictions, is the *capability classes,* which are described by Roman numerals. The smaller numerals have the fewest limitations for agricultural use. The capability classes can be found in Table 2.4.

The subclass would be a suffix on the class; for instance, IVe would be a class IV soil with erosion limitations. These classifications can aid in determining which areas of the watershed should be preserved or placed in permanent agricultural or conservation easements and will aid in determining those areas where building should be discouraged or encouraged. For instance,

TABLE 2.4   Soil Capability Classes

Class	Limitations
I	Few limitations that restrict their use.
II	Moderate limitations that reduce the choice of plants or that require moderate conservation practices.
III	Severe limitations that reduce the choice of plants, or that require special conservation practices, or both.
IV	Very severe limitations that reduce the choice of plants, or that require very careful management, or both.
V	Not likely to erode but have other limitations, impractical to remove, that limit their use.
VI	Severe limitations that make them generally unsuitable for cultivation.
VII	Very severe limitations that make them unsuitable for cultivation.
VIII	Limitations that nearly preclude their use for commercial crop production.

Capability Subclass	Limitation
e	Risk of erosion.
w	Water interferes with use.
s	Soil is shallow, droughty, or stony.
c	Climate limitation.

the Class I soils are usually primary agricultural soils and should be preserved. Soils with a risk of erosion subclass should also be highlighted as possible conservation land.

2.3.2 Soil Physical Properties

Soils play an important role in watershed characteristics and processes. Initially, the bedrock from which they are developed has an impact on the chemical and physical properties of the soils. Soils are comprised of mineral and organic matter. The ratio of each, and the density of soil, determines many properties of the soils, which affects how they function. Various characteristics of the soils, such as *permeability, hydraulic conductivity,* chemical characteristics, and so on, have a big impact on the hydrologic characteristics of watersheds. The physical, chemical, and biological breakdown of bedrock forms a wide range of particles from stones to gravel to sand to silt to clay. The sand, silt, and clay are termed *soil separates.* Each of these particles has a specific definition. Table 2.5 defines the size range for soil particles based upon the U.S. Department of Agriculture system (modified from Foth, 1990).

Soil *texture* is the relative proportion of sand, silt, and clay. The texture of a soil influences many of its characteristics, such as permeability, erodibility, and infiltration capacity. By obtaining the relative proportions of sand, silt, and clay, the soil texture can be determined through the use of the USDA Soil Textural Triangle, as shown in Figure 2.9. The relative proportion of sand, silt, and clay can be easily determined by the trained soil scientist in the field or can be measured more accurately using laboratory methods such as the hydrometer method (DeBarry, 1995).

Soil *structure* is the *aggradation* or arrangement of the sand, silt, and clay soil particles into larger secondary particles called *peds* or *aggregates.* Soils can be classified as having structure or being without structure (nonstructural). There are four major soil structure shapes: *speroidal (granular and crumb), platy, blocklike (blocky and subangular blocky),* and *prismlike (prismatic and*

TABLE 2.5 Characteristics of Soil Separates

Separate (cm²)	Diameter (mm)	Average Surface Area in 1 Gram (cm²)
Very coarse sand	2.00–1.00	11
Coarse sand	1.00–0.50	23
Medium sand	0.50–0.25	45
Fine sand	0.25–0.10	91
Very fine sand	0.10–0.05	227
Silt	0.05–0.002	454
Clay	<0.002	8,000,000

Source: Foth, 1990.

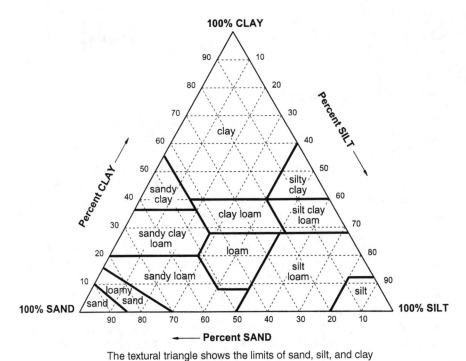

The textural triangle shows the limits of sand, silt, and clay
contents of the various texture classes.

Figure 2.9   USDA Textural Triangle (U.S. EPA, 1999, after Foth, 1990).

columnar), as shown in Figure 2.10. The most significant aspect of soil structure is the *interped spaces* (cracks or "cleavage planes"), formed through
physical pressures such as freezing and thawing, wetting and drying, and root
growth. Interped spaces that affect soil water percolation are much larger than
the pore spaces. Nonstructural soils do not form these cleavage planes between soil aggregates. As can be seen from Figure 2.10, the speroidal, blocklike, and prismlike structure will promote soil water percolation, whereas the
platy structure may actually form a barrier to downward water movement.
Platy structure is typical of clay soils.

Physical exertions also aid in forming dense layers of soils called *claypans,
fragipans,* or *hardpans*. These are tightly packed layers where pore spaces
are filled with clay and compacted. Downward water movement is impeded,
which is often the limiting factor in on-site waste disposal suitability and
affects the rate and direction of water movement in soils. (The subtle variations between these pans will not be discussed here.)

2.3.3   Soil Water Relations

The surface area of the soil particles has a direct influence on the capability
of the soils to retain water. The smaller the soil particles, the greater the total

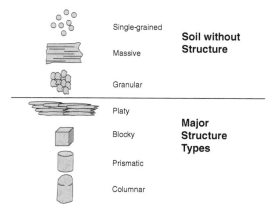

Figure 2.10 Soil structure: Structure types are based upon the congregation of aggregates. (Modified from Foth, 1990.)

surface area per unit quantity of soil. If one can envision a box that would hold one volleyball, then equating total surface area of a volleyball to its diameter, a string held around the volleyball's diameter would be approximately 28 inches long. However, if one were to fill the box with golf balls and take a string length equivalent to the diameter of all of these golf balls, the length would be approximately 750 inches long. This example illustrates that there is much more surface area, and in turn more *cohesive and adhesive forces*, with smaller soil particles. Cohesive force is the ability of the soil to "cling" to itself, whereas adhesive force is the ability to "adhere" or "stick" to other objects such as water.

Soil porosity can be described as:

$$\text{Percent Porosity} = (\text{Volume of pore space} \div \text{Volume of soil}) \times 100 \quad (2.4)$$

Soil density is its weight per unit volume, or:

$$\text{Density} = \text{Weight} \div \text{Volume} \quad (2.5)$$

and is important in understanding soil properties. The description of a soil's density includes *particle density* and *bulk density*. Particle density is the average density of just the soil particles, while bulk density is the bulk soil in its natural state, which includes both the particles and the pore space. Mineral soils composed primarily of mineral matter have a bulk density of about 1.3 g/cm^3 and a particle density of about 2.65 g/cm^3. Total pore space or porosity can be computed if the bulk density and particle density are known, as shown in equation 2.6.

$$\text{Porosity (\%)} = (1 - (\text{Bulk density} \div \text{Particle density})) \times 100 \quad (2.6)$$

Substituting the numbers for bulk density and particle density in equation 2.6 yields a natural porosity of about 50 percent for a typical mineral soil. Therefore, a typical undisturbed soil has about 50 percent pore space, which is occupied by some combination of air and water, the ratios of which will depend on many factors, including time elapsed since the last rain.

Soil pores provide a continuous interconnected system between the bedrock, soil column, and the atmosphere. Soil porosity is important in watershed management for aiding in analyzing the water budget (infiltration), determining the impacts of development, and understanding the rainfall/runoff process.

Sandy surface soils have less porosity than clayey surface soils. This is due in part to the fact that the clay soils may have a well-defined structure with interped spaces. Sands do, however, have larger pore spaces (*macropores*) than clays. Although soil porosity is important in water movement through a uniformly porous medium, water movement is more directly related to pore size. In saturated conditions, sands transmit water faster than clays due to the larger pore sizes (macropores). The *transmission rate* is the rate at which water moves in the soil and which is controlled by the soil's horizons. Pore space and arrangement of pores (structure) is the major factor in determining a soil's *permeability*, which is the capability of the soil to transmit water and air.

Percolation is the downward movement of *excess* water through the soil, while *infiltration* is the capability of the water to enter or "infiltrate" the surface of the soil. Infiltration is controlled by surface conditions. The maximum rate at which a soil is capable of absorbing water in a given condition is its *infiltration capacity*. Infiltration capacity of a soil is affected by many things, including:

- Antecedent rainfall conditions
- Antecedent soil-moisture conditions
- Compaction of the soil from raindrop forces, animals (cattle), development
- Inwash of fine material into soil pores
- Available macropores from burrowing animals, roots, insects, etc.
- Vegetative cover
- Entrapped air
- Surface slope
- Decreasing temperature (which reduces rainfall viscosity and freezes the surface)

If a soil is dry, wetting the top of it will create a strong *capillary* potential, or "pull," just under the surface, supplementing gravity. As soils become wetter, the infiltration rate decreases toward a minimum value. In some cases, after wetting, the clays in the soils swell, reducing the infiltration capacity shortly after a rain starts. Therefore, infiltration rate typically slows as the storm continues. If the precipitation rate is less than infiltration capacity, all precipitation is absorbed; if the precipitation rate is greater than infiltration capacity, water ponds on the surface and runoff may occur.

The driving force for water movement is the *water potential gradient* (f), or the difference in elevation between two points divided by the distance between those two points. The velocity of saturated flow (v) through soil is equal to the hydraulic conductivity (k) times the potential gradient, or:

$$v = kf$$

As mentioned, permeability is the soil's ability to transmit water or air. Soil permeability is measured as the saturated flow through soil columns and is expressed as the volume of water flow per unit of time. This information is published in the USDA NRCS Soil Survey Reports and is also attached to the Soil Survey Geographic (SSURGO) data discussed in Part B, Chapter 10. Soils can be grouped into permeability classes that provide a good general indication of their rate of water transmission, as shown in Table 2.6 (Hausenbuiller, 1985).

These classes can be displayed in a GIS map of the watershed being studied, as will be discussed in Chapter 10. Soil hydraulic conductivity (k) is the ability of the soil to transmit water and is related to the size and connectivity of the pores and spaces through which water moves.

Soil properties influence the runoff generation process. The USDA NRCS has established a criterion to determine how soils will affect runoff by placing all soils into Hydrologic Soil Groups (HSGs). HSGs are broken down into four categories (A through D) based on infiltration rate and depth as defined by NRCS (1993):

Group A (low runoff potential). Consists of deep, well- to excessively drained sands and gravels. "*A*" soils have a high infiltration rate, even when thoroughly wetted, and have a high rate of transmission.

TABLE 2.6 Soil Permeability Classes

	Range in Permeability	
Permeability Class	(in./hr.)	(cm/hr.)
Very slow	<0.06	<0.15
Slow	0.06–0.2	0.15–0.5
Moderately slow	0.2–0.6	0.5–1.5
Moderate	0.6–2.0	1.5–5.0
Moderately rapid	2.0–6.0	5.0–15.0
Rapid	6.0–20.0	15.0–50.0
Very rapid	>20.0	>50.0

Source: Hasenbuiller, 1985.

Group B. Moderately fine to moderately coarse-textured soils. They are characterized as having moderate infiltration rates when thoroughly wetted, and consist primarily of moderately deep to deep, moderately well- to well-drained soils that exhibit a moderate rate of water transmission.

Group C. Moderately fine to fine-textured soils that have slow infiltration rates when thoroughly wetted, contain a layer that impedes downward movement of water, and exhibit a slow rate of water transmission.

Group D (high runoff potential). Consists mainly of clay soils with a high swelling potential. These soils have very slow infiltration rates when thoroughly wetted and a very slow rate of water transmission. They typically exhibit a permanent high water table, a claypan, or clay layer at or near the surface, and/or shallow soils over a nearly impermeable material.

All the soil properties just mentioned are important to understand in watershed management for a variety of reasons. Land alteration compacts the soil and reduces porosity, permeability, and hydraulic conductivity, altering the natural hydrologic budget. The net effect is less infiltration and more runoff. Even where land is temporarily altered for construction purposes, where topsoil is stripped and eventually returned and reseeded, the subsoil is compacted by heavy equipment, reducing its permeability, as shown in Figure 2.11. Therefore, even if the area is revegetated, the soil profile does not function as it did in natural conditions. The NRCS is currently conducting research on the regeneration of the soil structure over time; it has found that some tend to become more permeable over time, particularly after several years of the freeze/thaw cycle (White, 2003).

Figure 2.11   Compacted soils from construction.

2.3.4 Soil Chemical Properties

A soil characteristic that must be understood due to its importance in watershed management is the *soil buffering capacity*. A soil's buffering capacity is its ability to resist changes in pH when an acid or base is applied. This is extremely important, particularly east of the industrial belt or Midwest, where acid rain with pH of 4 or less is now a common occurrence and where values as low as 1.5 are being recorded (Tchobanoglous and Schroeder, 1987). Soils formed from calcareous bedrock (limestone) have significant buffering capabilities, in contrast to soils derived from shales, which typically do not. The chemical composition of buffered soils allows inactivation of the added hydrogen ion (H^+) by adsorption to exchange sites or by a neutralization reaction between the H^+ and lime in the calcareous soils. The neutralization reaction takes on the form:

$$CaCO_3 + 2H^+ \rightarrow Ca^{2+} + H_2CO_3 \tag{2.7}$$

Calcium Carbonate + Acid $\rightarrow$ Calcium + Carbonic Acid

$$H_2CO_3 \rightarrow H_2O + CO_2 \tag{2.8}$$

Carbonic Acid $\rightarrow$ Water + Carbon Dioxide

This is the same reaction and reason that lime is applied to acidic soils. In unbuffered soils, the acid is not neutralized and flows through the soils and recharges stream baseflow at or near the acidic pH of the acid rain. In many eastern streams, the nonneutralization of the acid rain is ruining the biodiversity of the streams and, in particular, is damaging to sensitive species such as trout. Strongly acid soil conditions mobilize metal ions such as aluminum that can be toxic to organisms at high concentrations.

2.3.5 Filterability of Soils

Water molecules are electrically neutral; however, the charge is asymmetrically distributed, resulting in a polar attraction by hydrogen bonding between water molecules (see Figure 2.12). Soil particles have either negatively or positively charged sites, which attract water molecules (*adhesion*), as shown in Figure 2.13. If pollutants are contained within the water, they may be adsorbed to the soil particles with the water (Foth, 1990). This water has very little tendency to move and is affected by the thickness of the water film on the soil particle. The rate of movement will also depend on the pore size, which is directly related to soil texture. This hydrogen bonding is typical of sand and silt particles (Foth, 1990).

Most soils have small amounts of clay, either clays with oxide coatings, which have a net positive charge, or kaolinite, which have a net negative charge. (The difference between these two clays' physical makeup is beyond the scope of this text.) Humus is also abundant in most soils. Both clays and humus are present in the colloidal state and therefore expose a large amount

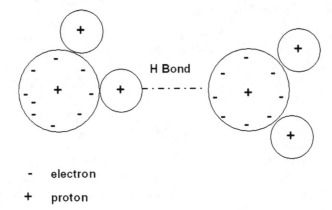

- electron

+ proton

Figure 2.12   Polar attraction of hydrogen bonding between water molecules.

of surface area for adsorption of water and ions. Ions will be adsorbed by a process known as *ion exchange*. Most clays and organic matter have a net negative charge, making them effective in "holding" positively charged particles (*cations*) such as Al, Fe, H+ and Ca. Cation holding depends on the size of ion and strength of charge. The colloid will adsorb a cation of opposite charge and release an ion held by the colloid, assuming the cation in solution has a greater number of charges than the ion already attached. The cation exchange capacity, measured in millequivalents of ion per 100 grams (me/ 100g), of a soil can be estimated by the following equation (Foth, 1990):

$$CEC = 2.5 \text{ (\% organic matter)} + 0.57 \text{ (\% clay)} \qquad (2.9)$$

The efficiency of an ion to be adsorbed is indirectly proportional to its size. The larger ions are more efficient in being replaced, so they will adsorb to water more quickly, forming a large hydrated radius, which reduces its ability to get close to the colloid's nucleus or micelle (Foth, 1990).

Most soils have a net negative charge from kaolinite clay and humus. This means that, in these soils, cations such as calcium will be adsorbed; and

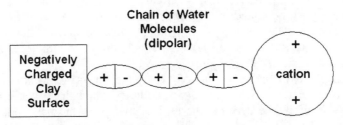

Figure 2.13   Hydrogen and soil particle bonding.

anions such as phosphate, chloride, and nitrate are soluble and could possibly percolate into the groundwater if sufficient positively charged colloids are not present (Foth, 1990).

Chemical contaminants, being in solution, tend to travel faster than biological contaminants, which are more easily filtered by silt and sand. Thus, the potential for pollution depends on the mobility of the contaminant, the accessibility of the groundwater reservoir, and the hydraulic characteristics of the reservoir.

2.3.6 Soil Erodibility

Undeveloped watersheds have natural vegetation cover, whether grass, meadow, or forest. Under natural conditions, soil erosion from wind and water is usually not significant and is a natural process. During rainfall events, the soil stays in place due to the protection of the soil from the impact of raindrops. The first thing a developer typically does when beginning the construction process is to clear the site of vegetation and "grub" the stumps, leaving the exposed soil. Often, the topsoil will be removed and either used elsewhere or stockpiled for reuse after construction. This leaves the bare soil exposed to the direct impact from raindrops. Although raindrops are small in scale, they provide a mighty blow to the exposed soil. Runoff is much higher in areas of bare soil versus vegetative cover because of the breakdown of aggregates and clogging of pores from the impact.

As humans alter the landscape through land development, deforestation, grazing practices, and agriculture, the rate of soil erosion significantly increases, a process termed *accelerated erosion*. This causes a chain reaction of *dislodgement, detachment, displacement,* and *transport,* and resultant deposition (usually in an unwanted location) as shown in Figure 2.14. *Detachment* is when the water or wind energy forces acting on the soil surface are stronger than the *cohesive forces*. Once the soil is exposed at the surface, detachment becomes more prevalent. It is important to note that fine-textured aggregated soils, or those with firm structure, are less likely to be detached than noncoherent sandy soils. However, once detached and broken apart, fine-textured soils will be transported more easily than the larger sand particles. The factors

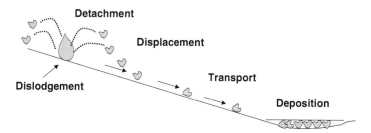

Figure 2.14 Soil dislodgement, detachment, displacement, transport, and deposition.

just described that influence soil infiltration, soil texture, structure, and organic matter determine the ease of particle detachment and soil erosion potential.

Dislodgement is when soil particles are moved from their original location by a raindrop splash and should not be confused with *displacement,* which is the lateral movement of soil from one place to another by mechanical forces such as equipment blades, vehicle traffic, or logs being yarded. The soil particles are then *transported* through raindrop "splash" or flowing water to where they are *deposited* when velocities can no longer support their weight.

Soil water erosion may occur when the rainfall rate exceeds the infiltration rate, producing runoff on exposed soil surfaces. Erosion from runoff, which forms larger and larger channels, begins as *sheet erosion,* leads to *rill erosion,* and can lead eventually to *gully erosion.* All forms of erosion may transport large volumes of sediment to waterways.

Construction sites can lose many tons of soil per year if not managed or regulated. The majority of this tonnage eventually ends up in our nation's waterways and bays.

Five major factors contribute to soil erosion and have been incorporated into the Revised Universal Soil Loss Equation (RUSLE), developed by the USDA Natural Resources Conservation Service (NRCS) for soil loss prediction in agricultural watersheds:

$$A = R \times K \times SL \times C \times P \tag{2.10}$$

Where:

A = predicted average soil loss (tons/acre/year)
R = a rainfall factor
K = a soil erodibility factor
SL = a slope factor based on the length of a slope in the area of interest
C = a cropping factor
P = a land use practice such as terracing or contour plowing

The RUSLE will be discussed in more detail in Chapter 15. All factors can be determined from the USDA Agricultural Research Service Agricultural Handbook Number 703 (Renard, 1996) or from Foth (1990). The equation is also useful in evaluating alternative land use practices to conserve soil. One important factor, the soil erodibility factor (K), measured as soil loss in tons per hectare per unit of rainfall erosion index, has been determined by NRCS for the soils across the United States by measurements on field plots. The higher the K value, the more erodible the soil. Knowing the K value provides the scientist a very good indication of erodibility potential, and when plotted with the GIS, a quick snapshot of potential problem areas can result, which can then be assessed (as will be discussed in Chapter 10).

Once washed into streams, soil particles undergo transport by streams and rivers. Streams and rivers tend to have flatter slopes and become wider as they progress downstream in distance and have increased contributing watershed area. As the slopes decrease, velocities decrease and the soil particles are no longer supported in suspension. When this happens, the soils are deposited; the accumulated sediment is called *alluvium* or *alluvial sediment.* Where steep mountain streams empty their deposits on flat plains, the deposited alluvium is termed an *alluvial fan* due to its fanlike shape. Where sediment is deposited on floodplains, it is termed *floodplain alluvium;* and when it is deposited where rivers empty into slow or still water, it forms a *delta.*

2.3.7 Hydric Soils

Another important soil characteristic to evaluate in watershed planning is whether the soil exhibits *hydric* soil characteristics. Hydric soils are typically associated with wetlands and exhibit conditions induced by *anaerobic* or *anoxic* conditions from long periods of saturation, flooding, or ponding during the growing season. Hydric soils support *hydrophytic vegetation* (U.S. Department of Agriculture (USDA) Soil Conservation Service (SCS), 1985, as amended by the National Technical Committee for Hydric Soils (NTCHS) in December 1986). Hydrophytic vegetation is also known as "water-loving" vegetation, or those plants that require saturated or near-saturated conditions to survive. During prolonged periods of flooded or inundated periods, the lack of oxygen allows the iron in soils to go from the *oxidized (ferric)* (Fe^{+3}) state, into a reduced *(ferrous)* (Fe^{+2}) state. This causes the soils to reflect the neutral color of the quartz, or the color of reduced iron, giving it a blue-gray appearance called *gleying*. In contrast, soils that are well aerated are typically uniformly brightly red, brown, or yellow colored due to various forms of oxidized iron, just as rusty iron is orange-brown colored.

Soils that are saturated for substantial periods of the growing season but not long enough to produce gleying often exhibit *mottles or redoximorphic features,* marked spots of contrasting oxidized soils (bright soil color value or orange color) and reduced soils (gleyed affect) developed through a fluctuating water table, producing alternate periods of aerobic and anaerobic conditions. Mottles are caused by the reduced iron becoming soluble, moving to other areas, and reoxidizing. The grayish areas are where the iron has been removed. Soils with mottling are shown in Figure 2.15.

2.4 TOPOGRAPHY

Topography, or *relief,* plays an important role in watershed processes and management. Topography is the three-dimensional boundary between the lith-

(a)
Extremely prominent mottling
in a clayey soil

(b)
Mottling in a loamy soil

(c)
Mottling in a sandy soil

(d)
Mottling inherited
from geologic processes

Figure 2.15 Examples of soil mottling: A, B, and C indicate seasonal soil saturation; D does not. (*Source:* Design Manual: On-site Wastewater Treatment & Disposal Systems, U.S. EPA, EPA 625/1-80-012, Washington, DC, October, 1980.)

osphere and the atmosphere, or more simply, the surface configuration or ground surface of the land. *Relief* is a broad term that relates the differences in elevation within an area; *microrelief* refers to small-scale differences in relief (USDA, 1951).

Topographic conditions are usually described in terms of *slope,* or the vertical rise in elevation divided by the horizontal run or distance between two points, and expressed as a percent (25 percent), horizontal-to-vertical ratio (4:1) or degrees (23.5 degrees). Many publications categorize slopes into general terms, such as flat, mild, moderate, and steep, or in terms of shape, such as level, undulating, rolling, hilly, and mountainous. Knowing the general slope of various areas of the watershed and stream will provide a good indicator of processes and management alternatives. (Utilization and manipulation of various topographic data sources will be described later in this text.) Steeper sloped areas typically have shallower soils, more erosion, faster times of concentration of surface runoff, different vegetation, and so on, than areas of flatter slope.

Watershed boundaries are defined by topographically high areas. Even when not used directly in a study, topographic data are often used in preparing visualization tools. Topography is shown on maps as *contours* or lines of equal elevation, *Digital Elevation Models* (*DEMs*), or other relief maps. The contours are typically spaced in even increments of elevation, or *bathymetry.* This "contour interval" helps to determine the vertical resolution of the contour map. Topographic maps often have point elevations, such as at mountain peaks, for *normal pool* elevations of lakes, or at other landmarks. Other implicit topographic information may include graphical sketches of terrain, as mountains are usually assumed to be topographically higher than their surroundings, while streams and lakes are usually considered to be topographically lower than their surroundings. Topography tells where both surface water and groundwater converge and concentrate or where water flow diverges and becomes less concentrated.

Topographic data have traditionally been presented as local topographic maps, in series of "rectangular" quadrangle sheets. Topographic maps may also be represented as more generalized maps, assigning colors to various ranges of elevations.

2.5 SUMMARY

The characteristics of watersheds are derived directly from the geology, soils, and landforms from which they originate. Having a thorough understanding of these three major physical factors will enable a better understanding and analysis of why certain features may exist in a watershed. A watershed assessment should begin with mapping and understanding the geology, soils, and relief of the watershed.

REFERENCES

Commonwealth of Pennsylvania Bureau of Topographic and Geologic Survey, "Sinkholes in Pennsylvania" (Harrisburg, PA: Commonwealth of Pennsylvania Bureau of Topographic and Geologic Survey), 1999.

DeBarry, Paul A. "Soils and Land Use–Lecture and Practicum Notes" (Wilkes-Barre, PA: Wilkes University), 1995.

Foth, Henry D. *Fundamentals of Soil Science,* 8th ed. (New York: John Wiley & Sons, Inc.), 1990.

Hausenbuiller, R.L. *Soil Science, Principles, and Practices,* 3rd ed. (Dubuque, IA: Wm. C. Brown Publishers), 1985.

NTCHS, National Technical Committee for Hydric Soils. "Hydric Soils of the United States." December 1986.

Renard, K.G., G.R. Foster, G.A Weesies, D.K. McCool, and D.C. Yoder. *USDA Agricultural Research Service, Agricultural Handbook No. 703—Predicting Soil Erosion by Water: a Guide to Conservation Planning with the Revised Universal Soil Loss Equation (RUSLE)* (Washington DC: U.S Government Printing Office), July 1996.

Strahler, Arthur N., and Alan H. Strahler. *Elements of Physical Geography* (New York: John Wiley & Sons, Inc.), 1976.

Tchobanoglous, George, and Edward D. Schroeder. *Water Quality* (Reading, MA: Addison-Wesley Publishing Co.), 1987.

USDA Natural Resources Conservation Service. *National Engineering Handbook* (Washington DC, U.S. Printing Office), 1993.

U.S. Department of Agriculture (USDA), Agricultural Research Administration. *Soil Survey Manual* (Washington, DC: U.S. Printing Office), 1951.

U.S. Department of Agriculture (USDA), Soil Conservation Service (SCS). "Hydric Soils of the United States," USDA-SCS National Bulletin No. 430-5-9, (Washington, DC: U.S. Printing Office), 1985.

U.S. Environmental Protection Agency (U. S. EPA). "Fundamentals of Soil Science as Applicable to Management of Hazardous Wastes," Burden, D.S., and J.L. Sims, EPA/540/S-98/500 (Washington, DC: U.S. EPA, Office of Research and Development), April 1999.

———. "Handbook: Ground Water" EPA/625/6-87/016 (Cincinnati, OH: Office of Research and Development), March 1987.

White, Ed, Pennsylvania State Soil Scientist, personal conversation, March 2003.

## CHAPTER 3

CLIMATE, PRECIPITATION, HYDROLOGIC CYCLE

3.0 INTRODUCTION

Watersheds begin with the hydrologic cycle, which begins with precipitation. If the rain were halted, streams would dry up, there would no longer be flooding or stormwater problems, and water supply would cease. Therefore, it is important to understand how precipitation affects the watershed and how the study of it can be utilized for watershed assessments.

3.1 CLIMATE CLASSIFICATION

Early climate classification systems divided the United States into regional climates, referred to as the Koppen system of classification (Koppen, 1931). The Koppen system was modified by G. T. Trewartha and others (1967) and Trewartha (1968), and is described and compared to ecoregion equivalents (discussed in Chapter 7) in Table 3.1 (USDA Forest Service, 2003) and shown in Figure 3.1. These climate classification systems became the precursor to today's ecoregion concept, which is described in Chapter 14. Knowing the general climate trends aids in determining which trends to expect in watershed analyses.

3.2 PRECIPITATION

Precipitation is a major factor in the formation of the landscape and *fluvial geomorphologic* processes, which will be discussed in Chapter 5. Understand-

TABLE 3.1 Climate Classification

Koppen Group and Types (Abbreviation*)	Ecoregion Equivalents (Class Code)
A—Tropical Humid Climates	Humid Tropical Domain (400)
Tropical wet (Ar)	Rainforest Division (420)
Tropical wet-dry (Aw)	Savanna Division (410)
B—Dry Climates	Dry Domain (300)
Tropical/subtropical semiarid (BSh)	Tropical/Subtropical Steppe Division (310)
Tropical/subtropical arid (BWk)	Tropical/Subtropical Desert Division (320)
Temperate semiarid (BSk)	Temperate Steppe Division (330)
Temperate arid (BWk)	Temperate Desert Division (340)
C—Subtropical Climates	Humid Temperate Domain (200)
Subtropical dry summer (Cs)	Mediterranean Division (260)
Humid subtropical (Cf)	Subtropical Division (230)
	Prairie Division (250)**
D—Temperate Climates	
Temperate oceanic (Do)	Marine Division (240)
Temperate continental, warm summer (Dca)	Hot Continental Division (220)
	Prairie Division (250)*
Temperate continental, cool summer (Dcb)	Warm Continental Division (210)
	Prairie Division (250)*
E—Boreal Climates	Polar Domain (100)
Subarctic (E)	Subarctic Division (130)
F—Polar Climates	
Tundra (Ft)	Tundra Division (120)
Ice cap (Fi)	

*Definitions and Boundaries of Koppen-Trewartha System:

Ar = All months above 64°F (18°C) and no dry season.

Aw = Same as Ar, but with two months dry in winter.

BSh = Potential evaporation exceeds precipitation, and all months are above 32°F (0°C).

BSk = Same as BSh, but with at least one month below 32°F (0°C).

BWh = One-half the precipitation of BSh, and all months above 32°F (0°C).

BWk = Same as BWh, but with at least one month below 32°F (0°C).

Cf = Same as Cs, but no dry season

Cs = Eight months 50°F (10°C) or higher; coldest month below 64°F (18°C); and summer dry.

Dca = Four to seven months above 50°F (10°C); coldest month below 32°F (0°C), and warmest month above 72°F (22°C).

Dcb = Same as Dca, but warmest month below 72°F (22°C).

Do = Four to seven months above 50°F (10°C); coldest month above 32°F (0°C).

E = Up to three months above 50°F (10°C).

Fi = All months below 32°F (0°C).

Ft = All months below 50°F (10°C).

A/C boundary = Equatorial limits of frost; in marine locations, the isotherm of 64°F (18°C) for coolest month.

B/A, B/C, B/D, B/E boundary = Potential evaporation equals precipitation.

C/D boundary = Eight months 50°F (10°C).

D/E boundary = Four months 50°F (10°C).

E/F boundary = 50°F (10°C) for warmest month.

**Koppen did not recognize the prairie as a distinct climatic type. The ecoregion classification system represents it as the arid sides of the Cf, Dca, and Dcb types (USDA Forest Service, 2003).

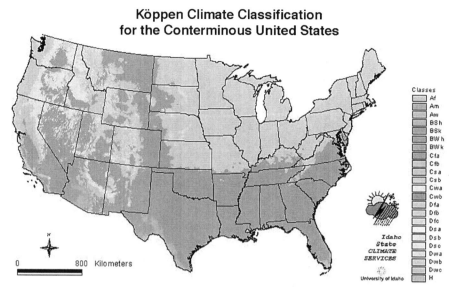

Köppen Climate Classification
for the Conterminous United States

Classes
- Af
- Am
- Aw
- BSh
- BSk
- BWh
- BWk
- Cfa
- Cfb
- Csa
- Csb
- Cwa
- Cwb
- Dfa
- Dfb
- Dfc
- Dsa
- Dsb
- Dsc
- Dwa
- Dwb
- Dwc
- H

Idaho State CLIMATE SERVICES

University of Idaho

0   800   Kilometers

Figure 3.1   Koppen climatic classification maps. (*Source:* Idaho State Climate Services. http://snow.ag.uidaho.edu/Clim_map/Koppen_usa_map.htm.)

ing annual precipitation patterns and variations is of utmost importance in being able to determine the water balance.

Note

There are many textbooks that contain detailed chapters on precipitation, so this section highlights only the important factors affecting watersheds.

Precipitation has two forms: liquid (rain and drizzle) and solid (snow, hail, and sleet). The percentage of these forms that falls in a watershed will dictate the amount of available soil moisture and much of the geomorphologic characteristics of the watershed. Drizzle has droplets typically less than 0.5 mm, whereas raindrops have diameters greater than 0.5 mm but typically between 1 and 5 mm (ASCE, 1996). Precipitation form, temporal distribution, seasonal distribution, frequency, intensity, and duration, as well as storm event distribution and mean annual rainfall, are all important factors to understand when performing a watershed assessment. A sample of monthly precipitation distribution can be found in Figure 3.2.

Raindrops typically have a terminal velocity (speed at which they hit the ground) of between 4 and 9 meters/second (Chow et al., 1988). This can have a tremendous impact on exposed soils. (Additional information on the impacts of rainfall on the watershed can be found throughout this book.)

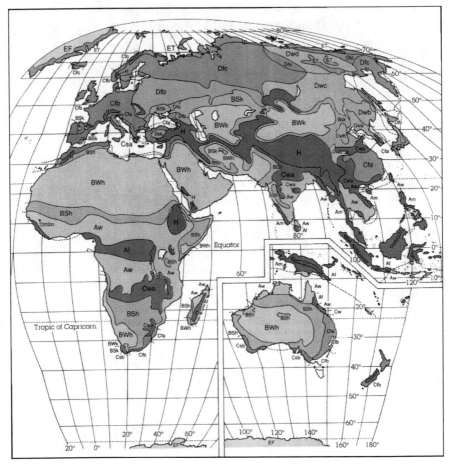

Figure 3.1   (*Continued*). (*Source:* Ahrens, 1991.)

3.2.1 Types of Precipitation

Precipitation may be classified by the conditions that produce the rising column of unsaturated air. *Convectional precipitation* is caused by uneven heating of the ground, causing the air to rise and expand, vapor to condense, and precipitation to occur, as can be seen in Figure 3.3. Typical convectional precipitation would be summer high-intensity, short-duration thunderstorms. Results are often flash flooding in small watersheds and a small portion of the precipitation infiltrating into the groundwater, and then only on a local level. *Orographic precipitation* is a result of topographic barriers such as mountain ridges forcing the moisture-laden air to rise and cool, resulting in precipitation, as can be seen in Figure 3.4. This is typical of the Pacific Northwest, where the Sierra Nevada cause the moist Pacific air to rise, re-

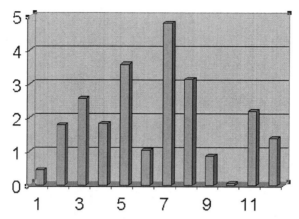

Figure 3.2 Monthly precipitation distribution in the United States. (*Source: Climatic Atlas of the United States,* U.S. Department of Commerce, 1979.)

sulting in about 100 inches of rainfall per year. *Cyclonic precipitation* is related to large low-pressure systems that take five to six days to move across the United States. These storms are usually low-intensity, long-duration storms lasting two to three days, as can be seen in Figure 3.5. These storms are the most beneficial for groundwater replenishment. Convectional and orographic storms are typically more localized, whereas cyclonic storms have a large spatial extent.

3.2.2 Recurrence Interval (Return Period)

The term "100-year storm" is often used to describe the storm that would have a magnitude and frequency of the theoretically largest storm that would

Figure 3.3 Convectional precipitation.

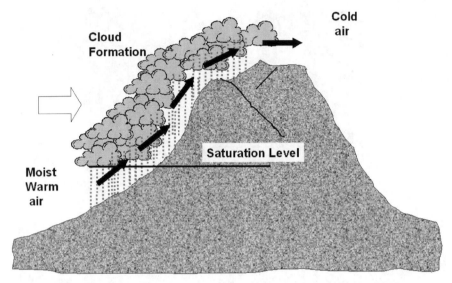

Figure 3.4 Orographic precipitation (modified from Strahler, 1976).

occur once every 100 years, often referred to as the *return period*. The 100-year storm may not necessarily create the 100-year flood. For instance, if a storm equivalent to the 100-year rainfall were to occur after a drought, the runoff may not produce a flood of the 100-year magnitude, due to greater-than-normal infiltration. However, a storm equivalent to the 25-year rainfall, falling on frozen ground or with a snow pack, might produce a 100-year flood. Therefore, the *x*-year storm should not be used interchangeably with the *x*-year flood.

This nomenclature—that is, the "100-year flood"—tends to comfort people who assume that they will not be flooded for another 100 years if they have just experienced a 100-year flood. In actuality, two storms of that magnitude could occur within the same year. The preferred nomenclature for the 100-year storm would be the storm with a 1 percent probability of occurrence in any one year, or the 1 percent exceedence-probability storm. The annual exceedence probability of a storm or precipitation (*P*) (or flood, *Q*) can be

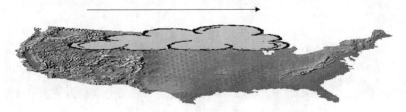

Figure 3.5 Cyclonic precipitation.

converted from the return period (T) using equation 3.1. The conversion from the old to the new nomenclature can be found in Table 3.2. The exceedence probabilities can be applied to storms or floods; however, it should be kept in mind that the two are not necessarily interchangeable, as just described.

$$P = 1/T \qquad (3.1)$$

Although the preferred method of describing storms or floods is the exceedence probability, the return period nomenclature is so ingrained in the terminology that it will continue to be used in this book. The 2.33–year storm shown in Table 3.2 is equivalent to the mean annual flood or the arithmetic average of all mean daily flows in the period of record (Dunne and Leopold, 1978).

3.2.3 Intensity/Duration/Frequency

Rainfall *intensity* is the rate of rainfall measured in depth over time (e.g., in/hr). The *duration* is the length of the storm. The *frequency* is how often a storm of a particular magnitude (intensity) and duration occurs. Intensity-duration-frequency (*I-D-F*) curves, as shown in Figure 3.6, have been developed throughout the world. These curves allow the user to determine the rainfall intensity for several return periods. As can be seen from Figure 3.7, for western Illinois, there is a 100-year, 24-hour storm (approximately 8 inches) or a 100-year, five-minute storm (approximately 1 inch) (Huff and Angel, 1989). The curves allow determination of the total rainfall amount for a particular duration and frequency.

Combining the values obtained for all regions of Illinois allows development of a *isohyetal* map of rainfall amounts for various return periods and durations, as shown in the 100-year, 24-hour example in Figure 3.7. This total amount, however, is typically not distributed uniformly over the 24-hour period. The USDA Natural Resources Conservation Service (NRCS), formerly

TABLE 3.2   Conversion of Return Period to Exceedence Probability

Return Period	Exceedence Probability
100	0.01
50	0.02
25	0.04
10	0.10
5	0.20
2.33	0.43
2	0.5
1	1.0

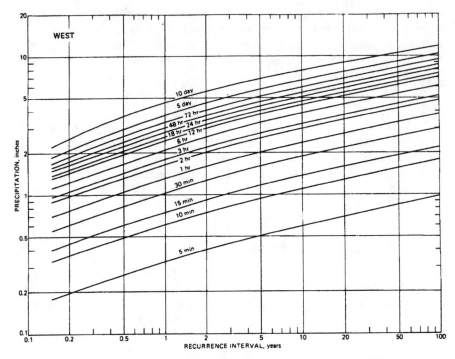

Figure 3.6   Rainfall Intensity-Duration-Frequency Curves (Huff, 1989; www.sws. uiuc.edu/atmos/statecli/pdf/chap3/fdhchri_Ch3all.PDF, 2003).

the Soil Conservation Service, has determined that a typical rainfall distribution follows an S-curve, as shown in Figure 3.8, based upon the regions shown in Figure 3.9 (USDA, 1986). NRCS's determination was from National Weather Service (NWS) duration-frequency data reported in Technical Paper Number 40 (TP-40) (Hershfield, 1961; Frederick et al., 1977). This was determined based upon NRCS research that concluded that "the highest peak discharges from small watersheds in the United States are usually caused by intense brief rainfall that may occur as distinct events or as part of a longer storm" (USDA, 1986). As can be seen from the near vertical portion of the curve, the high-intensity portion of the "synthetic" distribution occurs near the center of the storm.

3.3   HYDROLOGIC CYCLE, WATER BALANCE, AND WATER BUDGET

The *hydrologic cycle* can be described as the exchange of water between the earth's surface and the atmosphere, fueled by energy from the sun through processes such as condensation (cloud formation), precipitation, runoff, infiltration, evaporation, and transpiration (from plants and animals). The hydro-

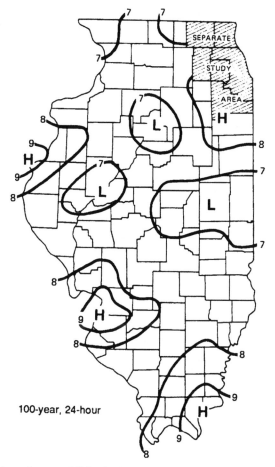

Figure 3.7 Isohyetal map of Illinois for the 100-year, 24-hour rainfall (Huff, 1989; www.sws.uiuc.edu/atmos/statecli/pdf/chap3/fdhchri_Ch3all.PDF, 2003).

logic cycle could be considered the complete cycle through which all water on earth is continuously moving. The hydrologic cycle often refers only to the physical parameters of water, but could also include the many chemical and biological processes. This would become a complex portrayal, which would vary by ecological region and chemical composition of geologic formations, for example; hence, it is beyond the scope of this book. It may be necessary, however, to determine the biological and chemical processes in a watershed assessment to explain chemical composition of the streams.

Under natural conditions, this cycle, called the *hydrologic* or *water balance,* is in equilibrium and accounts for the amounts of water in storage and in transit at any point in time within the hydrologic cycle. The water balance is actually conservation of matter applied to water within a watershed. The

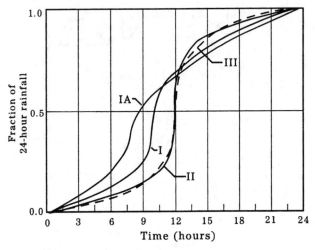

Figure 3.8   NRCS 24-hour rainfall distribution. (*Source:* USGS, 1986.)

term water balance was defined by C. Warren Thornthwaite as "the balance between the income of water from precipitation and snowmelt and the outflow of water by evapotranspiration, groundwater recharge, and streamflow" (in Dunne and Leopold, 1978). Comparatively, ASCE (1996) defines the term water or hydrologic budget as "an accounting of the inflow to (recharge),

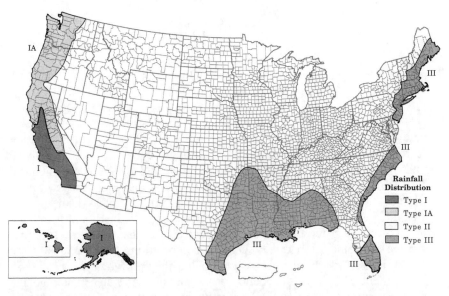

Figure 3.9   Approximate geographic boundaries for NRCS rainfall distribution. (*Source:* USGS, 1986.)

outflow from (discharge), and storage in, a hydrologic unit such as a drainage basin or aquifer."

By the definitions, it is hard to distinguish between the two terms, however, if one wants to differentiate them, one can think of the balance as being in balance with nature in the long term or in equilibrium conditions. In the long term, all water entering the watershed under natural conditions would balance water leaving the system, as shown in Table 3.3 (ASCE, 1996). The budget, in contrast, is described in the short term, where inflow and outflow may not balance due to a month, season, year, or several years of nonnormal precipitation. As with human accounting procedures alteration of the cycle may put the system or budget out of balance. Watershed planning would strive to put the system back into balance through management measures.

Note the term *consumptive use* in Table 3.3: Consumptive use is water that is taken out of the hydrologic system. Under the *natural hydrologic regime,* consumptive use would be water lost to evaporation or transpiration. Maintaining the natural water balance would aid in maintaining the natural hydrologic regime of the watershed, that is keep surface runoff, groundwater recharge, storage and supply, stream base flow, and evapotranspiration at values that occur under natural or undeveloped conditions.

The hydrologic cycle could be described as the *hydrogeologic cycle,* because surface water and groundwater are interconnected by recharge of groundwater through infiltration and stream *base flow* augmentation from groundwater. Stream base flow is the flow that is supplied to the stream by the groundwater system or *recharge of groundwater* or groundwater discharge. Figure 3.10 shows the hydrogeologic cycle schematically.

The hydro(geo)logic cycle can be described numerically by equation 3.2. The water budget can be summarized in many ways, depending on the goals and how detailed the breakdown of the particular constituents needs to be. One portrayal of the water budget where surface and groundwater divides coincide can be described by the hydrologic equation:

$$P = RO + Re + ET + S \qquad (3.2)$$

TABLE 3.3   Natural Hydrologic Balance*

Inflow	Outflow
Surface	Surface
Subsurface	Subsurface
Precipitation	Consumptive use
Surface storage	Increased surface storage
Soil moisture storage	Soil moisture storage
Ground water storage	Groundwater storage
INFLOW = OUTFLOW	

*Modified from ASCE, 1996.

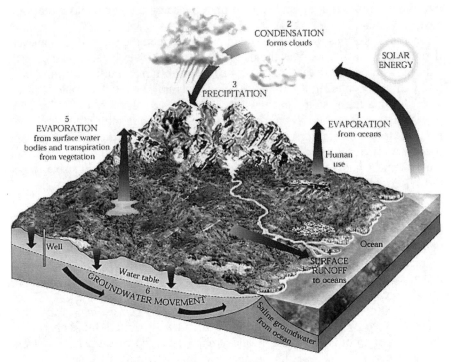

Figure 3.10 The hydro(geo)logic cycle. (*Source:* USGS, 2003, Illustration by John M. Evans USGS, Colorado District. http://ga.water.usgs.gov/edu/watercyclegraphic. html.)

Where:

    P = Precipitation
    RO = Runoff as stream flow (surface or direct runoff)
    Re = Recharge of groundwater (groundwater discharge to stream as base flow)
    ET = Evaporation and plant transpiration
    S = Storage in aquifers, streams and reservoirs

This equation is described many different ways in the literature, depending on how much detail is used to define the individual parameters. Of the parameters just cited, all but ET can be directly measured. Precipitation and stream flow can be determined from rainfall and stream flow gages, respectively. Groundwater and surface water storage can be measured through observation of well static water depth and storage changes in surface water impoundments, respectively. ET could therefore be calculated from Equation 3.2 to develop the watershed's water budget.

The ratios of the parameters in Equation 3.2 (runoff to recharge to evapo-transpiration loss) change significantly throughout the seasons in response to plant growth and temperature. Ratios may also vary considerably over time depending upon the intensity of rainfall, saturation of soil, and amount of water in aquifer storage. Figure 3.11 shows the input and output factors for the water balance in the eastern United States. Size of the arrows shows relative proportions of the water budget factors.

What is significant about Equation 3.2 is that, although it is usually applied to an annual water budget, it is also applicable for one-tenth of an inch of rainfall, or the 1 percent, 24-hour (100-year) storm (approximately 6 to 10 inches of rain in a 24-hour period, depending on location). The same equation would apply; however, the ratios of the parameters would be significantly different. For instance, there would be a greater proportion of surface runoff and very little ET in a 24-hour storm, due to the soil's infiltration capacity and storm duration, respectively. That said, when looking at an annual water budget for evaluation of long-term water supply, ET and groundwater recharge become the dominant values. Each parameter in Equation 3.2 is described in more detail in the following subsections.

3.3.1 Precipitation (*P*)

Precipitation includes rain, sleet, and snow. Precipitation (rain or melted snow), when it exceeds the infiltration capacity of the soils, will run off the land. The precipitation that infiltrates may be absorbed by plant roots, adsorbed by soil, or fill pores in shallow bedrock before a surplus exists that can reach the deep groundwater system or *water table*. Although humans have

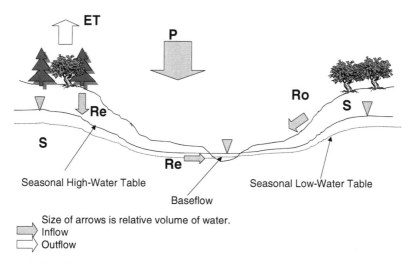

Figure 3.11 Cross section showing typical natural water balance for a humid climate.

the technology and ability to seed clouds to alter the precipitation pattern, it is not regularly performed for watershed management, and therefore is not discussed here.

3.3.2 Direct Runoff (RO)

Total runoff (*TRO*) or *total stream flow* is often used to denote the total amount of water flowing in a stream from *surface* or direct runoff (*RO*), also referred to as *surface discharge* plus *groundwater recharge* (*Re*) (*base flow*) sources. Direct runoff (*RO*) is that portion of the total runoff that flows on the surface of the earth (ASCE, 1996). More specifically, direct runoff is the portion of precipitation that is not absorbed by the soil or deep strata but finds its way into water bodies after meeting the demands of evapotranspiration, including interception. It is that portion of precipitation, sometimes referred to as *excess precipitation,* that flows over the surface during and shortly after a rain to directly or indirectly enter a stream, drainage ditch, swale, or storm sewer. Such runoff typically drains from a watershed within three to five days after a precipitation event, depending on soils, geology, land cover, and watershed size. Direct runoff can be thought of as *stormwater runoff.*

Direct runoff measurements can be made from direct stream flow data taken from a continuous recording station. Indirect, site-specific estimates can be made using empirically derived formulas such as the Natural Resources Conservation Service Curve Number Method or the Rational Method. The variety in surface materials, slopes, soil types, and rainfall intensities affect the total volume and rate of runoff. Stormwater runoff feeds the streams with a quick and short-lived supply of water, as opposed to the water supply that is essential to support aquatic life, the base flow supplied by groundwater discharge.

3.3.3 Recharge to Groundwater (Re) (Groundwater Discharge to a Stream)

Recharge, sometimes referred to as *groundwater runoff,* is the water that infiltrates the soils and that soaks into the aquifer system by natural methods. *Recharge* (*Re*) is also called groundwater discharge because it discharges base flow to the streams. This infiltration does not include very shallow waters that drain from the soil into ditches to feed streams immediately after a rain event. Water draining from saturated soils and bedrock, long after a rainfall, will feed the stream system as base flow. This value is usually calculated by knowing the change in groundwater level and aquifer storage coefficient. Infiltration from precipitation is very difficult to measure because it is affected by surface conditions, slope, soil, and bedrock, many of which can vary considerably. Therefore, the infiltration from precipitation is typically measured by the stream base flow, which is water leaving the aquifer. Because the natural system is in balance, stream base flow approximates the water flowing into the aquifer. Other factors, such as plant transpiration of groundwater, well consumption, and storage variances, must be considered to accurately determine the true infiltration value. This value may vary significantly over time and location. Recharge (Re) can be further divided into base flow (BF), change in groundwater storage (GWs) and groundwater evapotranspiration $GWet$, or:

$$Re = BF + \Delta GWs + GWet \qquad (3.3)$$

An important factor in understanding groundwater movement is stream *base flow*. Stream base flow is water that flows from the aquifer, allowing a stream to flow during dry periods long after a rain has ended, and thus is critical to the health of a stream. The amount of base flow entering the stream is an indication of the amount of precipitation that has entered the aquifer. If aquifer withdrawal by wells consistently exceeds base flow, the aquifer will experience deficit conditions, resulting in declining groundwater levels.

Stream base flow is affected by precipitation, evapotranspiration, well withdrawal, infiltration systems (sewage systems), aquifer storage, geology, and topographic considerations. The rate of stream base flow varies geographically; therefore, it is desirable to determine the stream base flow for a specific region. Stream base flow, which is measured in a variety of ways, is important to maintain stream temperatures and sustain aquatic life. Base flow varies throughout the year based upon precipitation and recharge.

3.3.4 Evapotranspiration (*ET*)

Evapotranspiration (ET) is water lost from the system by a combination of evaporation from surface water and transpiration from vegetation. According to the Pennridge Water Resources Plan, "*evaporation* is the direct movement of rainwater from ponded surface depressions, leaves, and branches into the air. *Transpiration* is the removal of water from the plant's vascular system via water loss from leaves. Tree transpiration differs from grass transpiration in that during dry periods, the deeply rooted trees can withdraw water from the entire soil profile down to bedrock and even from the shallow water table, thus removing water from the system long after the surface soils become dry" (BCPL, 2002). This is important, as wooded areas are cut for agriculture, forestry, parks, and residential development. One would think that deforestation would therefore aid in deep groundwater system replenishment, but the surface disturbance that is usually associated with these changes more than offsets the water withdrawal from roots by increasing surface runoff and having a net negative recharge effect.

ET can be calculated for the watershed using an indirect method that incorporates evaporative pan and soil moisture readings. Evaporative pan values are recorded at several National Oceanic and Atmospheric Administration (NOAA) weather stations. There are also standard empirically derived values that can be used to determine this factor, although they will vary significantly from reality as a result of the local effects of rainfall frequency, cloudiness, soil type, soil thickness, and plant type.

3.3.5 Storage (S)

The water that is temporarily stored in the saturated soils and bedrock is the *storage value*. The storage is typically in equilibrium, with inflow equaling outflow. Groundwater withdrawals (wells), prolonged droughts, or quarry pumping can all throw the equilibrium out of balance and change the available storage. Changes in aquifer storage from seasonal variations and wet-versus-dry precipitation years will ultimately balance out and may be considered as having no long-term effect on the equation.

Storage (S) could be broken down into subsurface soil (Ss), subsurface bedrock (Sb), and surface water (Ssw) storage, as shown in Equation 3.4.

$$S = Ss + Sb + Ssw \qquad (3.4)$$

The permanent stream channel, through its entire length, flowing at base flow conditions, stores a tremendous amount of water. This volume would be easy to determine knowing stream dimensions and the length of the study reach.

So, as one can see, the hydrologic budget can become extremely complex with many forms of Equation 3.5.

$$P = RO + BF + \Delta GWS + GWET + ET \pm Ss + Sb + Ssw \qquad (3.5)$$

On any water resource protection project, when trying to maintain the natural hydrologic "balance," keep this equation in mind.

3.4 ANTHROPOGENIC (MAN-MADE) CHANGES TO THE WATER BALANCE

Under natural, forested conditions, stormwater would fall on tree leaves and branches before flowing over the forest floor to enter small streams. Modifications in vegetative cover, through plowing, paving, and building rooftops, usually result in an increased total volume and peak rate of stormwater flow (i.e., runoff). Impervious surfaces such as pavement and rooftops are major culprits in altering the water balance by reducing infiltration and increasing stormwater runoff. Runoff in urbanized areas is also influenced by the lack of vegetation that intercepts the rainfall. The peak rate of stormwater runoff from land developments during and after storm events is, in some cases, controlled by *stormwater detention basins*. Over the past 15 to 20 years, the stormwater detention basins that can be seen as one passes residential and commercial developments have become commonplace. Unfortunately, that method of controlling stormwater runoff only reduces the peak rate of storm-water entering a stream; it does not reduce the volume. Because the creation of any new impervious surface reduces natural infiltration and increases the volume of stormwater runoff, there will be a net increase in the amount of stormwater leaving a land development site unless certain *best management* practices (BMPs) are installed to improve infiltration and reduce runoff volume.

Although land developers are required to install stormwater control structures for all new construction, many states do not have a formal mechanism in place at present to catalog or monitor the various stormwater controls as built. Such a lack of information makes it difficult to determine stormwater runoff rates and volumes for the wide range of storm events and landscape conditions that occur. (A more detailed discussion on urban hydrology can be found in Chapter 18.)

Many changes to the natural hydrologic balance occur due to land and water alteration and urbanization by humans. Land use has a significant impact on the hydrologic regime of the watershed. As mentioned previously, a natural watershed is in equilibrium. The human impact via land use throws this equilibrium out of balance. The impact of a variety of land use changes is presented in Table 3.4. Because the water balance as defined in this text includes water chemistry, the human impact on water quality is also discussed. Therefore, the previous table, Table 3.3, can be modified to show the human influence on the budget, which may affect the hydrologic balance one way or the other, as shown in Table 3.4.

The human influence on any of the factors in Tables 3.3 and 3.4 could put the budget out of balance. An example of human-induced influences on the water budget is consumptive use. A well withdraws groundwater, which is used and then discharged to a sanitary system, which bypasses the stream base flow system, as shown in Figure 3.12; or a diversion takes water out of one watershed and discharges it into another. Even if the stream base flow conditions are not significantly altered, the groundwater consumption could

TABLE 3.4   Factors of the Hydrologic Budget Modified by Humans*

Inflow	Outflow
Surface	Surface
Subsurface	Subsurface
Precipitation	Consumptive use
Imported watershed resources	Exported water
Decreased surface storage	Increased surface storage
Decreased ground-water storage	Increased groundwater storage
INFLOW may not equal OUTFLOW.	

*Modified from ASCE, 1996.

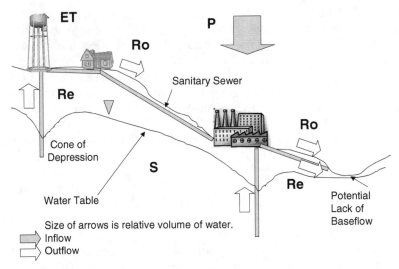

Figure 3.12 Several anthropogenic changes to the water cycle.

significantly affect aquifer supplies in times of drought. Increased storm surface runoff from an increased impervious area is actually a consumptive use as far as groundwater storage goes. When evaluating a holistic water resources management project and determining water balances, one must look at the various land use scenarios that affect input and consumption of water resources.

Runoff (from precipitation events and snowmelt) and infiltration replenish and restore surface water and groundwater resources. However, if adverse impacts such as excessive groundwater withdrawal (e.g., groundwater mining or pumping) or discharge of pollutants exceed the system's capability to recover, water availability and quality will decline. Stream base flow and groundwater aquifers will be reduced. With less base flow and a constant discharge of pollutants, the overall concentration of pollutants in a water body will increase. Damage to the natural environment (e.g., sensitive ecosystems) will result, and the costs for obtaining and treating water for human uses will increase. Inadequate or improper control of stormwater can also result in erosion and damage to stream channels, roads, bridges, and structures. Other anthropogenic (man-made) impacts to this balance (positive or negative) are sanitary sewer export, surface diversions, and change in recharge due to impervious areas. These should be considered when evaluating the balance of a watershed, and will be discussed in more detail in later sections of this chapter.

Human factors, such as impervious surfaces, stream discharges of treated wastewater, and withdrawals of surface water and groundwater, need to be incorporated into the hydrologic equation. Factors that influence groundwater discharge into streams as base flow also need to be considered. In addition to the natural infiltration from pervious surfaces, the effects of infiltration from sources such as on-lot septic systems, spray irrigation systems, and stormwater BMPs (e.g., infiltration basins) should be considered. Groundwater withdrawals from public wells that enter a public sewer system and are carried out of the watershed as treated wastewater effluent represent a net loss of stream base flow water from the system. The cumulative effect of individual, on-lot well pumping will be a negative factor in the equation, as it is removing water that would have ultimately entered the stream as base flow. In addition to the anthropogenic changes to the water balance just described, several additional issues will be discussed in the following subsections.

3.4.1 Urbanization (Land Development)

Perhaps the most studied land use as far as its effects on water quality is the urban setting. With urbanization comes an increase in impervious surfaces. Many studies have shown that as the percent imperviousness of a watershed increases, so does stream degradation. This is the result of many factors, including decreased infiltration and percolation of rainwater due to the impervious surface that results in decreased base flow; increased stormwater runoff peaks and volumes that lead to downstream and localized stormwater problems and flooding; increased stream temperature and contributions of nonpoint source pollution from parking lot or urban impervious area runoff, which decreases water quality; and pollutants from automobile emissions, lawn and garden chemicals, atmospheric deposition, and litter, which accumulate on impervious and man-made surfaces and get washed off into the waterways. The initial runoff carries with it the highest concentration of pollutants and is termed the *first flush*.

Streams in urban environments are stressed physically, chemically, and biologically. Physically, the increased magnitude and frequency of runoff increases the frequency of a stream reaching its critical erosive velocity, the point at which the channel begins to erode. This causes deepening, widening, straightening, and sedimentation problems. Most urban stream corridors have been straightened, enclosed, or channelized in the past. This increases the channel slope, as discussed in more detail in Chapter 5 and shown in Figure 5.1, which only serves to transport the problems downstream. It also removes habitat for essential aquatic species, thus degrading the biodiversity.

Rainfall hitting paved areas during the hot summer months produces increased temperatures in the runoff, termed the *thermal plug*. This, in turn, increases the temperature in streams, which could be lethal to sensitive aquatic organisms. Degraded water quality reduces its usability by humans and decreases aquatic diversity.

3.4.2 Wastewater

Sewage treatment plants (STPs) typically discharge treated wastewater to the surface water of streams. In many instances, water is withdrawn and collected, either from surface or groundwater sources, in one watershed and discharged via the STP in an adjoining watershed. These data can be obtained from sewage treatment plant records. Careful analysis is needed regarding the source of the water relative to watershed boundaries to be certain where the source water originated.

Wastewater discharges, usually from municipal sewage treatment plants, large businesses, and quarries, may provide consistent flows that influence natural stream base flow. Many chemical and temperature factors, as well as consistency of flows, ultimately determine whether such flows benefit aquatic life.

The most readily available source of information is from National Pollutant Discharge Elimination System (NPDES) permits, granted by the State Environmental Department. NPDES permits regulate any process water discharged onto the land surface, including businesses that discharge cooling waters, large and small-scale sewage treatment systems, and quarries that pump water from the pit to keep it dry.

Various water/wastewater land development scenarios exist with different implications to the water budget, as outlined here:

- On-lot septic and on-lot wells
- On-lot septic and community or public water supply
- Community or public sewer and on-lot wells
- Community or public sewer and community or public water supply

As one can see, the combination of water input and withdrawals could significantly alter a local water balance. For instance, a community withdrawing water from individual on-lot wells and discharging their effluent through community sewerage systems could conceivably take groundwater out of one watershed and discharge it as surface water in another watershed.

3.4.3 Infiltration Inflow (*I* and *I*)

Groundwater often infiltrates into the joints of the sanitary sewer system and flows to the STP to be discharged to the receiving stream, particularly in older sanitary collection systems in which the collection system was installed within the water table. This is a direct short circuit of the hydrologic cycle, with the groundwater being diminished, and can have severe consequences in times of droughts. I-and-I discharges will also affect the aquifer storage. Studies have shown I-and-I flows to be a significant portion of the wastewater discharge. A study performed on the Pennridge Wastewater Treatment Authority facility in Sellersville, Pennsylvania, estimated average annual infiltration rates from 1985 to 2000 that ranged from 28 to 48 percent of total sewage flow (Bucks County Planning Commission, 2002). This is water that is bypassing either the aquifer system or the stream system through travel within the sewage piping system. Attempts by municipal authorities to reduce infiltration, and thereby reduce treatment costs and exportation of groundwater from the watershed, typically are met with varying levels of success.

3.4.4 On-Site Septic Systems

On-site septic systems return treated septic system effluent to the aquifer via seepage from an in-ground or sand-mound drain field. Generally, lots using on-lot septic systems are rural in nature and utilize their own wells. Typically, approximately 10 percent of the water withdrawn from the wells is lost between the well and the septic system through evapotranspiration from the drain field and consumptive uses such as car washing.

3.4.5 Irrigation

Much of the irrigation water that originated from a potable water supply and is used to irrigate lawns, golf courses, or agriculture exits the system through plant transpiration and evaporation from ponded depressions. Irrigation water tends to accumulate salts as it continues to be circulated, a potential problem to the groundwater system. Wastewater spray irrigation systems allow wastewater to be recycled; however, they can experience significant evaporative losses.

3.4.6 Stormwater Retention Systems

Stormwater retention systems "retain" stormwater runoff, thus allowing it to soak into the ground and feed the aquifer system. Infiltration may occur through storage swales, basins, pervious pavement, or infiltration trenches.

3.4.7 Wells

Water wells extract groundwater out of the aquifer and bring that water to the surface, thus intercepting its natural flow. How that water is utilized and discharged has a big impact on the hydrologic budget. The potential effect of disrupting the natural flow is that the groundwater would have eventually entered the stream system as base flow. If this water is utilized, then discharged to a sanitary sewer that transports what is now wastewater many miles downstream before being treated in a sewage treatment plant and discharged back to the stream, that portion of stream from the withdrawal point to the discharge may be deprived of base flow.

For the purposes of this book, wells are categorized into four general categories, from smallest to largest:

1. Private home wells
2. Wells used for commercial (small business) establishments

3. Community water supply wells

4. Wells serving industrial activities

Groundwater may also be utilized by heat pumps (which commonly recirculate the water), industrial, commercial, and irrigation purposes. Public well data are typically reported on a monthly basis, with daily measurements also usually available. Individual small wells would require a survey or metering to determine their usage.

3.4.8 Basement Sump Pumps

Basement sump pumps, if discharging to a storm sewer or roadside, can remove a significant amount of water from the shallow groundwater source, diminishing shallow aquifer levels. If sump pumps are piped to a surface water conveyance system, the water that had been in the shallow aquifer is removed and rerouted for eventual discharge into streams and possibly exportation out of the watershed.

3.4.9 Agriculture

Perhaps the next most studied land use and its effect on watersheds is agriculture. Whereas the greatest impact of urban areas is on water quantity, agriculture significantly affects water quality. Agricultural cultivation processes disturb the surface and leave soil exposed, increasing sediment load to streams if not properly managed. In addition, application of fertilizers and pesticides increases nonpoint source pollution to streams during runoff events. Runoff from barnyards and pastures carries with it nutrients from the manure. Allowing cattle to graze and disturb the stability of stream banks is still a problem today.

3.4.10 Forestry

Timbering operations increase runoff on a short-term basis. If properties are reforested properly, timbering should have minimal long-term effects on the hydrology of a watershed. However, the short-term operations can cause substantial erosion from skidding and logging trails and exposed surfaces. In many states, logging practices may be exempt from many of the regulations that would protect water resources.

3.4.11 Mining and Quarries

Mining and quarries have a significant impact on the water balance and water quality if not properly managed. Unfortunately, much of our nation's mining activity occurred long before regulations were in place to ensure proper management of mined lands. Coal mining's biggest impact is from acid mine drainage. Surface mine cuts impound waters that collect minerals and may affect other surface waters and groundwater regimes and quality. Mine water is often pumped, extracting water from the groundwater system and discharging directly to the surface water system after being treated. To determine the impact on the water table both during the mining activity and after mining has ceased, it is necessary to understand groundwater flow systems.

Mines could intercept and either retain or detain surface runoff. Surface mines and quarries could act like a combination stormwater detention basin and well. The mine or quarry could have an influence on the adjoining aquifer, which may affect wells and stream base flow. Mines and quarries are extremely deep excavations, so they are typically filled with groundwater. This groundwater is pumped and discharged to surface streams to keep the pit free from water. Ultimately, the water enters the stream as it would from natural means; however, water is pumped from the ground and discharged at one point instead of flowing naturally from seeps in the ground into the stream. Ideally, the quarry should regulate large stormwater flows to some degree by collecting large rains and pumping them more slowly into the stream than would have occurred from direct surface runoff. Water that would have been transpired by plants from the soil into the air during the summer months will be captured and pumped into the stream. During periods of relatively low groundwater levels, water may be pumped into the stream when it may not have flowed under natural conditions, if the quarry is deeper than the stream channel.

Each quarry is unique in its design, current depth, rock type, and hydrogeologic setting. In general, quarries depress the local groundwater table and may thus adversely affect wells. This may also affect stream base flow upstream of the discharge point. Downstream of the discharge point, stream base flow may be more consistent and higher than prequarry conditions. Improperly placed spoil along stream banks could impede and obstruct flow and cause flooding. Erosion can lead to solids and sedimentation in waterways.

To predict water quality impacts, the chemical composition of the geologic material should be understood. In coal-mined areas, pyrite, oxygen (O_2), and water, with the aid of catalytic bacteria, chemically react to form acid mine drainage (AMD).

If planned out and managed properly, surface water impoundments within mined areas can provide habitat for aquatic and terrestrial wildlife, irrigation water, and recreation opportunities, and can be reclaimed as amenities for prosperous development sites. However, these recommendations do not ensure good water quality without extensive innovative solutions.

3.5 LOW FLOW

The stream flow during the dry season is the *low flow* that can be measured in a variety of ways. One measurement of low flow is the seven-day 10-year low flow or lowest seven-day consecutive discharge with a 10 percent probability of occurring every year (Q_{7-10}). The annual *seven-day minimum flow* is the lowest mean discharge for seven consecutive days for a calendar year, water year (October 1 to September 30), or climatic year (April 1 to March 31), and should not be confused with the Q_{7-10}. The *1-in-25-year low flow* is approximately half the rate measured for the 1-in-25-year annual low base flow rate. The *instantaneous low flow* is the minimum instantaneous discharge occurring for the designated period, typically the water year (Durlin and Schaffstall, 2002). Other low flow standards include the following two:

- 1Q20: Lowest one-day flow recorded over a 20-year period
- 30Q2: Lowest mean flow recorded over 30 consecutive days over a two-year recurrence interval

Stream base flow and low flow are typically measured in terms of flow per unit area, and commonly expressed as gallons per day per square mile of contributing drainage area (GPD/sm). Another useful stream flow statistic is the *annual mean flow,* which is the arithmetic mean of the individual daily mean discharges for the *water year.*

The amount of stream flow required for protection of a stream's fishery can be determined using the Instream Flow Incremental Methodology (IFIM), which was designed by a multidisciplinary United States Fish and Wildlife Service team in Colorado in the 1980s (Young, 2003). The methodology has been refined in many ways over the years, but in essence it involves the development of several computer models that are used to define the relationship between flow and fish habitat. The four basic types of models used are a hydraulic model, a biological model, a habitat model, and a water quality model.

The hydraulic model determines how water depth, velocity, cover, and substrate composition may change with flow at stream transects (cross sections) under various flows. The biological model determines what depths, water velocities, cover types (such as root wads, large boulders, undercut banks, or vegetation), and substrate types are most suitable for each species and life stage (spawning, fry, juvenile, and adult). The hydraulic model and the biological model are then combined to determine how habitat changes with flow for the described parameters.

The model can be used to estimate the amount of habitat loss due to flow reduction that a particular stream can undergo without damaging the aquatic species. Instream flow protection levels can then be determined; in turn, measures can be developed to maintain the hydrologic regime and water balance. A population loss of 5 percent is considered minimal or statistically nondetectable (Young, 2003).

In a natural system, where groundwater divides coincide with surface water divides, water enters the watershed as precipitation and leaves as evapotran-

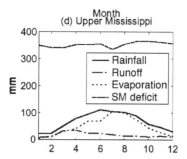

Figure 3.13   Annual precipitation, runoff, evaporation, and soil moisture (sm) deficit for the Upper Mississippi River Basin (Schaake, 2000).

spiration (ET) and stream flow, that is, the natural hydrologic cycle. However, when groundwater divides do not coincide with surface water divides, water also enters or leaves the hydrologic system as groundwater *underflow*.

3.6 SUMMARY

The complexity of understanding and managing the water balance and budget becomes more obvious as each component of the system is studied. The *natural flow regime* changes over time whether annually, monthly, or daily. Figure 3.13 shows the annual fluctuation in rainfall, runoff, ET, and soil moisture or storage deficit (sm) for the upper Mississippi River Basin (Shaake, et al., 2000). The change may be by orders of magnitude. Flows change almost daily on most streams. Fish are adapted to all flows in this naturally occurring flow regime, including floods and droughts, to a certain extent. Floods transport bed load, redistribute substrate, and shape the stream for critical fish habitat. Droughts allow instream habitat conditions for smaller fish species and fry to be protected temporarily from larger predator species, increasing biodiversity. However, when the natural flow regime is altered in some way, fish populations can be stressed.

Water budget studies attempt to define natural and human inputs and withdrawals in numeric terms for use in predictive regulatory and management models. In reality, quantitatively defining even one relatively simple factor such as drought can become a complex and controversial task. An understanding of the movement of water through the hydrologic system is important to the development of an accurate water budget.

In developing a sustainable water resources watershed plan, to include the goal of maintaining the natural hydrologic balance, the impact of land use changes by humans, such as land development, must be kept in mind. Historically, the impact of humans changes in land use—such as grazing, agriculture, forestry, mining, or development—has been, in certain instances, to diminish the benefits of a healthful, environmentally balanced hydrologic system. Proper management of the watershed should strive to maintain the water balance that existed prior to land use change by preventing water from being removed from the system at a greater rate than is required by people and stream-dependent life. This can be accomplished through proper water budget management and conservation, as will be discussed in later chapters.

REFERENCES

Ahrens, C. D. *Meteorology Today,* 4th ed. (Stamford, CT: West Publishing), pp. 473.

ASCE. *Hydrology Handbook,* 2nd ed., ASCE Manuals and Reports on Engineering Practice, No. 28 (New York: American Society of Civil Engineers), 1996.

Bucks County Planning Commission (BCPC), Borton-Lawson Engineering, and Phil Getty. Pennridge Water Resources Plan, 2002.

Brown, R.H. "Hydrologic Factors Pertinent to Groundwater Contamination," Public Health Service Technical Report W61-5 (Washington, D.C.: U.S. Public Health Service), 1961.

Cate, B. *Ground Water Pollution* (St. Louis, MO: Directory Press), 1972.

Chow, V.T., Dave R. Maidment, and L.W. Mays, *Applied Hydrology,* McGraw-Hill Book Co., New York, NY, 1988.

DeBarry, Paul A., and Dennis Livrone. "Innovative Intermunicipal Water Resources Protection and Management Plan Utilizing GIS," 2001 Pennsylvania Stormwater Management Symposium, Villanova University, October 17–18, 2001.

Dunne, T., and Luna L. Leopold. "Water in Environmental Planning" (New York: W.H. Freeman), 1978.

Durlin, R.R, and W.P. Schaffstall. Water Resources Data—Pennsylvania—Water Year 2001, vol.1. Delaware River Basin, Water Data Report PA-01-1, U.S. Geologic Survey, 2002.

Foth, H.D. *Fundamentals of Soil Science,* 6th ed. (New York: John Wiley & Sons, Inc.), 1978.

Frederick, R.H., V.A. Myers, and E.P. Auciello. "Five- to 60-Minute Precipitation Frequency for Eastern and Central United States," U.S. Department of Commerce, National Weather Service, National Oceanic and Atmospheric Administration Tech. Memo NWS HYDRO 35. Silver Spring, MD, 1977.

Fried, J.J. *Groundwater Pollution* (Amsterdam: Elsevier Scientific Publishing Co.), 1975.

Hershfield, D.M. "Rainfall Frequency Atlas of the United States for Durations from 30 Minutes to 24 Hours and Return Periods from 1 to 100 Years," Weather Bureau Technical Paper No. 40 (Washington, DC: U.S. Department of Commerce. Washington, DC), 1961.

Huff, F. A., and J. R. Angel. "Rainfall Distributions and Hydroclimatic Characteristics of Heavy Rainstorms in Illinois" (Bulletin 70), Illinois State Water Survey, 1989.

Idaho State Climate Services. http://snow.ag.nidaho.edu/Clim_map/Koppen_usa_map.htm.

Koppen, W. *Grundriss der Klimakunde* (Berlin: Walter de Gruyter Co.), 1931.

McLaren, A.D., and G.H. Peterson. *Soil Biochemistry* (London: Edward Arnold Publishers), 1967.

Penn State University, "Wetlands and Water Management on Mined Lands," Proceedings of a Conference, October 23–24, 1985.

Pettyjohn, W.A. *Water Quality in a Stressed Environment* (Minneapolis, MN: Burgess Publishing Co.), 1972.

Pruel, H.C. "Contaminants in Groundwaters near Waste Stabilization Ponds," *Journal of the Water Pollution Control Federation,* vol. 40, January 1968, pp. 659–661.

Pruel, H.C. and G.J. Schroepfer. "Travel of Nitrogen in Soils," *Journal of the Water Pollution Control Federation,* vol. 40, January 1968, pp. 30–44.

Reese, Geoffery, and Paul A. DeBarry. "The GIS Landscape for Wellhead Protection," Proceedings of the ASCE National Hydraulics Conference, San Francisco, CA, 1993.

Schaake, John C., Quingyun Duan, and Shuzheng Cong. "Retrospective Analysis of Water Budget for the Mississippi River Basin," Fifteenth Conference on Hydrology, AMS, Long Beach, CA, January 9–14, 2000.

Thornthwaite, C.W. and J.R Mather. "The Water Balance," *Laboratory of Climatology. Publication No. 8,* Centerton, NJ, 1955.

Trewartha, G. T. *An Introduction to Weather and Climate,* 4th ed. (New York: McGraw-Hill), 1968.

Trewartha, G. T., A. H. Robinson, and E.H. Hammond. *Physical Elements of Geography,* 5th ed. (New York: McGraw-Hill), 1967.

USDA, Soil Conservation Service. "Urban Hydrology for Small Watersheds—Technical Release 55 (TR-55)", 2nd edition, Washington D.C., June 1986.

USDA Forest Service Home Page, www.fs.fed.us/colorimagemap/images/app1.html (June 2003).

U.S. EPA. "Wellhead Protection: A Guide for Small Communities," EPA/625/R-93/002, Washington DC, February 1993.

U.S. Geological Survey, Water Science for Schools, Web Page, http://ga.water.usgs.gov/edu/watercycle.html, 2003.

U.S. Geological Survey, "Hydrogeology and Ground-Water Quality of Northern Bucks County, Pennsylvania," USGS Water-Resources Investigations Report 94-4109, Lemoyne, PA, 1994.

Walker, W.H. "Illinois Ground Water Pollution," *Journal of the American Water Works Association* (January 1969), vol. 61, p. 31–40.

Young, Leroy M. "Fish Habitat and Flow: What's the Connection?" Pennsylvania Fish Commission, Harrisburg, PA, http://sites.state.pa.us/PA_Exec/Fish_Boat/ma2001/habtflow.htm (2003).

CHAPTER 4

HYDROGEOLOGY

4.0 INTRODUCTION

Although watershed analyses typically concentrate on surface features, what goes on beneath the surface also affects the quality and quantity of stream flow. Contamination of groundwater begins at the surface, but affects water supply and surface water. Although there are many contaminants, it's the soluble chemicals such as nitrates that tend to travel most easily in groundwater; therefore, nitrogen contamination is used as an example in this chapter. Nitrates, in high concentrations in drinking water, cause *methemoglobinemia,* or *"blue baby,"* in infants, whereby the nitrates take the place of oxygen in the bloodstream, causing the babies to turn blue. It is important, then, to have a basic understanding of hydrogeology, or the study of groundwater and geology, to aid in determining various processes occurring in the watershed.

4.1 GROUNDWATER

Of the water that infiltrates, some replenishes the soil moisture deficiency if the soil is not saturated at the time of the storm. When saturated, this water is often referred to as the *shallow ground water system*. It then *percolates* through the soils and into the upper layer of bedrock until it reaches the saturated zone of bedrock called the *aquifer* or *deep groundwater system*. The upper water surface of this saturated zone, referred to as the groundwater, is called the *water table*. The portion of bedrock above the water table is typically not permanently saturated; it is called the *vadose* or *unsaturated zone*. Flow-through *aquifers* follow Darcy's law:

$$v = K(h_1 - h_2) \div l \tag{4.1}$$

Where:

v = Velocity
h_1 = Elevation of the higher point
h_2 = Elevation of the lower point
l = Horizontal distance between points h_1 and h_2
K = Hydraulic conductivity of the medium, as described in Chapter 2

Knowing the velocity and the distance, the time for a contamination plume to travel from point h_1 (a spill) to h_2 (a well) can be determined.

Aquifers may be in unconsolidated or consolidated materials. *Unconsolidated materials* are composed of loose rock, gravel, or mineral particles, such as alluvium. *Consolidated materials* are the hard bedrock, such as *sedimentary* (limestone, dolomite, shale, and sandstone), *metamorphic* (quartzite and gneiss), or igneous (granite and basalt).

Stream flow is composed of *groundwater discharge* (*base flow*) and *surface runoff* (*storm flow*). Where streams are fed by the shallow groundwater, they act as drains for the shallow groundwater system. This situation, known as *gaining streams,* is common in the humid Northeast and Northwest. However, in some instances, streams may lose water to the groundwater system, at which point they are known as *losing* or *ephemeral streams. Intermittent* streams, in contrast, alternate between gaining and losing streams, depending on water table elevation, which is dependent on time of year. All three situations are shown in Figures 4.1, 4.2, and 4.3, respectively. The situation can

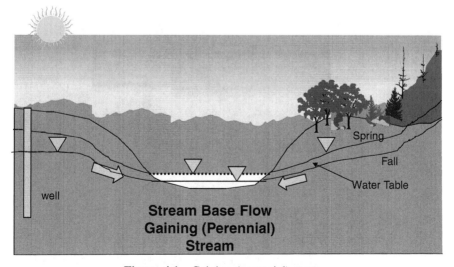

Figure 4.1 Gaining (perennial) stream.

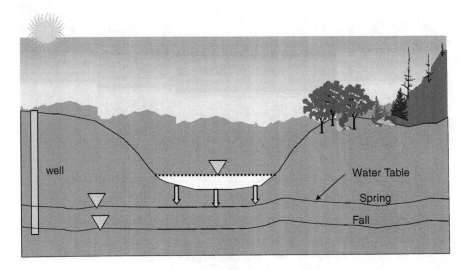

Stream Base Flow
Losing (Ephemeral)
Stream

Figure 4.2 Losing (ephemeral) stream.

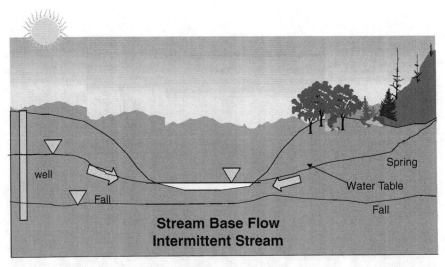

Figure 4.3 Intermittent stream.

be determined by measuring *static water levels* or elevations in wells and comparing them to the base flow surface water elevation of the stream, thus creating a *potentiometric map* of the area, as shown in Figure 4.4. In developing such a map, factors such as well water levels from confined versus unconfined conditions, timing of data collection, and previous precipitation conditions (e.g., droughts) need to be considered.

When the water level in the wells is at a higher level than the stream, the stream would most likely be a gaining stream. Conversely, if the water table were lower than the stream surface water elevation, it would be a losing stream. Streams may be gaining streams in the spring of the year when the water table is at its highest, directly after snowmelt, and losing streams during the end of the water year in late summer and fall. This would also apply to drought situations. Streams in carbonate geology are often losing streams. Studies have shown that water recharging the groundwater system from a losing stream could discharge to an adjoining watershed if the proper slope or *hydraulic gradient* exists (USGS, 1998).

Sedimentary rocks typically are the best source of groundwater supply, whereas the fine-grained rocks such as siltstone and shale are limited in this regard, due to low primary permeability. However, shale is often fractured, providing a very high *transmissivity*, the rate at which water is transmitted through a unit width of aquifer under a unit hydraulic gradient. It is equal to the product of hydraulic conductivity and saturated thickness of the aquifer (Todd, 1980), expressed in units of feet squared per day, or:

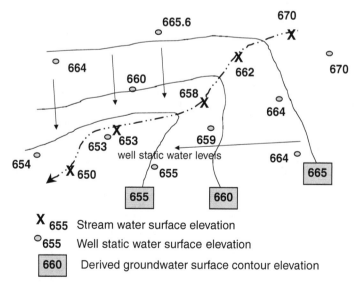

Figure 4.4 Potentiometric map showing contours derived from stream base flow elevations and well static water levels.

$$T = Kb \qquad\qquad (4.2)$$

Where:

T = Transmissivity $(ft^2)/day$
K = Hydraulic conductivity (ft/day)
b = Saturated thickness of the aquifer (ft)

This is important for water supply, and could have detrimental implications in contaminant transport. The water in these two strata is often also mineralized, with water from shale typically being "hard" and/or having a low pH. Carbonate rocks also possess fractures and other secondary openings that provide for high-yielding aquifers. Groundwater flow along these carbonate fractures and bedding planes dissolves the carbonates, the same process that forms caves. This provides rapid movement of groundwater. Where the dissolution reaches the surface, the groundwater is prone to possible contamination. Sandstone may have a porosity of up to 30 percent (U.S. EPA, 1987), but may also be fractured, providing even greater transmissivity.

Igneous and metamorphic rocks have very small primary pores. They are, therefore, poor water suppliers. The majority of the porosity and permeability of igneous and metamorphic rocks is in secondary openings formed by dissolution of certain minerals, faults, and fractures. These fractures drain quickly after a period of recharge from infiltration after a precipitation event, and if exposed at or near the surface, they are prone to contamination. Fractures, and in turn water and contaminant flow, are difficult to locate and determine their direction. Streams often form along these fracture zones because they provide the path of least resistance.

Unconsolidated aquifers typically have a much greater water supply yield, but are also more prone to contamination. Aquifers can range from a few acres wide and a few feet thick to thousands of miles wide and hundreds of feet thick. Porosities of up to 90 percent are possible in unconsolidated deposits (U.S. EPA, 1987); however, 40 to 50 percent is more common. The top of *unconfined aquifers* is atmospheric pressure, whereas *confined aquifers* have an impermeable or slowly permeable layer above them, essentially trapping the water and allowing pressure to build. The water level in wells tapped into a confined aquifer will rise above the aquifer; if it rises high enough to flow onto the land surface, it is known as an *artesian well,* as shown in Figure 4.5.

When undertaking a hydrologic study for a watershed assessment, one must fully understand the interaction between the land, surface water, and groundwater. In humid and semiarid regions, the water table generally conforms to the land surface topography. The hydraulic gradient, or slope, of the water table slopes away from the topographically high areas (mountain ridges and divides) toward adjacent low areas, typically streams and rivers. The high

Figure 4.5 Artesian well with water flow out of the well cap due to underlying pressure.

areas generally serve as groundwater recharge areas, while the low areas serve as groundwater discharge areas, as shown in Figure 4.6. Of importance is that the groundwater's chemical composition is altered by the minerals contained in the bedrock through which it passes.

The interrelation of surface and groundwater, and long-lasting effects of pollution, come to mind in the case where the people of the town of Crosby, North Dakota, boasted of "the best coffee in the state." Their water was drawn from a 38-foot deep-dug well. After testing the water that made this great coffee, it was found to contain 150 mg/l of nitrates and 2176 mg/l of dissolved solids. History showed that a horse stable had been operated from the late 1800s until 1930, the wastes of which provided the "body" for the best coffee in the state many years later (Pettyjohn, 1972).

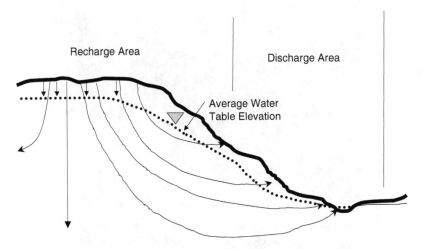

Figure 4.6 Cross-section of land showing recharge and discharge areas. Seasonal high water tables may provide discharges midslope. The water table generally follows the contour of the land.

4.2   FLOW THROUGH POROUS MEDIA

The most common cause of nitrogen contamination of groundwater is the discharge of wastewater near the ground surface. This wastewater originates from septic tanks, livestock feedlots, waste stabilization ponds or lagoons, waste treatment plant effluents, leaking sewers, and spray irrigation (Pettyjohn, 1972; Preul, 1968).

Groundwater pollution occurs mainly by two mechanisms: flow through porous granular material, or flow through dense rock with linear openings or fissures. The degradation of most pollutants is more efficient in flow through porous material. The pollutants are transported to the water table by the flow of percolating water containing the contaminant, which follows the path of least resistance, flowing readily through permeable zones and avoiding impermeable material. The pollutant flows under the influence of gravity from points of higher potential to lower potential (Pettyjohn, 1972).

4.3   FLUID MOVEMENT IN UNSATURATED MEDIA

It has been found that the rate of fluid movement in an unsaturated granular material is related to the fraction of pore space filled with fluid and can be expressed as (Brown, 1961):

$$K_r = \theta_1 - \theta_0 / 1 - \theta_0 \qquad (4.3)$$

Where:

K_r = Relative permeability, that is, ratio of permeability of unsaturated flow to saturated flow

θ_1 = Irreducible fraction of pore space that will retain liquid against the force of gravity

θ_0 = Fraction of pore space filled with liquid for any given regime or unsaturated flow

4.4 FLUID MOVEMENT IN SATURATED MEDIA

The water contained in saturated media is referred to as *groundwater,* with the upper boundary known as the *water table,* to distinguish the *unsaturated zone* from the *saturated zone.* Water will flow laterally in saturated zones, again following the path of least resistance, forming a hydraulic gradient or slope and eventually entering a stream in humid regions. In saturated media, flow is predominantly lateral (see Figure 4.4). Lateral flow in saturated media is often influenced by domestic or municipal discharge wells. With pumping, a cone of depression surrounds the well, influencing the groundwater flow toward the well (see Figure 4.7). The stream, if polluted, may not contaminate the aquifer. However, if a discharge well were located close enough to a

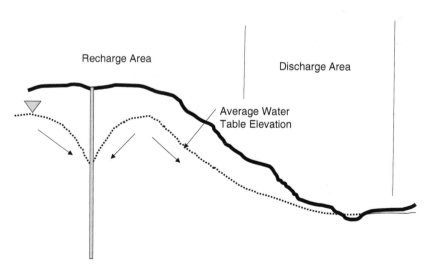

Figure 4.7   Cross section of land showing a pumping well and the cone of depression formed from overpumping.

stream, the water in the stream could be drawn out of the stream and enter the groundwater, contaminating the well. Wells should, therefore, not be drilled near a stream that is receiving sewage treatment plant discharges containing high amounts of nitrates (Walker, 1969).

4.5   FLOW THROUGH SECONDARY OPENINGS IN DENSE MATERIALS

Water and contaminants can flow relatively freely through secondary openings in dense rocks with little or no contaminant degradation. This is commonly the case where the aquifer is made of limestone or dolomite. The contaminated water will travel by gravity through the cracks and crevices in the aquifer, eventually reaching a discharge point or stream as shown in Figure 4.8. The contaminant will remain relatively unchanged since adsorption is minimal. Dilution is the only natural tool to diminish the contaminant concentration. Once in the aquifer mainstream, the contaminant could flow for miles relatively untouched.

4.6   NITROGEN POLLUTION

Many pollutants are of concern in regard to groundwater, too many to evaluate here; therefore, as mentioned in the introduction to this chapter, nitrogen is

Figure 4.8   Direct discharge of groundwater to the surface from fractured rock. Notice the groundwater seeps.

used as an example. In its nitrate form, nitrogen could cause methemoglo-binemia, also called "blue baby syndrome," in infants. Methemoglobinemia is a blood disorder caused when nitrite interacts with the hemoglobin in red blood cells, forming methemoglobin, which, unlike hemoglobin, cannot carry sufficient oxygen to the body's cells and tissues. Therefore, in the following subsections, nitrogen transport will be discussed, with many of the same principles applying to the range of soluble pollutants. Detailed information on a specific chemical's transport capabilities should be obtained if a watershed assessment determines that pollutant is a problem or potential problem.

4.6.1 Ammonia Nitrogen

Free ammonia in groundwater may be an indication of recent pollution and, in turn, contamination from other biological waste constituents such as fecal coliform bacteria. Ammonia nitrogen may be found in the groundwater as NH_3 or NH_4^+, depending on the pH; however, NH_4 is the predominant form (Preul, 1968). Because the ammonia ion (NH_4^+) is positively charged, it is adsorbed by the negatively charged soil particles. The adsorption capacity of the soil will eventually become exhausted as exposed negative sites are utilized. The process follows the empirical isotherm of adsorption proposed by Freundlich, and is expressed as follows:

$$A = KC_o^{1/n} \tag{4.4}$$

Where:

> A = Adsorption in micrograms of NH_4 per gram of dry soil or masses or substances adsorbed over the adsorbant
> C_o = Initial concentration of NH_4 in solution
> K and n = Empirical constants depending on soil surface area

Preul and Schroepfer plotted adsorption versus solution concentration for three soil textures and derived the graph shown in Figure 4.9 (Preul and Schroepfer, 1968).

Referring to the equation, it can be seen that adsorption is dependent upon the concentration of ammonia in solution and the surface area of the soil. The clay has the smallest particle sizes and, therefore, the largest surface area and, in turn, a greater adsorption capacity or ion exchange capacity.

Ammonium adsorption is also dependent on the relative abundance of other ions present. For example, potassium has similar ion-exchange characteristics as NH_4 and can, therefore, compete for ion exchange sites. Preul found that NH_4 adsorption did decrease with increasing concentration of the potassium cation present, as displayed in Figure 4.10. However, most wastewaters do not contain high concentrations of cations other than ammonia.

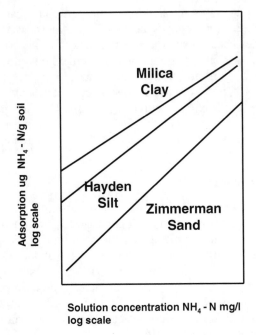

Figure 4.9 Adsorption isotherms for NH$_4$ on three soils (Preul, 1968).

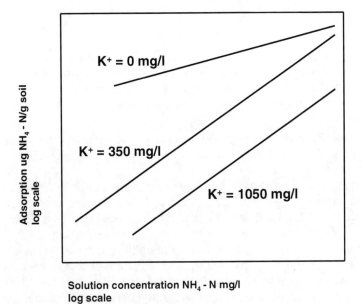

Figure 4.10 Adsorption isotherms for NH$_4$ with three concentrations of potassium cation (K$^+$) present. (Preul, 1968).

4.6.2 Nitrate-Nitrogen

As stated previously, nitrates have been found to cause methemoglobinemia in infants, which reduces the ability of the blood to carry oxygen. The control of nitrates in groundwater is therefore essential. Nitrate adsorption by soils is dependent on pH. At pHs below 7.0, nitrates may be adsorbed, whereas at neutral pH, nitrates will not be adsorbed.

Biological nitrification will change some influent NH_4 to nitrates under aerobic conditions. This will even occur in saturated flow if there is an abundance of dissolved oxygen in the waste influents (Preul, 1968).

Whether nitrates constitute a problem in groundwater depends greatly on their source. For example, ammonia nitrogen seeping through the bottom of a waste stabilization pond would remain unoxidized because oxygen would be at a minimum, whereas nitrates would become abundant in groundwater underlying a septic tank tile field (assuming a neutral pH) because the flow is near the surface (Preul, 1968). In one instance, nitrate levels in groundwater were found to be 22 times greater than normal groundwater levels just 65 days after a septic tank absorption field was installed (Cate, 1972). A diagram showing how septic tank effluent percolates through an unsaturated medium is shown in Figure 4.11 (Miller, 1980).

It has been found that nitrate concentrations will increase initially with distance from a sewage effluent percolation bed and will then diminish, due to nitrification and dilution, respectively (Cate, 1972). Under anaerobic conditions, nitrate serves as an oxygen source for certain bacteria, reducing nitrates to elemental nitrogen and oxides, or nitrogen. These gaseous forms could then fill soil pores (McLaren and Peterson, 1967).

4.6.3 Adsorption and Nitrification through the Unsaturated Soil Profile

Preul and Schroepfer (Preul, 1968) conducted a study in which they measured adsorption and nitrification of NH_4^+ as the waste percolated through a soil profile. They found that, initially, adsorption was the major inhibitor of nitrogen travel through soil. Then, as biological growth became established, specifically *nitrosomonas* and *nitrobacter,* nitrification became significant.

The concentrations of NH_4 versus NO_2 and NO_3, which existed on soil columns after 135 days of unsaturated flow, with an influent having 25 mg/1 $NH_4 - N$, are shown in Figure 4.12 (Preul and Schroepfer, 1968).

It can be seen that the $NH_4 - N$ was depleted from an influent of 25 mg/1 to less than 1 mg/1 after only 2.5 feet of travel. The depletion of 24 mg/1 was due to adsorption and nitrification. It can also be seen that the greatest amount of nitrification occurred within the first 2 feet of soil, and it occurred at a slower rate at greater depths due to the limited availability of oxygen. At 1.5 feet, adsorption was at its maximum (approximately 10 mg/1 $NH_4.N$), and decreased to about 5.6 mg/1 at the bottom of the column due

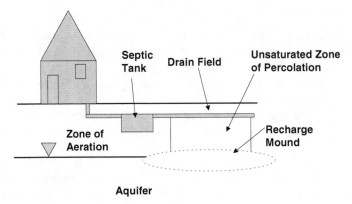

Figure 4.11 Percolation through the zone of aeration. Most of the natural removal or degradation processes function under these conditions (Deutsch, 1965).

to nitrification of the adsorbed NH_4 by microorganisms. Therefore, nitrification is the major influence on nitrogen transport through unsaturated soils. Different soil textures have different adsorption and nitrification rates; however, the same general trends occur. Under normal pH ranges and soil environments, practically no inhibition of nitrate movement in soils occurs, as discussed previously. This could cause serious nitrate problems in groundwater as the use of septic tanks becomes more abundant.

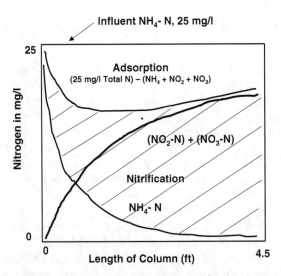

Figure 4.12 Combined effects of adsorption and nitrification in unsaturated column flow through sand (Preul, 1968).

4.6.4 Adsorption and Nitrification over Time for Saturated Flow

Preul and Schroepfer (1968) also experimented with nitrogen transport through saturated soils. In saturated soils, pollutants would be diffused and dispersed in the field; however, these aspects were eliminated in the columnar studies performed. Again, the influent nitrogen was in the form of NH_4 and the sand had a pH of 7.2.

The results showed that adsorption dominated for the first 50 days until the microorganisms became established. Adsorption followed an S-curve, showing a decreasing adsorption rate as ion exchange sites became saturated, creating an equilibrium.

Nitrification is dependent on the amount of oxygen contained in the influent, as oxygen would be limiting in the saturated soil. It is likely, then, that most or all of the nitrification occurred in the top layers of soil until the influent's oxygen was depleted. The tests did show a depletion of dissolved oxygen near the bottoms of the columns.

Under anaerobic conditions, as in the bottom of the columns, bacteria would reduce NO_3 to nitrogen gas, nitric oxide, or nitrous oxide (McLaren and Peterson, 1967). These forms of nitrogen would be represented in the adsorption section of the graph. This, along with less nitrification occurring, helps explain the smaller amount of NO_3 than in unsaturated flow. Both adsorption and nitrification of nitrogen are directly related to the concentration of influent NH_4, with oxygen as the limiting factor in nitrification (Preul, 1968).

Comparing the graphs of saturated flow versus unsaturated flow at day 135, the major difference is the nitrogen effluent reaching the groundwater in unsaturated flow is composed mainly of nitrates, whereas the effluent nitrogen in saturated media is mainly NH_4 (Preul, 1968).

Nitrogen, deposited in the soil mainly from wastewater application, occurs largely as the ammonium cation. This cation will be adsorbed by negatively charged clay particles as it percolates, until the negatively charged sites are saturated. In the presence of oxygen, ammonia will be converted by bacteria to nitrate (an anion), which will flow freely into the groundwater, causing a potential hazard. Once in the saturated material, the nitrates will be reduced to gaseous nitrogen at a minimal rate.

In unsaturated soils, most of the ammonia will be oxidized to nitrates, whereas in saturated flow, most of the ammonia will be unaltered. Once in the groundwater system, nitrates may become diluted by dispersion to allowable concentrations for drinking.

Nitrogen pollution will flow relatively untouched through crevices in solid material, such as limestone or dolomite, causing serious groundwater pollution problems.

4.6.5 Dispersion in Saturated Media

Dispersion of fluid flow in porous media is the disappearance of a transition zone between two fluids with different compositions, ending in a uniform

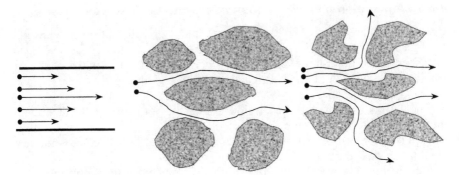

Figure 4.13   Components of dispersion (Fried, 1975).

chemical composition. Dispersion results from a combination of mechanical mixing and physiochemical phenomena.

Mechanical mixing is due to changes in velocity distribution from boundary effects of particles on the fluid flow. The fact that fluids are viscous gives them zero velocity at the soil profile, creating a velocity gradient in the flow. Stream lines will also fluctuate from the mean direction of the flow due to impeding soil particles, and variations in pore size will create velocity discrepancies (Fried, 1975). Figure 4.13 shows all three means of mechanical dispersion.

The physiochemical phenomenon consists of molecular diffusion, which is a function of a chemical potential gradient or concentration (Fried, 1975). If a diffusion gradient exists, ions will travel from high concentrations to low concentrations, seeking equilibrium. Dispersion is a major factor in keeping some groundwaters within the allowable limits of nitrate concentrations.

4.7   SUMMARY

Watershed processes and impacts to watersheds are often perceived as those that occur on the land surface. However, what happens underground is just as important, if not more so, to maintaining the hydrologic regime of a watershed. Groundwater and surface water comprise a single resource, intimately connected. Watershed assessments and management plans should include a hydrogeologic assessment based upon on the established goals of the watershed. Understanding chemical interactions and their movement is also important.

REFERENCES

Brown, R.H. "Hydrologic Factors Pertinent to Ground-Water Contamination," Public Health Service Technical Report W61-5 (1961).

Cate, B. *Ground Water Pollution* (St. Louis, MO: Directory Press), 1972.

Deutsch, M. "Natural Controls Involved in Shallow Aquifer Contamination," *Journal of Ground Water,* vol. 3, 1965.

Foth, H.D. *Fundamentals of Soil Science,* 8th ed. (New York: John Wiley & Sons, Inc.), 1990.

Fried, J.J. *Groundwater Pollution* (Amsterdam: Elsevier Scientific Publishing Co.), 1975.

McLaren, A.D., and G.H. Peterson. *Soil Biochemistry* (London: Edward Arnold Publishers), 1967.

Miller, David W., Ed. *Waste Disposal Effects on Ground Water* (Berkeley, CA: Premier Press), 1980, p. 188.

Pettyjohn, W.A. *Water Quality in a Stressed Environment* (Minneapolis, MN: Burgess Publishing Co.), 1972.

Preul, H.C. "Contaminants in Groundwaters Near Waste Stabilization Ponds," *Journal of the Water Pollution Control Federation,* vol. 40, January 1968, pp. 659–661.

Preul, H.C., and G.J. Schroepfer. "Travel of Nitrogen in Soils," *Journal of the Water Pollution Control Federation,* vol. 40, January 1968, pp. 30–44.

Todd, David Keith. *Groundwater Hydrology,* 2nd ed. (New York: John Wiley & Sons, Inc.), 1980.

U.S. EPA. "Handbook: Ground Water," EPA/625/6-87/016 (Cincinnati, OH: U.S. Environmental Protection Agency, Office of Research and Development), March 1987.

USGS. "Ground Water and Surface Water—A Single Resource," U.S. Geological Survey Circular 1139, Denver, CO, 1998.

Walker, W.H. "Illinois Ground Water Pollution," *Journal of the American Water Works Association,* January 1969, vol. 61, p. 31–40.

CHAPTER 5

HYDROGEOMORPHOLOGY (WATERSHED AND STREAM MORPHOLOGY)

5.0 INTRODUCTION

A *watershed* is defined as the "area of land that drains water, sediment, and dissolved materials to a common outlet at some point along a stream channel" (Dunne and Leopold, 1978). For this reason, watersheds are classified by size and complexity into other generalized terms such as *drainage basins* or *subwatersheds*. A drainage basin is often associated with a *river* as defined later, such as the Mississippi River Basin or Colorado River Basin. Often, a small watershed is a tributary to a much larger drainage basin. Watersheds are often broken into smaller subwatersheds for analysis purposes. Watersheds are therefore associated with streams. They are also referred to as *catchments*.

Natural drainage patterns vary significantly depending on topographic, geologic, morphologic, vegetative, soil, and climatic conditions. The ridges that separate the watersheds are referred to as *drainage divides* or *watershed boundaries*. The ridges are surface water divides, and as can be seen in Chapter 2, may not coincide with the subsurface flow divide due to strikes or dips in the geology. In other words, the subsurface flow may cross under surface watershed boundaries.

Watershed assessments, along with the specific watershed definition, are often defined to a particular *point*, often referred to for analysis' sake as the *point of interest*. Watershed drainage areas or watershed sizes are determined to a point of interest, or in many cases, several points of interest, depending on the scope of study or project. For the U.S. Army Corps of Engineers trying to keep open the Mississippi River for navigation at New Orleans, the point of interest would be where the Mississippi empties into the Gulf of Mexico, with a watershed drainage area of 1.25 million square miles. For the engineer designing a storm sewer system, the point of interest would be the inlet box with a watershed drainage area of perhaps one acre or less. For storm sewer design calculations, these small watersheds are often incorrectly referred to as *drainage areas,* when, in fact, the drainage area is actually the area in square feet, acres, or square miles of the watershed draining to the point of interest.

This book concentrates on coordinating the watershed management approach with stream restoration, and elaborates on the processes involved within the watershed even before the water reaches the stream, looking at a more comprehensive watershed management approach, that is, the bigger picture.

5.1 HYDROGEOMORPHOLOGY

Research revealed that the term *hydrogeomorphology* is defined differently in various references. In addition, terms like *headwaters* or *tributaries* are used rather loosely, with no consistency among disciplines, which may lead to misunderstandings. As noted in the Preface to this book, consistency in definitions is extremely important in a multidisciplinary science such as watershed management. Thus, a secondary goal of this chapter is to define the terms used by the disciplines that utilize watershed management principles to provide for future consistency.

Watershed analysis needs to examine two major branches of earth science: *geomorphology* and *hydrology,* which in this book are referred to collectively as hydrogeomorphology. Geomorphology is the study of landforms, and hydrology is the study of water. Hydrogeomorphology is, consequently, the study of the impact of hydrologic processes on the land. The key to watershed management is to truly understand these hydrogeomorphologic processes.

To begin to fully understand the interaction between the various fields of physical science that govern how watersheds, rivers, and streams function, we'll examine the terms that define them. For instance, what is the difference between *fluvial geomorphology* (*FGM*), *river morphology,* and *hydrogeomorphology?* We'll work backward: The suffix *-ology* means any science or academic field; *morph,* something that has a particular form, shape, or structure (e.g., morphology is defined as "the study of the structure of something"); and *geo-* (prefix) refers to earth or soil. Geomorphology, then, is the study of the formation and structure of the earth's surface features or the study of landforms and the nature of the materials underlying them. *Fluvial* refers to being produced by or found in a river or stream. *Fluvial geomorphology* is, therefore, the study of landforms and processes associated with rivers and streams, or a stream's definition based upon the climate, geology, soils (streambank material), vegetation, and topography of its watershed. *Hydro-* is the prefix for water, liquid, or moisture, and *hydrology* is the study of water. FGM is discussed in more detail in Chapter 13.

5.1.1   Fluvial Geomorphology

Because fluvial geomorphology (FGM) studies the processes and characteristics of streams and rivers, it is often applied specifically to stream restoration. However, if the watershed itself is out of balance, and the watershed changes over time due to urbanization, the restoration measure, just as dikes and levees of the past, will have a limited life span. These processes include rainfall/infiltration/runoff as they relate to the formation, functioning, and characteristics of streams, and are crucial in properly managing a watershed's water resources.

The fundamentals of fluvial geomorphology are an important factor in understanding the relationship between form and process in the watershed and past and present watershed dynamics. Knowledge of the stream's FGM enables the investigator to develop sound integrated management decisions, and, if assessed with hydrogeomorphologic principles, helps to determine the consequences of proposed restoration activities. Utilizing FGM principles to restore streams is termed *natural stream channel design* (Rosgen, 1996). Although most of Dave Rosgen's book concerns characterization, classification, and assessment of stream channels, it does include one chapter on design applications. As with any science, site- or region-specific factors should be evaluated, assessed, and incorporated into design. The need for region-specific applications, particularly in fairly urbanized eastern states' watersheds, has prompted some states/agencies to develop their own stream restoration guidelines. One such publication is "Guidelines for Natural Stream Channel Design in Pennsylvania," developed by the Keystone Stream Team (2003). This guideline recommends consistent procedures for problem identification, data collection and analysis, design options, permitting guidance, construction considerations, construction monitoring, and designer qualifications, recognizing and applying Rosgen's river restoration approach.

To summarize, the purpose of applying the FGM principles to natural channel design is to restore the channels to their natural configuration based upon future considerations of the watershed, so that future stability or equilibrium is maintained.

Note

In the discussions throughout this book, where principles apply to either streams or rivers, the two terms will be used interchangeably, even though they will be defined separately in Chapter 6.

5.1.2   Fluvial Processes

In analyzing hydrogeomorphic processes, one must think in geologic time, or the formation of hydrosystems over thousands and millions of years, in contrast to the natural tendency of people to think in human time, or the lifespan of a person. Often, as a result of the desecration or alteration by humans of the land surface, they accelerate the geologic processes, forcing the hydrosystem out of equilibrium, resulting in the man-made problems of erosion, sedimentation, flooding, and water quality problems. For instance, the process of geologic erosion is natural; it takes place over millions of years and creates our landforms, channels, and soil deposition and horizons. Humans, in contrast, while disturbing the soil profile for development activities, create an exposed surface and *accelerated soil erosion*. Rainstorms may detach and splash soil particles upwards of 2 to 3 feet in height and a radius of 5 feet, and as much as 100 tons of soil per acre may be detached from a single storm on a disturbed site (Beasley, 1972). This process exceeds the natural geologic erosion many times. Thus many states have developed erosion and sedimentation control guidelines and policies to minimize accelerated erosion.

Why is FGM important in watershed management? Hydrogeomorphic processes (watershed land use, natural stream channel formation, etc.), under equilibrium conditions, form stable conditions. Over a geologic time period, natural conditions (climate, vegetation life span, etc.) may cause streams to adjust and move. Over shorter time periods, a change wouldn't be noticeable unless something unnatural occurred to upset the equilibrium. This is especially true in alluvial rivers. The influence of humans (increased runoff, streambank erosion, etc.) or a unique high-intensity storm might accelerate these changes.

Streams, too, form through the hydrogeomorphic processes influenced by climate and physiography. Under natural conditions, stream patterns, shape, dimensions, and slope form until a "stable" condition is met. Over geologic time, the stream is then considered stable, and does not significantly accumulate sediment deposits or scour its stream banks, terms referred to as *aggrade* and *degrade*, respectively. Stream pattern or form is part of the "equilibrium" balance. If a meander is altered, the slope is changed. Some variation in meander length and radius of curvature is observed, but this is an allowable range of natural variability and an acceptable range of stability.

Stability is a function of whether a stream can transport its *sediment load*. When sediment load is less than a stream's *carrying capacity,* the stream will tend to scour its stream banks. When there is too much sediment, the stream will deposit this excess sediment, or aggrade. Sediment load will be discussed in further detail later in this chapter. Watershed changes or human-induced stream configurations alter this natural process, and streams will then "fight back" to get their dimensions, pattern, and profile back into equilibrium.

Contrary to what one would anticipate, streams carry a natural load of sediment. Sediment that is lifted from eddies and carried long distances by the stream is considered *suspended load* and its movement considered *transport by suspension*. Suspended sediment is typically less than 0.5 mm in diameter, depending on the stream. This would be equivalent to clay, silt, and up to medium sand, as described in Chapter 2. Sediment larger than the suspended sediment load particles that the stream flow bounces along the streambed is the *bedload;* its movement is by *traction.* Bedload is typically 5 to 50 percent of the total load (Dunne and Leopold, 1978). Sediment is oftentimes measured by its median grain size, or the diameter for which 50 percent of the particles are larger or smaller than this diameter, often referred to as the D_{50}.

Amounts of normal suspended load and bedload will vary with stream gradient, slope, dimensions, and so on, with each stream having its own equilibrium. If the stream flow has a sediment load less than its equilibrium load, then it will try to resupply this load by eroding the stream banks or channel. If the sediment load is greater than the equilibrium load, then the stream will try to get rid of the excess through *sedimentation* and *deposition.* These relationships are critical in understanding how to manage the watershed for erosion/sedimentation control in developing a watershed plan, and may explain cause-effect relationships.

This relationship can be summarized by the simple relationship of work:

$$\text{Available Power} = \text{Rate of doing work} \div \text{Efficiency} \qquad (5.1)$$

The available power is provided by the movement of mass from higher to lower elevations; therefore, the steeper the stream slope, the greater the power. Given a static power condition, the channel will adjust its geometry to increase its efficiency.

Relaying these principles to channel geometry, the effects of straightening a channel can be seen in Figure 5.1. Assuming the elevations at the top of the channel reach and the bottom of the channel reach for the natural me-

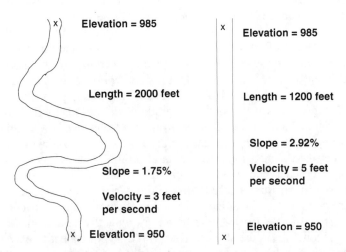

Figure 5.1 Typical effects of natural meandering channel (A) and channel straightened by humans (B).

andering channel as shown, and dividing by the meandering channel length, results in a specific slope. By straightening the channel, the flow length decreases, resulting in a steeper slope, resultant faster velocity, and, often, increased erosion, as described in more detail in Chapter 12, and more specifically in equations 12.5a and 12.5b.

Before a study of the specific fluvial geomorphologic characteristics of a stream or river is undertaken, a study of the hydrogeomorphology of the watershed should be performed. Watersheds have many unique identifying characteristics that will allow the scientist to better understand the reasons for fluvial characteristics. First, streams and watersheds can be classified into several "hydrogeomorphologic types," based upon the flow patterns of the streams and *tributaries* as described in the next section.

Many watershed management projects tend to concentrate simply on the problem at hand. For example, floodplain management plans concentrate on developing a peak 100-year flow from a watershed through the most cost-efficient, reliable method, such as regression equations. (These studies typically do not perform hydrologic modeling or detailed hydrologic budget calculations.) Floodplain hydraulics are then applied to determine the 100-year floodplain. Watershed stormwater management plans concentrate on the "hydrology" of the watershed, typically with hydrologic model analysis, or the rainfall-runoff processes as they relate to increased runoff from new development. Stream restoration projects evaluate the FGM of the streams, but often do not account for future changes in the watershed or *extreme event* protection. Extreme events are the most common causes of stream restoration project failure. Water supply/wellhead protection plans concentrate on the *water budget* or *water balance,* are typically more localized, and do not concentrate on the other factors. *Rapid bioassessment protocols (RBPs)* also look specifically at the biota in certain sections of the stream, but what is happening in the watershed could be assessed before even going out into the field. This book hopes to bridge the gap between all of these disciplines, to allow readers to obtain an overall picture before applying their specific disciplines, and to take a more comprehensive, coordinated watershed approach for long-term solutions.

5.2 WATERSHED CLASSIFICATIONS

Watersheds, when looked at from an overall aerial view, display a specific "footprint," which describes a lot about the watersheds' and the streams' characteristics. These footprints or patterns will be described in detail in the following subsections. Watersheds and streams with similar characteristics can be "grouped" or "categorized." Comparing a watershed to characteristics in its group makes it possible to ascertain further details on which characteristics might be encountered. Therefore, to simplify study objectives, streams and watersheds often are classified into generalized categories based upon similar characteristics. Classification systems allow the practitioner to quickly identify the characteristics of a stream and geomorphic processes of the watershed, consistency between practitioners, and cross-referencing between streams. As discussed elsewhere in this book, various disciplines have developed classification systems to serve the purpose of each particular discipline. This chapter discusses the various stream and watershed classification systems and how each may be coordinated.

Streams may be classified generally, based upon physical characteristics, or formally, according to a stream classification system such as the Rosgen Method. Watersheds are typically classified based on stream characteristics, and for that reason, classification names are often interchanged between streams and watersheds. In any study, however, stream classifications should be verified, as the high variability in stream dimensions may cause the stream, or sections of the stream, to fail "to fit the mold." Thus, caution should be used when any classification is applied, as the data do not always conform. Only experienced, well-trained professionals should utilize the classification methods. Both generalized and "official" classification systems will be discussed here.

5.2.1 Pattern Classification

Streams are classified based upon their form and patterns or networks they create. The term *stream networks* refers to the connectivity of the stream tributaries and is becoming an increasingly important concept because stream network connectivity is extremely important in the GIS hydrologic distributed modeling, as will be described in Chapter 17. Several watershed classifications are discussed in the following list, based upon their stream pattern.

Dendritic pattern. Streams showing a dendritic pattern form a treelike, or dendritic, arrangement of small streams or tributaries in the *headwaters* (branches) that flow in a variety of directions and continually join to eventually form the "major" stream or river. Dendritic patterns are common and are typical of regions of adjusted systems on erodible sediments, with no unusual geologic features, with horizontal strata, and with adequate rainfall. Dendritic watersheds are typically upside-down pear-shaped, with the stem of the pear being the outlet. The effects of any underlying geologic structure can become apparent when the main stem gets large. A typical dendritic watershed and stream pattern is shown in Figure 5.2a (Pidwirny, 2003).

Parallel pattern. The parallel pattern, in which tributary streams flow in the same general direction and usually join at small angles, is essentially an elongated variant of the dendritic pattern. Dunne and Leopold (1978) refer to this pattern as a "palmate dendritic pattern." A steep regional slope is the primary influence in developing a parallel drainage pattern,

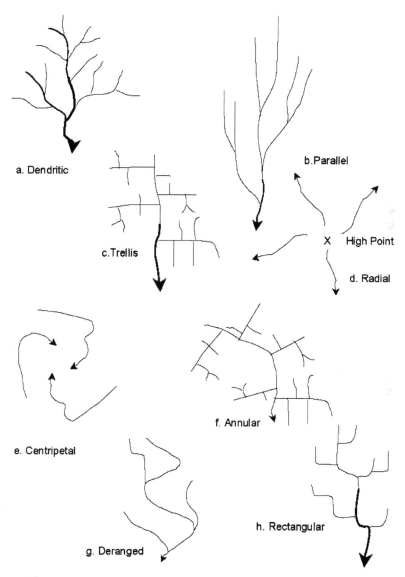

Figure 5.2 Examples of various drainage patterns or networks. (Figures 5.2a, b, c, modified from Pidwirny, 2003; d modified from Ford, 2003; g and h modified from Pidwirny, 2003.)

although any factor, such as noncohesive material that causes a stream to flow unusually far in one direction before merging with another, may cause a parallel system. A typical parallel watershed and stream pattern is displayed in Figure 5.2b (Pidwirny, 2003; Ford, 2003).

Trellis pattern. Streams with a trellis pattern typically have a main stream flowing parallel to bedrock structure, with the tributaries flowing into the main stem of the stream at right angles toward each other from opposite sides. It appears as a "squared-off" drainage pattern. Trellis patterns are typical where tilted sedimentary rocks form weakness planes through which the tributaries can flow until they eventually "break through" resistant layers, to form a *gap*. Trellis drainage patterns are also typical of fully dissected coastal plains. Trellis-style drainage networks dominate in folded rock terrains. A typical trellis watershed and stream pattern can be found in Figure 5.2c (Pidwirny, 2003), (Ford, 2003).

Radial pattern. The radial pattern is common in the vicinity of granite intrusions, salt domes, volcanic cones, and other localized uplifts and is a circular arrangement of streams that flow outward in all directions from the central high area. A typical radial watershed and stream pattern can be found in Figure 5.2d (Ford, 2003; ETE Team, 1997–2001).

Centripetal pattern. A centripetal pattern is uncommon but sometimes occurs in *karst* topography. It also could occur in deserts where *intermittent streams* flow toward a temporary salt lake or *basin*. It can be thought of as internal drainage, where streams flow to a central location. In a centripetal pattern, water flows inward from all directions toward the center of an area, forming a circular arrangement of streams (Ford, 2003; ETE Team, 1997–2003). A typical centripetal watershed and stream pattern is shown in Figure 5.2e.

Annular pattern. Annular drainage patterns can be considered to resemble a bent trellis; they are common on deeply eroded domes such as eroded volcanoes or uplifted sedimentary domes (Ford, 2003; ETE Team, 1997–2001). A typical annular watershed and stream pattern can be found in Figure 5.2f.

Deranged pattern. In areas recently disturbed by events such as volcanic deposition or glacial activity, the first stream patterns to emerge are called deranged stream patterns. These form by the water following the path of least resistance. As sediment gets transported, the stream adjusts its course accordingly over time. An example of a deranged stream pattern can be found in Figure 5.2g (Pidwirny, 2003).

Rectangular pattern. In a rectangular or gridlike drainage pattern, streams form angularly, near 90-degree turns, due primarily to following the fissures, tectonic faults, or joints in the bedrock, as shown in Figure 5.2h (Pidwirny, 2003).

5.2.2 U.S. Geological (USGS) Survey Hydrologic Unit Code (HUC)

In order to standardize the naming of stream categories, in the 1970s the United States Geological Survey (USGS) divided the United States into four levels of *hydrologic units:* regions, subregions, accounting units, and cataloging units, each based upon watershed divides. It is a hierarchical numeric coding system, or *hydrologic unit code (HUC)*, with two digits for each of the four levels. The first set of two digits is based upon 21 regions, as shown in Figure 5.3. The HUC generally serves as the backbone for the country's hydrologic delineation. The breakdown and general characteristics of the HUC are shown in Table 5.1. The unit is often referred to by the number of digits; for example, the eight-digit HUC in Table 5.1 would be referred to as HUC-8 watersheds.

Seaber and colleagues (1987) defined *subregions* to include "the area drained by a *river system,* a *reach* of a river and its tributaries in that reach, a closed basin(s), or a group of streams forming a coastal drainage area." The third level of classification subdivides many of the subregions into *accounting units*. These 352 hydrologic accounting units nest within, or are equivalent to, the subregions. The fourth and smallest element in the hierarchy of hydrologic units is the *cataloging unit* often called watersheds. A cataloging unit is a geographic area representing part or all of a surface drainage basin, a combination of drainage basins, or a distinct hydrologic feature. These units subdivide the subregions and accounting units into smaller areas. There are

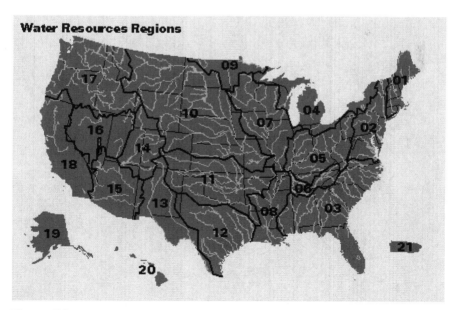

Figure 5.3 U.S. Geologic Survey HUC level 1 (HUC2) consists of 21 regions (USGS, 2003).

TABLE 5.1   Original USGS HUC Levels

Level	No. of "Units"	Typical Number	Term	Typical Range of Drainage Areas (SM)
Region	21	15	Region	>25,000
Subregion	222	1521	Basin	15,000–25,000
Accounting unit	352	152100	Subbasin	8,000–12,000
Cataloging unit	2,150	15210023	Watershed	500–2,000

2,149 or 2,150 eight-digit cataloging units in the nation, depending on the source, as shown in Figure 5.4.

A typical HUC classification system of up to eight digits is shown in Table 5.2 for Region 3.

Realizing the need for a smaller drainage area size and more detailed mapping, the Natural Resources Conservation Service (NRCS) further divided the eight-digit codes into smaller sizes and created digital GIS layers of the data, referred to as the *watershed boundary dataset* (*WBD*). NRCS has mapped units down to the 14-digit level (NRCS, 2003). The Federal Geographic Data Committee (FGDC) developed the FGDC Proposal, for federal standards for delineation of hydrologic unit boundaries, Version 1.0, March 1, 2002, providing guidelines for delineation of smaller hydrologic units (FGDC, 2002).

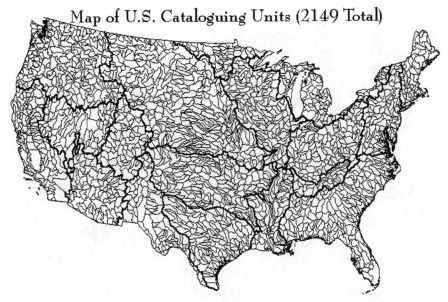

Map of U.S. Cataloguing Units (2149 Total)

Figure 5.4   The 2,149 USGS cataloging units (EPA, 2002).

TABLE 5.2 Typical HUC Region Classification

Region 03 South Atlantic-Gulf Region: The drainage that ultimately discharges into: (a) the Atlantic Ocean within and between the states of Virginia and Florida; (b) the Gulf of Mexico within and between the states of Florida and Louisiana; and (c) the associated waters. Includes all of Florida and South Carolina, and parts of Alabama, Georgia, Louisiana, Mississippi, North Carolina, Tennessee, and Virginia.

  Subregion 0301 Chowan-Roanoke: The coastal drainage and associated waters from and including the Back Bay drainage to Oregon Inlet. Virginia, North Carolina. Area = 18,300 square miles

  Accounting Unit 030101 Roanoke: The Roanoke River Basin. North Carolina, Virginia. Area = 9,680 square miles

    Cataloging Units
    03010101: Upper Roanoke. Virginia. Area = 2180 square miles
    03010102: Middle Roanoke. North Carolina, Virginia. Area = 1,750 square miles
    03010103: Upper Dan. North Carolina, Virginia. Area = 2,040 square miles
    03010104: Lower Dan. North Carolina, Virginia. Area = 1,240 square miles
    03010105: Banister. Virginia. Area = 590 square miles
    03010106: Roanoke Rapids. North Carolina, Virginia. Area = 590 square miles
    03010107: Lower Roanoke. North Carolina. Area = 1,290 square miles

The six different levels of hydrologic units are listed in Table 5.3 (FGDC, 2002). The FGDC proposal did not include the seventh level or 14-digit specifications.

In order to easily cross-reference streams and the watersheds in which they are contained, USGS, in conjunction with the U.S. Environmental Protection Agency (EPA), has developed the Reach File/National Hydrography Dataset

TABLE 5.3 Levels of Hydrologic Units per FGDC 2002 Proposed Standards (*Source:* FGDC, 2002)

Hydrologic Unit Level	Name	Digits	Size	Units
1	Region	2	Average: 177,560 square miles	21
2	Subregion	4	Average: 16,800 square miles	222
3	Basin	6	Average: 10,596 square miles	352
4	Subbasin	8	Average: 703 square miles	2,149
5	Watershed	10	63–391 square miles (40,000–250,000 acres)	22,000 (estimate)
6	Subwatershed	12	16–63 square miles (10,000–40,000 acres)	160,000 (estimate)

(RF/NHD). It is a Geographic Information System (GIS) database of streams, rivers, and water bodies (lakes) that is cross-referenced with the HUC system. The Reach Files are organized by USGS cataloging units. The U.S. EPA utilizes the Reach Files and the HUC as the units for its analyses, including the EPA's Surf Your Watershed Web-based service that is designed to help citizens locate, use, and share environmental information about their state and watershed. In Surf Your Watershed, EPA uses the eight-digit cataloging unit as the basis for information exchange. The HUC is part of the Federal Information Processing Standards (FIPS), which are the standards that processes should follow.

5.3 STREAM CLASSIFICATIONS

Streams or rivers can be classified as watersheds are, into various categories, as discussed in this section.

5.3.1 Generalized Stream Classifications

Rivers and streams are formed over thousands of years through natural erosion, until they establish a *bankfull* flow that represents the equilibrium mean conditions. Under natural watershed conditions, some natural variation in channel geometry and shape due to response from various storms larger or smaller than the bankfull flow can still be considered as a *stable* channel. Rivers and streams reach equilibrium and function as a consequence of the various physical features that comprise them, including climate, hydrology, geology, soils, and landscape. Streams balance erosion, sediment transport, and sedimentation to reach their equilibrium state. If one or more of these factors is altered, the streams will adjust to accommodate this change. For instance, the net effect of urbanization is increased storm flows. This increase in frequency and duration of storm flows erodes channels both vertically and horizontally and increases sedimentation or bed load. The net effect is an unstable channel that physically alters itself to accommodate the changed condition.

The human response to a change such as this has typically been to invoke a structural solution such as channelization, dikes, levees, or upstream impoundments, ignoring the equilibrium conditions of the stream in its natural state. These structural solutions have only a limited life span, with dikes eventually being overtopped and channelization causing increased velocities and downstream sediment deposition. A good approach would, therefore, dictate that equilibrium conditions be replicated as discussed by Leopold et al. (1964) and Rosgen (1996). However, just as channel restoration needs to mimic static stream channel conditions, if the watershed continually changes, or leaves those static conditions such as urbanization, forestry, or agricultural land use practices, the restoration efforts will be ineffective. One must implement a comprehensive, long-term watershed management plan to maintain the watershed in equilibrium conditions, or take the watershed back to its previous equilibrium state, to keep the restoration efforts intact.

This "plan" should not address the goals of the individuals or special-interest groups that are most active in the watershed, but should be structured to evaluate the natural conditions of the watershed and establish goals to bring or maintain the watershed in equilibrium conditions.

5.3.1.1   *Alluvial versus Nonalluvial Channels.* Alluvial channels are those that flow on deposited *alluvial material* (see Chapter 2), and most continually shift in horizontal and vertical location. Alluvial material is *unconsolidated mineral* matter or *transported regolith* that has been transported by streams and deposited in the valley floors. *Nonalluvial* channels are those running through nondeposited materials such as bedrock, and are more constant in their flow path. Streams in alluvial deposits sometimes can be considered as stable due to vegetation and natural energy dissipation in pool/riffle complexes—at least over an engineering life span. Stable conditions allow for deposition/aggradation in response to different storms, but the bankfull event correlates to equilibrium conditions. Similar variability is tolerated when *regional curves* are developed to the 95 percent confidence interval, which still leaves some room for variation centered on the regression equation line.

5.3.1.2   *Perennial, Intermittent, Ephemeral.* *Perennial* streams are those that flow continuously, whereas *intermittent* streams appear to "dry up" when the flow has the potential of being totally absorbed by the bed and underlying material. Intermittent streams may flow continuously during "wet" years. Streams are often intermittent in the headwaters, and become perennial as the watershed drainage area supplies enough base flow to support continued flow. Highly impervious watersheds can cause a perennial stream to become intermittent due to reduced recharge. Ephemeral streams flow only during or shortly after a rainfall event and are often referred to as channels.

5.3.1.3   *Influent versus Effluent.* Effluent streams are those that are re-charged by base flow; they are also referred to as "gaining" streams; conversely, influent streams, also referred to as "losing" streams, are those that recharge the aquifer. This is usually determined through stream gauge analysis at multiple points.

5.3.1.4   *Consequent, Subsequent, Insequent, and Obsequent.* Streams can be classified as consequent, subsequent, insequent, and obsequent based upon their formation or origin. These classifications, defined in the next section, are important for understanding stream morphology and in stream assessment. Consequent streams are those whose course is controlled by the initial slope of the land surface and flow parallel to the dip direction. Consequent streams are the first streams to begin in newly uplifted landscapes

and simply follow the random contours of the rock, flowing parallel to the tilt of the rock. Subsequent streams develop along a zone of weak rock or fissure, follow the pattern of rock exposure, and therefore flow parallel to the direction of strike. Subsequent streams follow weak planes in the rocks and later predominate because they can cut down rapidly and capture other streams. Obsequent streams flow opposite to the dip direction and are generally short and steep, whereas consequent streams are longer. Obsequent streams flow in a direction that is perpendicular to the tilt of the rocks. Insequent streams are those that do not follow a flow path related to tilt or soft places in rocks. A typical placement of sequent stream sequence is shown in the trellis pattern in Figure 5.5.

5.3.2 Formal Stream Classifications

Stream classifications arise out of a particular discipline's need to standardize analytic procedures. Different disciplines look at stream classifications in different ways, each to meet the specific purpose at hand. The unfortunate consequence of this is that a classification developed by one discipline may not be helpful to those in another. The major groups of stream classifications are based on *stream order* and fluvial geomorphology. For instance, the stream order classification does little to help the fluvial geomorphologist. However, future research could find a direct relationship between stream order and FGM class. As multidiscipline practitioners continue to work together to accomplish multifaceted goals, relationships between stream and watershed classification systems will begin to emerge.

One way to begin to try and formulate this relationship would be for each practitioner to classify the streams into all of the major "formal" classifica-

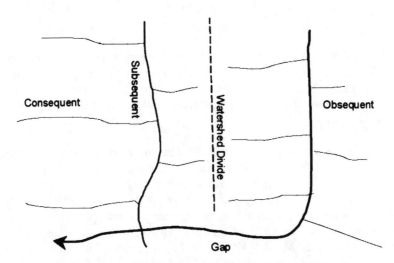

Figure 5.5   Trellis pattern (Ford and Hipple, 2003).

tions outlined in Table 5.4. Eventually, trends would begin to emerge. The point would be to have one classification that the biologist, ecologist, engineer, and fluvial geomorphologist could all reference so that they could understand and utilize each other's data, and cross-reference stream/river locations.

5.3.2.1 *Stream Orders.* Water or runoff, when flowing over the land, follows the path of least resistance in soil and bedrock. As the water accumulates, it forms larger and larger channels until a stream is formed. Streams become larger and larger as more and more drainage area contributes to the flow, eventually reaching the status of *river*. The form or pattern of this *drainage network,* or branching, will vary, depending on the physical features of the watershed, such as bedrock bedding planes.

Horton (1945) developed a method to classify the hierarchy of streams according to the number of tributaries upstream. A *tributary* is simply defined as a stream that flows into another stream. Strahler (1957) later refined this system of stream ordering. As with other classifications, classifying and mapping stream networks by stream orders allows for the comparison of the behavior of one stream to those with similar settings and drainage areas. It is useful in identifying hydrologic modeling points of hydrograph combination, model calibration points, and water quality sampling or study sites. As mentioned in an earlier chapter, points just above and below stream confluences are often referred to as *points of interest* for model calibration or sample collection. Individual tributary contributions to the three points of interest (above the confluence on both streams and one point below the confluence) can therefore be determined using this methodology.

The stream order naming convention uses numbers. In this system, the *first-order* streams are the initial, smallest tributaries; the stream located just below where two first-order streams combine would be considered a second-order stream; the stream located below the confluence of two second-order

TABLE 5.4 Major Classification Systems Source

Classification/Source	Purpose
Horton's law of stream order, slope and length (1945)	FGM
Stream Order, Strahler (1957)	FGM Dependent on scale of map used
Stream Order, Yang (1971)	FGM
Schumm (1977)	Morphology/Alluvial Streams, Sediment, Transport
River Continuum Concept, Vannote, et al. (1980)	Biological Activity
Rosgen (1996)	FGM
Montgomery and Buffington (1993)	Morphology
Others	

streams would be a third-order stream; and so on. Wherever two segments of the same order join downstream, the order of the downstream segment is increased by one; two third-order streams join to form a fourth-order stream.

One common mistake in naming stream orders is to name a stream below, for instance, the confluence of a first-order and a second-order stream as a third-order stream, which is incorrect, as can be seen in Figure 5.6. In this case, the stream would still be a second-order stream.

All that said, complicating the identification process is the fact that exactly what constitutes a first-order stream is somewhat subjective, and thus is the major difficulty in determining stream order. In other words, where the numbering starts may not be well defined, and hence will affect the entire ordered network. The Exploring the Environment (ETE) Team (1977–2000) describes stream-ordering requirements as starting with perennial streams: One should "start by identifying the smallest streams, those that have no permanently flowing tributaries. First-order streams are perennial streams, which carry water all year." USGS (2002) describes a first-order stream as "not determined by the presence or absence of flowing water, but by the shape of the land surface, which determines where flow will be concentrated when water is present." This is consistent with Leopold and others (1964) who defined a first-order stream as the smallest unbranched tributary that appeared on a USGS 7.5 minute topographic map on which intermittent streams are shown. Unfortunately, this discrepancy in stream order determination could result in two scientists ordering the same network differently, with the stream numbering consistently off by one or more. Clearly, this is another area where standardization would benefit users, and the National Hydrography Dataset

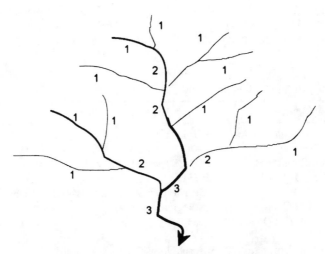

Figure 5.6 Sample stream order diagram. The main stem would not become a fourth-order stream until two third-order streams came together.

(NHD) is the first step in the right direction. (Note: Irrigation canals and other "artificial" systems are not included in the stream order.)

There is a direct relationship between stream order and the relative size of streams. As watershed drainage area and additional tributaries combine, they lead to larger and larger stream order numbers. Naturally, the width, depth, and *flow,* or *discharge,* of streams also generally increase with stream order. General slope of the streambed also tends to decrease as the stream order number increases. However, as with any classification systems, due to multiple variable factors, such as climate, geology, and so on, the statistical variance of the correlations is large.

First-, second- and third-order streams are classified as *headwater* streams, and they make up over 80 percent of the total length of earth's rivers and streams (USDA, 1998). Fourth- through sixth-order streams are considered "medium" streams. Streams can be termed "rivers" when they become seventh order or larger, thus the definition of river. The Amazon River, the largest in the world, is a twelfth-order stream. Mapping and ordering stream networks provides a useful stream order diagram, as shown in Figure 5.7.

A *tributary* is a stream or river that flows into a larger one. Headwaters are the small streams that are the sources of a river. What is the difference between a stream, creek, crick, and brook, and how do they relate to stream order? Many people use these terms interchangeably, but according to Webster's, a *brook* is defined as a small stream. Another name for a brook is a

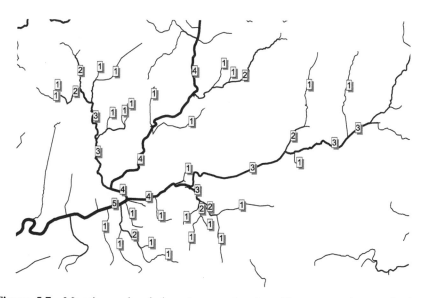

Figure 5.7 Mapping and ordering stream networks with stream orders results in a useful diagram, as shown in this GIS data layer.

run. A *creek* is defined as "a small stream, somewhat larger than a brook." Most brooks and creeks are lumped into the stream category; however, it may be possible to equate a brook to a first-order stream, a creek to a second-order stream, and a stream to a third-order stream. A *crick,* used to refer to flowing water, is slang, hence, simply, is incorrect. A *channel* can be considered a dry *swale* for passing storm flows or the *bed* of a river. A *swale* is also a dry channel for passing storm flows. The biologists performing rapid bioassessment protocols (described in Chapter 14) typically utilize stream order numbering systems in their work.

5.3.2.2   *Federal Information Processing Standards (FIPS): Water Feature Classification.* Under the Information Technology Management Reform Act (Public Law 104-106), the Secretary of Commerce approves standards and guidelines that are developed by the National Institute of Standards and Technology (NIST) for federal computer systems. These standards and guidelines are issued by NIST as Federal Information Processing Standards (FIPS) for use governmentwide. NIST develops FIPS when there are compelling federal government requirements, such as for security and interoperability, and no acceptable industry standards or solutions.

The FIPS has made available six categories of geographic or spatial data. These six categories are:

Land Features
Political Features
Populated Places
Postal Codes
Structural Places
Water Features

The FIPS has further broken down the water features database into the following categories:

Canals: Abandoned Canal, Canal, Canal Bend, Drainage Canal, Irrigation Canal, Navigation Canal(s), Section of Canal, Underground Irrigation Canal(s)

Channels: Channel, Cutoff, Lake Channel(s), Marine Channel, Navigation Channel, Ravine(s), Reach, Seachannel(s), Tidal Creek(s), Watercourse

Glaciers: Arctic Land, Cirque(s), Glacier(s), Icecap, Icecap Depression, Icecap Dome, Moraine, Nunatak(s), Snowfield

Inlet: Former Inlet, Inlet

Lagoon: Lagoon(s), Section of Lagoon

Lake: Asphalt Lake, Crater Lake(s), Intermittent Lake(s), Intermittent Oxbow Lake, Intermittent Salt Lake, Lake Bed(s), Lake Channel(s), Lake(s), Oxbow Lake, Salt Lake(s), Section of Lake, Underground Lake

Ponds: Fishponds, Intermittent Pond(s), Intermittent Pool, Intermittent Salt Pond(s), Pan(s), Pond(s), Pool(s), Salt Pond(s)

Reef: Reef(s), Section of Reef, Shoal(s)

Reservoir: Intermittent Reservoir, Reservoir(s), Water Tank

Seas: Apron, Bank(s), Borderland, Deep, Fracture Zone, Hole, Moat, Ocean, Polder, Rise, Sea, Seachannel(s), Seamount(s), Section of Bank, Shelf, Shelf Edge, Shelf Valley, Slope, Sound, Tablemount(s) or Guyot(s), Tongue, Trench, Trough

Spring: Geyser, Hot Spring(s), Spring(s), Sulphur Spring(s)

Straits: Strait

Stream: Abandoned Watercourse, Anabranch, Arroyo, Bar, Bend, Canalized Stream, Confluence, Current, Delta, Distributary(-ies), Dry Stream Bed, Estuary, Ford, Gut, Headwaters, Intermittent Stream, Levee, Lost River, Narrows, Rapids, Reach, Section of Intermittent Stream, Section of Stream, Stream Bank, Stream Bend, Stream Mouth(s), Stream(s), Weir(s), Whirlpool

Swamps: Bog(s), Gut, Hammock(s), Intermittent Wetland, Mangrove Swamp, Marsh(es), Moor(s), Peat Cutting Area, Quicksand, Salt Marsh, Swamp, Wetland

Wadi: Section of Wadi, Wadi Bend, Wadi Junction, Wadi Mouth, Wadi(es)

Wells: Abandoned Well, Overfalls, Section of Waterfall(s), Sinkhole, Waterfall(s), Waterhole(s), Well(s), Whirlpool

A further description of the streams database includes, for example, Global Streams databases, with latitude and longitude distance calculations. It contains officially recognized place names, physical locations, geographical coordinates, administrative regions, countries, and additional information. It also includes streams (bodies of running water moving to a lower level in a channel on land); bars (shallow ridges or mounds of coarse, unconsolidated material in a stream channel, at the mouth of a stream, estuary, or lagoon and in the wave-break zone along coasts); guts (sloughs, places of deep mud or mire; bayous; creeks; and water passages); rapids (turbulent sections of a stream associated with a steep, irregular streambed); and more.

5.4 WATERSHED HYDROLOGY

Watershed hydrology is the science of the water of the earth, its cycle and processes. Although hydrology is closely related to the many disciplines covered in this text, including climatology, soils, hydrogeology, geology, geomorphology, ecology, chemistry, and hydraulics, it is most often related to surface runoff processes. Understanding watershed hydrology is essential in being able to develop a comprehensive watershed management plan. A watershed responds to rainfall events. As explained in Chapter 3, the proportion of rainfall that is recharged and contributes to base flow, versus evapotranspiration and surface runoff, depends on soils, geology, time since the last rainfall (antecedent moisture conditions), rainfall amount, anthropogenic changes, and watershed size and shape.

Stormwater runoff entering a stream as surface runoff produces a *hydrograph,* a plot of discharge (cubic feet per second) over time. The hydrograph can be classified into the *rising limb, peak,* and *falling* or *recession limb,* as shown in Figure 5.8. Before the storm hydrograph, the flow in Figure 5.8 is base flow. After the recession segment, barring any new precipitation, the hydrograph gradually recedes back to the original base flow condition described by what is called the *depletion curve.* The hydrograph includes surface runoff and groundwater recharge of base flow. Methods have been developed from gauged data to separate base flow from surface runoff.

The effect of watershed shape on the two watersheds of the same contributing drainage area can be seen in Figure 5.9 (USDA SCS, 1985). Notice that the elongated watershed tends to reduce the flood peak and increase the time to peak due to the increased time it takes for the surface runoff to reach the mouth (travel time of the watershed).

The various aspects of watershed hydrology are intertwined throughout the text of this book.

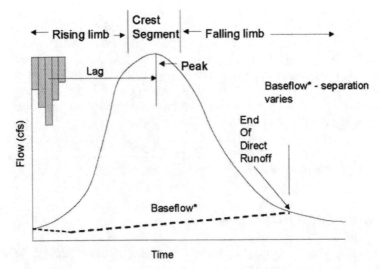

Figure 5.8 Typical hydrograph and its components (modified from Viessman et al., 1977).

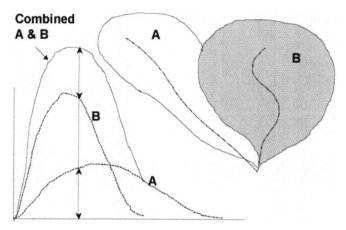

Figure 5.9   Effect of watershed shape on the individual watershed hydrograph and the cumulative hydrograph. (USDA SCS, 1985).

5.5 SUMMARY

Understanding the drainage pattern of streams aids in determining how streams were formed and what geophysical processes to expect in a watershed. Classifying streams and watersheds, and comparing them to a reference system, or what to expect, aids in determining how the system should be performing.

REFERENCES

Beasley, R.P. *Erosion and Sediment Pollution Control* (Ames, IA: Iowa State University Press), 1972.

Dunne, T., and Luna L. Leopold. *Water in Environmental Planning* (New York: W.H. Freeman), 1978.

EPA. Surf Your Watershed, www.epa.gov/surf/ (accessed 2002).

ETE Team, Wheeling Jesuit University/NASA. "Classroom of the Future, 1997–2001," www.cotf.edu/ete/modules/waterq3/WQintermpuzzle.html (accessed March 19, 2003).

FGDC Proposal. "Federal Standards for Delineation of Hydrologic Unit Boundaries," Version 1.0, www.ftw.nrcs.usda.gov/HUC/HU_standards_v1_030102.doc (March 1, 2002).

Ford, Robert E., and James D. Hipple. "Streams and Rivers Lab: An Introduction to Fluvial Geomorphology," www.usra.edu/esse/ford/ESS205/fluvial/fluvial.html (June 2003).

Horton, R.E. "Erosional Development of Streams and Their Drainage Basins— Hydrophysical Approach to Quantitative Morphology," *Bulletin of the Geological Society of America,* 1945, vol. 56, pp. 275–370.

Keystone Stream Team. "Guidelines for Natural Channel Design for Pennsylvania Waterways," Alliance for the Cheasapeake Bay, March, 2003

Leopold, L.B., M.G. Wolman, and J.P. Miller. *Fluvial Processes in Geomorphology* (San Francisco: Freeman), 1964.

NRCS. Watershed Boundary Dataset (WBD), National Cartography and Geospatial Center, www.ftw.nrcs.usda.gov/huc_data.html (June 2003).

Pidwirny, Michael J., PhD. "Fundamentals of Physical Geography: Introduction to Geomorphology, The Drainage Basin Concept," Department of Geography, Okanagan University College, www.geog.ouc.bc.ca/physgeog/contents/11k.html (accessed 2003).

Rosgen, Dave. *Applied River Morphology, Wildland Hydrology* (Pagosa Springs, CO: 1996).

Seaber, P.R., F.P. Kapinos, and G.L. Knapp. "Hydrologic Unit Maps," U.S. Geological Survey Water-Supply Paper 2294, 1987.

Strahler, A.N. "Quantitative Analysis of Watershed Geomorphology," *Trans. Am Geophys. Union,* 1957, vol. 38, pp. 913–920.

USDA Forest Service. "Chesapeake Bay Riparian Handbook: A Guide for Establishing and Maintaining Riparian Forest Buffers," Chesapeake Bay Program, June 1998.

USDA Soil Conservation Service (SCS). *National Engineering Handbook,* Section 4: "Hydrology" (Washington, DC: Engineering Division, Soil Conservation Service, USDS), March 1985.

USGS. water.usgs.gov/nawqa/protocols/OFR-93-408/habitp15.html, 2002.

———. "Water Resources, Hydrologic Unit Maps: What are Hydrologic Units?" Adapted from Seaber, et al., 1987, "Hydrologic Unit Maps," U.S. Geological Survey Water-Supply Paper 2294, 2003.

CHAPTER 6

LAKES, RESERVOIRS, STREAMS, AND WETLANDS

6.0 INTRODUCTION

Lakes and reservoirs have an impact on the hydrology, water quality, and flooding potential of watersheds. Most lakes in the United States are man-made and actually store water during flooding events, thereby reducing flows downstream. A typical inflow and outflow hydrograph from a particular lake or reservoir can be seen in Figure 6.1. Lakes and reservoirs also impact water quality. Most lakes stratify into distinct temperature regimes during the summer and winter months. Whether water is withdrawn from the bottom or the surface of a lake in a typical spillway will affect the temperature and aquatic life downstream of that lake or reservoir. Properly managing a lake and its watershed will have a cumulative impact on the entire watershed in which that lake is located. Minimizing nutrients and erosion leading into a lake will help not only in maintaining the lake depth and water quality but also in maintaining downstream water quality. Minimizing lakeshore erosion by maintaining buffers and lakeshore protection will also prolong the life of the lake and maintain its depth, which in turn maintains lake quality. Lake management should be a component of all watershed management plans.

One must know basic ecological processes to be able to assess whether a water body is polluted or could potentially become polluted. The study of freshwater ecosystems, lakes, reservoirs, streams, and rivers, is termed *limnology*. Freshwater systems can be classified into *lentic* (standing) or *lotic* (flowing) water. Lentic systems such as reservoirs, lakes, and ponds are more susceptible to pollution because they act as sinks, retaining pollutants. Flowing systems such as streams and rivers have more of a tendency to "flush" pollutants downstream, the result being an accumulation of pollutants in

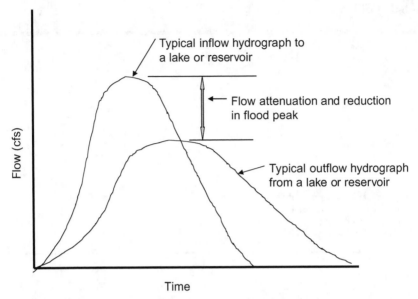

Figure 6.1 A typical inflow and outflow hydrograph from a particular reservoir. Notice the flood flow attenuation and reduction in peak flow.

downstream estuaries and bays. Impoundments or dams with an outlet structure that can regulate flows have the capability to release some pollutants from the reservoir, albeit these structures are typically used more for water quantity regulation than water quality control. When assessing water quality for its suitability for viable aquatic life, systems are often classified into *impounded water* and *flowing water* divisions. An impounded water body is defined as one with a 14-day or greater detention time under average annual flow conditions. A flowing water body is defined as a water body with fewer than 14 days detention time under average annual flow conditions

6.1 LENTIC WATER (PONDS, LAKES, RESERVOIRS)

Ponds are typically smaller and shallower than lakes. Sunlight can typically reach the pond's bottom, thus plants requiring photosynthesis can grow and the water temperature throughout the *water column* varies little. The water column is the profile of the water from the surface to the bottom of the lake. Lakes are typically deeper than ponds with a *stratified temperature gradient,* which will be discussed later. The line drawn between ponds and lakes is not exact, as subjectivity is involved in their classification.

There are several differences between *reservoirs* and natural lakes that affect water quality. As a rule, lakes are geologically older. Reservoirs are usually dammed rivers or stream valleys fed by a major tributary, and are typically longer and narrower than lakes, which are more circular. Reservoirs typically have more irregular shorelines. These physical differences affect the physical, chemical, and biological characteristics of the water bodies. Table 6.1 summarizes several differences between lakes and reservoirs that are important for watershed management, and Figure 6.2 shows views of a pond, lake, and, reservoir, respectively.

To better understand and be able to evaluate lake conditions, one must have an understanding of natural lake succession. Lakes naturally, over geologic time, fill with sediment. Sediment tends to deposit concentrically from the outer edges (where velocities are the lowest) toward the middle.

6.1.1 Lentic Zones

A pond, lake, or reservoir can be categorized into the following zones and as shown in Figure 6.3:

Marginal zone. The area where the land meets the water.

Littoral zone. The area from the shoreline lakeward, to where rooted plants can no longer be supported.

Pelagic zone. The zone of open water, from the littoral zone to the center of the lake.

Profundal zone. The mass of water and sediment occurring near the bottom of the lake below light penetration.

Benthic zone. The bottom stratum of the lake.

Rooted aquatic vegetation, when adequate depth of sediment accumulates and water depth is shallow enough to support it, tends to follow the same

TABLE 6.1   Comparison of Characteristics of Natural Lakes and USACE Reservoirs*

Variable	Natural Lakes ($N = 309$)	Reservoirs ($N = 107$)
Drainage area, km^2	222.0	3228.0
Surface area, km^2	5.6	34.5
Maximum depth, m	10.7	19.8
Mean depth, m	4.5	6.9
Hydraulic residence time, years	0.74	0.37
Drainage area/surface area	33.0	93.0
Phosphorus loading, g P m^{-2} year^{-1}	0.87	1.70
Nitrogen loading g N m^{-2} year^{-1}	18.0	28.0
Transparency, m	1.4	1.1
Total phosphorus, mg ℓ^{-1}	0.054	0.039
Chlorophyll a, μg ℓ^{-1}	14.0	8.9

*Modified from Cooke and Kennedy, 1989.

Figure 6.2 Pond (a), lake (b), and reservoir (c). Ponds are typically smaller than lakes, but water depth and corresponding stratification also play an important role.

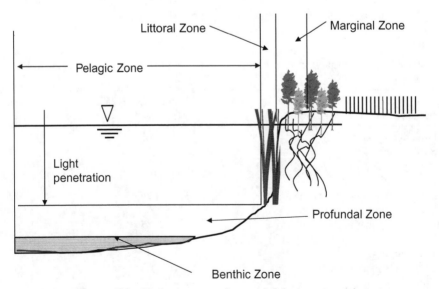

Figure 6.3 Various zones of a pond, lake, or reservoir.

succession as the sediment deposition, progressively encroaching toward the center of the lake. *Macrophytes,* or large aquatic rooted plants such as alligator weed, smartweed, and cattails, occur in a band along the water's edge. Next in succession are floating leaved emergent plants such as water lilies and American lotus, which maintain roots in the sediment. These grow in water depths from 0 to 5 feet. Next in line toward the lake/pond center are the submerged rooted weeds such as watermilfoil, pondweed, and coontail, which will grow in water depths of up to 10 feet. The open water region is comprised of unrooted macroscopic floating species such as duckweed and microscopic floating species, or *plankton.*

6.1.2 Lentic Succession

Lentic systems can be classified as *oligotrophic* (young, low productivity), *mesotrophic* (middle-aged, medium productivity), and *eutrophic* (old, high productivity). As a general rule, oligotrophic lakes are newly formed or are in unproductive watersheds; they are characterized by low levels of nutrients and low numbers of aquatic plants and animals (*biological productivity*) and are often crystal clear. Consequently, organisms high on the food chain, such as game fish, are low in numbers.

Mesotrophic lakes have average nutrients per unit volume of water, are more productive, and therefore have an abundance of organisms high on the food chain. A mesotrophic lake condition is preferred for recreational, water quality, and game fishing reasons. The natural succession process changes an oligotrophic lake to a mesotrophic to a eutrophic lake over geologic time. The human influence with excess nutrients and sediment shortens this time span.

As sediment and nutrients in excess of what the natural biomass can assimilate enter the lake, the succession process is enhanced, a process called *eutrophication.* Eutrophication in lakes is an overabundance of silt, organic matter, and nutrients, the result of which is increased production of algae and rooted plants and decreased reservoir volume. Increased nutrients cause increased plant growth. As the plants complete their life cycle and die-off occurs, the decomposition of these plants by bacteria and other organisms consumes oxygen. The oxygen depletion could cause fish kills.

Phosphorus is typically the *limiting nutrient* in the ecological system. That is, nitrogen is typically in abundance and plant growth and productivity are controlled by the amount of available phosphorus. However, in systems where phosphorus is in sufficient supply, such as in areas with alluvial sediment or limestone, nitrogen may become the limiting nutrient. Once the limiting nutrient is supplied, plant growth and reproduction will explode, often causing algal blooms, or an overabundance of algae. The result of these blooms is an overutilization of oxygen by the bacteria degrading the algae, causing odors and oxygen depletion for other aquatic organisms such as fish, often causing fish kills. Decaying organic matter produces hydrogen sulfide, causing a "rotten egg" odor to occur. Water quality is typically unfit for recreational purposes. Phosphorus management should be part of any good watershed management plan. The typical pH range of lakes should be between 6 and 9. Lakes with pH values less than 6 are considered acidic, and most lifeforms cannot survive in waters with a pH of 4 or lower.

The consequences of eutrophication are weed-choked shallow areas, algal blooms, oxygen-depleted deep waters, degradation of potable water supplies, limitations on recreation use, degraded fisheries, reduced storage capacity, and disruption of downstream biological communities.

6.1.3 Stratification and Lake Overturn

Nutrients stimulate aquatic plant growth, thus a measure of the lake's biomass (aquatic vegetation) can serve as an indirect indicator of nutrient levels. As a rule, as plant numbers increase, the number of aquatic animal organisms also increases. The measured biomass of aquatic plants and animals in a water body at a given time is its *biological productivity*. Photosynthetic activity takes place in the lighted region of the lake, or the *photic zone*. The photic zone extends down to a depth where plant growth (photosynthetic oxygen production) is balanced by respiratory uptake and is dictated by the amount of light available. This depth is referred to as the *compensation limit*. The compensation depth is generally assumed to correlate with the depth at which at least 1 percent of the surface light penetrates. Below this depth there will be no net phytoplankton growth (USEPA, 1988).

Lakes and reservoirs undergo annual thermal changes to their storage pools, which have an impact on the water released and downstream conditions. Knowing a lake's *stratification* is helpful in understanding downstream biotic communities, hence should be a part of any watershed management plan. In the spring of the year, the temperature of a lake's water is fairly evenly distributed and wind mixing of the entire water column is fairly common. As the energy of the sun in the summer months heats the surface, the less dense, warmer surface water "floats" on top of the cooler, denser water, and stratification occurs. In the upper water, termed the *epilimnion*, photosynthesis exceeds respiration. Plants use carbon dioxide from the water and produce oxygen. Gas exchange with the atmosphere also occurs in this layer. In the lower layer, the *hypolimnion*, water is fairly stagnant; respiration may exceed photosynthesis, consuming oxygen; and nutrients and organic sediments accumulate and may be rereleased from organic matter decomposition and sediment. Due to its cooler temperatures, the hypolimnion maintains an adequate supply of oxygen. Fish typically stay in this oxygenated water during the hot summer months and only come to the surface at night to feed. The zone between the epilimnion and the hypolimnion is an area of sharp temperature change or thermal gradient called the *mesolimnion* (also called the *metalimnion*) (USEPA, 1988). The plane through the point of maximum temperature change is the *thermocline*. Figure 6.4 shows a cross section of a typical summer stratified lake.

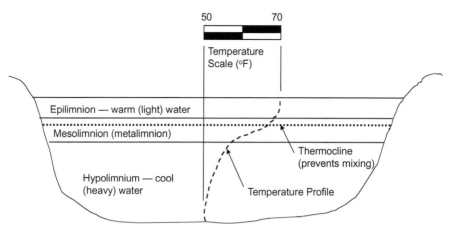

Figure 6.4   Cross-section of a typical summer stratified lake (modified from U.S. EPA, 1988).

In the autumn, the heat from the epilimnion is lost to the atmosphere, and wind mixing creates an evenly distributed temperature or *isothermal* lake where the water is uniformly dense. This mixing of the "limnions" is called the *turnover* and is common in the spring and autumn. The stratification process is affected by such factors as regional location, lake depth, and orientation in accordance to prevailing wind direction. Lakes in the south may have the entire column exhibit epilimnetic conditions. This whole overturn is what distributes the nutrients and dissolved oxygen in a lake.

This concept is important in watershed management for several reasons. The temperature of the incoming water, whether base flow or produced from rainfall events, will travel through the lake at a level equivalent to its density until it becomes mixed. For instance, if the incoming water has a lower temperature and density than the lake water, when the incoming velocities subside, the incoming water will "plunge" beneath the surface, with extensive mixing possible. If the incoming water is nutrient-laden, these nutrients will be mixed with the lake water. If it is high in organic matter, which would settle in the hypolimnion, the microbial metabolism could deplete the oxygen supply. Conversely, the discharge of the lake, whether from a surface spillway or outlet pipe at the bottom of the lake, will have a major influence on the temperature of the lake, the depth of the mesolimnion, the water temperature of the discharge, and the receiving stream. This could affect water quality and oxygen content and, in turn, the aquatic species makeup of the receiving stream.

Another important feature that affects a watershed is the lake's or reservoir's *hydraulic residence time,* sometimes called the *exchange rate.* This is the time it takes for the volume of water in storage to replace itself with an equal volume of water. This factor affects the storage and exchange of nutrients and other pollutants. A lake with a short exchange rate would have a better chance of flushing itself should a pollution event occur. Large lakes and reservoirs may have an exchange rate of one year, whereas streams have a constant flushing rate. The methodology used to identify a lake's or reservoir's hydraulic residence time is to first determine the system's water budget, or determination of water inflow and outflow. The general equation for lake or reservoir water budgets can be expressed as:

$$\text{Water input} = \text{Water output} \pm \text{Water stored in the lake} \qquad (6.1)$$

Water input is precipitation, stream inflow, and groundwater inflow; outputs include surface discharge (outflow), evaporation, losses to groundwater, and man-made withdrawals such as irrigation. Lakes that are formed at the intersection of the groundwater table are supplied primarily by the groundwater and are called *seepage lakes*. The water level of seepage lakes fluctuates with the rising and falling water table. Lakes fed primarily by inflowing streams are called *drainage lakes*.

The outflow is approximately equal to the inflow, except in periods of drought, when a water level may be below the spillway or evaporation is great. The hydraulic residence time can be calculated by determining a lake's inflow, outflow, and water storage volume based on average daily stream flow. The general equation would be:

$$\text{Hydraulic residence time} = \text{Volume of storage} \div \text{inflow or outflow}$$

For example, a lake with a 1,000-acre-foot storage volume and an average inflow of 10 acre feet per day would have a hydraulic residence time of:

$$1,000 \text{ ac-ft} \div 10 \text{ ac-ft/day} = 100 \text{ days}$$

The more detailed the actual inflow and outflow data are, the more accurate the residence time computations will be. Inflow or outflow from groundwater sources should also be determined.

The shorter the residence time, the less time algal biomass has to accumulate and reproduce in the lake. Longer residence times (more than 100 days) and adequate nutrients will allow algae to accumulate. If nutrients are limiting, the algal production will be kept in check. An excess supply of nutrients, such as from leaking septic systems or sewage treatment plant discharges, will allow the algal populations to explode. Therefore, the nutrient inflow or nutrient loading needs to be examined in conjunction with the hydraulic residence time, as it can become a good management tool. The EPA's "Lake and Reservoir Restoration Guidance Manual" (U.S. EPA, 1988) is recommended reading to evaluate a lake or reservoir's influence in a watershed. For *total maximum daily load* (*TMDL*) calculation purposes, lakes can be defined as surface waters with a hydraulic residence time of 14 days or more, based upon an average daily stream flow and *normal pool* volume.

(TMDLs will be discussed in more detail in Chapter 15.) The normal pool is the level of the water at the spillway crest. In the absence of actual data, an average annual daily discharge rate of 1.5 cubic feet per second (cfs) per square mile can sometimes be used.

6.2 THE GREAT DAM DEBATE

As early as 1901, John Muir and his newly created Sierra Club opposed the Hetch Hetchy dam and reservoir, which was proposed by the San Francisco City directors for water supply and hydroelectric power (Old News, 2003). The debate was about the effect of the dam and reservoir on the natural ecological system. After years of conflict, President Woodrow Wilson signed a bill allowing San Francisco to build the dam on December 19, 1913. The debate continues today.

There are definitely two completely opposite points of view in this regard. Dams, and the reservoirs they create, have advantages and disadvantages, as outlined in Table 6.2 (modified from The Heinz Center, 2002). For streams or rivers that originally carried migratory fish such as shad or salmon, dams create an impediment to their migration. The created pool of water also warms the water that is discharged and could affect downstream *salmonid* populations, which are a cold-water species. As discussed earlier, streambed load under natural conditions is typically in balance, whereas dams trap this sediment, making the stream downstream of the dam "hungry" for sediment, causing downstream stream bank erosion. Conversely, dams attenuate flood peaks, and provide a variety of habitat for species such as blue and green herons, ducks and geese, muskrats and beavers, and the variety of warm-water macro-invertebrates and fish species. The millions of bass fishermen who have had many hours of fishing pleasure and can see the smiles on their

TABLE 6.2 Reasons for and against Dams and Reservoirs

For	Against
Clean hydroelectric power	Economic obsolescence
Water supply	Structural obsolescence
Irrigation	Safety considerations
Flood suppression	Lotic ecosystem and watershed restoration
Recreational opportunities (hiking, hunting, swimming)	Return of stream stability
	Restoration of downstream riparian habitat
Boating/sailing	Water quality
Shoreline and aquatic habitat for many species	Unregulated flow recreation
	Return of the sediment balance
Warm-water fisheries	
Groundwater recharge	

children's and grandchildren's faces when they hook a lunker, or even a six-inch sunny, would definitely argue for the advantages of having such facilities.

The decision to build a new dam, or removing an old dam, should carefully weigh the scientific, environmental, social, economic, regulatory, and policy long-term issues of the entire watershed and its streams, not just the localized impacts. (Note that the category of political issues was left off of the list.) Emotions should be left out of the decision-making process.

The effects of dams/dam removal on the watershed and its components should be incorporated in any watershed assessment.

6.3   LOTIC WATER (STREAMS AND RIVERS)

As with lakes, streams have a natural life cycle. A stream is not static, but is a delicately balanced system responding to natural events and man-made changes in its watershed. Youthful streams cut deeply through their parent soil material in a fairly straight path (Terrell and Perfetti, 1989). Commonly found in hilly terrain, youthful streams form V-shaped valleys with steep-sloped banks. As the stream matures, its path begins to meander, cutting into adjacent banks and widening the valley. In old age, the stream has a broad floodplain through which it meanders back and forth.

Although not static, streams can be stable. A stream in a well-balanced "ideal" condition will be in equilibrium with its surroundings. The most extreme examples are those that have smooth, stable, well-vegetated banks, free from erosion or failure. The channel bed will not be in a scour or a sediment deposition mode. A stream responds to environmental changes by changing its banks or bed or both. If sediment exists in excess of the stream's natural carrying capacity (or *bedload*), the stream will tend to deposit this excess sediment, a process termed *aggradation*. If bedload is less than the natural carrying capacity, the stream will tend to degrade, or scour its bed and bank. Both situations are unstable and can lead to problems. If the streambed is rising due to deposition, the carrying capacity of the channel is reduced, and the next large rainstorm could cause significant flooding. A stream will also attempt to adjust by scouring out its bed and eroding its banks. A degrading stream may erode its banks, causing obstructions to flow, and the eroded sediment from the bank will be carried and deposited downstream. In nature, the ideal condition will seldom occur. A stream can be in a stable condition when its banks and bed are in equilibrium and it is undergoing its natural succession.

The human influence has altered all but a few of the world's streams. Urbanization, grazing, deforestation, and mining have all influenced the natural balance of streams. Increased flows, velocities, and sediment have caused stream erosion, fallen trees, and deposition. These changes to streams cause many ecological changes, not only at the problem area, but in a domino effect downstream.

6.4 PHYSICAL PROPERTIES OF RECEIVING WATERS

Streambed bottom materials (*substrate*) have a major influence on the type and quantity of aquatic organisms present. The materials vary by location in the watershed, regional geology, and topography. *Headwaters,* which typically have steeper slopes and swiftly flowing waters (higher velocities), tend to have deep channels free of smaller sediments. There are exceptions, however, where boulders cause a roughness that significantly slows the velocities. In contrast, streams in level terrain with lower velocities typically have shallow beds and substrate composed mainly of smaller sediment. Velocities therefore determine the size and substrate available for aquatic organisms. Velocities can be classified as slow, moderate, and swift, as delineated in Table 6.3.

Several velocity/sediment relationships are important to recognize, as shown in Table 6.4.

As can be seen from these relationships, the various disciplines (soils, hydrology, geomorphology, biology) are closely interrelated.

Watercourses with swift velocities that have cobble and gravel beds have the greatest invertebrate biodiversity. The cobble or gravel bottom is stable in times of high flows and provides habitat and hiding places for the bottom-dwelling animals. Typically, these streams are characteristic of alternating *riffles* (shallow, fast-moving waters) and *pools* (deep, slow-moving waters). Greatest insect production occurs in riffles with cobbles of 6 to 12 inches (Marriage and Batchelor, 1983). The fluvial processes of stream formation were discussed in more detail in Chapter 5.

6.5 WETLANDS

Wetlands are areas of high biological diversity; they are complex and sensitive ecosystems, and are among the most biologically productive natural ecosystems in the world. They provide unique habitat to a large variety of plant and animal species and are often host to threatened and endangered species. They

TABLE 6.3 Stream Velocity Classification*

Velocity Class	Velocity (ft/sec)	Velocity (m/sec)
Slow	<0.8	<0.1
Moderate	0.8–1.6	0.25–0.5
Swift	>1.6	>0.5

*Modified from Terrell, 1989.

TABLE 6.4   Velocity/Sediment Relationships

Relationship	Velocity (ft/sec)	Velocity (m/sec)
Silt and organics deposit	<0.7	<0.2
Sand settles	0.8–3.9	0.25–1.2
Gravel settles	3.9–5.6	1.2–1.7
Erosion of sand and gravel riverbeds occurs	>5.6	<1.7

are defined by vegetation, hydrology, and soils conditions: vegetation adapted to wet conditions, a water table at or near the land surface during the growing season, and soil conditions that exhibit characteristics of anaerobic conditions from saturation for some period during the year. Wetlands include swamps, marshes, bogs, and similar areas. Typical wetlands such as marshes and bogs are fairly easily identified by the general public; however, many specific wetland types are more difficult to identify. Some examples of these are *vernal pools* (pools that form in the spring rains but are dry at other times of the year), *playas* (areas at the bottom of undrained desert basins that are sometimes covered with water), and *prairie potholes*.

Each wetland has a unique water budget and plays an important natural function in stormwater and flood flow attenuation, acting like natural sponges for storing stormwater, then slowly releasing it. Wetlands actually perform the same flood flow storage and peak flow attenuation functions as stormwater detention basins are designed to do, although not to quite the magnitude. Wetlands also improve water quality by filtering pollutants and sediment and by "treating" nutrients and organic wastes. Some wetlands often promote groundwater recharge, which supports or enhances stream base flow and groundwater supply. The magnitude of this function depends on the type, size, and location of the wetland in the watershed.

When the uppermost part of the soil is inundated and saturated with water during the growing season, oxygen is consumed by soil organisms, producing conditions unsuitable for most plants. Such conditions also develop *hydric* soils, or soils under reduced conditions. A hydric soil is one that is saturated, flooded, or ponded long enough during the growing season to develop anaerobic conditions that favor the growth and regeneration of hydrophytic vegetation (U.S. Department of Agriculture (USDA) Soil Conservation Service (SCS), 1986). Plants such as marsh grasses that can grow in wetland conditions are called *hydrophytes* or *hydrophytic vegetation*. Hydrophytic vegetation is defined by the U.S. Army Corps of Engineers as "the sum total of *macrophytic* plant life that occurs in areas where the frequency and duration of inundation or soil saturation produce permanently or periodically saturated soils of sufficient duration to exert a controlling influence on the plant species present" (U.S. Army Corps of Engineers, 1989). The presence of water by ponding, flooding, or soil saturation is also a good indicator of wetlands, although these conditions do not have to be present throughout the year to be considered wetland hydrology. Temporary inundation can also qualify as wetland hydrology. An area that meets the hydric soils, wetland hydrology, and hydrophytic vegetation definitions constitutes a wetland.

The *hydrologic regime* plays a key role in wetland ecosystem processes, type, structure, and growth. Wetlands have their personalized water budget, an important *geoindicator,* which balances groundwater, runoff, precipitation, and physical forces such as wind and tides (inputs) with drainage, recharge, and evapotranspiration (outputs). Wetlands may also exhibit a seasonal pattern of water levels, or the rise and fall of surface and subsurface water known as a *hydroperiod.* Changes in the range of water levels, whether annual or seasonal, affect surface biota, decay processes, *detritus* accumulation rates, and gas emissions. External factors, such as fluctuations in water source (stream diversions, tile drainage, groundwater pumping), climate, or land use (forest clearing, land development) can alter the wetland balance and makeup (U.S. Global Change Research Information Office, 2003).

Waters flowing out of wetlands are chemically distinct from inflow waters, because a range of physical and chemical reactions take place as water passes through the organic materials in wetlands, causing some elements such as heavy metals to be entrapped, and others such as humic acids to be mobilized.

6.5.1 Marshes, Estuaries, and Bays

Marshes may be found inland (freshwater) or along the fringes of bays (saltwater). In general, freshwater marshes have larger and more diverse plant populations than saltwater marshes. Irregularly flooded salt marshes exhibit the fewest species of plants. Ecosystem services supplied by marshes include removal of sediment, absorption of the erosive energy of waves, absorption of excess nutrients, and provision of an abundant plant community, which in turn provides food and shelter to many species of animals.

According to the EPA (2003), an *estuary* is "a partially enclosed body of water formed where freshwater from rivers and streams flows into the ocean, mixing with the salty sea water. Estuaries and the lands surrounding them are places of transition from land to sea, and from fresh- to saltwater. Although influenced by the tides, estuaries are protected from the full force of ocean waves, winds, and storms by the reefs, barrier islands, or fingers of land, mud, or sand that define an estuary's seaward boundary." The National Estuarine Research Reserve System (2003) points out that estuarine environments are among the most productive on earth, creating more organic matter each year than comparably sized areas of forest, grassland, or agricultural land. Their preservation is, therefore, a top national priority.

Bays are extremely important as they are typically the "sink" for all that takes place in the watershed. Thus, when performing a watershed assessment in which the stream or river flows to a bay, the bay should also be assessed. An overabundance of nitrogen (from urban runoff, sewage treatment outfalls, and leaking septic systems) and phosphorus (from fertilizer, and that is attached to sediment) often creates eutrophication problems in the lower portions of the watershed or the upper portions of the estuaries and bays.

6.5.2 Regulatory Aspect of Wetlands

Recognizing the potential for continued or accelerated degradation of the nation's waters, the U.S. Congress enacted the Clean Water Act (hereafter referred to as "the act"), formerly known as the Federal Water Pollution Control Act (33 U.S.C. 1344). The objective of the act is to maintain and restore the chemical, physical, and biological integrity of the waters of the United States. Section 404 of the act authorizes the Secretary of the Army, acting through the Chief of Engineers, to issue permits for the discharge of dredged or fill material into the waters of the United States, including wetlands. In 1987, the EPA and the Corps developed the *Corps of Engineers Wetlands Delineation Manual* to systematically define wetlands for the Clean Water Act, Section 404, permit program.

The definition of wetlands used by the U.S. Army Corps of Engineers (Corps) and the U.S. Environmental Protection Agency (EPA) since the 1970s for regulatory purposes is, "areas that are inundated or saturated by surface or groundwater at a frequency and duration sufficient to support, and that under normal circumstances do support, a prevalence of vegetation typically adapted for life in saturated soil conditions." Ralph Tiner (date unknown) provides a comprehensive listing of regulatory and nonregulatory definitions of wetlands.

The 1987 manual was briefly replaced by the 1989 *Federal Manual for the Identification and Delineation of Jurisdictional Wetlands,* which remained in use until proposed changes to the manual in 1991 and the consequent National Academy of Sciences study of wetlands definition. The 1987 manual was then clarified and readopted by the EPA and Corps of Engineers. It organizes environmental characteristics of a potential wetland into the three categories described previously: soils, vegetation, and hydrology. The manual contains criteria for each category. With this approach, an area that meets all three criteria is considered a wetland.

The Section 404 regulations require that new wetlands be created if natural wetlands are filled or dredged. However, natural wetlands provide much better ecosystem services than man-made wetlands. Years after being put into place, man-made wetlands do not show functional equivalence to natural wetlands.

6.5.3 Wetland Classifications/Categories

The Environmental Protection Agency (EPA) identifies six categories of special aquatic sites in its Section 404 b.(1) guidelines (Federal Register, 1980):

- Sanctuaries and refuges
- Wetlands

- Mudflats
- Vegetated shallows
- Coral reefs
- Riffle and pool complexes

Although all of these special aquatic sites are subject to regulation by the Clean Water Act, the U.S. Army Corps of Engineers *Wetlands Delineation Manual* considers that wetlands are vegetated. Thus, unvegetated special aquatic sites such as mudflats are not covered in the manual.

There are five general types of wetlands (Cowardin, et al., 1979) as shown in Figure 6.5 and listed below:

- Marine (open ocean and associated coastline)
- Estuarine (salt marshes and brackish tidal water)
- Riverine (rivers, creeks, and streams)
- Lacustrine (lakes and deep ponds)
- Palustrine (shallow ponds, marshes, swamps, sloughs)

Riparian forests are delineated on the U.S. Fish and Wildlife Service's National Wetland Inventory mapping as part of riverine wetland systems. Riparian forests protect water quality, reduce water temperature during summer for coldwater fish populations, and create leaf litter in the stream that becomes the base organic material to maintain macroinvertebrate populations for a healthy aquatic ecosystem.

Due to natural topographic conditions, wetlands are often found in valleys or depressions, ironically the most hydrologically advantageous location for stormwater detention or flood control structures. Although the Section 404 regulations prohibit dredging or filling of wetlands without a permit, the language is not as clear as to whether wetlands can be flooded or inundated for stormwater control structures if dredging or filling of the wetlands does not take place.

Note

This text does not address the argument of whether natural wetlands could be used for stormwater detention; rather, it attempts to apply an unbiased approach to this issue.

Wetlands can be classified by their importance for flood control and ecological value. One option for evaluating whether wetlands could be flooded for stormwater detention would be to develop a ranking system for wetlands based on their value for flood control, ecological significance, aesthetics, recreation, and habitat. This would allow a cost-benefit ratio of not utilizing such areas for flood control to be determined. For example, if they harbor endangered plant and/or animal species or provide recreational value, they should

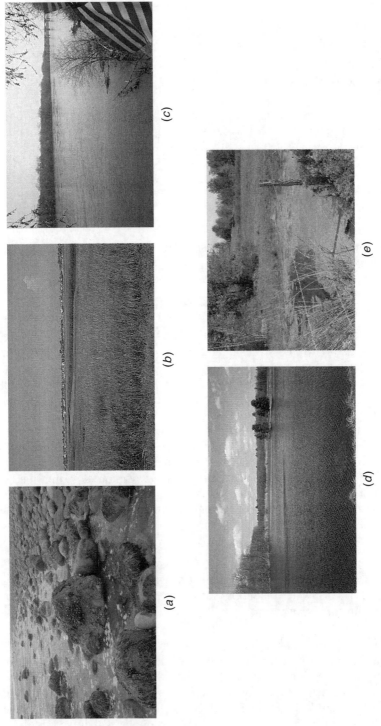

Figure 6.5 Typical marine (a), estuarine (b), riverine (c), lacustrine (d), and palustrine (e) wetlands. (*Source:* a and b: Ferren, Jr. et al. (1996); www.mip.berkeley.edu/wetlands/figures/V22toend.html; c, d, and e; DeBarry, 2003.)

be maintained in their existing conditions. Conversely, a wetland with no unique plant or animal species may be considered for flood control and/or stormwater management measures. This is a dangerous question: Who gets to decide? A wetland without unique species may still be performing important ecosystem functions, including supporting more common species that are very important in the food chain. It also depends on whether one is talking about created wetlands or natural wetlands. Again, this could potentially open a big can of worms.

In order to evaluate a wetland's functions, one may use the Wetland Evaluation Technique (WET) or WETII procedure, or the HEP procedure described below. Adamsus and others (1991) developed WET and WETII to assess wetlands functions related to the 404 and other regulatory management and planning programs. These procedures address 11 wetland functions, as listed in Table 6.5.

The system ranks each function as low, medium, or high probability. The values should then be evaluated cumulatively. WET also provides a procedure to evaluate habitat suitability for 14 waterfowl species groups, four freshwater fish species groups, 120 species of wetland-dependent birds, and 133 species of saltwater fish and invertebrates (Adamus et al., 1991).

The Habitat Evaluations Procedure (HEP) was developed by the U.S. Fish and Wildlife Service to be used to evaluate changes in an area's habitat value as a result of alterations to the system, and to assess the habitat value of an area for particular fish and wildlife species (Hill, 2003). Other procedures are also available.

6.5.4 Wetlands Spatial Data (Mapping)

Wetland spatial data for the GIS mapping has been generated by the U.S. Fish and Wildlife Service (FWS) to meet the agency's mandate to map the wetland and deepwater habitats of the United States, the product being the National Wetland Inventory (NWI). The purpose of the survey was not to map all wetlands and deepwater habitats of the United States, but rather to use aerial photo interpretation techniques to produce thematic maps that show the larger wetlands and types that can be identified by such techniques. The objective was to provide better geospatial information on wetlands than found

TABLE 6.5   WET Categories

Groundwater recharge	Groundwater discharge
Flood flow alteration	Sediment stabilization
Sediment/toxicant retention	Nutrient removal/transformation
Production export	Wildlife diversity/abundance
Aquatic diversity/abundance	Recreation
Uniqueness/heritage	

on the U.S. Geological Survey topographic maps. It was not the intent of the NWI to produce maps that show exact wetland boundaries comparable to boundaries derived from ground surveys. Boundaries are therefore generalized in most cases (U.S. Fish and Wildlife Service, 2003).

It should be noted that wetland areas exist that may not appear on the NWI maps. These wetlands may be too small to be seen in the aerial photos used to create the NWI maps, or they do not exhibit characteristics that allow

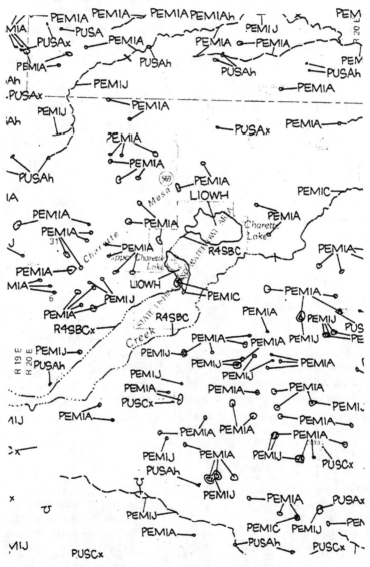

Figure 6.6 NWI wetlands delineation map.

interpretation from the air. On-site wetland delineations are needed to accurately establish the presence and extent of wetlands on the ground. A sample NWI wetland map can be found in Figure 6.6.

Wetlands should be located, and their function and importance evaluated, in any watershed plan to determine flood control capabilities, groundwater recharge potential, and habitat preservation.

6.6   SUMMARY

Lentic water (lakes, ponds, reservoirs), lotic water (streams and rivers), and wetlands are all integrated in the watersheds hydrologic system. Understanding their function, and how they affect the biology, chemistry, and hydrology of the watershed should be an integral part of any watershed assessment and plan.

REFERENCES

Adamus, P.R., E.J. Clairain, R.D. Smith, and R.E. Young. "Wetland Evaluation Technique (WET)," *Volume II: Methodology* (Vicksburg, MS: Department of the Army, Waterways Experiment Station, NTIS No. ADA 189968), 1987.

Adamus, P.R., L.T. Stockwell, E.J. Clairain, M.E. Morrow, L.D. Rozas, and R.D. Smith. "Wetland Evaluation Technique (WET)," *Volume I; Literature Review and Evaluation Rationale.* Technical Report WRP-DE-2 (Vicksburg, MS: U.S. Department of the Army, Waterways Experiment Station), 1991.

Cooke, G. Dennis, and Robert H. Kennedy, "Water Quality Management for Reservoirs and Tailwaters; Report 1, in "Reservoir Water Quality Management Techniques," Technical Report E-89-1 (Vicksburg, MS: U.S. Army Corps of Engineers Waterways Experiment Station), 1989.

Cowardin, L.M., Virginia Carter, F.C. Golet, and E.T. LaRoe. "Classification of Wetlands and Deepwater Habitats of the United States," U.S. Fish and Wildlife Service Report FWS/OBS-79/31, U.S. Government Printing Office, Washington, D.C., 1979.

Federal Register (45), Regulation versus Guidelines, 85343 December 24, 1980.

Ferren, Jr., Wayne R., Peggy L. Fiedler, Robert A. Leidy. "Wetlands of the Central and Southern California Coast and Coastal Watersheds—A Methodology for Their Classification and Description" Final Report Prepared for United States Environmental Protection Agency, Region IX, San Francisco, CA, February 6, 1995 (Revised August 1996).

The Heinz Center. *Dam Removal—Science and Decision Making* (Washington, DC: The H. John Heinz III Center for Science, Economics and the Environment), 2002.

Hill, P.S. Nelson, and P. Mason (eds.). "Chesapeake Bay Wetlands Research Recommendations and Program Descriptions," CRC No. 150b, www.chesapeake.org/stac/pubs/wrkshops/abstracts/crc150bab.html (2003).

Marriage, L.D., and R.F. Batchelor. "Ever Changing Network of Our Streams, Rivers, and Lakes," *Using Our Natural Resources, Yearbook of Agriculture* (Washington, DC: U.S. Government Printing Office), 1983.

National Estuarine Research Reserve System, Information on Estuaries, http://inlet.geol.sc.edu/nerrsintro/nerrsintro.html (accessed May 2003).

Old News, July and August 2003.

Terrell, Charles R., and Patricia B. Perfetti. *Water Quality Indicators Guide: Surface Waters* (Washington, DC: U.S. Department of Agriculture, Soil Conservation Service), September 1989.

Tiner, Ralph W. "Technical Aspects of Wetlands: Wetland Definitions and Classifications in the United States," National Water Summary on Wetland Resources United States Geological Survey Water Supply Paper 242 (date unknown), at http://water.usgs.gov/nwsum/WSP2425/definitions.html.

U.S. Army Corps of Engineers. *Corps of Engineers Wetlands Delineation Manual,* Technical Report Y-87-1, U.S. Army Corps of Engineers, Vicksburg, MS, 1987.

U.S. Army Corps of Engineers. *Federal Manual for the Identification and Delineation of Jurisdictional Wetlands,* U.S. Army Corps of Engineers, Vicksburg, MS, 1989.

U.S. Department of Agriculture (USDA) Soil Conservation Service (SCS). "Hydric Soils of the United States," USDA-SCS National Bulletin No. 430-5-9, Washington, D.C., 1985, as amended by the National Technical Committee for Hydric Soils (NTCHS), December 1986.

U.S. Environmental Protection Agency (EPA). "National Estuary Program," www.epa.gov/nep (May 2003).

———. *The Lake and Reservoir Restoration Guidance Manual,* 1st ed., EPA 440/5-88-002 (Washington, DC: Criteria and Standards Division), February 1988.

U.S. Fish and Wildlife Service. "Wetlands and Deepwater Habitats Classification: National Wetlands Inventory Mapping Code Description," www.nwi.fws.gov/atx/atx.html (May 2003).

U.S. Global Change Research Information Office. *Wetlands extent, structure and hydrology.* www.gcrio.org/geo/wetlands.html (May 2003).

CHAPTER 7

ECOLOGY/HABITAT

7.0 INTRODUCTION

One of the primary goals in sustainable watershed management is to maintain the species *biodiversity* and the aquatic and terrestrial habitat that supports them. It is therefore important to understand the interaction between organisms and their habitat, those species that indicate the relative health of an ecological system, and how to properly manage for species preservation. This chapter provides a general background of why each of these factors is important in watershed assessment.

7.1 ECOSYSTEMS

An *ecosystem* is a community of organisms and their physical and chemical environment. Aquatic ecosystems such as those found in streams, rivers, ponds, lakes, wetlands, estuaries, and bays are complex and diverse and vital to life in many ways. Each species has a unique *niche* within that community defined by those physical and chemical parameters (or *abiotic* parameters) and biological interactions. The *abiotic* (nonbiological) parameters include:

- Water quality
- Temperature
- Salinity
- Nutrients
- Substrate

- Light
- Oxygen
- Habitat

It is important to understand a particular ecosystem's *structure* and *dynamics*. Ecosystem structure is the spatial organization of the various communities, or where in the ecosystem each community resides. *Ecosystem dynamics* refers to the interaction of the community species and patterns of change.

The *food chain* or *food web* is the series of organic material or organisms that contribute to the diet of the top consumer, or the links among species in an ecosystem, as shown in Figure 7.1. The base of a food web is occupied mostly by vegetation (producers) and fine organic debris (decomposers). Herbivores (primary consumers) and carnivores (secondary consumers) occupy the higher levels. Omnivores occupy an intermediate level in the food web. Food webs are complicated by the fact that many species feed at various levels. The understanding of the food chain is essential in determining how pollutants may be *bioaccumulated* (concentrated) up the food chain into the species that are consumed by humans. Human changes in the environment affecting one level of the food chain can change the entire ecosystem.

The complexity of organisms ranges from communities based on the *predator-prey* relationship to others that rely on mutually beneficial or symbiotic relationships, or *mutualism*. Symbiotic relations rely on each other for

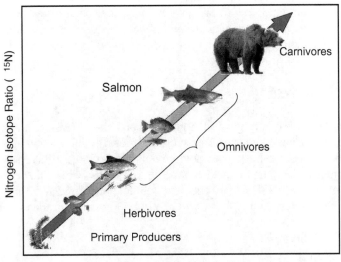

Figure 7.1 Typical food chain or web. (*Source:* USGS, How High on the Food Web are You, wwwrcamnl.wr.usgs.gov/isoig/projects/fingernails/results/foodchain.html, accessed, Dec 26, 2003.)

survival. For example, mycorrhizal fungi in soils penetrate the roots of some plants, particularly legumes, and aid in dissolving nutrients (nitrogen), thus making more nitrogen available to the plant than would be otherwise be available. Other examples of mutualism include the plant-pollinator relationship, the ant *Pseudomyrmex,* which grooms the acacia plant in return for food and protection; and marine fish, which clean parasites from other larger marine fish in return for protection. Lichens illustrate the mutually beneficial relationship between algae and fungi. All organisms are interrelated in one aspect or another, even if it's primarily through the food chain rather than symbiotically (Ricklefs, 1978).

Another type of ecological relationship includes *parasitism,* wherein a parasite resides either on the surfaces or inside the bodies of its prey. Worms such as the roundworm and tapeworm are examples. Eggs from the trematoad worm can be passed from out of human or animal intestines into water, which then develop into free-swimming larvae, which then infect snails and are then transformed into free-swimming cercariae, which then can again penetrate the human skin (Ricklefs, 1978).

For the most part, the entire food chain is dependent on energy produced from the sun. Energy from the sun is transformed into food through *photosynthesis* in green plants. This is an important concept to remember because it indirectly affects all watershed processes. In some cases, organisms obtain their energy from chemical constituents, a process called *chemosynthesis.* The energy transformation process is an extremely inefficient one. For instance, of the total *biomass* produced as phytoplankton, only a small portion of that energy is transferred to the next upper level, or ingested by the zooplankton. The efficiency of transfer of the carbon/energy from one *trophic level* to the next is only approximately 10 percent. The term *trophic* refers to a feeding level, or a group of organisms within the food chain that secures food in the same general way. For instance, all the animals that derive their energy directly from eating green plants, such as grasshoppers, mice, and deer, are in one trophic level. The assemblage of the trophic levels is the *trophic structure.* Typical ecosystems have three to six trophic levels, through which energy, organic material, and nutrients flow.

The food chain is complex, with the primary food web dependent on dead organic matter (detritus). *Primary producers* are organisms that occupy the first trophic level in the *grazing food chain.* These organisms are primarily *photosynthetic autotrophs,* which produce food molecules *inorganically* by using light and the chemical process of photosynthesis. Plants are the dominant photosynthetic autotrophs on the earth. This organism does not require outside sources of organic food energy for survival.

The aquatic food chain begins with *phytoplankton* (microscopic one-celled plants occurring in colonies), commonly known as *algae.* They are then eaten by the *zooplankton* (tiny one-celled animals). Extremely small phytoplankton are called *nannoplankton.* The general term including both phytoplankton and zooplankton is *plankton.* The zooplankton are eaten by small fish, which are in turn eaten by larger fish. This food chain includes many species consumable by humans, such as blue crabs, oysters, clams, and, eventually, game fish such as striped bass, flounder, and weakfish.

Excess sediment smothers the primary food producers. Excess nutrients induce undesirable plant growth. What humans do in the watershed that contributes sediment to the bays, such as river channel alterations and agriculture, directly affects the bays, and hence, eventually, humans themselves.

Maintaining or improving terrestrial and aquatic habitat is an integral part of a watershed plan that can be implemented through wetlands preservation or mitigation, riparian buffers, and floodplain management. Two important principles of habitat preservation are *habitat succession* and *edge*. Habitat succession is a recognizable pattern of progression of plant and animal communities over time. For instance, a field left uncut will succeed into, initially, *primary trees* and, eventually, *secondary forest growth*. A pond may silt in and eventually become a marsh, then fill with *emergent vegetation* and shrubs to become a wetland, while the animals that inhabit each habitat also change over time. It has been found that species diversity is greatest at habitat edges, or the boundaries between two different habitat types, such as the hedgerow between a farmer's field and the woodland. Gradual edges provide more integrating habitat, and in turn better biodiversity than abrupt edges. Any riparian habitat restoration/replacement or wetlands restoration effort should consider these relationships.

7.2 BIODIVERSITY

"Biodiversity is the variety and variation of all species of plants, animals, fungi, and microbes, including their genetic makeup, their ecological roles, and their interrelationships in biological communities throughout the world ecosystems" (Kim, 2002). Biodiversity has three components. The first is the variety of species; the second is the variety of genetic makeup within a species, called *genetic diversity* or *gene pool;* the third is the complexity of the community assemblage within the ecosystem, called the *ecosystem* or *community diversity*. The interactions and activities of thousands of organisms maintain the earth's atmosphere, soils, and water, and aid in filtering and breaking down pollutants, storing and filtering water, pollinating crops and flowers, and preventing disease. Each individual organism can be thought of as a strand in the spider web of life, supporting one another for strength and sustainability through the ecosystem interactions just described.

Aquatic biodiversity includes the phytoplankton, zooplankton, aquatic plants, insects, fish, birds, and mammals, and their interaction with their environment. Their environment includes lakes, ponds, reservoirs, rivers and streams, groundwater and wetlands (freshwater ecosystems), oceans, estuaries, salt marshes, seagrass beds, coral reefs, kelp beds, and mangrove forests (marine ecosystems). The banks of streams or riparian areas are also important areas associated with maintaining biodiversity of freshwater systems.

7.3 STRESSORS

Stressors are physical, chemical, or biological factors that place the aquatic ecosystem under stress. Physical stressors include temperature, flow, mixing, or habitat destruction. Chemical stressors include conventional and toxic pollutants. Toxic pollutants can be further classified as chronic and acute. *Chronic toxicity* would be any harmful effects to organisms with long-term exposure to specific substances or mixtures. This is most important if this exposure occurs during a sensitive period of the life cycle. Early life stages or reproductive toxicity tests may be used to determine chronic impacts. Therefore, most chronic exposure indices are expressed as time-weighted average (TWA) values. The nature of chronic effects is such that most can be anticipated and/or monitored, and occupational safety measures can be taken to minimize their impacts.

Acute toxicity refers to the impacts of single exposures, often at high concentrations. It suggests the possible risk levels for the consequences of accidental releases, such as from a spill.

One biological stressor that is becoming an ever-increasing problem is the introduction of *exotic invasive species,* organisms (plants, fish, insects, viruses) that are introduced into an environment in which they did not originate and may not have a regional natural predator to keep their populations in check. Invasive species often overtake naturally occurring species, food, or habitats, resulting in decreased biodiversity. If growth is unrestricted, they can kill off the native diverse population, thus creating a *monoculture.* Flowing water provides a means of seed and organism dispersal, and coupled with periodic flooding, allows ample opportunity for invasive species to become established. The problem species will vary by region and the local list should be obtained before conducting a watershed assessment. Below is a sampling of several invasive plants (Cole et al., 1996):

Tree of heaven	Canada thistle
Autumn olive	Japanese knotweeed
Multiflora rose	Phragmites
Purple loosestrife	Poison hemlock
Japanese stiltgrass	Mile-a-minute vine

Figure 7.2 shows purple loosestrife chocking out the natural vegetation of a wetland. There are also native species, such as reed canary grass and cattails, that are aggressive invaders of moist soils and should be kept in check. "Invasive plants threaten the habitat of two-thirds of all threatened and endangered wildlife species" according to Larry Kuhns, Professor of Ornamental Entomology ("Strange Invaders," *Penn State Agriculture,* Penn State Univ., Fall, 2003).

One example of the ecological significance of invasive species is the "dead zone," a large region of depleted oxygen where natural life cannot live, that is occurring in Lake Erie. Early indications are that it is being caused by the

Figure 7.2   Purple loosestrife, an invasive species, dominating a wetland.

invasive zebra and quagga mussels, deposited in the lake from the bilges of oceangoing ships and whose population has exploded. The mussels filter large amounts of organic matter, and in turn deposit it on the bottom of the lake. The organic matter then decomposes, creating a large demand for oxygen.

Another example of an invasive species problem is with the hemlock woolly adelgid, *Adelges tsugae,* a small insect with a cotton-like outer coating that sucks the juices from the eastern and Carolina hemlock to a point where the tree dies, as shown in Figure 7.3 (U.S. Forest Service, 2003). Hemlocks frequently inhabit ravines and stream valleys, and their thick mass of needles prevents the impact of raindrops from eroding the steep slopes. Elimination of these hemlock stands can leave the soil exposed for soil displacement and erosion. An invasive species inventory may be a valuable task in a watershed assessment.

All stressors can cause a loss of sensitive species, thus changing the biodiversity of the aquatic community. A cumulative impact of physical, chemical, or biological stressors often has more of an effect on the biotic community than a single stressor. Thus, the biodiversity of the biotic community is a good indicator of water quality and the presence of stressors.

7.4   BIOINDICATORS

Indicator species are those organisms found in particular environmental conditions; they are also referred to as *bioindicators* or *ecological indicators.* Indicators provide a means to determine watershed health, measure progress, and evaluate program effectiveness. The EPA defines an ecological indicator as "a measure, an index of measures, or a model that characterizes an eco-

Figure 7.3 The woolly adelgid kills eastern hemlock by sucking its sap and weakening the tree. Erosion can occur on dead hemlock stands where the soil is no longer protected from rainfall impacts. (*Source:* http://www.fs.fed.us/na/morgantown/fhp/hwa/hwasite.html.)

system or one of its critical components" (U.S. EPA, 2003). Coal miners took canaries into the mines, and although mines are not the birds' natural environment, they indicated when lethal gases were present. Canaries are more sensitive to toxic gases than humans, so when they perished, they provided a warning system to the miners to seek fresh air.

Bioindicators are very sensitive to changes in their environment. Biological organisms, or the lack of their presence, are, therefore, good indicators of the quality of the environment in which they live.

7.4.1 Aquatic Bioindicators

7.4.1.1 Periphyton. *Periphyton* and algae are primary producers and are most directly affected by chemical and physical factors. Periphyton are sedentary, have rapid reproduction rates and short life cycles, and hence are excellent indicators of short-term impacts. They are more sensitive to some pollutants than the other assemblages. Methods of evaluation include biomass and chlorophyll measurements (Barbour, 1999).

Periphyton are not the only primary producers. Aquatic systems also include *macrophytes* such as mosses, rooted aquatics, and *emergents,* which also provide primary production.

7.4.1.2 Macroinvertebrates. *Macroinvertebrates,* or large (nonmicroscopic) animals without backbones, are also excellent indicator organisms. *Benthic macroinvertebrates,* or *benthos,* are those that dwell on the bottom, or *substrate,* of streams, rivers, lakes, or bays, and are therefore excellent indicator species. They can be classified based upon their functional feeding designations and their habit/behavior designations, as shown in Tables 7.1 and 7.2 (Barbour, 1999).

Benthic macroinvertebrates are good indicators of localized water quality conditions because their migration is limited to a small area. They typically have a life cycle of one year or more, and sensitive stages will respond quickly to stress. They are easy to identify by an experienced biologist, and are made up of a variety of species with variable tolerances to pollutants. Therefore, by identifying the major macroinvertebrates, general stream health can be assessed (Barbour, 1999).

Clean streams typically contain larval insect species such as stoneflies, caddis flies, and mayflies in the cobbles and gravel of riffle areas. Although the presence of these organisms represents clean water, the absence of these organisms does not necessarily indicate polluted water. A prolonged drought or a recent flood may have eradicated the larval insect species temporarily and the populations may not have had a chance to recolonize. One must look at the history of the stream before making a determination of a stream's health based upon the organisms present.

Streams with sandy bottoms support very few macroinvertebrate species due to small pore size for cover and instability of the shifting sand to provide a stable strata. *Aggrading,* or depositing, streams and rivers typically have silt or mud bottoms. These support a different class of invertebrate species such as bloodworms, midges (chironomids), tube-building worms, burrowing may-flies, mussels, and clams. Silty beds are also very unstable.

W.M. Beck (1955) classified macroinvertebrates into *intolerant* (sensitive) to pollution, which are assigned a value of 2, and *facultative,* which can tolerate some pollution and are assigned a value of 1. The higher the index value, the healthier the stream (U.S. EPA, 1990). Benthos can also be classified as tolerant to pollution. Although originally developed for Florida conditions, this classification system can be adapted to other regions. Each class is described in further detail here:

- *Class I Organisms* (*Sensitive or Intolerant*). These organisms are highly sensitive to changes in their aquatic environment, including pollution, and are either killed or driven away from their home by these changes. A

TABLE 7.1   Benthic Macroinvertebrates Functional Feeding Designations

PA = Parasite	FC = Filter/collector
PR = Predator	SC = Scraper
OM = Omnivore	SH = Shredder
GC = Gatherer/collector	PI = Piercer

Source: Barbour, 1999.

**TABLE 7.2   Benthic Macroinvertebrates Habit/
Behavior Designations**

cn = Clinger	sw = Swimmer
cb = Climber	dv = Diver
sp = Sprawler	sk = Skater
bu = Burrower	

Source: Barbour, 1999.

significant drop in populations would indicate a change in their environmental factors.

- *Class II Organisms (Facultative).* These organisms can adapt to varying environmental conditions and are more tolerant to changes in their environment. Most can survive where organic pollution is causing eutrophication. An example of these organisms is the facultative anaerobes, bacteria that can survive in oxygenated or anaerobic (devoid of free oxygen) conditions.

- *Class III Organisms (Tolerant).* These organisms are the most tolerant to stressed environmental conditions; they can survive and reproduce under these conditions. If one were to equate the macroinvertebrate indicator organisms to vertebrate indicators, the trout would be intolerant, susceptible to slight changes in oxygenated conditions, temperature, and pollution, whereas the carp would be tolerant, able to live in muddy, slow-moving, oxygen-poor waters. The macroinvertebrate organisms that fall into these categories can be found in Figure 7.4 (Terrell and Bytnar, 1989).

Beck's Biotic Index (BBI) would be computed as the addition of two times the number of Class I species identified in the sample plus the number of Class II species, or:

$$BBI = 2n_{I} + n_{II} \qquad (7.1)$$

Where:

BBI = Beck's Biotic Index
n_{I} = Number of Class I species identified in the samples
n_{II} = Number of Class II species identified in the samples

Using this approach (Terrell and Bytnar, 1989; Beck, 1955), several trends in indicator organism diversity can be observed:

- An undisturbed community will contain the majority of groups in Class I but also representative numbers of Classes II and III.

1. Intolerant (sensitive) to pollution:

C
L
A
S
S

1

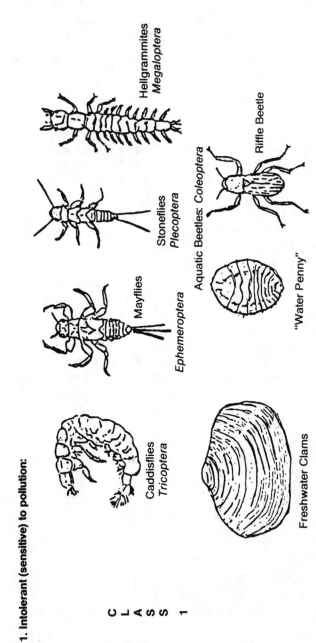

Caddisflies
Tricoptera

Mayflies

Ephemeroptera

Stoneflies
Plecoptera

Hellgrammites
Megaloptera

Aquatic Beetles: *Coleoptera*

Riffle Beetle

"Water Penny"

Freshwater Clams

Figure 7.4 Macroinvertebrates according to Beck's Biotic Index Classes (Terrell, 1989).

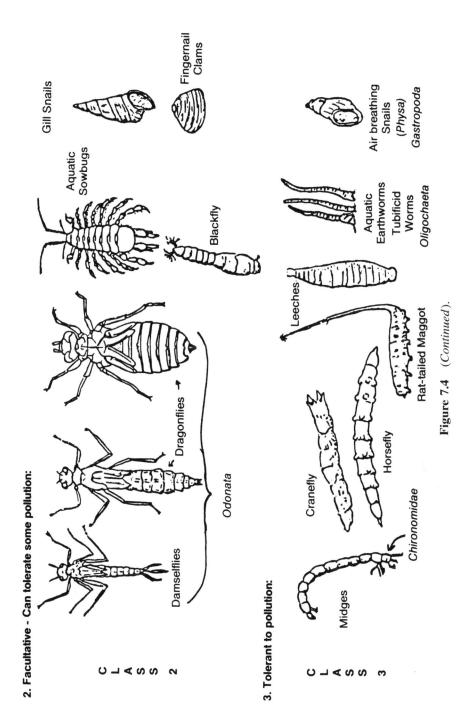

2. Facultative - Can tolerate some pollution:

Gill Snails

Fingernail
Clams

Aquatic
Sowbugs

Blackfly

Air breathing
Snails
(*Physa*)
Gastropoda

Dragonflies →

Damselflies

Odonata

C
L
A
S
S

2

Aquatic
Earthworms
Tubificid
Worms
Oligochaeta

Leeches

3. Tolerant to pollution:

Cranefly

Horsefly

Rat-tailed Maggot

Midges

Chironomidae

C
L
A
S
S

3

Figure 7.4 (*Continued*).

145

- A community with natural limits (drought, homogeneous substrate) or that is impacted by human activities will contain mainly Class II organisms.
- Waters dominated by Class III organisms are adversely impacted by organic pollution. Organic pollution affects both the diversity of the species and the number of species per unit area. Toxic pollution, such as acid mine drainage (AMD) or pesticides, has essentially the same effect.
- Waters with a sediment pollution problem tend to show a greater reduction in density of organisms than in the diversity of organisms.
- As can be seen from the preceding trends, by evaluating species type, diversity, and populations (richness), one can obtain a general type of pollution.

7.4.1.3 Fish. Although some fish, such as grass carp, are *grazers* (feed on vegetation), and others are *detritivores* (feed on *detritus* or dead organic matter), most fish are at the top of the aquatic food chain. Fish are relatively long-lived and are therefore a good indicator of long-term effects and habitat conditions. And because they are consumed by humans, fish play an important role in assessing contamination. They can be classified by their trophic designations and tolerance designations, as shown in Tables 7.3 and 7.4 (Barbour, 1999).

At the extremes, the presence of species such as trout would be a good indicator of cold-water fisheries that are relatively free of pollution and well oxygenated, whereas the presence of carp would indicate warmer, less oxygenated waters.

7.4.2 Terrestrial Bioindicators

Certain aquatic organisms, as we have seen, are excellent bioindicators for the aquatic environment. There is increasing evidence that songbirds are an actual terrestrial bioindicator of a region's ecological conditions. Species differ in their tolerance for physical, biological, and chemical stressors, including the availability of appropriate food, shelter, and breeding conditions. During the gypsy moth invasion of the Pennsylvania oak forests in 1988 and 1989, spraying was not supposed to affect bird or mammal species, but dead animals were encountered throughout the forests. It was observed that birds' songs were absent for up to five years afterward. Here again, it has been found that

TABLE 7.3 Fish Trophic Designations

P = Piscavore	F = Filter feeder
H = Herbivore	G = Generalist Feeder
O = Omnivore	V = Invertivore
I = Insectivore (including specialized insectivores)	

Source: Barbour, 1999.

TABLE 7.4   Fish Tolerance Designations (Relevant to Nonspecific Stressors)

I = Intolerant
M = Intermediate
T = Tolerant

Source: Barbour, 1999.

undisturbed areas have substantially better species diversity than disturbed areas. A study performed by Jackson and colleagues (2000) related the types of bird species found to the surrounding land cover (urban, residential, agriculture, forest, etc.) to create the Bird Community Index (BCI). The BCI ranges from 0, for poor to fair ecological conditions, to 2, representing good to excellent ecological conditions.

The Canaan Valley Institute (CVI) worked with the U.S. Environmental Protection Agency's Office of Research and Development and the Penn State Cooperative Wetlands Center to apply the BCI to the Mid-Atlantic Highlands. Mapping the BCI within a GIS, the ecological conditions of a watershed can be implied. Mapping the BCI over time can determine when watersheds become stressed. The BCI study showed that stressed habitats such as agriculture, mining, timber harvesting, and residential and urban land uses had lower BCIs. Forested watersheds had the highest BCIs, as shown in Figure 7.5. Lower BCI habitats included American crows, starlings, and blue jays. Forested habitats, in contrast, housed species such as the scarlet tanager, Louisiana waterthrush, and cerulean warbler, neotropical migratory birds that are more sensitive species. These species spend their winters in the subtropics but breed along streams and within deep forests. Although the BCI cannot be used as the only indicator of a watershed's health, due to the many intricacies and interrelationships within a watershed, it is a valuable tool that may be added to any watershed assessment (Claggett, 2001).

7.5   FISHERIES

Maintaining reproductive fisheries is essential to human survival. Water bodies can also be classified by the type of fisheries they support. Three classes include cold-water, cool-water, and warm-water fisheries. Cold-water fisheries typically support the family *Salmonidae* (trout and salmon) and other flora and fauna indigenous to cold-water habitat. Trout are primarily insect feeders that thrive in well-oxygenated, swift waters, whose temperatures range between 50 and 60 degrees Fahrenheit. The cool-water class includes species such as the smallmouth bass, which feeds in lower-stream reaches that are warmer and slower than the cold water and are marginal for trout. These waters still contain the riffles and deep pools. Typical food for smallmouth bass includes insects, crayfish, and minnows. Warm waters are typical of

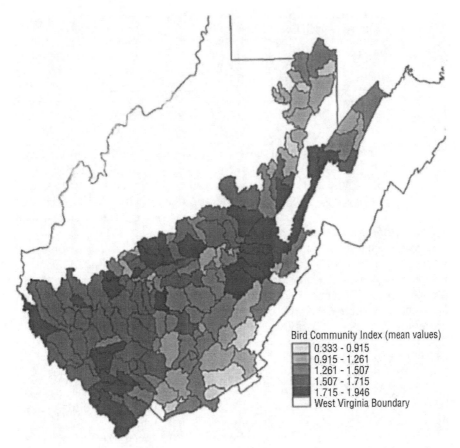

Figure 7.5 Mean Bird Community Index (BCI) values for watershed in southern West Virginia.

lakes, ponds, and lower reaches of rivers where water is still and slow flowing. Warm-water species include largemouth bass, bluegill, perch, crappie, and catfish, and other flora and fauna indigenous to warm-water habitats.

Many states have developed further classifications, thus state or regional specifications should be consulted before applying the standards. For example, in Pennsylvania, cold-water fisheries are classified as either high quality (HQ) or exceptional value (EV) for the purposes of regulating uses and maintaining their pristine qualities. HQ streams are those that have a high quality that exceeds levels necessary to support propagation of fish, shellfish, and wildlife. They have a high-quality aquatic community based upon peer-reviewed biological assessment procedures that consider physical habitat, benthic macroinvertebrates, or fishes such as described in "Rapid Bioassessment Protocols for Use in Wadeable Streams and Rivers: Periphyton, Benthic, Macroinvertebrates, and Fish" (Barbour et al., 1999). EV waters are those of high water quality that meets one or more of the following specific conditions:

- A surface water with exceptional ecological significance
- A surface water with exceptional recreational significance
- Water located in a national wildlife refuge or state propagation and protection area (PADEP, 2003)

Streams can also be classified based upon their cold-water fisheries or trout waters. For instance, a Class A wild trout water, which supports a population of naturally reproducing trout of sufficient size and abundance to support a long-term and rewarding sport fishery, may be classified by a state's fish commission based upon species-specific biomass standards. Individual state definitions and regulatory requirements should be consulted for each watershed.

The EPA has classified streams into several categories based upon their water quality, as shown in Figure 7.6. These include impaired or unimpaired and unhealthy or healthy/sustainable (U.S. EPA, 2003a).

7.6   THREATENED AND ENDANGERED SPECIES

A watershed plan should identify *threatened* and *endangered* species and their habitat, and develop goals to preserve both. Threatened species as defined by

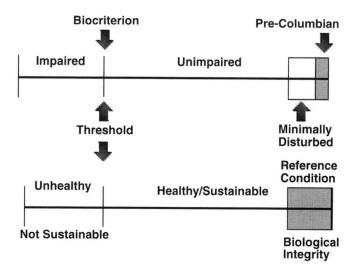

Figure 7.6   EPA classifications of streams (USEPA, 2003a).

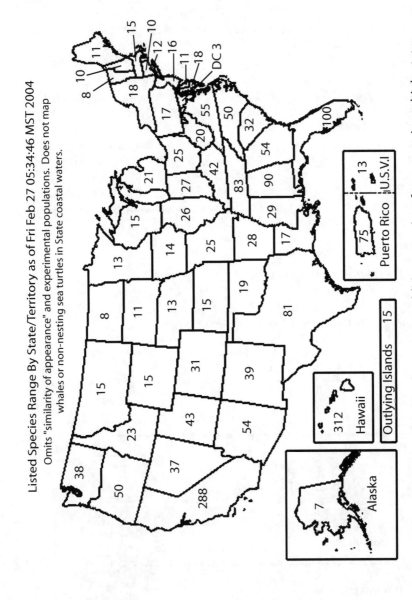

Listed Species Range By State/Territory as of Fri Feb 27 05:34:46 MST 2004

Omits "similarity of appearance" and experimental populations. Does not map whales or non-nesting sea turtles in State coastal waters.

Total U.S. Species is 1258. Numbers are not additive, a species often occurs in multiple states.

Figure 7.7 Number of threatened and endangered species by state (*Source:* U.S. Fish and Wildlife Service, 2003).

the Federal Endangered Species Act of 1973 (87 Stat. 884, as amended; 16 U.S.C. 1531 *et seq.*) are species that are likely to become endangered within the foreseeable future throughout all or a portion of their range. An endangered species is a species in danger of extinction throughout all or a significant portion of its range (other than insects that are a threat to humans, as defined by the Secretary of the Department of Interior or Commerce).

Information about threatened and endangered (T&E) species can be obtained from a state's Natural Diversity Inventory (NDI) or from the appropriate federal or state resource agencies. Species inventories are typically organized by computer, map, and manual files that describe locations of endangered, threatened, and rare species, and that sometimes include the most outstanding examples of a state's natural community and ecologic features. In some cases, current field surveys have found that historically documented species were lost due to habitat destruction. In other cases, plant and animal species have not been included because past records are not specific enough to precisely document sites. Although historical records may not completely reflect the known range of a particular species, they are the best documentation available to estimate the former distribution of threatened and endangered plants and animals.

Species data are available in a variety of formats: for example, GIS-compatible format, composite maps in Adobe Acrobat .pdf, or other electronic formats. These data are often published on the World Wide Web. The maps should be examined to determine whether threatened or endangered species are present or were historically present in or near the watershed. Maps that show species in the general region can be georeferenced and then used to produce the threatened and endangered species area GIS layer. Other threatened and endangered species may be present in the watershed but are not mapped. The number of T&E species by state is shown in Figure 7.7.

Preservation of T&E species is dependent on their terrestrial (land) or aquatic (water) habitat. *Critical habitat* is defined as habitat where certain physical, chemical, and biological features in the environment are essential to the conservation and support of one or more life stages of a species.

TABLE 7.5 Nature Conservancy Rare and Endangered Species Classification

GX:	Presumed extinct (not located despite intensive searches)
GH:	Possibly extinct (of historical occurrence; still some hope of rediscovery)
G1:	Critically imperiled (typically 5 or fewer occurrences or 1,000 or fewer individuals)
G2:	Imperiled (typically 6 to 20 occurrences or 1,000 to 3,000 individuals)
G3:	Vulnerable (rare; typically 21 to 100 occurrences or 3,000 to 10,000 individuals)
G4:	Apparently secure (uncommon but not rare; some cause for long-term concern; usually more than 100 occurrences and 10,000 individuals)
G5:	Secure (common; widespread and abundant)

Source: Barbour, 1999.

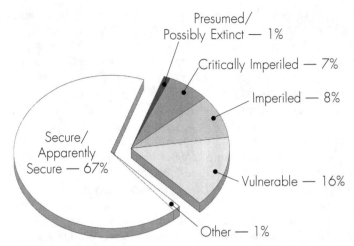

Figure 7.8   Breakdown of presumed/possibly extinct, critically imperiled, imperiled, vulnerable, and secured/apparently secured species in the U.S. (EPA, 2003b).

The Nature Conservancy classified rare and endangered species according to Table 7.5 (U. S. EPA, 2003b), and a breakdown of species in the United States can be found in the pie chart in Figure 7.8.

7.7 SUMMARY

Maintaining the fisheries of aquatic systems and biological integrity of aquatic and terrestrial systems should be a goal of any watershed plan. Stressors should be identified and goals developed to minimize them. Biological indicators are an efficient means to determine environmental integrity of systems. Threatened and endangered species located in the watershed, and their habitat, should also be identified; and preservation of the habitat and species should be incorporated into the watershed plan.

REFERENCES

Bailey, Robert G. "Hierarchy of Ecoregions at a Range of Scales" (map) (Washington, DC: U.S. Department of Agriculture, Forest Service), hwww.fs.fed.us/land/ecosysmgmt/colorimagemap/ecoreg1_provinces.html (1994).

———. "Description of the Ecoregions of the United States," Miscellaneous Publication 1391 (Washington, DC: U.S. Department of Agriculture), 1980.

———. Personal communication.

Bailey, Robert G., Peter E. Avers, Thomas King, and W. Henry McNab, W, eds. "Ecoregions and Subregions of the United States (map) (Washington, DC: U.S.

Geological Survey), 1994. (Note: The scale of this colored map is 1:7,500,000. It is accompanied by a supplementary table of map unit descriptions that were compiled and edited by McNab and Bailey, and prepared for the U.S. Department of Agriculture, Forest Service.)

Barbour, Michael T., Jeroen Gerritsen, Blaine D. Snyder, and James B. Stribling. "Rapid Bioassessment Protocols for Use in Streams and Wadeable Rivers: Periphyton, Benthic Macroinvertibrates, and Fish," 2nd ed., EPA 841-B-99-002 (U.S. EPA, Office of Water), July 1999.

Beck, W.M. "Suggested Method for Reporting Biotic Data," in Sewage and Industrial Wastes, 27:1193–1197, 1955.

Claggett, Peter. "What Birds Tell Us," in *Echo* (Davis, WV: Canaan Valley Institute), 2001.

Cole, Charles A., Thomas L. Searfass, Margaret C. Brittingham, and Robert P. Brooks. "Managing Your Restored Wetland," Penn State University, Cooperative Extension Service, University Park, PA, 1996.

Common Ecological Region National Interagency Technical Team. ECOMAP: National hierarchical framework of ecological units. Unpublished administrative paper (Washington, DC: U.S. Department of Agriculture, Forest Service), 2002.

Jackson, L.E., T.J. O'Connell, and R.P. Brooks. *MAIA project summary: Birds indicate ecological condition of the Mid-Atlantic highlands.* EPA/620/R-00/003. USEPA–Office of Research and Development, Washington, D.C., 2000.

Kim, Ke Chung. "Biodiversity, Our Living World: Your Life Depends on It!" (University Park, PA: Penn State University, Center for Biodiversity Research), 2002.

McNab, W. Henry, and Peter E. Avers. "Ecological Subregions of the United States" WO-WSA-5. Prepared in cooperation with Regional Compilers and the ECOMAP Team of the Forest Service, July 1994. *http://www.fs.fed.us/land/pubs/ecoregions/*

Omernik, J.M. "Ecoregions of the Conterminous United States," *Annals of the Association of American Geographers,* vol. 77, no. 1 (1987), pp. 118-25.

Omernik, J.M. and G.E. Griffith. "Ecological Regions Versus Hydrologic Units—Frameworks for Managing Water Quality," *Journal of Soil and Water Conservation,* vol. 46, no. 5 (1991), pp. 334–340.

Omernik, J.M. Ecoregions: "A Spatial Framework for Environmental Management." In *Biological Assessment and Criteria: Tools for Water Resource Planning and Decision Making,* W. Davis and T. Simon (Eds.; Boca Raton, FL: Lewis Publishers), 1995, pp.49–62).

PADEP—Pennsylvania Department of Environmental Protection, 28 Pa. Code § 17.51 Title 25, Environmental Protection Chapter 93. Water Quality Standards General Provisions–Antidegradation Requirements, 2003

Ricklefs, Robert E. *Ecology* (New York: Chiron Press), 1978.

Stein, Bruce A., Lynn S. Kutner, and Jonathan S. Adams, eds. Precious Heritage: *The Status of Biodiversity in the United States,* compiled for The Nature Conservancy and Association for Biodiversity Information (Oxford: Oxford University Press), 2000.

Terrell, Charles R., and Patricia Bytnar. "Water Quality Indicators Guide: Surface Waters" (Washington, DC: USDA, SCS), September 1989.

U.S. EPA. Proceedings of the 1990 Midwest Pollution Control Biologist Meeting (EPA905/9-90-005), Chicago, IL, April 1–13, 1990.

————. "Biological Indicators of Watershed Health," *http://www.epa.gov/ bioindicators/html/about.html* (February 17, 2003).

U.S. Forest Service. "Ecological Subregions of the United States," www.fs.fed.us/ land/pubs/ecoregions/toc.html (February 17, 2003).

U.S. Forest Service. "Hemlock Woolly Adelgid," *http://www.fs.fed.us/na/ morgantown/fhp/hwa/hwasite.html,* December 2003.

Wiken, E.B. (compiler). "Terrestrial Ecozones of Canada," Ecological Land Classification Series No. 19: Environment Canada, Hull, Quebec, 26 pp and map, 1986.

CHAPTER 8

WATER QUALITY: NONPOINT SOURCE POLLUTION

8.0 INTRODUCTION

Maintaining or improving the quality of groundwater and surface water-courses is of primary concern for future sustainability. Abuse of our water resources has far-ranging environmental and economic consequences, among them degradation of water quality, habitat, fisheries, and recreation uses. Pollution comes from a variety of sources, including point and nonpoint sources. Management or control of those sources can only be accomplished after learning to understand the mechanisms of the pollutant buildup-washoff processes.

8.1 WATER QUALITY

Naturally, maintaining excellent water quality, whether groundwater or surface water, should be a main goal of any watershed assessment and management plan. This is often referred to as *antidegradation,* which, in a regulatory context, sets the goal of maintaining and protecting high-quality waters. Understanding indicators of water quality health is, therefore, a reasonable process to determine the overall health or stress of a receiving water body. The following will provide a concise, qualitative surface water quality indicator guide, utilizing geologic, hydrologic, biologic, and ecologic principles. These approaches could be supplemented with quantitative chemical testing to confirm the results.

As with the geomorphic processes, streams under natural conditions have the ability to naturally cleanse themselves, with the inputs of parameters such as pesticides, nutrients, and sediment equaling the outputs. However, human influence from agriculture, urbanization, or mining overloads this equilibrium, resulting in polluted streams. Overloaded streams are in stressed conditions, with sensitive organisms being the first to succumb to the effects of the imbalance. The end result is a decrease in biodiversity.

The quality of water in watercourses or bodies of water is affected by many factors, including flow rate, water depth, amount of suspended sediment, amount of sunlight reaching the water, nutrient loads, and other pollution. The amount of sunlight reaching the water body's bed, which supplies energy to aquatic vegetation, can be affected by sediment, canopy, and depth of water. Atmospheric deposition also plays a role in water quality. Before looking at the quality of surface and groundwater, the quality of the source of this water, precipitation, should be studied.

8.2   CHEMICAL PROPERTIES OF PRECIPITATION

The quality of surface water and groundwater is influenced by the quality of precipitation itself. As pollutants are discharged into the air through car exhaust and industrial smokestacks, these particles either settle to the earth or are carried back to the land surface in raindrops. One of the major concerns in the midwestern and eastern United States is acid precipitation.

Acidity is the capacity to neutralize bases. *Alkalinity,* conversely, is the capacity of water to neutralize acid. Water exists in the following relationship (Tchobanoglous, 1987):

$$H_2O \rightarrow H^+ + OH^- \tag{8.1}$$

The H^+ and OH^- are essentially water in its ion form.

Levels of acidity and alkalinity, or base, are measured by the pH scale. The hydrogen ion determines the pH. The pH scale ranges from 0 to 14, with 0 being the most acidic and 14 being the strongest base value. A pH of 7 is neutral. Optimum conditions for most aquatic life in the waters is a pH of 7 to 9. The pH can be defined as:

$$pH = -\log[H^+] \tag{8.2}$$

The equilibrium expression (K) would, therefore, be:

$$K = [H^+] [OH^-] \div H_2O \tag{8.3}$$

Because the molar concentration of water is essentially constant, the expression can be simplified to:

$$K_w = [H^+] [OH^-] \tag{8.4}$$

Where:

K_w = Equilibrium constant of water

When air contaminants having ionizable H^+ or OH^- dissolve in water, the equilibrium between the H^+, OH^-, and H_2O shifts. If the H^+ increases, the pH decreases (increased acidity). The burning of fossil fuels introduces sulfur dioxide (SO_2) into the atmosphere. As these gases dissolve in water droplets, sulfurous acid is produced:

$$SO_2 + H_2O \rightarrow H_2SO_3 \qquad (8.5)$$

Other chemical reactions may raise the hydrogen ion concentration by as much as three orders of magnitude.

Acid rain with pH values of 4 has been recorded in the northeastern United States, Canada, and Europe. The Midwest industrial belt is the major source of the pollution in the northeastern United States.

Acid rain percolates through the soils and eventually feeds the stream as *base flow*. Soils derived from acidic bedrock typically do not have a buffering capacity; however, soils derived from calcareous material such as limestone typically have a good buffering capability. *Buffering* is the ability of the soil to resist change in pH when supplied with an acid or base. Well-buffered soils inactivate added H^+ and OH^- ions and thus bring the percolated water closer to a neutral pH. The chemical reaction would be:

$$CaCO_3 + 2H^+ \rightarrow CA^{2+} + H_2CO_3 \qquad (8.6)$$

which is the same reaction as when lime is applied to soil to increase its pH for growing conditions. Streams in watersheds with soils that are unbuffered, then, are more susceptible to the effects of acid precipitation than those in watersheds with calcareous soils.

Winter conditions typically cause precipitation to fall in the form of snow and ice. Snow and ice deposits contain acidity similar to rain; however, the precipitation is stored in a solid form until spring thaw occurs. During surface ice and snow thaw conditions, percolation through soils does not occur due to direct surface runoff over frozen ground. Therefore, buffering does not occur until the soils thaw and unbuffered surface water runoff flows directly to the stream. Extended periods of low pH runoff typically occur during the spring thaw, which can cause extensive damage to aquatic life.

A common trend in streams is that pH levels may increase as the stream leaves the mountainous terrain and enters the lowlands, where farming is a typical land use. This can be attributable to several factors. As reported in the *Mehoopany Creek Watershed Assessment,* (BLE, 2003)

as the stream enters the lowlands, reaching deeper soils and longer percolation characteristics, the precipitation receives longer contact time with the soils during precipitation events. Soils in the lowland areas are typically deeper and naturally higher in pH than the mountainous soil series, thereby creating a better capability to buffer the acid precipitation. Additionally, the deeper, richer lowland soils are used for agriculture more often, and lime is added to improve pH levels for crop production. The addition of lime improves the pH and alkalinity of precipitation as it leaches through the agricultural areas and, eventually, improves the stream pH.

Limestone addition programs to streams have been successful in raising the pH and enhancing the sport fishing of the streams, and have shown that fish and other aquatic organisms increased in numbers and diversity (Weigmann et al., 1993).

In performing a watershed assessment, a source of data for *deposition pollution* would be the National Atmospheric Deposition Program (NADP, 2003), which includes a network of monitoring stations throughout the United States, as shown in Figure 8.1, and identifies the quantities of air pollutants deposited. Samples of the maps produced from its data for pH and the ammonium ion can be found in Figures 8.2 and 8.3, respectively.

In addition to the pH and the ammonium ion data, the NADP also supplies annual spatial data for the parameters listed in Table 8.1, and total annual precipitation (see Figure 8.4).

These data can aid in explaining trends found in water quality sampling. In addition to the data, however, water samples should be collected and a

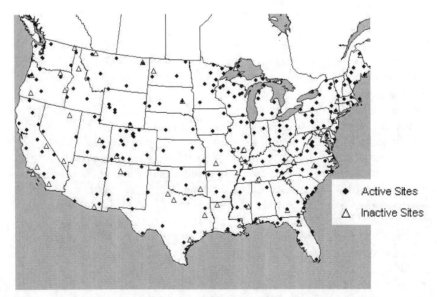

Figure 8.1 National Atmospheric Deposition Program (NADP) data collection sites (NADP, 2003).

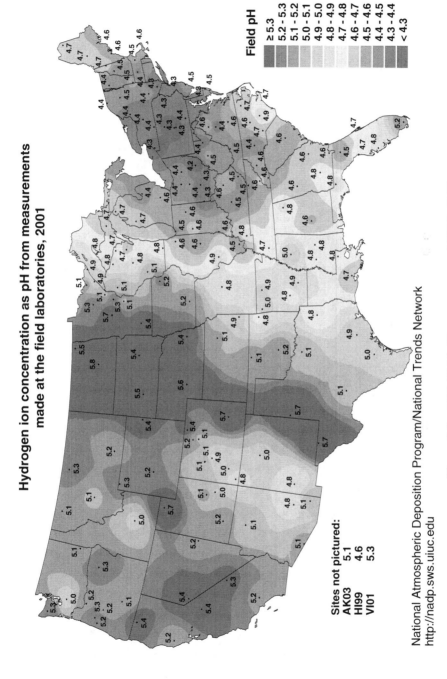

Hydrogen ion concentration as pH from measurements made at the field laboratories, 2001

Sites not pictured:
AK03 5.1
HI99 4.6
VI01 5.3

Field pH

≥ 5.3
5.2 - 5.3
5.1 - 5.2
5.0 - 5.1
4.9 - 5.0
4.8 - 4.9
4.7 - 4.8
4.6 - 4.7
4.5 - 4.6
4.4 - 4.5
4.3 - 4.4
< 4.3

National Atmospheric Deposition Program/National Trends Network
http://nadp.sws.uiuc.edu

Figure 8.2 2001 Field pH measurements (NADP, 2003).

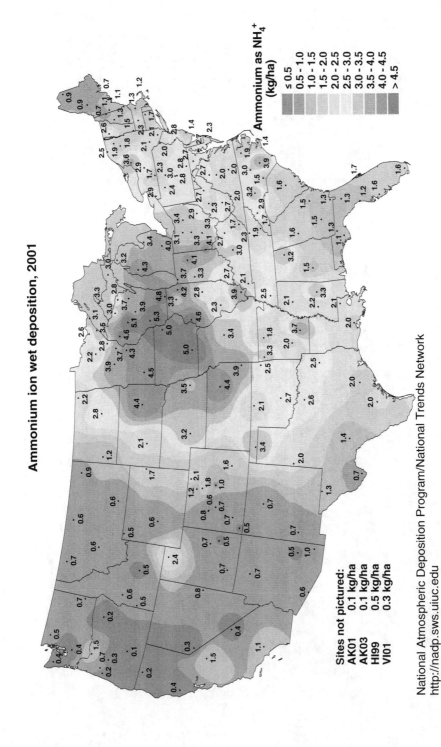

Figure 8.3 2001 Wet deposition of the ammonium ion (NADP, 2003).

TABLE 8.1 Parameters Measured by NADP (2003)

Lab pH (measured at the Central Analytical Laboratory)
Lab H Deposition (kg/ha) (measured at the Central Analytical Laboratory)
Field pH (measured at the field sites)
Field H Deposition (kg/ha) (measured at the field sites)
(Sulfate) SO_4^{-2} Concentrations (mg/L)
SO_4^{-2} Deposition (kg/ha)
Nitrate (NO_3) Concentrations (mg/L)
NO_3 Deposition (kg/ha)
Ammonium (NH_4^+) Concentrations (mg/L)
NH_4 Deposition (kg/ha)
Calcium (Ca) Concentrations (mg/L)
Ca Deposition (kg/ha)
Magnesium (Mg) Concentrations (ug/L)
Mg Deposition (kg/ha)
Potassium (K) Concentrations (ug/L)
K Deposition (kg/ha)
Sodium (Na) Concentrations (ug/L)
Na Deposition (kg/ha)
Chlorine (Cl) Concentrations (mg/L)
Cl Deposition (kg/ha)
Nitrogen (N) Deposition (from NO_3 and NH_4) (kg/ha)
Measured Precipitation (cm)

water quality analysis performed to verify pH differences and other water quality parameters in stream reaches. Use of pH data is critical in identifying the amount of lime required to adjust pH values of the stream to improve conditions for trout survival and reproduction success. It is critical to collect field data on the pH of the streams in the entire watershed to be able to determine those areas where the problem originates and to assess the need for lime addition in those areas.

8.3 CHEMICAL PROPERTIES OF RECEIVING WATERS

In addition to precipitation, the chemistry of a water body is influenced by its physical surroundings, such as soils and internal reactions. Naturally, the soils surrounding the water influence its pH, acidity, alkalinity, hardness, and nutrients. Land uses also have a major influence on the chemical composition of receiving waters. Sediment causes *turbidity* in the water, which affects sunlight penetration. Phosphorus, typically the limiting nutrient, attaches itself to clay particles due to the negative ionic charges of the particles, and then settles with the solids. This phosphorus may then become soluble and available for excess plant growth. Nutrients, pesticides, and other pollutants are also chemically and biologically detached and/or altered in the aquatic sys-

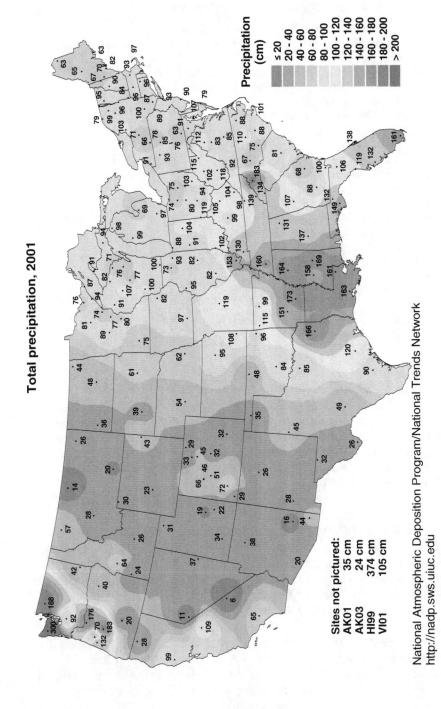

Figure 8.4 2001 Total precipitation in centimeters (NADP, 2003).

National Atmospheric Deposition Program/National Trends Network
http://nadp.sws.uiuc.edu

tem, depending on a variety of factors including pH, oxygen content, and temperature.

8.4 POINT VERSUS NONPOINT SOURCE POLLUTION

Nonpoint source (NPS) pollution is pollution occurring from locations other than from a pipe or point source. However, surface runoff that collects in storm sewers and discharges via a pipe is still considered NPS pollution, since it originates as diffuse runoff from the land surface. Point sources are visible, discrete, and easily identifiable. By contrast, nonpoint sources are diffuse and often hard to trace. These sources typically include stormwater runoff, construction site drainage, feedlot drainage, malfunctioning septic systems, highway deicing, and farm field runoff. Studies have found that nonpoint source pollution contributes approximately 45 percent of the total pollution to estuaries, 76 percent to lakes, and 65 percent to rivers, as shown in Table 8.2 (U.S. EPA, 1989).

NPS pollution presents a complex problem to surface water and groundwater, which can accumulate as the problem moves downstream. NPS pollution to surface water stems from storm events, contrasting sharply with the relatively even, continuous discharges encountered from point sources of pollution.

Point sources are "end-of-the-pipe" discharges that are easy to regulate with federal and state permits. NPS pollution, in contrast, is the direct result of land use patterns, so many of the solutions lie in finding more efficient, rational ways to manage the land and stormwater runoff. Federal and state government does not play a major role in local land use decisions; therefore, local governments must be responsible for maintaining water quality through NPS controls.

Pollution can be expressed as a concentration in milligrams/liter (mg/l) or a loading in pounds/acre/year (lb/ac/yr). Expressing pollutant concentrations in mg/l is important in urban areas where trace metals and toxicants may impact the environment when their ambient concentration exceeds a critical

TABLE 8.2 Percentage of Pollutant Sources to Estuaries, Lakes, and Rivers

	Estuaries	Lakes	Rivers
Combined Sewer Overflow	4%	1%	1%
Natural Causes	3%	12%	6%
Industrial Point Sources	8%	1%	9%
Municipal Point Sources	22%	8%	17%
Nonpoint Sources	45%	75%	65%
Other/Unknown	18%	3%	2%
	100%	100%	100%

threshold over some period of time. In rural watersheds, where trace metals and toxicants are relatively low, the loadings of parameters that directly affect the biology of the receiving waters (total nitrogen, total phosphorous, BOD, fecal coliforms, and sediment) are more meaningful when expressed as total annual loads, or pounds/acre/year. Loadings can be calculated as an average annual unit area (lbs/acre/year) for each subwatershed to compare against other subareas and EPA or state standards.

The land development process and urbanized areas substantially increase *impervious surfaces* such as roads, sidewalks, parking lots, roofs, and driveways. Impervious surface can be defined as any land cover that prevents the infiltration of water into the soil (Arnold and Gibbons, 1996). A major factor in causing impairment to our nation's waters is this increase in impervious area (U.S. EPA, 2002). Impervious areas reduce the amount of rainfall that infiltrates into the ground and accumulate pollutants, which are then washed off during rainfall events and produce quick runoff responses to rainfall events. Although urban runoff is a major impairment to our nation's waters, the net result of increased imperviousness is far worse, as delineated in Table 8.3.

As can be seen from the contents of the table, the entire hydrologic process is a cause-and-effect relationship, in which each effect leads to exacerbation of the next effect. For example, increased flows and velocities contribute to stream bank erosion, which in turn causes sediment pollution.

Therefore, impervious cover has a direct effect on a watershed's hydrology, aquatic habitat, water quality, and biodiversity. The percent imperviousness

TABLE 8.3 Effects of Impervious Area on Water Resources

Impervious Area
  Rapid Runoff
    Earlier and Higher Peak Flows
      Increased flooding
      Higher velocities in streams
    Higher Flow Volumes
      Increased frequency and duration of bankfull flows
      Increased stream bank erosion
      Decreased geomorphological stability
  Nonpoint Source Pollution
    Impaired water quality
    Decreased aquatic diversity
  Decreased Groundwater Recharge
    Decreased stream base flow augmentation
    Impaired water quality
      Decreased aquatic diversity
  Increased Stream Temperatures
    Impaired water quality
      Decreased aquatic diversity

of a watershed or to a point of interest is a useful indicator to measure the impacts of land development. There is a strong correlation between percent imperviousness of a watershed and stream health (Schueler, 1995). Schueler reported that stream degradation occurs when a watershed becomes about 10 percent impervious, and degradation becomes unavoidable at 30 percent imperviousness. Therefore, streams can be classified, based on these percent impervious area thresholds, as *protected* (less than 10 percent), *impacted* (10-30 percent), and *degraded* (greater than 30 percent). Therefore, watershed percent impervious area can be utilized as an environmental indicator and planning tool. (Arnold and Gibbons, 1996). This relationship is shown graphically in Figure 8.5.

Keeping these thresholds in mind, community comprehensive planning and individual site development plans should be designed around these thresholds. Although one cannot limit a watershed to 10 percent impervious surface and then stop growth, one can maintain the natural hydrologic regime through proper management of development and by minimizing the effects of impervious area through innovative management of stormwater.

8.5 NONPOINT SOURCE (NPS) POLLUTANTS

It is important to understand the basic characteristics of the pollutants themselves in order to understand their effects on water quality. Many of the pollutants are interrelated. For example, phosphorus tends to bind to sediment. Although this text will not go into great detail, a general description of each pollutant category follows (Schueler, 1987).

8.5.1 Solids

Solids in surface waters are generally attributable to surface runoff and can be exacerbated by erosion from construction sites. When transported to wa-

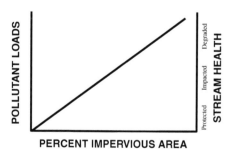

Figure 8.5 Impact of impervious surface on stream health and pollutant loads (modified from Arnold, 1996).

TABLE 8.4 Approximate Suspended Sediment Load by Land Use (Marsh, 2003)

Land Use Sediment Load	Approximate Suspended (pounds/acre/year)
Low-density residential	300
Agricultural land in rotation crops	1,000
Agricultural land in row crops with management practices	1,200
Medium-density residential	3,000
Highways and freeways	3,100
Agricultural land in row crops without management practices	5,600
Areas under development	8,800

tercourses, sediment-laden runoff increases stream turbidity, which reduces penetration of light, smothers desirable aquatic plants, and settles onto gravel stream bottoms, choking aquatic life and fish eggs. Sediment is the most common nonpoint source pollutant type. The American Farmland Trust (1984) has estimated that approximately 1.6 billion tons are deposited in our nation's waterways each year. Approximate suspended sediment loads by land use can be found in Table 8.4 (Wisconsin Department of Natural Resources, undated).

Sediment accumulation reduces channel, bridge, culvert, and storm sewer capacity. Sediment that settles to the bottom of lakes decreases water depth. This allows sunlight to reach the root zone much further out into the water body, which may allow undesirable aquatic vegetation to take root. Water with a high concentration of suspended solids affects water supply, which causes a laxative effect when ingested and increases water treatment costs. Sediment also carries with it attached or adsorbed phosphorus, which can add to the eutrophication process. Sediment accumulation also inhibits navigation, commerce, and recreational activities.

The amount of sediment in the water can be determined by measuring the *total suspended solids* (*TSS*) and *total dissolved solids* (*TDS*). The TSS are all suspended solids including both dissolved and undissolved. TDS consist mainly of small amounts of organic matter, inorganic salts, and dissolved gases, and are dissolved in water.

8.5.2 Nutrients

Nutrients are those chemicals (nitrogen, phosphorous, and potassium) that provide plants with the essential elements for survival. Although plants require nutrients, an overabundance of them can cause eutrophication, or undesirable plant growth (algae and other aquatic organisms) in waterways. Excessive plant growth causes oxygen depletion, which causes stress on aquatic species (including fish). Algal blooms and proliferation of macrophytes (water plants) reduce light and may release toxic compounds into the water, killing aquatic organisms. An overabundance of nutrients affects swimming, boating, fishing, and recreational activities.

Phosphorus is typically the *limiting nutrient* in surface water, meaning that plant growth is regulated by the available phosphorus. An excess concentration of phosphorus triggers excessive plant growth. Phosphorus, in the form of phosphate ($H_2PO_4^-$) from fertilizer, detergents, and organic wastes, becomes *adsorbed* (attached) to sediment, which is carried to streams during the erosion/sedimentation process. This excess phosphate then equalizes itself with the existing phosphate in the water, thus increasing the phosphate in the water (Foth, 1990). Nitrogen can be found in many forms (organic nitrogen, ammonia, nitrite, and nitrate) and is easily converted from one form to the other. In most healthy freshwater systems, ammonia is present in low concentrations. In highly eutrophic water bodies, particularly those that become devoid of oxygen (*anoxic*), much higher ammonia concentrations can prove lethal to organisms if the pH of the water body is high. Nitrogen originates from the decomposition of organic matter or from sewage or septic pollution and can be carried to streams during storm events.

8.5.3 Organic Matter (Oxygen-Demanding Pollutants)

One of the most important measures of the quality of a water body or stream is its dissolved oxygen level, as many of its primary functions, such as the ability to sustain a thriving aquatic community, are dependent upon the presence of dissolved oxygen at sufficient concentrations. Biological or organic pollution are those pollutants that, while undergoing decomposition, place an oxygen demand on the water. Although stormwater runoff typically does not carry lethal levels of these parameters, organics are a concern, particularly in slow-moving waters characteristic of the lower portion of the watershed and lakes. The decomposition process also releases nutrients and contributes to eutrophication.

Biochemical Oxygen Demand (BOD_5), Chemical Oxygen Demand (COD), and Total Organic Carbon (TOC) are tests to measure a water's ability to consume oxygen due to organic pollutants present. BOD_5 is the amount of oxygen used in the metabolism of biodegradable organics under aerobic conditions at 20 degrees centigrade during a five-day test. When organic matter enters surface water, bacteria feeding on the organic matter multiplies, consuming the oxygen that is required by the more desirable plant and animal growth. COD is a chemical test to determine the organic content of natural water, wastewaters, and industrial wastes. TOC is the total amount of carbon in the water.

8.5.4   Microorganisms

Water that flows on the earth's surface is exposed to many microorganisms that can cause diseases, which are then carried into waterways. Water flowing downstream and into lakes gathers plant particles and animal matter in various states of decay and other biological impurities. Runoff from yards and roads can carry animal wastes to streams and lakes. Leaking and/or malfunctioning septic tanks and cesspools are also a major source of fecal coliform contamination to water bodies and receiving streams. In most instances, where fecal coliforms are present, the water is also receiving nutrients, which may overly enrich the lake and cause eutrophication.

Microorganisms such as *fecal coliforms* and *pathogens* are a concern from nonpoint source pollution. Fecal coliforms and *fecal streptococci* are bacteria found in the feces of mammals and are commonly used to measure possible contamination from human or animal waste. Fecal coliforms are naturally present in humans, and are not harmful to them; however, they are *indicator organisms* that more harmful organisms may be present. The extent to which fecal coliforms are present in the water can indicate the general quality of that water and the likelihood that the water is contaminated by human or animal wastes. The ratio of fecal coliform (FC) to fecal streptococci (FS) counts can be used to determine if the source of contaminated water is from human or animal wastes. It has been found that an FC/FS count of around 4 is indicative of human waste, whereas an FC/FS count of less than 1 usually indicates domestic animal waste (Tchobanoglous, 1987).

The detection of coliforms in water may also signal the presence of contaminants that may cause disease. Waters that are polluted may contain several different disease-causing organisms, commonly called pathogens. Enteric pathogens (i.e., those that live in the human intestine) can carry or cause a number of infectious diseases. Swimmers in sewage-polluted water could contract any illness that is spread by ingestion of fecal-contaminated water.

8.5.5   Heavy Metals

The most common heavy metal pollutants are cadmium, copper, lead, and zinc; they are also referred to as *trace inorganics*. Metals are of concern for their impact on aquatic life and drinking water, as they can be toxic in high enough concentrations. The primary sources of heavy metals are atmospheric deposition, heavy industry, and exhaust and tire deposition from automobiles. Heavy metals tend to accumulate in bottom sediments, which then can be assimilated into organisms through ingestion.

8.5.6   Synthetic Organic Components

Synthetic organic components are man-made chemicals; they include pesticides and herbicides, petroleum hydrocarbons (oil and grease, benzene, toluene, xylene, and ethyl benzene), solvents, and household and industrial chemicals. Petroleum hydrocarbons are a major concern due to bioaccumulation, but they can also cause acute toxicity to aquatic organisms at low concentrations. Sources of petroleum hydrocarbons include leaking underground storage tanks (USTs), runoff from roads and parking lots (car emissions and leaking oils and grease), as shown in Figure 8.6, and improper disposal of waste oil. Pesticides can also reach toxic levels in receiving streams.

8.5.7   Chlorides

Chlorides are salts such as sodium or magnesium chloride, which are applied for deicing purposes during the winter months. Runoff can produce salt concentrations in receiving streams that are damaging to freshwater aquatic life, particularly during spring thaw and runoff events. Runoff from highway maintenance sheds can also be a source of chloride pollution if not properly managed. And because chlorides are soluble, they also percolate into the groundwater.

8.5.8   Temperature

One pollutant that is typically overlooked is the temperature of runoff. Impervious surfaces absorb a high level of the sun's heat energy. Much of this heat is then transferred to the stormwater runoff passing over the impervious surfaces. The temperature change in a stream from a sudden thunderstorm in an impervious watershed on a hot summer day can be dramatic, and possibly lethal to aquatic life. Of significance are impervious areas adjacent to receiving streams and water bodies, where the heated runoff flows directly to the water without mixing and cooling from contributing runoff. A study per-

Figure 8.6   Petroleum hydrocarbons are a common pollutant from parking lots. (Modified from Scheuler, 1995.)

formed by Galli (1990) for the Metropolitan Washington Council of Governments (MWCOG) has shown the temperature of a stream increases 0.14 degrees Fahrenheit per 1 percent increase in imperviousness. Galli found that a developed watershed can have average stream temperatures up to 16 degrees Fahrenheit higher than undisturbed natural reference streams. Other factors that generally increase the temperature of streams in urban areas include:

- Decreased base flow
- Increased air temperature from blacktop
- Reduced vegetation on stream banks
- Lack of buffers
- Industrial/municipal discharges with elevated temperatures

Variability in levels of dissolved oxygen corresponds to fluctuations in the water temperature, with the warmer water able to support less oxygen than the colder water. Therefore, there is less oxygen available for respiration by aquatic organisms in the warmer water. In addition, increased water temperatures increase aquatic organism metabolism and respiration rates. It has also been found that increases in stream temperature can change the algal community from the desired diatoms to the undesirable blue-green algae. This occurs at a temperature of approximately 86 degrees Fahrenheit, which can be common in urbanized watersheds.

Sensitive cold-water species such as salmon and trout, which typically inhabit headwater streams, require lower temperatures for spawning and egg hatching. Therefore, impervious areas in headwaters will have more of a detrimental effect on aquatic organisms than in lower streams.

Most stormwater management measures do not improve the temperature of runoff. However, best management practices, which are typically designed for filtering sediment pollution, can be modified to decrease the temperature of the discharge. One such modification would be to provide temporary storage of runoff underground, allowing the water to cool. In critical areas of a watershed, thermal pollution must be addressed. This can be accomplished through infiltration and underground storage to help reduce the temperature of the outflow.

8.6 BUILDUP AND WASHOFF IN URBAN AREAS

Between rain events, pollutants accumulate on impervious surfaces from atmospheric deposition (*fallout*), general human activity, brake and tire wear, vehicle emissions, automobile fluids, sediment carried from the wind, road deicing compounds, salts and cinders, undissolved phosphates, nitrates, petroleum-based organics, and heavy metals, a term called pollutant *buildup*. The American Public Works Association (APWA) conducted one of the earliest studies to determine street surface accumulation of "Dust and Dirt" (DD)

(APWA, 1969). Dust and dirt can be defined as anything passing through a quarter-inch mesh screen. Typical accumulation values of dust and dirt can be seen in Table 8.5, while typical pollutant values (milligrams) for each gram of dust and dirt can be seen in Table 8.6 (APWA, 1969; James et al., 1999). The APWA study found that buildup of these pollutants is nearly linear with time and can be measured in terms of length of curb.

Amy and colleagues (1974), AVCO (1970), Lager and colleagues (1977), Pitt (1979), Pitt and Amy (1973), Shaheen (1975), James et al. (1999), and Manning (1977), are several excellent sources on pollutant accumulation and/or buildup of dust and dirt data. During the precipitation event, the pollutants "wash off" the impervious surfaces and get carried into the waterways, causing degradation of the receiving watercourse. Physical features such as percent impervious area, land use, soils, and slopes impact the pollutant washoff process. Several water quality computer models such as the Storm Water Management Model (SWMM4) (James et al., 1999) include a buildup and washoff routine.

The majority of the pollutants (close to 80 percent) get washed off during the early stages of rainfall or the first half-inch of rain, a term called the *first flush* (Lee, 2000). The first flush is when the accumulated pollutants are at their maximum concentration. The longer the time period between runoff events, the greater the pollutant accumulation from fallout and other sources, and in turn, the higher the concentrations of pollutants washed off the surfaces. Not coincidentally, approximately 80 percent of storms are one inch of rainfall or less. Rainfall intensity and duration and the interval between storms, and that amount of snowmelt, will also affect the concentration of pollutants in the runoff.

Measurement of these concentrations over time will produce a *pollutograph*. Just as a hydrograph can be developed to model runoff response, a comparable pollutograph can be modeled and used to analyze the concentration or mass flux of a particular constituent over time. Despite the ability to model such a pollutograph, there is still much to be understood about the nature of various stormwater constituents and how they affect the quality of receiving waters. When plotted against the hydrograph of the same drainage

TABLE 8.5   Measured Dust and Dirt (DD) Accumulation in Chicago by the APWA

Type	Land Use	Pounds DD/dry Day per 100 ft-curb
1	Single-family Residential	0.7
2	Multi-family Residential	2.3
3	Commercial	3.3
4	Industrial	4.6
5	Undeveloped or Park	1.5

Source: APWA 1969; James, 1979.

TABLE 8.6   Milligrams of Pollutant per Gram of Dust and Dirt (Parts per Thousand by Mass) for Four Chicago Land Uses from 1969 APWA Study*

Parameter	Single-Family Residential	Land Use Type Multifamily Residential	Commercial Industrial	Industrial
BOD$_5$	5.0	3.6	7.7	3.0
COD	40.0	40.0	39.0	40.0
Total Coliforms**	1.3×10^6	2.7×10^6	1.7×10^6	1.0×10^6
Total N	0.48	0.61	0.41	0.43
Total PO$_4$ (as PO$_4$)	0.05	0.05	0.07	0.03

*Source: James, 1979.
**Units for coliforms are MPN/gram.

area, one can see the first flush phenomenon, as shown in Figure 8.7. The data from the figure came from a one-acre impervious test plot. Several parameters were tested for the first flush phenomenon; however, for display purposes, only nitrates are shown.

Therefore, the initial runoff from impervious areas will be the most polluted; the runoff will become "cleaner" as the storm progresses. It should be noted that although the concentration of pollutants is highest in the first flush, the first flush may not necessarily deliver the highest volume of pollutants. The storm duration will dictate the final volume of pollutants delivered. For this reason, although capturing the first flush will remove the majority of the pollutants, the volume of runoff should also be considered as a management measure.

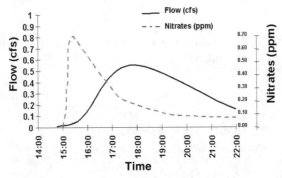

Figure 8.7   Typical pollutograph showing nitrates first flush (DeBarry, 1996).

Utilization of methods such as the Revised Universal Soil Loss Equation (RUSLE) to predict sediment load and computer simulation models such as SWMM (James, 1999), Generalized Watershed Loading Function (GWLF) (Haith, 1992), and the EPA's Better Assessment Science Integrating Point and Nonpoint Sources (BASINs) (U.S. EPA, 2001) to predict nonpoint source pollution will be discussed in Chapter 15. Quantification of the source of nonpoint source pollutants from urban runoff is difficult due to the diffuse nature of the sources, in spite of the available computer models.

8.7   RUNOFF POLLUTANT SOURCES

Stormwater systems in urban areas were typically designed and installed to efficiently remove stormwater from the streets and parking areas. They consist of streets with curbs and gutters, catch basins, and storm sewer networks. Pollutants are transported in urban runoff directly to streams or through storm sewers during storm events, from sanitary or industrial discharges, or a combination thereof. Often, in the older urban areas, storm sewer systems are combined with sanitary sewer systems (*combined sewers*), which convey the combined storm and sewage flows to a sewage treatment plant (STP). During storm events, the STPs are hydraulically overloaded, allowing untreated sewage to directly enter the streams. These types of discharges are referred to as *combined sewer overflows* (*CSOs*). In contrast, the preferred *municipal separate storm sewer system,* or *MS4,* conveys only stormwater; however, in most cases it discharges directly to streams without stormwater treatment. As will be discussed in Chapter 16, the National Pollution Discharge Elimination System (NPDES) program intends to provide the incentive to treat the MS4 discharges by forcing dischargers to develop a stormwater management program and implement the use of innovative BMPs. Pollutants in runoff and their sources can be categorized into several categories, as indicated in Table 8.7.

8.8   SUMMARY

In general, nutrient sources are animal waste, fertilization of urban lawns, malfunctioning septic systems, and sewage treatment plants. Nutrients include phosphorous and nitrogen compounds such as ammonium, nitrates, and the like. Heavy metals such as iron, zinc, and lead come from brake linings, industrial emissions, and tires. Solids pollution, which would include total suspended solids and dissolved solids, can come from atmospheric or wind deposition, litter, erosion from construction sites, or from in-stream erosion. The biological pollutants come from animal waste or from overflowing sewage treatment plants, sanitary sewers, or failing septic systems. A further breakdown of pollutant sources can be found in Table 8.8.

TABLE 8.7 Pollutant Categories and Their Sources

	General Pollutant Source								
Pollutant Type	Residential	Commercial	Industrial	Construction	Highways	Agriculture	Forestry	STP's	Stream Bank Erosion
Nutrients	x			x		x	x	x	x
Total Kjeldahl Nitrogen (TKN)									
Ammonia Nitrogen (NH_3)									
Nitrates (NO_3)									
Nitrites (NO_2)									
Total Phosphorus (TP)									
Soluble Phosphorus (SP)									
Solids	x	x	x	x	x	x	x	x	x
Total Suspended Solids (TSS)									
Total Dissolved Solids (TDS)									
Organic Matter (Oxygen-demanding)	x			x		x	x	x	x
Biochemical Oxygen Demand (BOD_5)									
Chemical Oxygen Demand (COD)									
Total Organic Carbon (TOC)									
Biological	x					x	x	x	x
Fecal Coliforms									
Fecal Streptococci									
Pathogens									
Heavy Metals		x	x	x	x			x	
Lead (Pb)									
Copper (Cu)									
Zinc (Zn)									
Iron (Fe)									
Synthetic Organic Compounds	x		x	x	x	x	x	x	x
Herbicides									
Pesticides									
Petroleum Hydrocarbons									
Oil and Grease									
Chlorides	x		x	x	x				
Temperature		x	x	x	x			x	x

TABLE 8.8   Sources of Primary Pollutants of Concern in Point and Nonpoint Discharges to Receiving Waters (PaDER, 2000)

The following parameters must be considered at a minimum when evaluating the question of measurable change for the specific wastewater type under which they are listed. Additional parameters should be considered on a case-by-case basis depending upon what is known or anticipated about the wastewater characteristics of an existing or proposed discharge.

Nonmunicipal Sewage

C-BOD$_5$	Total Phosphorus
Total Suspended Solids	Volatile Organic Compounds as
Ammonia Nitrogen (NH$_3$-N)	measured by EPA Method 624
Nitrate Nitrogen (NO$_3$N)	Fecal Coliform density
Nitrite Nitrogen (NO$_2$N)	

Municipal Sewage

C-BOD$_5$	Hardness
Total Suspended Solids	Total Iron
Ammonia Nitrogen (NH$_3$-N)	Total Manganese
Nitrate Nitrogen (NO$_3$-N)	Total Zinc
Nitrite Nitrogen (NO$_2$-N)	Total Copper
Total Phosphorus	Total Nickel
	Total Aluminum
	Total Lead
	Fecal Coliform Density

Volatile Organic Compounds as measured by EPA Method 624

Industrial Wastes
(including noncoal mining)

See attached EPA tables for general guidance. Site-specific parameter lists can be developed on a case-by-case basis in cooperation with the appropriate DER Regional or District Office.

Oil and Gas Extraction Activities

pH	Total Cadmium
Total Suspended Solids	Total Chromium
Total Dissolved Solids	Total Copper
Total Chlorides	Total Iron
Oils and Grease	Total Lead
Total Phenolics	Total Lithium
Total Aluminum	Total Nickel
Total Arsenic	Total Silver
Total Barium	Total Zinc
Total Berylium	Total Benzene
Ammonia Nitrogen (NH$_3$-N)	
Osmotic Pressure	
MBAS	

Coal Mining Activities

pH	Total Copper
Total Dissolved Solids	Total Iron
Total Suspended Solids	Total Manganese
Total Sulfate	Total Nickel
Total Aluminum	Total Zinc

TABLE 8.8 (*Continued*)

Waste Management Facilities

Municipal Wastes (All metals expressed as total)

pH	Arsenic
Ammonia Nitrogen (NH_3-N)	Barium
Nitrate Nitrogen (NO_5-N)	Beryllium
Nitrite Nitrogen (NO_2-N)	Boron
Total Phosphorus	Cadmium
C-BOD$_5$	Chromium (Total & Hexavalent)
Total Suspended Solids	Cobalt
Total Dissolved Solids	Copper
Fecal Coliform Density	Iron
Total Chlorides	Lead
Total Sulfate	Manganese
Total Organic Carbon	Mercury
Total Alkalinity	Nickel
Fluoride	Selenium
Free Cyanide	Silver
Total Phenols	Thallium
Aluminum	Tin
Antimony	Vanadium
	Zinc

Volatile and semivolatile organic compounds and pesticides as measured by EPA Methods 624, 625, and 608.

Residual Wastes

All parameters listed for municipal wastes plus:

Bicarbonate	Total Calcium	Total Magnesium
COD	Total Sodium	Total Potassium

Table A Treatment plant discharge should be analyzed for all of the following parameters.

Parameter	Units
COD	mg/l
Color	STD Units
Oil and Grease	mg/l
Alkalinity (as $CaCO_3$)	mg/l
Sulfate (as SO_4)	mg/l
Sulfide (as S)	mg/l
Sulfite (as SO_3)	mg/l
Aluminum (total)	mg/l
Iron (total)	mg/l
Manganese (total)	mg/l
Surfactants (MBAS)	mg/l
Temperature (Summer)	F*
Temperature (Winter)	F*

TABLE 8.8 *(Continued)*

Table B Analyze for those pollutants believed present in the industrial discharges to the system: Report as mg/l.

Barium
Boron
Cobalt
Fluoride
Magnesium
Molybdenum
Tin
Titanium
Antimony
Arsenic
Beryllium
Cadmium
Cyanide (Tot)
Copper
Lead
Mercury
Nickel
Phenol (Tot)
Selenium
Thallium
Zinc

Table C Analyze for those pollutants believed present in the industrial discharge of the system: Report as mg/l or μg/l.

Volatile Organics

1V. Acrolein
2V. Acrylonitrile
3V. Benzene
5V. Bromoform
6V. Carbon Tetrachloride
7V. Chlorobenzene
8V. Chlorodibromomethane
9V. Chloroethane
10V. 2-Chloroethylvinyl Ether
11V. Chloroform
12V. Dichlorobromomethane
14V. 1,1-Dichloroethane
15V. 1,2-Dichloroethane
16V. 1,1-Dichloroethylene

17V. 1,2-Dichloropropane
18V. 1,2-Dichloropropylene
19V. Ethylbenzene
20V. Methyl Bromide
21V. Methyl Chloride
22V. Methylene Chloride
23V. 1,1,2,2-Tetrachloroethane
24V. Tetrachloroethylene
25V. Toluene
26V. 1,2-trans-Dichloroethylene
27V. 1,1,1-Trichloroethane
28V. 1,1,2-Trichloroethane
29V. Trichloroethylene
31V. Vinyl Chloride

GC/MS Fraction—Acid Compounds

1A. 2-Chlorophenol
2A. 2,4-Dichlorophenol
3A. 2,4-Dimethylphenol
4A. 4,6-Dinitro-O-Cresol
5A. 2,4-Dinitrophenol
6A. 2-Nitrophenol

7A. 4-Nitrophenol
8A. P-Chloro-M-Cresol
9A. Pentachlorophenol
10A. Phenol
11A. 2,4,6-Trichlorophenol

177

TABLE 8.8 (*Continued*)

Table C (*Continued*)

GC/MS Fraction—Base/Neutral Compounds

1B. Acenaphthene
2B. Acenaphthylene
3B. Anthracene
4B. Benzidine
5B. Benzo (a) Anthracene
6B. Benzo (n) Pyrene
7B. 3,4-Benzofluoranthene
8B. Benzo (ghi) Perylene
9B. Benzo (k) Fluoroanthene
10B. Bis (20Chloroethoxy)
  Methane
11B. Bis (2-Chloroethyl) Ether
12B. Bis (2-Chloroisopropyl) Ether
13B. Bis (2-Ethylhaxyl) Phthalate
14B. 4-Bromophenyl Phenyl Ether
15B. Butyl Benzyl Phthalate
16B. 2-Chloronaphthalene
17B. 4-Chlorophenyl Phenyl Ether
18B. Chrysene
19B. Dibenzo (a,h) Anthracene
20B. 1,2-Dichlorobenzene
21B. 1,3-Dichlorobenzene
22B. 1,4-Dichlorbenzidine
23B. 3.3′-Dichlorobenzidine

24B. Diethyl Phthalate
25B. Dimethyl Phthalate
26B. Di-N-Butyl Phthalate
27B. 2,4-Dinitrotoluene
28B. 2,6-Dinitrotoluene
29B. Di-N-Octyl Phthalate
30B. 1,2-Diphenylhydrazine (as
  Azobenzene)
31B. Fluoroanthene
32B. Fluorene
33B. Hexachlorobenzene
34B. Hexachlorobutadiene
35B. Hexachlorocyclopentadiene
36B. Hexachloroethane
37B. Indeno (1,2,3-cd) Pyrene
38B. Isophorone
39B. Naphthalene
40B. Nitrobenzene
41B. N-Nitrosodimethylamine
42B. N-Nitrosodi-N-Propylamine
43B. N-Nitrosodiphenylamine
44B. Phenanthrene
45B. Pyrene
46B. 1,2,4-Trichlorobenzene

GC/MS Fraction—Pesticides

1P. Aldrin
2P. α-BHC
3P. βBHC
4P. γ-BHC
5P. ζBHC
6P. Chlordane
7P. 4,4′-DDT
8P. 4,4′-DDE
9P. 4,4′-DDD

10P. Dieldrin
11P. α-Endosulfan
12P. β-Endosulfan
13P. Endosulfan Sulfate
14P. Endrin
15P. Endrin Aldehyde
16P. Heptachlor
17P. Heptachlor Epoxide
25P. Toxaphene

**Table D Analyze for those pollutants believed present in the industrial discharge
  to the system: Report as mg/l or μmg/l.**

18P. PCB-1242
19P. PCB-1254
20P. PCB-1232
21P. PCB-1248
22P. PCB-1260
23P. PCB-1260
24P. PCB-1016

TABLE 8.8 (*Continued*)

Table E Analyze for those pollutants believed present in the industrial discharge to the system: Report as μCi/ml or pCl/ml.

Radioactivity
(1) Alpha, Total
(2) Beta, Total
(3) Radium, Total
(4) Radium 226, Total

Source: Pennsylvania Department Environmental Resources, 1992.

The majority of the pollutants entering the waterways are from nonpoint sources, or those that are diffuse and do not come from end-of-pipe discharges. Understanding the types of nonpoint sources of pollution, their sources, and their impact on receiving water bodies aids in developing a plan to minimize or abate their impacts.

The sudden appearance of Pfiesteria shows the impact of mismanagement by humans and how that can upset the ecological balance. *Pfiesteria piscicida* is a free-swimming, single-celled organism or dinoflagellate that is toxic and fatal to fish, causing unsightly sores as shown in Figure 8.8 and has also been linked to short-term memory difficulties and respiratory problems in humans. It has been a problem in North Carolina but didn't appear in the Chesapeake Bay until 1997 (U.S. EPA, 1999). It was linked to over 50,000 dead fish in Maryland's eastern shore in August and September 1997 alone. Its population explosion has been linked to the increase in nutrients in stormwater runoff entering waterways. It appears to be worse during dry summers, when flows are low and, in turn, nutrient concentrations are high.

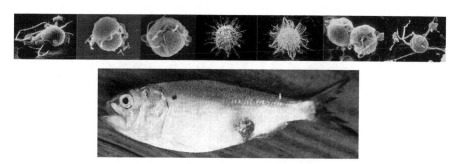

Figure 8.8 Pfiesteria and its impact on fish. (*Source:* North Carolina State University Center for Applied Aquatic Ecology, http://www.pfiesteria.org/imagearchives/lesioninamges.html.)

REFERENCES

American Farmland Trust. "Soil Conservation in America," (Washington, DC: American Farmland Trust), 1984.

American Public Works Association. "Water Pollution Aspects of Urban Runoff" (Washington, DC: Federal Water Pollution Control Administration, contract WP-20-15), 1969.

Amy, G., R. Pitt, R. Singh, W.L. Bradford, and M.B. LaGraff. "Water Quality Management Planning for Urban Runoff," EPA-440/9-75-004 (NTIS PB-241689) (Washington, DC: Environmental Protection Agency), December 1974.

Arnold, Chester L., and C. James Gibbons. "Impervious Surface Coverage: The Emergence of a Key Environmental Indicator," *Journal of the American Planning Association,* vol. 62, no. 2 (Spring 1996), pp. 243–258.

AVCO Economic Systems Corporation. "Storm Water Pollution from Urban Land Activity," EPA 111034FKL07/70 (NTIS PB-195281) (Washington, DC: Environmental Protection Agency), July 1970.

Beasley, R.P. *Erosion and Sediment Pollution Control* (Ames, IA: Iowa State University Press), 1971.

Borton-Lawson Engineering. "Mehoopany and Little Mehoopany Creek Watersheds Assessment," Borton-Lawson Engineering Report (Wilkes-Barre, PA: Borton-Lawson Engineering) March 17, 2003.

Cook, G. Dennis, and Robert H. Kennedy. "Water Quality Management for Reservoirs and Tailwaters," Report 1, "In-Reservoir Water Quality Management Techniques," Technical Report E-89-1 (Vicksburg: MS: U.S. Army Corps of Engineers Waterways Experiment Station), January 1989.

DeBarry, P.A. Tobyhanna Creek Watershed Management Plan, Monroe County Conservation District, Monroe County Planning Commission, Monroe, PA, 1996.

———. "G.I.S. Applications in Nonpoint Source Pollution Assessment." Proceedings from ASCE's 1991 National Conference on Hydraulic Engineering.

Foth, Henry D. *Fundamentals of Soil Science,* 8th ed. (New York: John Wiley & Sons, Inc.), 1990.

Galli, J. "Thermal Impacts Associated with Urbanization and Best Management Practices, Metropolitan Washington Council of Governments" (Washington, DC: Maryland DEP), 1990.

Haith, D.R., R. Mandel, and R.S. Wu. "GWLF: Generalized Watershed Loading Functions User's Manual," version 2.0 (Ithaca, NY: Cornell University), 1992.

James, W., W.C. Huber, R.E. Dickinson, and W.R.C. James. *Water Systems Models—HYDROLOGY* (Guelph, Ontario, Canada: Computational Hydraulics International), December 1999.

Kumble, P. and T. Schueler (Eds). "Mitigating the Adverse Impacts of Urbanization on Streams: A Comprehensive Strategy for Local Government," in *Watershed Restoration Sourcebook,* Publication #92701 of the Metropolitan Washington Council of Governments, Washington, DC.

Lager, J.A., W.G. Smith, W.G. Lynard, R.F. Finn, and E.J. Finnemore. "Urban Stormwater Management and Technology: Update and User's Guide," EPA-600/8-77014

(NTIS PB-275264) (Cincinnati, OH: Environmental Protection Agency), September 1977.

Lee, G.F. "Assessing and Managing Water Quality Impacts of Urban Stormwater Runoff," presented at International Conference on Urban Drainage, via Internet, May 2000.

Manning, M.J., R.H. Sullivan, and T.M. Kipp. "Nationwide Evaluation of Combined Sewer Overflows and Urban Stormwater Discharges, Volume III: Characterization of Discharges," EPA # EPA600/277064c, NTIS # PB.272107, U.S. EPA, Cincinatti, OH, August 1977.

Marsh, D. "When a Watershed Is Gripped by Nonpoint Source Pollution, All Riled Up: Nonpoint Source Pollution; The Point of No Return" (Madison, WI: Wisconsin Department of Natural Resources), undated.

National Atmospheric Deposition Program (NADP), http://nadp.sws.uiuc.edu/ (April 25, 2003).

New York State Department of Environmental Conservation. "Reducing the Impacts of Stormwater Runoff from New Development," 1992.

Pennsylvania Department of Environmental Resources. *Special Protection Waters Implementation Handbook,* 2nd ed., Harrisburg, PA, March 1, 2000.

Pennsylvania Department of Environmental Protection, Pennsylvania Code, Title 25 Environmental Protection, Chapter 93, Water Quality Standards, December 15, 2001.

Pitt, R. and Amy G. "Toxic Materials Analysis of Street Surface Contaminants," EPA-R2-73-283 (NTIS PB-224677) (Washington, DC: Environmental Protection Agency), August 1973.

Pitt, R. "Demonstration of Non-Point Pollution Abatement Through Improved Street Cleaning Practices," EPA-600/2-79-161 (NTIS PBSO-108988) (Cincinnati, OH: Environmental Protection Agency), August 1979.

Sawyer, Clair N., and Perry L. McCarty. *Chemistry for Environmental Engineering* (New York, McGraw-Hill), 1978.

Schueler, T.R. "Controlling Urban Runoff: A Practical Manual for Planning and Designing Urban BMPs," Metropolitan Washington Council of Governments, 1987.

———. "Site Planning for Urban Stream Protection," Prepared for Metropolitan Washington Council of Governments, Water Resources Publications, LLC, December 1995.

Shaheen, D.G. "Contributions of Urban Roadway Usage to Water Pollution," EPA-600/2-75-004 (NTIS PB-245854) (Washington, DC: Environmental Protection Agency), April 1975.

Tchobanoglous, George, and Edward D. Schroeder. *Water Quality* (Reading, MA: Addison-Wesley), 1987.

USDA, Soil Conservation Service. "Water Quality Indicators Guide: Surface Water," SCS-TP-161, USDA-SCS, Washington, D.C., September, 1989.

United States Environmental Protection Agency (U.S. EPA), Office of Water. BASINs [Better Assessment Science Integrating Point and Nonpoint Sources], Version 3.0, EPA-823-C-01-003, United States Environmental Protection Agency, June 2001.

———. "Preliminary Data Summary of Urban Storm Water Management Best Management Practices," EPA 821-B-99-012, 1999.

————. StormWater. "Phase II Final Rule—An Overview," EPA 833-F-00-001, 2000.

————. Office of Wetlands, Oceans, and Watersheds. "National Management Measures to Control Nonpoint Source Pollution from Urban Areas–Draft," EPA 842-B-02-003, Washington, DC, July 2002.

————. Final report of the National Urban Runoff Program (NURP), Water Planning Division, Washington, DC, 1983.

————. *Federal Register,* Part VI–40 CFR Part 9 et al. "Revisions to the Water Quality Planning and Management Regulation and Revisions to the NPDES Program in Support of Revisions to the Water Quality Planning and Management Regulation: Final Rules," July 13, 2000.

————. *Lake and Reservoir Restoration Guidance Manual,* 1st ed., EPA 440/5-88-002 (Washington, DC: U.S. Environmental Protection Agency), February 1988.

————. Watershed Protection: TMDL Note 2 Bioassessment and TMDLs, EPA 841-K-94-005a (Washington, DC: U.S. Environmental Protection Agency), 1994.

————. "What You Should Know About Fish Lesions." Office of Water, Washington, DC, U.S. Environmental Protection Agency, 1999.

U.S. Geological Survey. "Effects of Urbanization on Stream Ecosystems," Fact Sheet 042-02, May 2002.

Weigmann, Diana L., Louis A. Helfrich, and Daniel M. Downey. *Guidelines for Liming Acidified Streams and Rivers* (Blacksburg, VA: Virginia Water Resources Center, Virginia Polytechnic Institute and State University), 1993.

PART B:

WATERSHED ASSESSMENT

CHAPTER 9

WATERSHED ASSESSMENT: DATA COLLECTION

9.0 INTRODUCTION

One can think of a watershed assessment as utilizing a microscope to eventually identify an organism. Watershed assessments are usually triggered by a certain problem, usually human-induced, somewhere in the watershed. The most common mistake, however, in performing watershed assessments is that the person or team organized to correct the problem concentrates specifically on that problem. One example, which is a common occurrence, is the case of a stream bank erosion problem: Typically, significant time and money are spent to evaluate, design, and construct a fluvial geomorphology (FGM) stream restoration project, but little or no attention is paid to what may happen with future development in the watershed upstream of this project. As with flood control projects, these projects are often designed for the flows and sediment load deliveries of existing conditions and do not take into account future land use changes; hence, their life span is limited.

The other drawback to this approach is that significant funds may be spent to collect data targeted at that particular problem, whereas, if an expanded scope were considered, other funding opportunities might be available or several goals achieved with minimal additional expense. Likewise, in targeting a specific problem, a team of individuals may be organized to concentrate specifically on that problem; in the process, they may not consider the overall picture, hence may overlook the long-term, holistic solution.

This chapter will provide a review of the watershed assessment procedures and introduce some of the concepts, such as water quality sampling, that will be described in more detail in later chapters. Watershed assessments should address the cumulative impact of *stressors,* which include chemical, biological, and habitat degradation, that would result in a loss of biological diversity. They should include a stormwater analysis (hydrologic modeling), floodplain analysis, survey of obstructions, problem areas, existing and proposed stormwater and flood control structures, point and nonpoint source pollution sources, and so on.

9.1 PROJECT INITIATION

The watershed assessment must be a methodical, comprehensive, and interdisciplinary team approach. The process should begin with establishing the team that will meet to establish the goals and objectives.

Watershed assessments are typically initiated to address a recognized problem, as part of a planned watershed management or restoration program, or to meet a specific regulatory objective. Problems that initiate a watershed assessment could be flooding, stormwater-related problems, stream bank erosion, poor water quality, problems in a lake or reservoir, and so on. An example of a planned watershed management program might be the Adirondack Water Supply program. Examples of regulatory objectives might include Pennsylvania's Act 167 Storm Water Management program or the Clean Water Act's National Pollution Discharge Elimination System (NPDES) program, whereby total maximum daily loads (TMDLs) may have to be established.

9.2 ESTABLISH THE WATERSHED ASSESSMENT TEAM

The watershed team should consist not only of the specialist in the field of the primary problem, but also include individuals with expertise in addressing the secondary and tertiary problems; these members may include specialists in watershed management unrelated to the identified problems who may be able to point out the interaction between watershed management and the problems.

The assessment should be a cooperative effort, and team members should consist of engineers, planners, conservationists, biologists, soil scientists, fluvial geomorphologists, geologists, economists, environmental attorneys, citizens, and so on. The team should also include consultants; county planning commissions; conservation districts and conservation organizations; federal, state, and local environmental and regulatory agencies; university professors and students; local high schools, and other stakeholders (citizens with a direct interest in the livelihood of the watershed).

9.3 INVOLVE CITIZENRY

In development of any watershed plan, it is essential to involve the stake-holders or local citizenry because the preservation and protection of the water resources will benefit them the most. In addition, they typically have an intimate knowledge of its characteristics and problem areas. Involving citizens can be initiated through the development of a watershed plan advisory committee, which should meet periodically throughout the watershed planning process to obtain input on goals and objectives of the local citizenry.

The formation of a watershed association, if not already formed, is another way to get the citizens involved. Often, groups such as watershed associations will volunteer their time to aid in watershed sampling and data collection. Obviously, a training program in sample collection and storage should be provided to ensure the integrity of the results. Associations such as these can often become organized and, thereafter, obtain 501(c)(3) tax-exempt status to apply for grant applications to further the study and promote the educational portion of the ultimate goals and objectives of the study, such as BMPs and stormwater controls and workshops.

The advisory committee should be made up of local citizens, organizations (such as Trout Unlimited and the Nature Conservancy), local municipal officials, Conservation District members, Planning Commission members, and local universities or high school teachers. Frequently, universities and schools are looking for programs such as this to provide their students with an opportunity to collect data and be involved in a watershed project, which is another inexpensive way to provide data collection efforts. This also provides training to prospective environmental professionals in collection of data for watershed analysis and is an avenue for long-term monitoring of the watershed, as grants or funds typically have a contract period, at the end of which the funds cease.

Educational efforts typically put forth by volunteer citizen groups include development of educational pamphlets or brochures for distribution to their mailing lists. Topics for such print media should include what a homeowner can do, nonpoint source pollutants, BMPs, and what the citizens and/or municipal officials can do to improve water quality in their watershed and/or to prevent stormwater problems. A Web page is also an excellent means of disseminating information to the stakeholders. Field days for citizens for stream walk programs should also be organized. Children's programs can be developed in a festive atmosphere to show general water quality sampling procedures and to educate the youth, our future stewards. In addition, a stream watch and monitoring program should be developed to continually monitor the streams at critical locations to ensure their nondegradation.

Watershed associations should be comprised of a sampling of the general population, to include environmentalists, developers, businesses, representatives of industry, members of the educational community, and other professionals.

9.4 SELECT A CONSULTANT

It is amazing how many "experts" appear once a project with funding is made public, claiming to be watershed specialists, but who in fact lack formal training. In addition, now that scientific information is readily available on the Internet, many "Internet consultants" appear, who have not had a formal university education, but who can regurgitate examples found online. Thus, the most common mistake that watershed groups make is to hire a consultant without thoroughly researching his or her qualifications. A consultant may look good on paper, but the hiring group must look into his or her past history of successful projects and contact the references listed for those projects. One should research not only the most recent projects, but older projects as well. Often, previous clients did not rehire the consultant for a reason. Likewise, so-called technicians take a few continuing education courses in FGM and proclaim themselves as FGM experts.

One problem often encountered in watershed management plans is the lack of understanding of the mechanics of rivers and streams in formulating solutions; this lack frequently leads to project failure. In any watershed assessment, a complete understanding of river and stream mechanics is required, so the team requires an engineer who has a thorough understanding of meteorology, hydrology, hydraulics, fluid mechanics, and soil mechanics.

9.5 IMPLEMENT QUALITY ASSURANCE/QUALITY CONTROL (QA/QC)

Quality assurance refers to the standards and methodologies to be followed throughout the assessment process. *Quality control* refers to the checks, balances, and reviews of the data, analysis, and final report, to ensure they are accurate and true. Early in the watershed assessment process, the team should establish a QA/QC program comprising the protocols to be used for data collection and consistency, responsible party, and so on. A QA/QC manual should be developed and followed by all parties involved in the assessment process.

Quality assurance/quality control should be established for water quality sampling, transport, and storage, according to EPA's Standard Methods (U.S. EPA, 2003).

9.6 IDENTIFY GENERAL PROBLEMS IN THE WATERSHED

In order to develop comprehensive goals and objectives for a watershed management plan, problems, their causes, and proposed solutions to correct those problems must be identified. Problems can be classified as *general* (i.e., poor water quality, AMD affected) or *site-specific*. The general problems should be identified first, to establish the goals and objectives of the watershed plan.

During the watershed assessment process, the site-specific problem areas can be identified through citizen surveys, then validated through field reconnaissance surveys. Proposed corrections to the problems can then be determined, as discussed in more detail in Section 9.7. Problems often translate directly to the goals and objectives of the watershed plan. Typical problems in a watershed that should be identified are listed in Table 9.1.

In the watershed assessment, one should first specifically classify the primary, secondary, and tertiary problems through field reconnaissance or public survey. One should be realistic in developing the goals, realizing that not all desired results may be attainable, at least initially, based upon time and monetary constraints. Once these problems have been categorized, one should step back and begin collecting information on the entire watershed.

9.7 DEVELOP GOALS AND OBJECTIVES

The goals and objectives need to be established early in the watershed assessment process, with the understanding that they may change as new information and/or problems are uncovered. In performing watershed assessments, several goals and objectives may be determined. These should

TABLE 9.1 Typical Watershed-Related Problems

- Flooding (see Figure 9.1)
- Stormwater-related problems (see Figure 9.2)
- Obstructions
- Diminished stream base flow (see Figure 9.3)
- Stream bank erosion (see Figure 9.4)
- Sediment deposition
- Water quality degradation (see Figure 9.5)
- Habitat degradation
- Aquatic life degradation
- Fisheries degradation
- Wastewater discharges
- Water supply contamination
- Water supply sustainability
- Urbanization
- Point source pollution (Sewage Treatment Plant (STP), illicit discharges, industrial waste outfalls)
- Storm sewer outfalls (see Figure 9.6)
- Nonpoint source pollution (NPS)
- Improperly designed stormwater management facilities (see Figure 9.7)
- Concentrations of geese (see Figure 9.8)
- Concentrations of farm animal operations
- Agricultural runoff
- Impact on a need (e.g., recreation)
- Land use issues (urbanization, mining; see Figure 9.9)

Figure 9.1 Development in the floodplain often leads to obvious problems, even if following current regulations of elevating the first floor one and a half feet above the floodplain. This is indicative of a regional or watershed problem. (Photo courtesy of the Delaware County Planning Department.)

Figure 9.2 Undersized storm drains or clogged inlets create localized flooding. (Photo courtesy of Borton-Lawson Engineering.)

Figure 9.3 Diminished base flow is a common problem in urbanized or disturbed watersheds. (Photo by Paul DeBarry; courtesy of Borton-Lawson Engineering.)

Figure 9.4 Stream bank erosion from increased flows from urbanization is common. (Photo courtesy of Borton-Lawson Engineering.)

Figure 9.5 Water quality problems, such as this AMD stream on the left, are not always obvious and may require water quality testing to determine their extent. (Photo by Paul DeBarry; courtesy of Borton-Lawson Engineering)

then be prioritized so that time and resources can be used most effectively. They should be prioritized based upon the following: magnitude of environmental improvements that would be made to the watershed, budgetary constraints, social and economic factors, and so on. Less significant objectives may not be achievable, at least in the first phase of the assessment process. One may be surprised to discover how environmental, physical, or man-made factors located in an extreme portion of the watershed are contributing to the problem. This will then lend itself to the holistic management approach to correct the problem.

An example of how goals and objectives may change in the assessment process is the Wysox Creek Watershed, in Bradford County, Pennsylvania. In developing a comprehensive watershed stormwater management plan, the field survey located several stream bank erosion problems where farmers' barns were being undercut by the lateral movement of the stream bank and falling into the water. A detailed look at the soils with the GIS showed that the soils were indeed highly erodible. Soil samples determined the critical velocity of the streams at which the soils would begin to erode. Modeling determined the flow at which the critical velocity was exceeded, which coincidentally happened to be the bankfull flow and the one-and-a-half-year storm. The stormwater management plan was tailored to reduce the frequency

Figure 9.6 Storm sewer outfalls should be surveyed. Flow three days after a rain may indicate an illicit discharge to that storm sewer. (Photo by Paul DeBarry; courtesy of Borton-Lawson Engineering.)

at which the flows exceeding the one-and-a-half-year storm or critical velocity would be reached.

9.8 OBTAIN GRANT(S)

Once the goals and objectives have been defined and prioritized, it's time to pursue grants to obtain funding to correct the problems or establish a watershed management, protection, or restoration plan. To help with this effort, the EPA has developed an online resource called "Catalog of Federal Funding Sources for Watershed Protection," at: www.epa.gov/watershedfunding. It allows a search by type of assistance required (grant or loan), type of organization applying, whether a match is required, and keyword; it also provides

Figure 9.7   Standard detention basins with low flow channels allow the first flush of pollutants to flow through the basin untreated. (Photo by Paul DeBarry; courtesy of Borton-Lawson Engineering.)

a link to the applicable funding source. In addition, state programs should be explored; typically, these can be found through the state agency responsible for environmental protection and/or conservation.

9.9   HOLD PUBLIC MEETINGS AND CONDUCT SURVEYS

No one knows the ins and outs of the watershed as well as its stakeholders; therefore, a citizens' survey should be a part of any successful watershed assessment program. Much required assessment data—such as location of problem areas, type of problems, frequency of occurrence, potential causes—can be gathered from citizens and municipal officials within the watershed. A series of public meetings should be planned, and the goals and objectives of each meeting should be established early in the process. Data collection forms can be handed out at public meetings, with a deadline set for their return.

A strict QA/QC program is essential when collecting data from volunteers (i.e., nonprofessionals) to ensure reliability. Quantitative data that may affect the outcome of the study should be obtained from trained professionals only.

Figure 9.8 An ever-increasing problem is large numbers of geese due to the concentration of nutrients their droppings contribute to the waterways. Planting the shorelines of lakes and ponds with natural high vegetation discourages geese, as it inhibits their ability to see predators. (Photo by Paul DeBarry; courtesy of Borton-Lawson Engineering.)

Figure 9.9 Typical area of disturbance: mined land. (Photo courtesy of Borton-Lawson Engineering.)

9.10   COLLECT DATA

Data collection is perhaps one of the most important aspects of the watershed assessment. Knowing what data to obtain, how and when to efficiently obtain it, how to organize it, and how to analyze it are crucial to the success of the assessment and plan. Data on stream flow, land use, streams with negatively impacted water quality, macroinvertebrates, and other rapid bioassessment protocol data, flooding and stormwater problems, and more should all be collected early in the study to establish the plan of action. Data on the physical features of the watershed such as ecoregions, physiographic provinces, soils, geology, land use/land cover, and topography (DEMs), and so on, as discussed in Chapter 10, should all also be collected as the study begins.

After evaluating these macro influences, one should evaluate the watershed in more detail, collecting information on capacities of bridges and culverts, problem areas, existing and proposed stormwater collection, and control facilities. Each is described more fully in the following subsections.

Field verification of data is a very important aspect of the watershed assessment. Once map and survey data have been collected, field collection of data should begin. The locations to visit in the watershed should be selected based upon the public surveys and maps. A field assessment trail can be mapped in the GIS and then brought along in the field. Digital photographs should be taken of the sites visited and "hyperlinked" to the location point in the GIS. Field surveys can be performed through what is known as a *window survey*, an automobile drive through the area to quickly evaluate the conditions, with brief stops at each site to record the required information. Physically hiking the entire stream length or to identified sites not accessible by vehicle is recommended if time, budget, and watershed size permit. An aerial survey recording the flight with a video camera is also a good way to catalogue potential problem areas and stream conditions. The assessment process should be completed for the field through one or a combination of the above techniques, again depending on the goals and objectives of the study.

Data can be collected by a consultant, state agency, municipality, or members of the Watershed Advisory Committee (WAC). All available design data from federal and state agencies should be obtained before collecting field data. If volunteers collect the data, they should be properly trained, and a quality assurance/quality control plan should be in place and implemented. Data collection forms should be well thought out and devised ahead of time so that all the necessary data will be collected during one site visit. The forms can then be completed at the site by field measurement and locations can be collected through the GPS.

Realizing data gaps is also crucial to the success of the plan, particularly with respect to the implementation phase. Data gaps are those areas where data is required but is not readily available, and whose absence may hinder the full assessment process. Data gaps should be identified and new data collected to fill in these gaps if conceivably possible within the time frame of the study. If long-term data is required to make sound decisions, then a plan to collect the missing data should be a component of the assessment. For instance, if one primary goal of a watershed assessment is to replenish groundwater recharge through stormwater management, and stream gage data is missing and an accurate water budget cannot be determined, a stream gauge network should be established. After installation, it will take several years to collect enough data to be usable. The short duration data cannot be used to predict flood flows, but should be sufficient to determine base flow conditions.

If a hydrologic computer model is being utilized, data collected and assimilated should be consistent with hydrologic computer model input requirements. If several goals are to be met, the data collected should be able to be utilized by the models applicable to the analyses required.

9.10.1 Existing Mapped Data

Although the GIS is an extremely useful tool in watershed assessment, older paper maps provide an excellent history of previous watershed conditions. Older United States Geological Survey (USGS) topographic quadrangles, older SCS soils survey maps (which are based on aerial photography), Agricultural Stabilization and Conservation Service (ASCS) conservation assessment aerial photographs, and other municipal or county maps can provide a snapshot of watershed conditions at some previous date. Comparing these historic data to new mapped or digital data could provide a clue as to the how the changes in the watershed are affecting the streams, flooding, water quality, and many other factors.

9.10.2 GIS Data

As mentioned several times here, and discussed more fully in Chapter 10, GIS data is an invaluable tool for watershed assessment and analysis. Thus, all possible sources of digital spatial data should be obtained and evaluated. Where there are multiple sources of the same data, the most accurate source should be incorporated into the watershed assessment. Accuracy and scale of data will be discussed in Chapter 10. However, one should remember not to just replicate the data layers to produce the map of that particular data layer, but to analyze and/or combine them to determine trends and patterns and to develop solutions. For instance, the digital geology data could be used to produce a geologic map of the watershed; however, utilizing the attributes of the database, one could produce maps of pollution vulnerability, aquifer yield, etc. that would be much more useful for a watershed assessment.

9.10.3 Regulatory Agency Survey and Existing Reports

Although the citizens' survey is invaluable to the watershed assessment, the public does not typically have knowledge of a great deal of information that is collected by federal, state, and local regulatory agencies. For example, whereas surface-water-related data is often readily observable by the common citizen and is, therefore, more readily obtained from the citizen survey, information related to groundwater quality is not as easy to observe. Such data, then, is best obtained from the regulatory agencies that gather and maintain it. Any watershed-specific reports available from the U.S. Army Corps of Engineers, USGS, FEMA, USDSA-NRCS, river basin commissions, universities, watershed associations, county planning commissions, or state or local engineering, planning, or biological groups should be obtained and perused before the data collection effort begins. These reports will often lead to even more appropriate references. Local representatives of these agencies and associations should also be contacted. To begin a search for data, www.science.gov is a good place to start. The website categorizes the myriad of data and agencies into specific subject groupings for quick reference.

A matrix of the available data such as that listed in Table 9.2 should be developed. Data gaps should also be identified from these reports, then targeted in the assessment in order to obtain the most comprehensive, reliable set of data.

9.10.4   Rainfall and Stream Flow Data

Rainfall and stream flow data is necessary for hydrologic modeling, to determine the watershed's water budget and stream base flow, to develop regional curves of bankfull flow, and possibly to help explain some of the problems encountered in the watershed. (Rainfall and stream flow data collection is described in detail in Chapter 11.)

9.10.5   Problem Areas

One can develop and utilize a survey form such as the one in Table 9.3 to collect data on the problems previously identified and to determine the type of problem (flooding, stormwater-related, water quality, stream bank erosion, pollution source, etc.). These problem areas should be identified and their locations placed on a map, to be transferred to the GIS. This data can be further augmented by reviewing any County Conservation District's (CCD) local government records regarding stormwater complaints and requests for assistance.

Problem areas can be grouped or classified into one or several of the following categories:

- Regional flooding
- Localized flooding
- Erosion and sedimentation (E&S)
- Stream bank erosion
- Water quality problems

TABLE 9.2 Data Matrix

Report/Data	Responsible Agency	Date	Date Obtained
Problem Areas	Stakeholders/Consultant	3-Jan-03	3-Oct-04
Obstructions	DOT, Consultant	3-Mar-03	3-Oct-04
Hydrologic and Hydraulic Studies	U.S. Army Corps of Engineers	22-Mar-87	3-Oct-04
Flood Studies	FEMA/USACE	22-Sep-98	3-Oct-04
Ground Water Report(s)	USGS	28-Apr-90	3-Oct-04
Surface Water Reports	USGS	18-Mar-97	3-Oct-04

TABLE 9.3 Watershed Problem Areas

WATERSHED PROBLEM AREAS—FORM X.

SHEET ____ OF ____

FORM COMPLETED BY

Before Filling Out Form,
See Instructions on Back

WATERSHED _____

Name: _____

Municipality: _____   Telephone: _____   Comments:

County: _____   Date: _____

MAP NO. *	X-1	X-2	X-3	X-4	X-5	X-6	X-7	X-8	X-9	X-10	X-11	X-12
Types of Problems												
Flooding												
Stormwater related												
Obstructions												
Diminished base flow												
Stream bank erosion												
Sediment deposition												
Water quality degradation												
Habitat degradation												

Aquatic life degradation
Fisheries degradation
Wastewater discharges
Water supply contamination
Water supply sustainability
Urbanization
Point source pollution
Storm sewer outfalls
Nonpoint source pollution
Improperly designed facilities
Geese concentrations
Farm animal concentration
Agricultural runoff
Impact on a need (i.e., recreation)
Land use issues
Other (explain)
Explanation Line No. (on back)

- Insufficient summer base flow
- Wells drying up
- Other

Additional detail should be collected on each problem's cause, frequency of occurrence, magnitude, duration, damage, and source (if determinable). Information on drainage problems and proposed solutions can be obtained by submitting the forms to the Watershed Advisory Committee (WAC) and requesting that they solicit input from people in their local areas. A detailed list tailored to the watershed should be developed. Evaluating the results of the survey will often show the problem, or magnitude of the problem, that most frequently occurs. This will help shape the goals and objectives of the watershed plan. One or more engineers from the watershed assessment team should visit the problem areas identified in the public survey and devise potential solutions. Sketches may be required. The possible source of the problem should be determined while visiting the problem site.

9.10.6 Areas of Disturbance

Areas of disturbance, which may also be a problem area, include mined areas (see Figure 9.9), quarries, junkyards, construction sites, and highly urbanized areas, and could include forestry or agricultural operations. Known areas of disturbance such as surface mines, new construction sites, and highly urbanized areas should be mapped and targeted for inspection to determine possible pollution sources. These can often be determined from new digital aerial photographs, as shown in Figure 9.10, if available; however, input from local citizens and County Conservation Districts are also excellent sources of these locations. As with the other survey items, areas of disturbance should be visited and their potential for pollution determined. Likewise, acid mine drainage (AMD) discharges should be visited and their impact on the receiving stream determined. Additional water quality data may be required. The impact of the AMD discharge should be followed downstream to determine how far it is felt. Locations for possible AMD treatment sites should be located. AMD discharge flow should be obtained to determine the size of the potential treatment facility.

9.10.7 Animal Feeding Operations (AFO)

Livestock and poultry operations (LPOs), animal feeding operations (AFOs), and concentrated animal feeding operations (CAFOs) should be identified (see Section 16.2.1). Solutions to the problems should be devised as part of the watershed plan.

Unbuffered stream reaches in pasture areas are responsible for many detriments to stream health. They open direct sunlight to the streams, which

Figure 9.10   Area of disturbance identified from digital aerial photography.

warms the water. They provide direct access for the nutrients contained in manure to enter the stream during rainfall-runoff events and allow livestock to trample the stream banks, causing accelerated stream bank erosion, as shown in Figure 9.11.

Identifying these areas is critical to any watershed assessment, so that funding can be obtained to fence natural stream buffers. Once fenced, the natural

Figure 9.11   Unbuffered stream banks not only give nutrients from manure direct access to the streams, but allow trampling and warming of the water from direct sunlight, resulting in channel instability. The second photo has a compounded problem of land development just upstream from unprotected stream banks. (Photo courtesy Borton-Lawson Engineering.)

vegetation root mass will quickly provide stream bank stabilization and its canopy, shading. Often, much money is spent importing vegetation; in some cases, supplemental plantings with species such as black willow may be warranted. Many millions of dollars have been spent to restore these eroded stream bank areas using natural channel design and fluvial geomorphologic procedures; in fact, this money would have been better spent had a riparian buffer been maintained in the first place. In addition, if the upstream watershed is managed properly in a proactive, planned manner, the channels will eventually restabilize themselves. This is perhaps the most critical issue affecting the Chesapeake Bay and the Susquehanna River watershed; the situation should be analyzed in order to direct funds in the most cost-effective, environmentally beneficial manner.

9.10.8   Stream Water Quality Sampling Data

Water quality sampling for the parameters that may be of concern in the watershed should be obtained, as well as the locations of sampling points for future input to the GIS.

Water quality testing is expensive; therefore, testing parameters should be determined *after* the GIS analysis has been performed to pinpoint which parameters to test. Although many citizens' groups collect water quality data, their collection process often is not performed under strict quality assurance/quality control conditions; rather, it tends to be random-grab sampling, generally using less sophisticated equipment, procedures, and methodologies. Moreover, the sampling is not necessarily coordinated with required flow periods, such as base flow conditions. For example, a citizens' group may collect one set of data 10 days after a rainstorm and the next set of data three days after a rainstorm. Without collecting flow data, this information is of very little use, although it may show a general trend and coincident pollution events. In contrast, state or federal data, which is tied to flows, is typically very reliable.

Water quality samples should be collected based upon the goals and objectives of the project, to fill in any data gaps. The type of sampling procedures (e.g., grab versus continuous, base flow versus storm flow) required for the intended purpose of the study should be determined and the appropriate samples collected. The collection of existing water quality and quantity data in the streams and tributaries to establish the TMDL waste load allocation is important, if this has not been performed previously. Inflow and outflows from lakes should be obtained to determine hydraulic retention time and nutrient budget.

9.10.9   FGM Channel Reaches and Designations

Fluvial Geomorphologic Assessment data should be collected. (This is described in more detail in Chapter 13.) This survey would include the stream bank erosion problems (shown in Figure 9.11) encountered during the FGM assessment.

Figure 9.12   Obstructions include culverts, bridges, and fallen trees, and their limited capacity could cause flow to be diverted and cause flooding. (Photos by Paul DeBarry, courtesy of Borton-Lawson Engineering.)

9.10.10   Biological Data

The biological condition of the watershed and stream should be determined through a scientifically valid assessment procedure, such as the EPA's Rapid Bioassessment Protocol, RBP (U.S. EPA, 1999).

9.10.11   Obstructions to Flow

Obstructions are structures, whether man-made or natural, that may obstruct the flow or hydraulic capacity of the channel, as shown in Figure 9.12. This may cause a *backwater* and flooding. Examples are bridges, culverts, buildings, fallen trees, and earth slumps.

Information on obstructions and their capacities can be obtained by several methods.

The dimensions required to determine capacities should be obtained; these include diameter, width, depth, cover or head, material, skew, or any other factors that may affect the capacity of the opening, including sediment deposition. Capacities can then be calculated using accepted hydraulic methods and design charts. The Federal Highway Administration's Hydraulic Design of Highway Culverts (2001) is particularly useful.

Next, the obstruction capacities can be compared to the peak flow derived through the modeling process obtained at that location for each design storm duration and frequency. Flood-frequency relationships can then be developed for each obstruction and recorded in tabular format, suitable for attachment as "attributes" to a GIS database. From these flood-frequency relationships, significant obstructions can be determined. A *significant obstruction* is defined as any structure or assembly of materials that would impede, retard, or cause ponding or diversion of stormwater runoff or erosion of surrounding land or stream banks. Significant obstructions can be classified into categories based upon their hydraulic capacity from available modeling data as follows:

- Those obstructions able to pass the 100-year, 24-hour storm and greater without obstructing the flow.
- Those obstructions not able to pass the 100-year, 24-hour storm and greater without obstructing the flow.
- Those obstructions not able to pass the 50-year, 24-hour storm and greater without obstructing the flow.
- Those obstructions not able to pass the 25-year, 24-hour storm and greater without obstructing the flow.
- Those obstructions not able to pass the 10-year, 24-hour storm and greater without obstructing the flow.
- Those obstructions not able to pass the 5-year, 24-hour storm and greater without obstructing the flow.
- Those obstructions not able to pass the 2.33-year, 24-hour storm and greater without obstructing the flow.

Another often overlooked effect of bridges and other obstructions is that they often force into a constricted opening the flow that would have naturally passed in the floodplain. By taking the natural flow and forcing it into a smaller area, it increases the velocities over those which would occur under natural conditions. The net affect is an "unstable" channel reach, with stream bank erosion and sediment transported downstream as shown in Figure 9.13. A watershed plan should require new bridge designs to consider bridge openings to allow flow in the floodplain, as shown in Figure 9.14.

9.10.12 Existing and Proposed Stormwater Collection Facilities (Storm Sewers)

Stormwater collection facilities are those structures such as storm sewers, diversions, and swales that would channel flow to a discharge point. It is important to collect information on these stormwater collection facilities as they sometimes alter or divert the flow path, and typically provide a single discharge point for sampling or flow measurements. Municipalities' storm sewer maps should be obtained to pinpoint collection areas and discharge points. This will aid in the nonpoint source pollution assessment. Storm sewer inlets and outfalls should be field-verified from available maps, or catalogued and identified if this data is not available. This would also be a good time to perform the inlet stenciling, describing that the stormwater flows to a stream, if funds are available to do so. Impacts of the discharge should be noted and photographed, and water quality samples collected.

Figure 9.13 Bridge constrictions force floodplain flows into a smaller area, often causing downstream stream bank erosion.

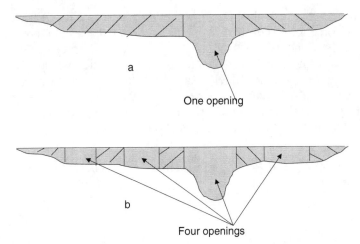

Figure 9.14   (a) Typical bridge opening design, and (b) one that allows natural flood-plain flow.

9.10.13   Existing and Proposed Stormwater Management Facilities

Stormwater control facilities include dams, reservoirs, lakes, detention basins, retention and infiltration basins, best management practices (BMPs), or any facility that would alter or attenuate the flows and have an impact on the hydrology of the watershed. Locating these facilities ahead of time through the public survey will increase the efficiency of data collection. These should then be evaluated as to their efficiency for treating the water quality of stormwater runoff; subsequently, retrofit suggestions should be made to improve the pollutant removal efficiency of these facilities. The public survey most likely will not locate all the stormwater management structures, so before going into the field, one should obtain the latest digital aerial photography and locate overlooked basins utilizing the GIS, as shown in Figure 9.15.

Many existing detention basins were designed for quantity control only; in other words, they were designed to reduce a postdevelopment peak flow for a specified design storm, down to the predevelopment peak for that same design storm. This does little to treat for water quality improvement or infiltrate potential recharge. In fact, many basins even have concrete low-flow channels running through their bottoms to prevent them from staying wet (refer back to Figure 9.7). However, this allows the first flush of pollutants to pass through the basin uncaptured and untreated. All stormwater management structures should be visited, followed by development of potential BMP retrofit options to treat water quality and promote groundwater recharge. Sketches of solutions should be drawn and design grants pursued. The structures should be ranked based upon those that would benefit the watershed the most by a retrofit. In other words, those that may be passing large amounts

Figure 9.15 Digital aerial photography aids in the location of detention basins for retrofit potential analysis.

of untreated nonpoint source pollution but could be treated with a proper retrofit.

Conversion of a standard detention basin to a water quality BMP (shallow wetland) is shown in Figures 9.16a-c. Notice that the existing detention basin allowed the polluted stormwater to pass directly to the outlet, whereas the retrofit allows extended detention times and filtering of pollutants by aquatic vegetation.

In evaluating the retrofit options, one must keep in mind that the basins were designed for quantity control, and a retrofit typically takes away some of the volume required to control the peak rates of flow; therefore, a detailed routing of the new retrofit must be performed to ensure the basin will continue to function safely for its peak rate control.

Figure 9.16a Photograph of dry (standard) stormwater detention basin that allows pollutants to flow right through the basin.

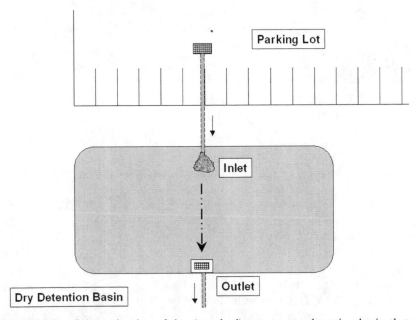

Figure 9.16b Schematic plan of dry (standard) stormwater detention basin that allows pollutants to flow right through the basin.

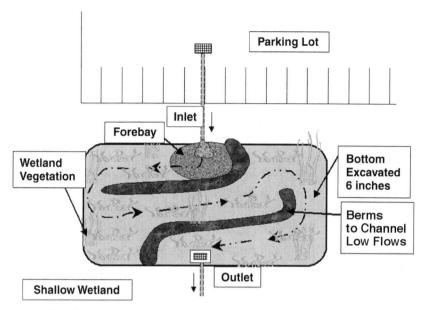

Figure 9.16c Schematic of the same dry (standard) stormwater detention basin retrofitted to a shallow wetland with baffles to provide extended detention and filtering and bioassimulation of pollutants by aquatic vegetation.

9.10.14 Existing and Proposed Flood Control or Flow Alteration Facilities

Flood control or flow alteration facilities or projects include levees, dikes, floodwalls, diversions, pump stations, or other flood protection or flood management facilities that would contain or alter flood flows. Before performing the field survey, design data should be obtained from the U.S. Army Corps of Engineers or other agency to determine capacities, design storm, velocities, environmental impact, and so on. The structures should then be visited and, if design data is not available, the aforementioned information determined. Upstream and downstream effects of the structure should be determined, as well as the distance of the impacted area.

9.10.15 Projected and Alternative Land Development Patterns in the Watershed

In addition to the existing land use/land cover data, a future land use scenario should be developed in the GIS; further, the hydrologic model should be rerun based upon future growth, to determine the impact on the waterways from that growth.

9.10.16   Illicit Discharges

Utilizing the NPDES MS4 guidelines, locations of illicit discharges should be obtained, and the source of flow and pollutant type to those discharges determined.

9.10.17   Industrial Discharge Points

Locations of industrial discharges should be solicited and any existing water quality data collected and analyzed. These locations should field-verified. Additional water quality sampling data and bioassessment may be required to better determine the impact of the discharge on the receiving stream's water quality. All data should be placed into the GIS database.

9.10.18   Sewage Treatment Plant (STP) Outfalls

Sewage treatment plants and their outfalls should be located in the GIS. Any outfall water quality data should be collected and analyzed for its impact on the receiving stream's water quality. Sewage treatment plants and their outfalls should be visited to confirm the discharge location. Additional water quality sampling data and bioassessment may be required upstream and downstream of the discharge to better determine the impact of the discharge on the receiving stream's water quality. All data should be placed into the GIS database.

9.10.19   Sewage Service Areas

Sewage service areas should be collected and mapped, if not already mapped by the authority and entered into the GIS.

9.10.20   Water Service Areas

Water system service areas should be collected, identified, and mapped, if not already mapped by the water authorities and entered into the GIS. These, in addition to the sewer service areas, are particularly important when developing localized water budgets and evaluating consumptive versus nonconsumptive uses.

9.10.21   Potential Contamination Sites

An inventory of potential contamination sites, such as gas stations, chemical factories, and chemical storage areas, should be determined. Most state regulatory agencies have an inventory of this information. Other industries that might be potential contamination sites, such as dry cleaners, may not be in the database, hence would have to be collected by some other means such as tax parcel land use data or field survey.

9.10.22 Well Data

USGS and some state environmental regulatory agencies obtain data on new and existing wells. The information attributed to these wells is, however, often derived from the well drillers, so its accuracy should be field-verified. Important information such as *static water table level* or *potentiometric surface,* water quality, and potential contamination may be obtained from this data. Field verification of data is a very important aspect of the watershed assessment.

9.11 CHECK LOCAL ORDINANCES

Many changes in the watershed are due to disturbances and land use changes, and most land use controls are governed by local land development or zoning ordinances, or lack thereof. Ordinances should, therefore, be collected and analyzed for consistency with the goals and objectives of the proposed watershed assessment and management plan. Further, these ordinances should be checked to be sure they include watershedwide stormwater management criteria, water quality criteria, and groundwater recharge criteria. Some of the inconsistencies to look for can best be described with a few examples:

- *Stormwater ordinance-requiring detention of only the 100-year storm.* Although most likely written to be conservative, detaining only the 100-year storm would result in an outlet structure that would pass flows unimpeded up to and including the 100-year storm. This would allow the increased flows from postdevelopment conditions for storms smaller than the 100-year storm to be passed through the basin undetained, causing increased flows and possibly stormwater-related or flooding problems downstream.
- *Wellhead protection ordinance prohibiting recharge in the zone of contribution.* Although intended to prevent contaminants from entering the groundwater supply, recharge of "clean" runoff—that is, from roofs— would be beneficial, if not mandatory, for sustained water supply.
- *A conservation design ordinance prohibiting land development on hydrologic soil groups C and D.* It was suspected that this was written to avoid soils that have a high water table or a slow percolation rate, which could be remedied through soil testing and septic system design to account for the percolation rate. One problem with this ordinance was that 80 percent of the municipality was composed of C and D soils. The other problem with this concept is that, for sustainable groundwater and stream

base flow supply, the stormwater should be recharged in the most permeable A and B soils, and building should occur on the C and D soils, which are least desirable for groundwater recharge.

9.12 ASSIMILATE DATA

Once all the data has been collected, the task of keeping track of it all can be overwhelming. One way to do this is to summarize it into matrices. Excel spreadsheets are convenient for this purpose, particularly if the identifiers are kept consistent with those in the GIS. Then the spreadsheet essentially becomes a database, and the tables can be joined so that the database can be displayed in the GIS, as shown in Figure 9.17.

9.13 PROCEED WITH ANALYSIS: ASSESSMENT AND MODELING

Once the data has been collected, analyzed, and assimilated, the watershed assessment can begin, utilizing the criteria and procedures established in this book. The GIS should be implemented to determine background watershed conditions, to identify patterns and trends, and to evaluate scenarios. The established goals should be consulted to prioritize capital and resources. Modeling efforts may focus on flood control, stormwater management, TMDLs, water quality, groundwater flow, or stream bank erosion. In any case, the data collected and assimilated should be consistent for this assessment or any model that may be required in the future. The assessment should determine the impact of sources of pollution and other problems and should include reaches of impacted stream; how this impact was measured and determined; and impacts on existing and proposed stream/lake uses and existing and planned land uses. A TMDL should be established for any parameter that does not meet water quality standards in the watershed. The allowable load for each point and nonpoint source should be determined using the Waste Load Allocation Process (WLAP). Included in this analysis should be the percentage of contribution from each source. The percent reduction required from each contributing source should be determined and methods chosen to achieve it.

Potential restoration/retrofit projects should be identified and cost estimates made to design and construct those projects. Prioritization should be based on the objectives of the assessment, in addition to measurable environmental benefits, budgetary constraints, social and economic concerns, environmental constraints, land ownership, and physical conditions of the site.

9.14 WRITE THE PLAN REPORT

The plan report should follow this general outline:

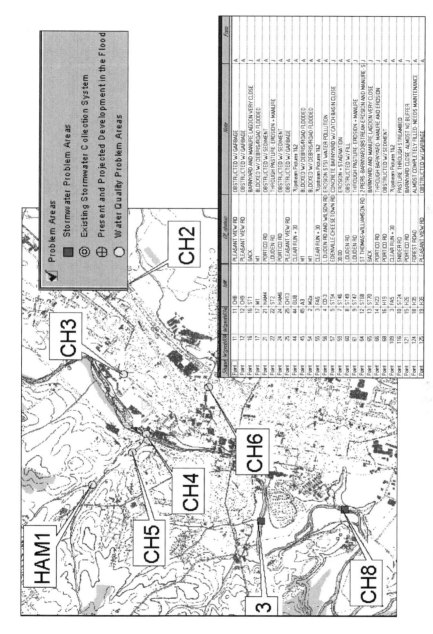

Figure 9.17 Excel spreadsheet database of problem areas in a watershed, converted to GIS attributes (by Paul DeBarry; courtesy of Borton-Lawson Engineering).

Executive Summary: Goals, Objectives, Action Items, Schedule for Implementation.

  I. Description of the watershed (physical, social, economic considerations)
 II. GIS analysis (data, sources, analysis, final product)
III. Data collection efforts (data gaps, collection procedures, QA/QC)
 IV. Problem areas (types, locations, field evaluation)
  V. Analysis (modeling results, parameter analysis)
 VI. Results of assessment (findings, priorities)
VII. Implementation (individual site priorities)
VIII. Summary—conclusion

The report should be concise, with extraneous material placed in a separate technical document.

9.15 SUMMARY

The assessment process should involve all interested parties, and well-defined, unbiased goals and objectives should be determined early in the process. These goals and objectives should then be tailored and prioritized to the specific watershed. That is not to say that items of lesser importance should be ignored and the data not collected for them. Keep in mind long-term trends and potential studies and collect data that will be usable for a variety of studies and purposes. A strict quality assurance/quality control (QA/QC) program should be developed and followed. Spatial data should be stored within the GIS. Local ordinances should be reviewed and made consistent with the goals of the plan. Political and social influences also must be considered in the assessment and plan. The report should be concise and to the point, with recommended actions prioritized for implementation. Grants should be pursued to obtain funding to accomplish the goals of the plan.

REFERENCES

U.S. Department of Transportation, Federal Highway Administration. "Hydraulic Design of Highway Culverts," Hydraulic Design Series No. 5, 2nd ed. NHI-01-020, National Highway Institute, Arlington, VA, September 2001.

U.S. EPA. "Rapid Bioassessment Protocols for Use in Wadeable Streams and Rivers, Periphyton, Macroinvertebrates, and Fish," Second Edition, EPA 841-B-99-002, July 1999.

U.S. EPA. "Standard Methods," www.epa.gov/waterscience/methods/, September 2003.

## CHAPTER 10

GEOGRAPHIC INFORMATION SYSTEMS

10.0 INTRODUCTION

Geographic Information Systems (GIS) were originally developed in the 1960s by Roger Tomlinson, who used computers to model land inventory for the Canadian government on the large mainframe computers. With the advent of personal computers and their increased speed, GIS became more readily available to the common practitioner. Subsequently, the invention of Microsoft Windows and the graphical user interface (GUI) made it much easier for any government agency, engineering firm, planning commission, or watershed organization to train and be up and running with GIS in a short time.

GIS is comprised of five components: people, policies and procedures, data, hardware, and software. GIS software functionality is not only a computerized mapping tool but a spatial database, with the tabular data, or *attributes,* attached to the spatial data. (Recently, GIS has been commonly referred to as a geospatial database.) It is, therefore, a highly valuable query and analysis software package. GIS software typically provides the capabilities for data management (data collection and maintenance), spatial analysis (modeling), and mapping (visual display and plotting).

10.1 USE OF GIS FOR WATERSHED ASSESSMENT

GIS data can be visualized as a volume of encyclopedias stacked on their sides, as shown in Figure 10.1. Each volume is a layer in the mapped product, such as roads, streams, topography, and so on. However, different GIS soft-

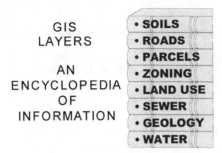

Figure 10.1   GIS is like an encyclopedia of spatial data in which the layers can be overlaid, and each volume contains a wealth of data or attributes attached to that layer.

ware programs define the "layers" differently, with some of the most common terms being *layers, coverages, themes,* and *features;* and each software program is distinct.

Note

In spite of the technical differences between various GIS software programs, this text refers generically to all of them when referencing layers so as to not promote any particular type.

It's important to point out that the "layer" is just not the spatial data; it contains large volumes of information on that particular data layer. This volume of information, or data, contains the attributes attached to each piece of spatial data. The Geographic Information System is an excellent tool for gathering and analyzing geographic data related to the watershed. Because each watershed's characteristics are unique, the GIS makes it possible to overlay and manipulate the various data layers specific to the needs and goals of a particular watershed. The data are the most important part of a GIS system, and often the most costly. Data generation, projection, and manipulation can require many hours, but once established can increase analysis and updating efficiency, thereby saving long-term costs.

10.2   UNITS, SCALE, AND ACCURACY

Various agencies develop their map material in different units, whether they be meters or feet, so the appropriate units to use should be determined early in the project. The next item the user should be concerned with before beginning the project is data accuracy. One must be sure that the data being obtained is valid for its intended purpose. Data accuracy is usually substantiated using national map accuracy standards. Table 10.1 provides a good reference source for scale-dependent projects.

TABLE 10.1 Scale According to National Map Accuracy Standards

Scale Range	Scale	1 inch = X feet	Horizontal Accuracy	Example Use
Large Scale	1:1,200	100	3.33	Engineering Design
	1:2,400	200	6.66	Engineering Design
	1:4,800	400	13.33	Concept Design, Master Planning
	1:12,000	1,000	33.33	Regional Planning
	1:24,000	2,000	40.00	Watershed Planning
	1:63,360	5,280	105.60	Transportation Planning
Small Scale	1:100,000	8,333	166.66	Regional Basin Planning

Scale represents the ratio between distance or area on the map and the corresponding distance or area on the ground. The original scale of the data source, as input into the GIS, dictates to what detail an analysis can take place. The output scale is simply the portrayal of the data at a "size" that can be read and understood by the chosen audience; in actuality, it is usually dictated by the size of the paper on which it will be displayed. For instance, a spatial data set generated within the GIS can be displayed at an output scale equivalent to an architectural D-size paper (or 24 by 36 inches); or that same spatial data set can be displayed on an 8½-by-11-inch paper for a standard report.

The larger the scale, the smaller the area covered. For instance, a typical scale for local watershed assessment and analyses within a county is 1:24,000. This may be considered too large a scale for a river basin commission whose watershed spans several states. Data developed from 1:100,000-scale sources may be more appropriate here.

Map detail, scale, and accuracy are all interrelated. A large-scale map is not necessarily more accurate than a small-scale map; however, it will show more detail. Maps cannot be more accurate than the original data source. Therefore, enlarging maps will not necessarily result in more detailed or accurate results than the original map. Vertical accuracy should also be considered when utilizing data with elevations or Z-data (as opposed to the horizontal X, Y values) to ensure accurate results.

10.3 COORDINATE SYSTEMS, DATUMS, AND PROJECTIONS

A *coordinate system* is a reference system used to measure horizontal and vertical distances on a *planimetric* map, often confused with *projections* and *datums*. Coordinate systems are projections of the working surface onto the earth's surface; they are defined by a projection, datum, spheroid of reference, and parallel data. *Geographic coordinate systems* (*GCS*) use three-dimensional spherical surfaces to define locations on the earth. A *datum* is a set of parameters that define the coordinate system, and the control points are either calculated or measured. The datum is also defined by the earth's spheroid and its relationship to the center of the earth. Early datums were determined based upon theories of the earth's surface. More recently, the use of Global Positioning Systems (GPS) has enabled a refinement of the earth's definition. Two datums used in North America are the North American Datum of 1927 (NAD27) and the revised North American Datum of 1983 (NAD83). The origin of NAD27 is a point at Meade's Ranch in Kansas, which is the approximate centroid of the continental United States. The datum is based upon a Clarke spheroid to represent the shape of the earth. Due to limitations in ability to accurately calculate information in those days, many errors were introduced into this datum.

However, with the advent of new technologies such as GPS satellites, Very Long Baseline Interferometry, electronic theodolites, and Doppler systems, a more accurate portrayal of the United States could be determined, and thus the NAD83 system. The origin of this datum is the earth's center of mass; it uses the GRS80 spheroid, and is based on new calculations made with the use of satellite information systems and earth observations. The most recently developed and widely used datum is the World Geodetic System of 1984, or WGS84 (Kennedy and Kopp, 2000).

A *projection* is a transformation of spheroidal or three-dimensional data onto a flat or two-dimensional map. Software projection routines can transform a location on a sphere to a flat surface. However, just as peeling an orange and trying to flatten the rind to make it horizontal, projections cause distortions in the shape, area, and distance of the mapped data. Different projections will distort the spheroidal data differently. (For further reading on map coordinate systems, datums, and projects refer to ESRI's *Understanding Map Projections* [Kennedy and Kopp, 2000].)

To establish a watershed GIS, the first step is to determine the coordinates at which the watershed should be located. There are many projections and coordinate systems available, and a lot of data is being developed in the various coordinate systems, as summarized in Table 10.2 (U.S. Department of the Interior, 2000). Table 10.2 represents the best projections utilized for a specific area of analysis.

Several types of coordinate systems utilized include: State Plane Coordinate System (not a projection), Universal Transverse Mercator (UTM) projection system, conical projection system, Albers, and others. Spatial data can be transferred from one projection to the other through standard projection equations, which are now incorporated into most GIS software. Coordinate systems and final projections should all be checked before utilizing spatial data. Typically, for watersheds ranging from 1 to 300 square miles, or that do not span states or state plane zones, it is best to utilize a State Plane Coordinate System. For larger regional basin watershed studies that span several states, it is best to utilize a UTM coordinate system.

A comparison of Latitude Longitude, State Plane, and UTM coordinate system grids and Mercator, Lambert Conformal Conic, and Unprojected Lat-

TABLE 10.2 Projections Best Utilized for Watershed Assessment

Summary of Areas Suitable for Mapping with Projections
Key: * = Yes x = Partly

Projection	Type	World	Hemisphere	Continent/Ocean	Region/sea	Medium scale	Large scale
Globe	Sphere	*					
Mercator	Cylindrical	x			*		
Transverse Mercator	Cylindrical			*	*	*	*
Oblique Mercator	Cylindrical			*	*	*	*
Space Oblique Mercator	Cylindrical						*
Miller Cylindrical	Cylindrical	*					
Robinson	Pseudocylindrical	*					
Sinusoidal Equal Area	Pseudocylindrical	*		*			
Orthographic	Azimuthal		x				
Stereographic	Azimuthal		*	*	*	*	*
Gnomonic	Azimuthal				x		
Azimuthal Equalidistant	Azimuthal	x	*	*	*		
Lambert Azimuthal	Azimuthal		*	*	*		x
Equal Area							
Albers Equal Area Conic	Conic			*	*	*	
Lambert Conformal Conic	Conic			*	*	*	*
Equidistant Conic	Conic			*	*		
Polyonic	Conic					x	x
Bipolar Oblique Conic Conformal	Conic			*			

itude and Longitude projections are shown in Figures 10.2a and 10.2b, respectively (Dana, 1999).

10.4   GIS DATA TYPES

There are three main types of spatial data that can be utilized in watershed assessment. The most common spatial data is *vector data,* which includes

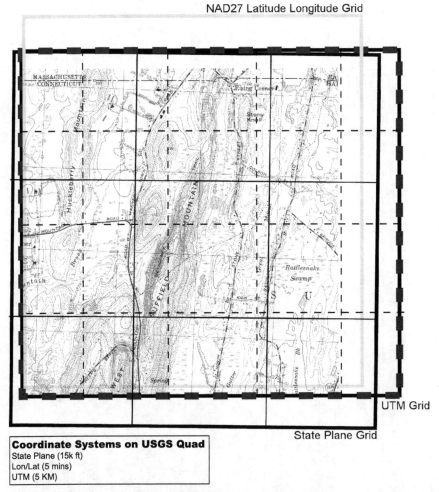

Figure 10.2a   Comparison of NAD27 Latitude Longitude, State Plane, and UTM Coordinate Systems. (Adopted from Peter H. Dana, The Geographer's Craft Project, Department of Geography, The University of Colorado at Boulder.)

Three Map Projections Centered at 39 N and 96 W

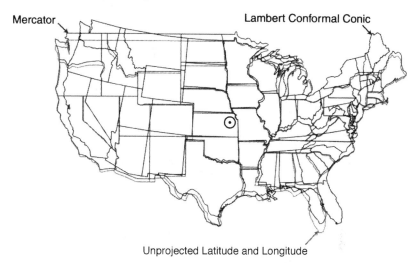

Figure 10.2b   Comparison of three projections for the United States. (*Source:* Peter H. Dana, The Geographer's Craft Project, Department of Geography, The University of Colorado at Boulder, 1999.)

points, lines, and *polygons.* Examples of these three types of vector data used in watershed analysis are wells, streams, and soil polygons, respectively. Vector data should have *topology,* which refers to the structure of a set of features and the relationships between connecting or adjacent features in a geometric network such as a stream. Topology should not be confused with *topography,* or the elevations and lay of the land. In *planar* or polygon spatial data, topology would be the relationships between shared borders (Zeiler, 1999).

The second form of data used in watershed analysis is *raster* or *grid data,* a series of grid cells with assigned values, whether referring to elevation, land use, or soil type, as shown in Figure 10.3. The smaller the grid cell size, generally, the more refined the analysis and the display.

The third type of data useful in watershed assessment is *image data,* which is typically also in a raster-type format. An example of this data would be digital aerial photographs, as shown in black and white and infrared in Figure 10.4. Infrared photographs may show features, such as thermal pollution plumes (which are important to watershed analysis), better than black-and-white photographs.

Each set of data has its advantages and disadvantages for use in the application at hand. Discrete features and data summarized by area, and line and point data, are usually represented using the vector model. Vector data is also utilized for spatial reference, such as creating base maps, stream maps, and so on. The raster data represents continuous numeric values and is often

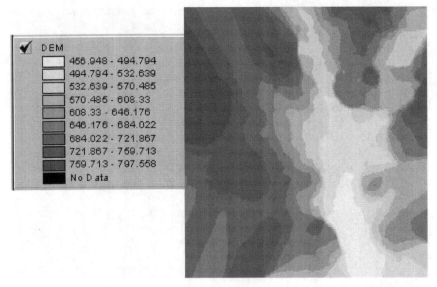

Figure 10.3 Typical raster dataset: a Digital Elevation Model (DEM).

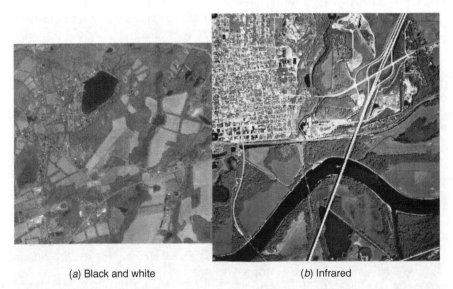

(*a*) Black and white

(*b*) Infrared

Figure 10.4 Typical image data: digital aerial photograph. (*Source:* USGS Digital Aerial Photograph Web site, ftp://ftp.emtc.usgs.gov/pub/gis_data/aerial_photos/1994 _illinois_river_72dpi/peoria/, 12/27/03.)

utilized for combining discrete features with other layers in a model processing of spatial data.

The *geodatabase* is a mechanism for storing vector and raster data, and is an object-oriented data model. It is a spatial database that resides within a relational database management system and includes validation rules, relationships, and topological associations (MacDonald, 2001).

10.5 METADATA

In addition to the three spatial data types, another important aspect of the GIS is the *metadata,* or data about data. Metadata is the descriptive information about the contents of a digital geospatial data file. Metadata describes the source, content, quality, projection, and other information about the GIS data. Metadata is extremely important in enabling a user to readily see the properties of the data, including its projection, accuracy, and source. Not capturing or recording accurate metadata could cause users to misinterpret the data and falsely apply the whole analysis. Although originally not readily accepted by GIS users due to the time-consuming process to prepare it, recent software enhancements have made it easier to develop metadata. Metadata can be as comprehensive as the user wishes and can be displayed in many different ways, including:

- Federal Geographic Data Committee (FGDC) Classic
- Federal Geographic Data Committee (FGDC)
- Federal Geographic Data Committee (FGDC)—Environmental Systems Research Institute
- Federal Geographic Data Committee (FGDC)—Frequently Asked Questions Format (FAQ)
- Federal Geographic Data Committee (FGDC)—Geography Network
- International Organization for Standardization (ISO)
- ISO—Geography Network
- eXtensible Markup Language (XML)

Metadata typically includes the following sections and supporting information:

Identification Information
- Publication date of digital files
- State/county name
- Bounding coordinates
Data Quality Information
- All source documents used to create the digital Q3 flood data

Spatial Data Organization Information
- Object counts (points, lines, and areas)

Spatial Reference Information
- Projection
- Datum

Entity and attribute information

Distribution information

Metadata reference information

10.6 GIS LAYERS AND DATA SOURCES USEFUL FOR WATERSHED ASSESSMENT AND ANALYSIS

Most watershed management plans—whether for flood control, stormwater management, open space or greenway planning, water supply and wellhead protection, lake management, rapid bioassessment protocols, or FGM—require much of the same type of data. Typical layers required for watershed assessment include land use/land cover, soils, geology, topography, relief, stream networks, water features, and base map materials, including municipal boundaries and roads. This list is by no means exhaustive, but practically provides the most commonly used data for watershed management. Each data source would need to be projected to the working coordinate system, via the appropriate functions in the GIS software being utilized.

The most common mistake in developing watershed assessments and management plans is to do a simple compilation of available GIS data, which are then "regurgitated" into individual "pretty" maps to display that data. In essence, this could also be accomplished by copying the map from a book. The goal of using GIS for watershed assessments should not be to produce a map showing a map layer. The goal should be to use the GIS data in combination with other data to evaluate interactions and trends and to provide a solution. In other words, the combination of data should tell a story and lead to educated decisions. The "map" is just a by-product of the evaluation and analysis process that aids in translating the complicated interactions. Thus the GIS data and their sources are described in this section, not to enable users to reproduce the layer's maps, but to help the user begin to think about watershed processes and interactions. Therefore, the term "map" is intentionally avoided in this text.

The place to start to look for GIS data is the www.geodata.gov Web-based portal, which offers one-stop access to maps, data, and other geospatial services. This site is part of the geospatial one-stop initiative, set up to accelerate the development and implementation of the National Spatial Data Infrastructure (NSDI) and enhance government efficiency. The geodata.gov portal includes information from state, local, and tribal governments, along with those from the private sector and academia. The site also allows users to publish their own data to the site. It provides easy access to data in the following data categories

- Administrative and Political Boundaries
- Agriculture and Farming
- Atmosphere and Climatic
- Biology and Ecology
- Business and Economic
- Cadastral
- Cultural, Society, and Demographic
- Elevation and Derived Products
- Environment and Conservation
- Geological and Geophysical
- Human Health and Disease
- Imagery and Base Maps
- Inland Water Resources
- Locations and Geodetic Networks
- Oceans and Estuaries
- Transportation Networks
- Utilities and Communication

Now that the second decade of GIS use on a local level has begun, much of the data required for water resources analysis is also becoming readily available. The following sections review the data sources required to establish a working watershed assessment project.

10.6.1 Base Data (Digital Geospatial Data)

The first step in developing a GIS database is to determine the *base map* from which most of the analysis will be performed and results displayed. A base map is a one whose features will not change in the watershed, hence will aid the user to determine the location of the assessment data. A base map typically includes data sets such as roads, streams, political data, and so on. Its data may be planimetric or raster type, as described earlier.

Map scale and resolution will play a part as to the data layers and sources from which data is obtained. Road centerline files, *hydrography* (streams, lakes), political boundaries, and so on are typical "layers" that make up the data suitable for base mapping. These are typically available from the State

Department of Transportation (DOT) on CD-ROM or downloadable from the Internet (most DOTs were aggressive in the early states of GIS data development). Another typical source of the road centerline data and streams is the USGS 1:24,000 topographic quadrangles, which were digitized and *mapjoined*. Therefore, the data matches the National Map Accuracy Standards, with an accuracy of plus or minus 40 feet horizontal. The scale and accuracy of the data to be utilized should match the required scale of the application. The next sections describe data sources that can be used for base data.

10.6.2   U.S. Census Bureau TIGER Files

TIGER is an acronym for the Topologically Integrated Geographic Encoding and Referencing system. Its files comprise a planimetric database developed at the U.S. Census Bureau to support its mapping needs for the Decennial Census and other Bureau programs. The *TIGER/Line* files are extracts of selected geographic and cartographic information from the Census Bureau's TIGER database, whose source materials were principally 1:100,000 U.S. Geological Survey (USGS) topographic maps scanned by USGS for the bureau and included: roads, railroads, hydrographic features, and miscellaneous transportation features layers. These layers were then vertically integrated into a *topologically* consistent dataset, meaning they are not stored as separate layers. For the 2 percent of the land area covered by the major urban areas, the Census Bureau substituted the GBF/DIME files that were created in the 1970s (with updates in 1981 to 1985) for more detail. The features were also geographically coded to identify the areas. The 2004 TIGER/Line files will be the first version of the files to reflect substantial updates from the Master Address File/Topologically Integrated Geographic Encoding and Referencing Accuracy Improvement Project (MAF/TIGER AIP). TIGER files are available at: www.census.gov/geo/www/tiger/tiger2002/tgr2002.html.

The TIGER/Line files contain data describing three major types of features/entities:

1. Line Features
   - Roads
   - Railroads
   - Hydrography
   - Miscellaneous transportation features and selected power lines and pipelines
   Political and statistical boundaries
2. Landmark Features
   - Point landmarks—schools and churches; sporadic coverage added on an as-needed basis
   - Area landmarks—parks and cemeteries; sporadic coverage added on an as-needed basis

3. Polygon Features
   - Geographic entity codes for areas used to tabulate the Census 2000 statistical data and governmental unit boundaries legally in effect as of the latest Boundary and Annexation Survey (BAS)
   - Locations of area landmarks

The TIGER/Line files are provided in ASCII text format only; therefore, users will have to convert or translate the files into a format compatible with their specific software package. A sample TIGER/Line file is shown in Figure 10.5.

10.6.3   U.S. Census Bureau's Cartographic Boundary Files (CBF)

The U.S. Census Bureau's *Cartographic Boundary Files* (*CBF*) are selected *generalized extracts* from the Census Bureau's TIGER geographic database designed for use in a Geographic Information System (GIS) or similar mapping system. They have been developed for various internal Census Bureau projects and have been made available to the general public. The generalized files are of a much smaller size than the original file extraction from the Census Bureau's TIGER database, resulting in faster download and processing times. The CBFs are primarily designed for small-scale, thematic mapping applications at a target scale range of 1:500,000 to 1:5,000,000. The files are *unprojected,* wherein the geographic coordinates are referenced by latitude and longitude (a.k.a. the "geographic" projection). The CBFs are available in ARC/INFO EXPORT (.e00) format, ArcView Shapefile (.shp) format, and ARC/INFO Ungenerate (ASCII) format at www.census.gov/geo/www/cob/index.html. A sample of the County CBF files for Alabama can be found in Figure 10.6.

Figure 10.5   Tiger/Line files showing northern Wyoming counties (left), and Colorado Urban Areas, and roads, streams, and urban areas (right). (*Source:* www.lib.ncsu.edu/stacks/gis/tiger.html, 12/27/03.)

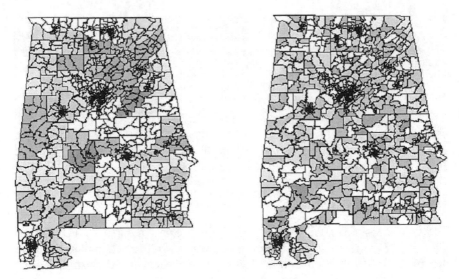

Figure 10.6 Typical County Cartographic Boundary File (CBF) for Alabama, showing county boundaries and census tracts. (*Source:* www.census.gov/geo/www/cob/tr2000.html, 12/27/03.)

10.6.4 Geographic Names Information System (GNIS)

USGS (2003a) has developed the *Geographic Names Information System* (*GNIS*), which has three databases: the National Geographic Names Data Base, the USGS Topographic Map Names Data Base, and the Reference Data Base. These databases can be downloaded and used as labels, text, or annotation to aid in identifying features on the base data set. Information on these three datasets can be found at http://geonames.usgs.gov/index.html.

10.6.5 USGS National Map

The U.S. Geological Survey is in the constant process of developing the "National Map," a consistent framework for nationally current, accurate, consistent geospatial digital data, and topographic maps derived from those data. Working with the various governmental and other partners, its goal is to develop a seamless integrated set of data and standardize symbols, and so on, thus the National Map.

The National Map makes it possible to access, integrate, and apply this geospatial data (USGS, 2003). Available online is The National Map (TNM) data viewer, which allows access to seamless national datasets and high-resolution orthoimagery of selected urban areas from USGS and its partners. Unique about the TNM data viewer is that it simultaneously accesses data layers from multiple sources to display them onto the user's system.

Some of the seamless data includes:

- Urban Areas High-Resolution Orthoimagery
- National Elevation Dataset (NED)
- National Land Cover Dataset (NLCD) 1992
- Shuttle Radar Topography Mission (SRTM)

The TNM data viewer is shown in Figure 10.7.

The *National Hydrography Dataset* (*NHD*), discussed later in this chapter, is one product from this effort. The USGS Digital Terrain Elevation Dataset Level 1 and the USGS National Elevation Dataset are represented as shaded relief (color and gray-scale) maps; and the USGS National Land Cover Dataset—also discussed later—is the source of land cover information. Part of the effort is to allow users to access mapped data through the standard Web browser without the need for GIS software.

The National Map program also obtains data from the new *National Atlas of the United States* updates, and digitizes a large, bound collection of paper maps that was published in 1970. Data available from the National Atlas of the United States can be found at www.nationalatlas.gov/, 2003. The interactive screen can be found in Figures 10.8 and 10.9.

10.6.6 Streams and/or Hydrography

Hydrography is the scientific study of seas, lakes, and rivers, especially the charting of tides and changes in coastal bathymetry or the measurement and recording of river flow. In addition to the hydrography data mentioned earlier, several hydrography data layers may be obtained, typically on a statewide

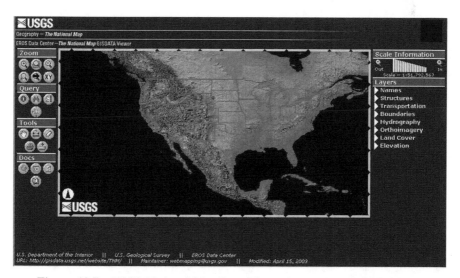

Figure 10.7 USGS National Map Data Viewer home page (USGS, 2003).

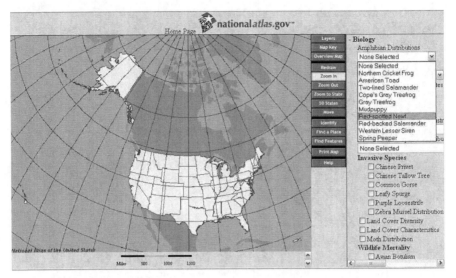

Figure 10.8 Interactive National Atlas of the United States: display.

level; the layers may vary depending on individual goals and objectives of the watershed plan. A stream coverage could also be developed through standard routines in the GIS from digital elevation models (DEMs).

10.6.6.1 National Hydrography Dataset (NHD). One comprehensive source of digital spatial surface water data (lakes, ponds, streams, rivers,

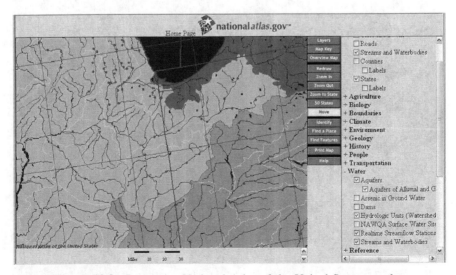

Figure 10.9 Interactive National Atlas of the United States: results.

springs, and wells) is the aforementioned National Hydrography Dataset (NHD) (USGS, 2003b) developed jointly by the USGS and EPA. Initially based on 1:100,000-scale data, the latest NHD data is based upon the USGS Digital Line Graph (DLG) hydrography dataset and is integrated with EPA's reach-related information from the Reach File Version 3 (RF3) (U.S. EPA 2003), categorized by *cataloging units*. The cataloging unit is a geographic area that subdivides the accounting units within *hydrologic units (HUC)*.

Within the NHD, surface water features are combined to form "reaches," which provide the framework for linking water-related data to the NHD surface water drainage network. From RF3, the NHD acquires hydrographic sequencing, upstream and downstream navigation for modeling applications, and reach codes. The reach codes provide a way to integrate data from organizations at all levels, by linking the data to this nationally consistent hydrographic network. The feature names are from the Geographic Names Information System (GNIS). These linkages enable the analysis and display of these water-related data in upstream and downstream order. A sample of the NHD data is shown in Figure 10.10.

According to the USGS, the characteristics of the National Hydrography Dataset are as follows (http://mac.usgs.gov/isb/pubs/factsheets/fs10699. html, 2003:

- It is a feature-based dataset that interconnects and uniquely identifies the stream segments, or "reaches," that make up the nation's surface water drainage system.

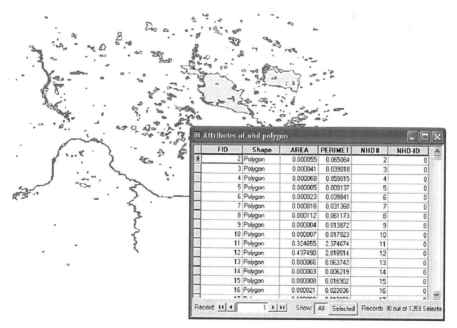

Figure 10.10   NHD water feature polygons and attributes.

- Unique reach codes (originally developed by the U.S. EPA) are provided for networked features and isolated water bodies.
- The reach code structure is designed to accommodate higher-resolution data.
- Common identifiers uniquely identify every occurrence of a feature.
- It is currently based on the content of the USGS 1:100,000-scale data, giving it accuracy consistent with those data.
- Data are in decimal degrees on the North American Datum of 1983.
- Names with GNIS identification numbers are included for lakes, other water bodies, and many stream courses.
- It provides flow direction and centerline representations through surface water bodies.

Because the reaches are "connected"—that is, each reach is coded to realize which are upstream and downstream from it—hydraulically connected modeling is possible, which will become even more powerful in the future as models are developed to utilize this feature. NHD data is available by *sub-basin* (formerly known as cataloging unit) and can be downloaded at no charge from: http://nhd.usgs.gov/data.html.

10.6.6.2 HYDRO1k. HYDRO1k is a geographic database providing consistent, comprehensive, and seamless (by continent) global coverage of topographically derived, hydrologic-related, georeferenced datasets at a resolution of 1 kilometer (1km). HYDRO1k is developed at the U.S. Geological Survey's (USGS) Earth Resources Observation Systems (EROS) Data Center, and is derived from that agency's recently released 30 arc-second digital elevation model (DEM) of the world (GTOPO30). The HYDRO1k package provides six raster and two vector datasets and many of the common derivative products used in hydrologic analysis, as delineated in Table 10.3 and shown in Figure 10.11.

TABLE 10.3 HYDRO1k Hydrologic Datasets

Raster	Vector
Hydrologically correct DEM	Streamlines
Derived flow directions	Drainage basins
Flow accumulations	
Slope	
Aspect	
Compound topographic (wetness) index	

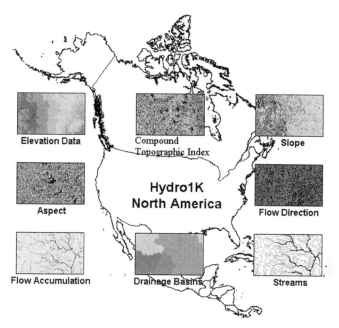

Figure 10.11   HYDRO1k datasets (USGS, 2003).

The North American HYDRO1k data can be downloaded from: http://edcdaac.usgs.gov/gtopo30/hydro/.html, and detailed descriptions of the techniques used in the development of the HYDRO1k dataset can be found in the Readme file on the Web site.

10.6.6.3 Environmental Protection Agency Reach File. The EPA Reach File (RF3, Version 3.0), developed from USGS 1:100,000-scale hydrography data by the Environmental Protection Agency, is also a national hydrologic database. The RF3 interconnects and uniquely identifies 3.2 million stream segments, or "reaches," that comprise the country's surface water drainage system (Dewald and Olsen, 1994). Each stream segment is assigned a unique reach code, which determines the upstream/downstream relationships of each reach. This provides a common nomenclature for reporting surface water conditions under the Clean Water Act and provides a nationally connected hydrologic transport network. The connected stream network also makes it possible to hydrologically model waterborne pollution associated with both point and nonpoint sources.

The compilation aspect of RF3 production involves combining relevant portions of Reach File Versions 1.0 and 2.0 (RF1 and RF2, respectively), the USGS Geographic Names Information System (GNIS) database, and the 1988 USGS 1:100,000-scale Digital Line Graph (DLG) hydrography (Dewald, et al., 1985). The RF3 aids EPA's STOrage and RETrieval (STORET) program, which is the EPA's national water quality data management system. RF3 files may be retrieved by HUC at the BASINS Web site at www.epa.gov/waterscience/ftp/basins/gis_data/huc/.

10.6.7 Digital Raster Graphics (DRG)

DRGs are scanned digital images of the standard USGS 7½-minute topographic quadrangles. They are an image data source saved as a Tag Image File Format (TIFF) and are extremely useful in analyzing drainage areas, terrain, slopes, stream network verification, and watershed delineation, and sometimes serve as an excellent base map for watershed analysis. DRGs can be found through links from the USGS GIS Web site, www.usgs.gov, or the ftp site, ftp://ftp.emtc.usgs.gov/pub/gis_data/. The data is now commonly distributed at each state's digital geographic data distribution Web site or state university. The digital raster graphics are in an Universal Transverse Mercator (UTM) grid coordinate system. A sample DRG is shown in Figure 10.12.

10.6.8 Digital Elevation Models (DEMs)

DEMs are extremely useful in watershed analysis, as they are records of terrain elevations at regularly spaced horizontal intervals placed in a raster or grid-based GIS data source, as shown in Figure 10.13. The most common DEMs are based on USGS topographic 7½-minute quadrangles with a 30-meter resolution and UTM projection. Most DEMs give single values for each location. These values are often considered to represent the Earth's surface

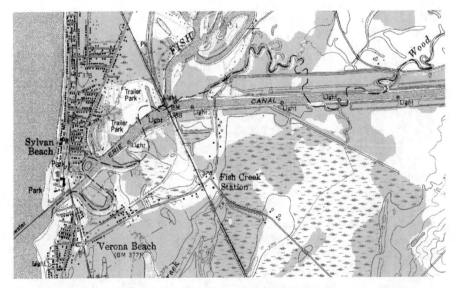

Figure 10.12   Portion of a USGS DRG.

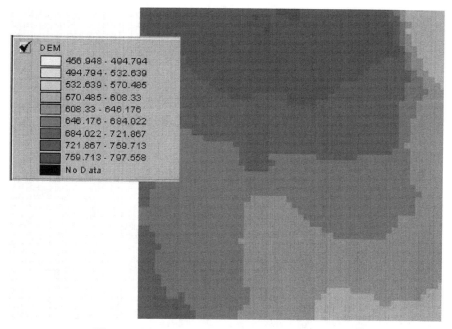

Figure 10.13 DEM showing individual grid cells.

at that "point." Each cell has an average elevation of that 30-meter cell. Elevation units are in meters or feet relative to National Geodetic Vertical Datum of 1929 (NGVD 29) in the continental United States. Utilizing the DEMs, a variety of hydrologic and terrain analyses can be performed, such as slope analysis, flow direction, and accumulation and three-dimensional (3-D) modeling. These will be described further in Chapter 17. The USGS is in the process of developing 10-meter DEMs, which are accessible from the Web site, www.usgs.gov, as they become available. The 10-meter DEMs provide much more accurate processing, as they contain nine cells of differing elevations for the previous one 30-meter DEM cell. A typical 3-D image of the watershed terrain is shown in Figure 10.14.

Vertical accuracy of DEM data is dependent upon the spatial resolution (horizontal grid spacing), quality of the source data, collection and processing procedures, and digitizing systems. As with horizontal accuracy, the entire process, beginning with project authorization, compilation of the source datasets, and the final gridding process, must satisfy accuracy criteria customarily applied to each system. Each source dataset must qualify to be used in the next step of the process. Errors have the effect of compounding for each step of the process; therefore, production personnel are directed to account for each production step leading to the final DEM.

10.6.8.1 *GTOPO30.* The U.S. Geological Survey's Earth Resources Observation Systems (EROS) Data Center, in conjunction with the National Aer-

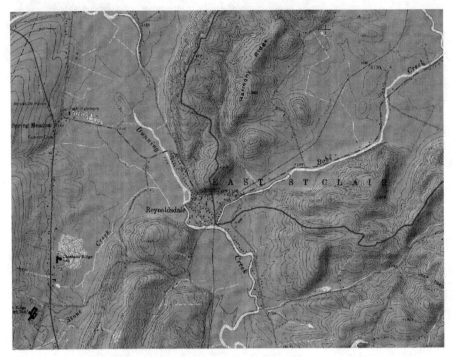

Figure 10.14 Watershed DEM with a DRG overlaid on top.

onautics and Space Administration (NASA) and other national and international agencies, completed the GTOPO30 global digital elevation model (DEM) in late 1996 (USGS, 2003). GTOPO30 was derived from several raster and vector sources and has a horizontal grid spacing of 30 arc seconds (approximately 1 kilometer). The vertical accuracy of GTOPO30 varies by location according to the source data; however, the areas derived from the USGS DEM raster source data have a vertical accuracy of plus or minus 30 meters linear error at the 90 percent confidence level. It is too broad for localized watershed assessment, but may be helpful in large-scale basin analysis. GTOPO30 can be downloaded by tiles, as shown in Figure 10.15 from the USGS's anonymous ftp site, ftp://edcftp.cr.usgs.gov/. More specific information on GTOPO30 may be found in the Readme file at the Web site.

10.6.9 Digital Line Graphs (DLGs)

Digital Line Graphs (DLG) are vector files containing planimetric data, such as roads and streams, digitized from USGS topographic maps and related sources. They are, therefore, a digital cartographic representation of the USGS 7½-minute quadrangles, which, depending on scale, consist of U.S. Public Land Survey System, Boundary, Transportation, Hydrography, Hypsography, and other layers as listed in Table 10.4.

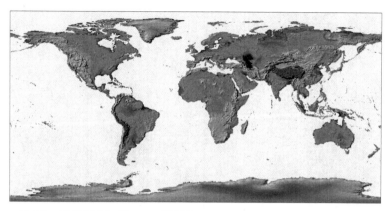

Figure 10.15   GTOPO30. (*Source:* USGS, http://edcdaac.usgs.gov/gtopo30/gtopo30. html, access 12/27/03.)

The products are archived and available through the USGS and offer a full range of attribute codes; they are accurate and are topologically structured, which makes them ideal for use in GIS. DLGs are available in Optional format, written as ANSI-standard ASCII characters in fixed-block format. A sample DLG is shown in Figure 10.16.

10.6.10   Digital Orthophoto Quarter Quadrangles (DOQQs)

DOQQs are scanned color-infrared or black-and-white digital images, 1-meter ground resolution, orthophoto-rectified photos of one-quarter of a USGS 7½-minute quadrangle (or 3.75 minute). They are also a product of the United

TABLE 10.4   DLG Layers and Feature Types

Layer	Feature Type
Public Land Survey System (PLSS)	Township, range, and section lines
Boundaries	State, county, city, and other national and state lands such as forests and parks
Transportation	Roads and trails, railroads, pipelines, and transmission lines
Hydrography	Flowing water, standing water, and wetlands
Nonvegetative features	Glacial moraine, lava, sand, and gravel
Survey control and markers	Horizontal and vertical monuments (third order or better)
Man-made features	Cultural features, such as buildings not collected in other data categories
Vegetative surface cover	Woods, scrub, orchards, and vineyards

Source: USGS, http://edcwww.cr.usgs.yor/products/map/dlg.html.

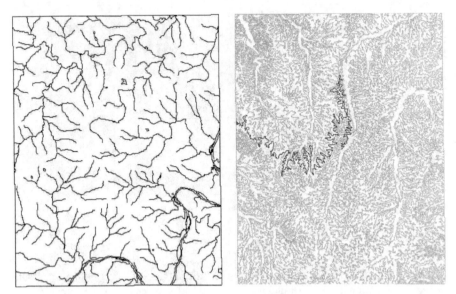

Figure 10.16 USGS Hydrology and Hypsography Digital Line Graphs (DLG), based on 7½ minute quadrangles. (*Source:* www.webgis.com/terr_pages/MO/dlgutm/miller.html, 1/27/03.)

States Geological Survey (USGS, 2003). The rectification removes the displacements caused by the camera and the terrain. A DOQ combines the image characteristics of a photograph with the geometric qualities of a map; and because it is a photographic image, the digital orthophoto displays features that may be omitted or generalized on other cartographic maps. This makes the digital orthophoto valuable as a layer in a GIS base map or for updating other map materials such as DLGs and topographic maps. It is also an excellent source for a watershed's land use and land cover. Compressed DOQ county files are distributed in JPEG format on CD-ROM. An example of a DOQQ at the resolution to be viewed and zoomed to show the pixelization can be seen in Figure 10.17.

10.6.11 Satellite Images and Satellite-Derived Data

The USGS has substantial satellite coverage of the United States available, including satellite photographs from NASA; digital satellite images, among them Advanced Very-High Resolution Radiometer (AVHRR) images, Land Satellite (Landsat) Thematic Mapper (TM), Landsat 7 Enhanced Thematic Mapper (ETM+) Data, and Multispectral Scanner (MSS) digital data. Photo products derived from airborne and satellite sensor data also include color and black-and-white photographic products generated from Side-Looking Airborne Radar (SLAR), and Systeme Probatoire d'Observation de la Terre

Figure 10.17 Digital orthophoto quarter quadrangles (DOQQ) and pixelization.

(SPOT) imagery digital data (USGS). Several commonly used products are summarized in the following subsections.

10.6.11.1 *Landsat 7 ETM+ and Thematic Mapper Data.* Landsat 1 was launched in 1972 and, since then, the use of *remotely sensed digital images* has been steadily increasing. Landsat 7 satellite data came online in July 1999. Landsat land cover data is remotely sensed (satellite) image data of the Earth's surface. It provides the spatial information and synoptic view that allows interpretation of land use changes or conflicts, such as between development and the environment (wetlands, prime agricultural land, wildlife habitat, steep slopes). For instance, loss of prime agricultural land to urban sprawl can be tracked. Landsat 7 ETM+ data monitors ecologic, geographic, geologic and hydrographic, and land use change (urbanization) processes. The differences in spectral reflection allow identification of building rooftops, parking lots, concrete surfaces, and asphalt roads, and makes it possible to visually determine suspended sediment plumes in streams, rivers, and bays caused by wash-off from earth disturbance, as shown in Figure 10.18. It is also useful for modeling purposes.

Landsat data are *remotely sensed multispectral data,* meaning they have several bands of color. Multispectral image data are digital image data taken of a given scene in a number of different spectral bands at once. The Thematic Mapper of the U.S. Landsat series of satellites has seven spectral bands. Other

Figure 10.18 Landsat ETM+ Data of the San Francisco Bay Area showing urbanization (USGS, 2003).

examples of existing sources of such data are from the Multispectral Scanner (four spectral bands) and from the French SPOT (three spectral bands), which is discussed a little later. Color bands can be turned on or off to analyze the scene in a variety of different spectral combinations. Certain features, such as suspended sediment in streams, may be able to be seen under varying spectral combinations better than others.

Another advantage of remotely sensed satellite images and data is that they allow a multitemporal analysis for change detection (Chavez and MacKinnon, 1994). In other words, they provide the user with the ability to look at image data taken several years apart to determine environmental impacts, as shown in Figure 10.19.

10.6.11.2 Advanced Very High Resolution Radiometer (AVHRR) (Global Land Cover Characterization Background). Advanced Very High Resolution Radiometer (AVHRR) data was collected from April 1992 through September 1996 (Eidenshink and Faundeen, 2003) and includes 1-kilometer resolution global land cover characteristics. Since the data was collected globally, it is on the Interrupted Goode Homolosine and Lambert Azimuthal Equal Area projections. The U.S. Geological Survey, with NASA and the Earth Resources Observation System (EROS) Data Center and many other national and European agencies, have partnered to develop this "global land cover database." The major uses of the AVHRR data are related to the study of surface vegetation cover. The global land cover characterization data has been manipulated to represent the following thematic maps (USGS, 2003h):

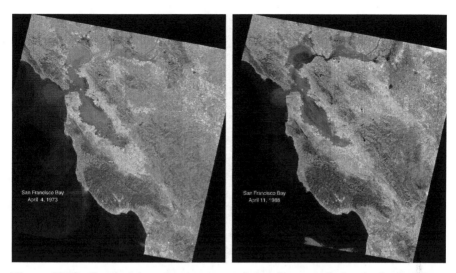

Figure 10.19 Landsat data showing suspended sediment pollution in the San Francisco Bay (USGS, 2003g). Landsat satellite images are used to identify and map wetlands and water quality changes that have occurred during the past 20 years in and around the San Francisco Bay. (Satellite image supplied by Pat Chavez, USGS.)

- Seasonal land cover regions
- Global Ecosystems (Olson, 1994a, 1994b)
- International Geosphere Biosphere Programme Land Cover Classification (Belward, 1996)
- USGS Land Use/Land Cover System (Anderson et al., 1976)
- Simple Biosphere Model (Sellers et al., 1986)
- Simple Biosphere 2 Model (Sellers et al., 1996)
- Biosphere-Atmosphere Transfer Scheme (Dickinson et al., 1986)
- Vegetation Lifeforms (Running et al., 1995)

The North America Land Cover Characteristics Data Base, Version 2.0 (http://edcdaac.usgs.gov/glcc/glcc.html), is shown in Figure 10.20; the global land cover database is shown in Figure 10.21.

10.6.11.3 *Digital Globe Data.* The three-dimensional images of Baghdad during Operation Iraqi Freedom were from Digital Globe, a commercial vender that has its own satellite, the Quickbird. This data, too, can be utilized to pinpoint types of sources of pollution, as shown in Figure 10.22. This imagery is high-resolution satellite 61-centimeter (2 feet) panchromatic or 2.44-meter (8 feet) multispectral data.

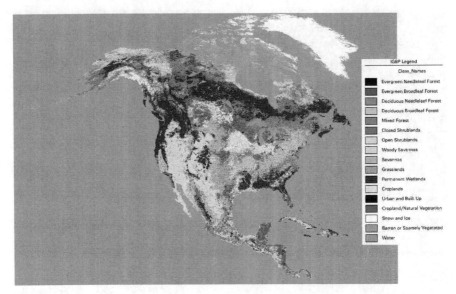

Figure 10.20 The North America Land Cover Characteristics Data Base, Version 2.0. (*Source:* http://edcdaac.usgs.gov/glcc/glcc.html, 12/27/03.)

10.6.11.4 Systeme Probatoire d'Observation de la Terre (SPOT) Imagery Digital Data.
The SPOT Satellite System provides high-quality medium resolution imagery, as shown in Figure 10.23. The SPOT 5 satellite, launched in May 2002, enhanced the SPOT system by offering higher-resolution products while retaining the wide field of view, excellent operational capacity, and image quality of its predecessors. SPOT 5 offers 2.5-meter

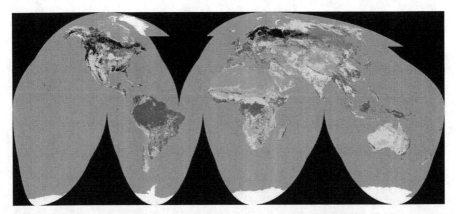

Figure 10.21 Global land cover database. (*Source:* http://edcdaac.usgs.gov/glcc/global2_img.html, 12/27/03.)

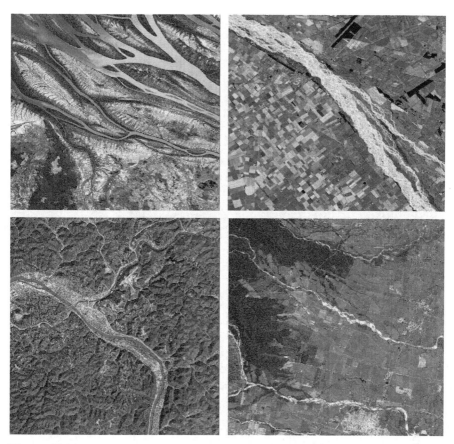

Figure 10.22 Digital Globe image data showing suspended sediment pollution in Madagascar, flooding relief, and flow patterns (QuickBird imagery courtesy of Digital Globe).

and 5-meter resolution in *panchromatic* (black and white) mode, 10-meter resolution in multispectral (color) mode in bands B1 (green), B2 (red), and B3 (near IR), and 20-meter resolution in the short-wave IR band and 10-meter resolution panchromatic stereopairs (Digital Globe, 2003; SPOT, 2003). The data comes in 60 kilometer by 60 kilometer tiles or 60 kilometer by 120 kilometer, when the two High Resolution Geometric (HRG) instruments are used. The *location accuracy* is better than 50 meters, without ground control points, allowing their use for applications that require scales from 1:25 000 to 1:10 000.

10.6.12 Aerial Photography

Aerial photographs are an important tool in watershed assessments and can be utilized by a number of disciplines, including hydrologists, biologists, land

Figure 10.23   SPOT digital imagery. (*Source:* SPOT, 2003.)

use planners, real estate developers, environmental specialists, and many other professionals that rely on detailed and timely aerial photographs. Aerial photographs currently supply a more detailed resolution than satellite images; however, their digital density is much greater, often making them cumbersome to work with due to regeneration time on the computer screen. An aerial survey company would be hired to fly over and photograph the area (if the area had not already been flown), then the photographs would be converted into a digital format. There are times, however, where the detail of aerial photographs will be worth the additional expense.

Until 1975, there was no systematic approach for locating an aerial photograph, or series of photographs, quickly and easily. In 1975, the U.S. Geological Survey (USGS) initiated the Aerial Photography Summary Record System (APSRS), which describes aerial photography projects that meet specified criteria over a given geographic area of the United States and its territories. Each listing is a summary of aerial photography projects within a 7.5-by-7.5-minute quadrangle area, corresponding to the USGS 7.5-minute map series. Entries are sorted by project date and describe the scale, project code, file type, cloud cover, and camera focal length. The entries also give the name of the holding agency or firm.

Aerial photography vendors can deliver high-resolution digital aerial photographs at a scale and resolution specified by the user or client. When performing a large-scale watershed assessment project, one may have the watershed flown at a higher level and a product developed at a lower resolution scale than for a small-scale watershed assessment, for example, a borough storm sewer assessment.

10.6.13 Existing Land Use

One important aspect of any watershed assessment is its land use and/or land cover. Land use is the present use of the land and is defined by Clawson and Stewart (1965) as "man's activities on land which are directly related to the land." Land cover is the type of vegetation or man-made material covering the land surface. These terms are often combined to provide the Land Use Land Cover (LULC) classification system. Land use is important in watershed assessments to determine urban growth, determine land suitability for new development, project future growth patterns, determine percent imperviousness, evaluate surface roughness for hydrologic modeling, aid in determining infiltration, determine hydrologic travel time, and evaluate possible sources of stream reach stress. Land use can be derived by one of several methods depending on the intended use—if, for example, one is looking primarily at an impervious area, one needs land use classifications according to the TR-55 methodology for hydrologic modeling or other purposes.

Land use or land cover can be derived from interpretation of the satellite image data and aerial photography data described previously, or obtained from already digitally published formats. The LULC data provides a key component during watershed land surface analysis in addressing planning issues such as:

- Which portion of the watershed is eligible to absorb population growth?
- Which portion of the watershed is susceptible to acid mine drainage?
- Which portion of the watershed is suitable for future development based on the presence of natural resources such as level terrain or water?
- Which portion of the watershed should be zoned for nondevelopment?
- Which portion of the watershed is most vulnerable to urbanization?

10.6.13.1 USGS (Anderson). The USGS has been the national and international leader in LULC innovation in mapping, and part of that effort has been to address the standardization of the LULC classification system with the publication of paper no. 964 (Anderson et al., 1976). In this paper, James R. Anderson and his colleagues laid the framework for a standardized LULC classification system for use on a national level, but left it open-ended so that state, regional, and local agencies could develop more detailed LULC maps. Note: This paper was an update to USGS Circular 671 (Anderson, 1972), therefore the classification system that uses paper no. 964 is often referred to as the "modified Anderson classification system." As one can imagine, on a spatial basis, classifying land use is somewhat subjective. The modified Anderson method categorizes the types of land use and land cover according to land capability and its vulnerability to certain management practices.

The evolution of this system is interesting. In June 1971, the Conference on Land Use Information and Classification was held in Washington, D.C., at which Anderson and others presented their ideas for standardizing LULC classifications. Anderson's proposal used remote sensing as the primary source for LULC development. Others, from the New York State Land Use and Natural Resources Inventory at Cornell University, used aerial photography and had well-established classification systems. The team of Anderson, Ernest Hardy, and John Roach came together to modify the New York system to be adaptable on a national level; the result: Circular 671.

The modified Anderson classification system, as described in paper 964, utilizes a hierarchical classification, which is shown in Table 10.5.

The use of each level of data is dependent on the user's budget, project area, and required resolution. Thus, levels I and II would be for large-sized projects on a statewide level and larger; levels III and IV would be more appropriate on a watershed level, for regional studies, and by counties and municipalities. The original Anderson land classification system looks like the partial data given in Table 10.6 (Anderson, 1976), while the newer modified Anderson land classification system, which breaks the LULC into six more detailed classifications, appears in Table 10.7 (USGS, 1997).

As a part of its National Mapping Program, the United States Geological Survey (USGS) originally developed Land Use and Land Cover (LULC) datasets and associated maps. The EPA converted the USGS's Land Use/Land Cover data, Anderson levels I and II, into ARC/INFO digital format at 1:250,000 scale. It is now available in two formats, vector and raster, or grid, cell. The spatial data is distributed through the Geographic Information Retrieval Analysis System (GIRAS). The grid cell data is available in a Composite Theme Grid (CTG) format.

The (LULC) datasets describe the water, vegetation, cultural, and natural surface features on the land surface. Each original feature, with the smallest being about 10 acres (4 hectares) in size, has coded attribute data to distinguish it. The original source of data was developed through manual interpretation and digitization of NASA's high-altitude aerial photographs. This process has been supplemented with other existing land use maps and field surveys. And because the data was digitized, it follows polygon topology. The original data conforms to the 1:250,000-scale accuracy, and the original data is in the UTM projection. The interpreted data is compiled onto the 7½-

TABLE 10.5   Anderson Classification Levels

Level	Data Characteristics	Flight Height	Scale
I	Landsat (formerly ERTS)		
II	High-altitude	40,000 ft (12,400 m) or above	<1:80,000
III	Medium-altitude	10,000–40,000 ft (3,100–12,400 m)	1:20,000–:80,000
IV	Low-altitude	<10,000 ft (3,100 m)	>1:20,000

TABLE 10.6 Anderson Hierarchical Land Use Classifications

Level I	Level II	Level III
1 Urban or Built-up Land	11 Residential	111. Single-Family Units
		112. Multifamily Units
		113. Group Quarters
		114. Residential Hotels
	12 Commercial and Services	
	13 Industrial	

Classification Codes: First- and Second-Level Categories

 1 Urban or Built-Up Land
11 Residential
12 Commercial and Services
13 Industrial
14 Transportation, Communications, and Utilities
15 Industrial and Commercial Complexes
16 Mixed Urban or Built-up Land
17 Other Built-up Land
 2 Agricultural Land
21 Cropland and Pasture
22 Orchards, Groves, Vineyards, Nurseries
23 Confined Feeding Operations
24 Other Agricultural Land
 4 Forest Land
41 Deciduous Forest Land
42 Evergreen Forest Land
43 Mixed-Forest Land
 5 Water
51 Streams and Canals
52 Lakes
53 Reservoirs
54 Bays and Estuaries
 6 Wetlands
61 Forested Wetlands
62 Nonforested Wetlands
 7 Barren Land
72 Beaches
73 Sandy Areas Other Than Beaches
75 Strip Mines, Quarries, and Gravel Pits
76 Transitional Areas

TABLE 10.7 Modified Anderson Hierarchical Land Use/Land Cover Classifications and Category Definitions

Level 1	Level 2	Level 3	Level 4	Level 5	Level 6
1.0 Water	1.1 Open Water	1.11 Stream/river 1.12 Canal/ditch	1.121 Lined canal/ditch 1.122 Unlined canal/ditch		
		1.13 Lake/pond 1.14 Reservoir 1.15 Bay/estuary 1.16 Sea/ocean			
	1.2 Perennial Ice/ Snow	1.21 Snowfield 1.22 Glacier			
2.0 Developed	2.1 Residential	2.11 Single-family residential 2.12 Multi-family residential			
	2.2 Non-residential Developed	2.21 Commercial/ Light Industry	2.211 Major Retail 2.212 Mixed/Minor Retail and Services 2.213 Office 2.214 Light industry		
		2.22 Heavy Industry	2.221 Petro-chemical Refinery		
		2.23 Communications and Utilities			
		2.24 Institutional	2.241 Schools 2.242 Cemeteries		

2.25 Agricultural Business
  2.252 Confined feeding
2.26 Transportation
  2.261 Airport
2.27 Entertainment/ Recreation
  2.271 Golf Course
  2.272 Urban Parks
2.3 Mixed Urban

3.0 Bare
  3.1 Transitional
  3.2 Quarries/Strip mines/Gravel pits
  3.3 Bare Rock/ Sand
  3.4 Flats
  3.5 Disposal

4.0 Vegetated
  4.1 Woody
    4.11 Forested
      4.111 Deciduous
      4.112 Evergreen
      4.113 Mixed

TABLE 10.7 *(Continued)*

Level 1	Level 2	Level 3	Level 4	Level 5	Level 6
		4.12 Scrubland	4.121 Deciduous 4.122 Evergreen 4.123 Mixed 4.124 Desert scrub		
		4.13 Orchards/vineyards/groves	4.131 Irrigated orchard/vineyards/groves		
	4.2 Herbaceous	4.21 Natural Herbaceous	4.211 Natural grassland		
		4.22 Planted/cultivated	4.221 Fallow/bare fields		
			4.222 Small grains	4.2221 Irrigated small grains	
			4.223 Row crops	4.2231 Irrigated row crops	
			4.224 Planted grasses	4.2241 Pasture/hay	4.22411 Irrigated Pasture/hay
				4.2242 Other grass	4.22421 Irrigated Other grass
			4.225 Irrigated planted/cultivated		
	4.3 Wetlands	4.31 Woody wetlands 4.32 Emergent wetlands			

minute topographic quadrangle basis, and merged if required for a particular watershed project or urban area. Each quadrangle of land use data may have a different representative date. Date ranges from the mid-1970s to the early 1980s are common. When joined together, these quadrangles probably will not match along edges due to differences in interpretation and time coverage. Likewise, edges of each map file were manually digitized and so may not join neighboring maps.

USGS's new initiative, through its National Mapping Division (NMD) has an ongoing Land Cover Characterization Program (LCCP) whose purpose is to develop high-resolution LULC data. The four components of the program are global (1:2,000,000 scale), national (1:100,000 scale), urban (1:24,000 scale), and special projects that may map at a resolution higher than 1:24,000 scale. The LULC features are often determined from 1-meter resolution digital orthophoto quadrangles (NASA high–altitude missions) and supplemented with interpretation from USGS topographic quadrangles, available aerial photographs, and field data sources. Original interpretation was developed by manual digitization of scanned images. A typical USGS LULC map can be found in Figure 10.24.

A summary of the various land use data and the appropriate use is shown in Table 10.8.

The LULC data is available for download from an anonymous File Transfer Protocol (FTP) account at http://mrdata.usgs.gov or from http://edcwww.cr.usgs.gov/glis/hyper/guide/1_250_lulcfig/states100k.html.

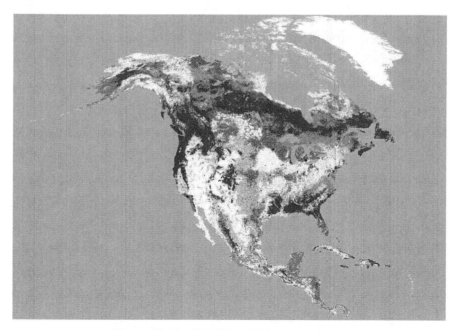

Figure 10.24 USGS Land Use/Land Cover.

TABLE 10.8   Land Use, Format, and Applicability

Source	Format	Applicability
Planimetric digitized from aerial photography	Vector	Hydrologic Modeling
LandSAT	Raster	Hydrologic Modeling
USGS LULC	Raster	Hydrologic Modeling

The USGS's Rocky Mountain Mapping Center (RMMC) is currently developing high-resolution LULC data for the urban and special projects components. The RMMC's primary involvement with the Land Cover Characterization program, prior to the commencement of the USGS Director's Initiatives in fiscal year 1996 (FY96), was for the USGS Water Resources Division National Water Quality Assessment (NAWQA) program. This LULC requirement was for Intermediate LULC data interpreted from Landsat Thematic Mapper (TM) source imagery and Large-Scale (high-resolution) LULC data interpreted from USGS digital orthophoto quadrangle (DOQ) source imagery. Large-scale LULC data collected in support of the NAWQA program employed a 2.5-acre minimum mapping unit (MMU) for water, a 5-acre MMU for urban areas, and a 10-acre MMU for rural areas. The high-resolution LULC data compiled for NAWQA used the DRAFT digital line graph (DLG-3) dual classification schema within urban areas. For example, an urban polygon received both a land use code as well as a land cover code. For the new initiative, up to six levels have now been added to this LULC data since Anderson's original paper. As a matter of comparison, Level 1 has four categories, whereas Level 6 has more than 75 categories. An example of this system can be found in Table 10.7. High-resolution features include a minimum polygon width of 125 feet and a minimum mapping unit of between 2.5 and 5 acres.

10.6.13.2 *Natural Resources Conservation Service (NRCS) or SCS Land Use Classification.* Another example of a land use classification system is described in the Soil Conservation Service's (SCS) Technical Release 55 (TR-55) (SCS, 1986) as shown in Tables 12.6a and 12.6b, Chapter 12. In this particular case, the ultimate outcome is curve numbers for hydrologic modeling; therefore, specific categories of land use have to be classified. A good source of land use for TR55 classification would be development of the land use coverage directly via screen digitizing of the USGS digital orthophoto quadrangles, or, if available, more accurate digital orthophoto information. Many land use maps can be developed from aerial photography that can be obtained either by hiring an aerial photography consultant or by obtaining digital satellite data as described earlier. As has been pointed out, land use can be classified in several ways, depending again on the purpose of the assessment. This system and the modified Anderson method should be coordinated so that hydrologists and engineers who need a watershed's land use defined for hydrologic modeling purposes can utilize the Anderson map.

10.6.13.3 Tax Parcel Land Use Source. Many counties and municipalities classify land use based on the tax parcel identification or SIC code. Caution should be used when utilizing a tax parcel base for land use/land cover, as a tax parcel database reflects land use better than land cover. This procedure may classify large tracts of land that are only partially developed as commercial or industrial areas and where the majority of the parcel might be forested. For instance, an industry may own 25 acres, 5 of which are developed with the remaining 20 wooded. However, the tax parcel in the GIS database may show it as industrial. Thus, displaying a land use map based on this attribute would not provide a correct reflection of the actual land cover. For residential land developments, a tax parcel database is typically suitable. However, the road right-of-way as shown in a tax parcel database includes the roads, the shoulder, and undeveloped portions of the right-of-way that may be wooded or grassed or be another pervious area. When examining an impervious area in the watershed, this must all be taken into account when designating percent impervious areas for right-of-ways.

10.6.14 Physiographic Provinces

Physiographic provinces data are supplied by USGS and can be downloaded from http://water.usgs.gov/lookup/getspatial?physio (USGS 2003d). This data is a polygon coverage in ArcInfo export format; the coverage is the Physiographic Divisions in the conterminous United States, which is based on eight major divisions, 25 provinces, and 86 sections representing distinctive areas having common topography, rock types and structure, and geologic and geomorphic history, as developed from Fenneman and Johnson's "Physical Divisions of the United States" 1:7,000,000-scale map (1946). It is recommended that the local USGS or EPA regional office be contacted to determine if there is more accurate region data coverage. The Physiographic divisions of the United States is shown in Figure 10.25.

10.6.15 Ecoregions

Ecoregions are regionalized areas exhibiting common physical geography, geology, soils, vegetation, land use, wildlife, and climate. The two systems of ecoregion classification are the U.S. Forest Service system (Bailey, 1995) and the U.S. EPA system (Omernik 1987, 1995). The U.S. Forest Service Ecoregion database can be obtained from www.fs.fed.us/institute, and is shown in Figure 10.26; the EPA ecoregions can be found on the BASINS CD or Web page (www.epa.gov/docs/ostwater/BASINS/), and are shown in Figure 10.27.

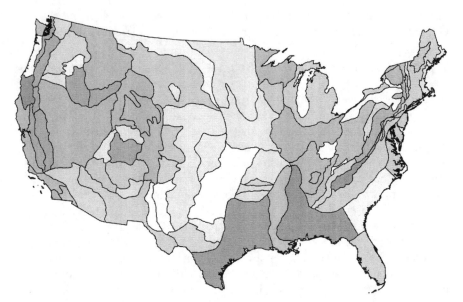

Figure 10.25 Physiographic provinces.

Divisions

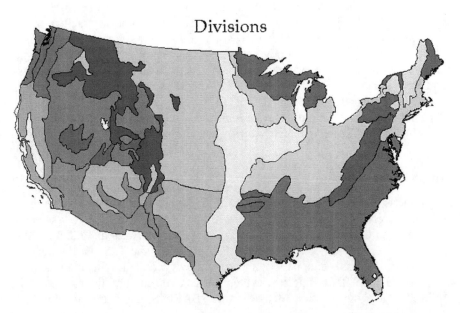

Figure 10.26 U.S. Forest Service ecoregions. (*Source:* Baily, 1975.)

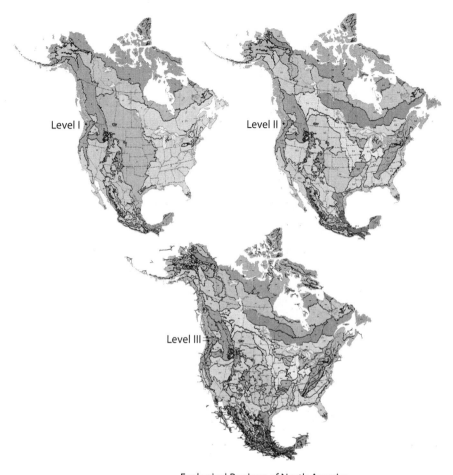

Ecological Regions of North America

Figure 10.27   U.S. EPA ecoregions (Levels I–III).

10.6.16 Geology

Digital geologic databases typically can be obtained through the state geologic surveys Web page or possibly through the USGS. A geologic map of the watershed should be plotted to look for various important features, such as limestone geology or karst terrain or strip-mined areas in alluvial or glacial deposits. Implementing a watershed plan where infiltration measures are proposed will have to be handled differently in karst terrain as opposed to other geologic formations. Therefore, a typical karst terrain map would be an important feature. It doesn't appear that there is a national clearinghouse for digital geologic data, so the user is advised to search the states' GIS clearinghouse for geologic data.

10.6.17 Soils

No watershed analysis is complete without first evaluating the soils. This book emphasizes soils because so much can be learned about a watershed by evaluating the soils data within the GIS.

There are essentially two sources of soils data: one, through field measurement, and, two, through the United States Department of Agriculture Natural Resources Conservation Service (NRCS), formally the Soil Conservation Service (SCS) Soil Survey. The soil survey reports provide a wealth of information on a county-by-county basis for the entire United States. It is recommended that the soil survey be obtained from County Conservation Districts and consulted prior to performing fieldwork to obtain a preliminary indication of the soil properties. Naturally, the goals and objectives of the particular watershed plan should be developed prior to analysis of the soil survey report. Particular properties could then be targeted and evaluated from the available information.

The NRCS developed most soil surveys in the 1960s. They investigated soils in the field and determined soils series boundaries by use of augers or shovel excavation and evaluation of landforms. Most of the data contained in the soil survey reports is now also contained in the Soil Survey Geographic (SSURGO) datasets, which are explained shortly.

The NRCS established three soil geographic databases representing different intensities and scales of mapping (USDA NRCS, 1995). Each database has a common link to a set of attribute data files for each map unit component and provides the descriptive data for each spatial database. The soils data is collected as part of the National Cooperative Soil Survey (NCSS), which has set the standard for soil mapping and classification. The data was stored in a national database named the Soil Interpretations Record (SIR); however, it is now stored in the National Soil Information System (NASIS) database.

The three soil geographic databases, in order of increasing detail, are the National Soil Geographic (NATSGO) database, the State Soil Geographic (STATSGO) database, and the Soil Survey Geographic (SSURGO) database.

NATSGO soils are on a national level and at a scale of 1:2,000,000; therefore, they should only be used for very preliminary analyses.

The STATSGO database provides coverage of the conterminous United States at a scale of 1:250,000 in a Universal Transverse Mercator (UTM) projection. This database was designed primarily for regional, multicounty, river basin, state, and multistate resource planning, management, and monitoring purposes, and is not detailed enough to make interpretations for local areas within a county. The data has been generalized or aggregated from the published soil surveys or the SSURGO maps and are similar to the "General Soil Map" in published soil surveys. The dominant features of the aggregated component data in STATSGO soils data can be developed from a weighted average of the soils' components and their properties. The STATSGO soil survey product may be used as a reference source or wide area analysis but is not designed for use in determining specific soil properties or interpretations of a site because of its multicomponent data aggregation. The STATSGO spatial data can be linked to the soil interpretations and soil properties attribute database.

SSURGO is the most detailed level of soil mapping performed by the Natural Resources Conservation Service (NRCS); it is constructed using field-mapping methods based on national standards, and are based on the published soils survey maps. Mapping scales generally range from 1:12,000 to 1:24,000, and compilation and digitizing were performed for national map accuracy standards for 1:24,000 7½-minute USGS quadrangles or plus or minus 40 feet horizontal. This level of mapping is designed for use by landowners, engineers, townships, and county natural resource planning and management agencies. The SSURGO soils are typically managed at the State Office of the NRCS; the coordinates are derived from the North American Datum of 1983 reference system, which is based on the Geodetic Reference System of 1980 on a county or soil survey area basis. The SSURGO spatial data can be linked to the NASIS attribute database. The database can be equated to the tables of data in the County Soil Surveys; it contains data on the soils' properties and interpretations, including depth-to-water table and bedrock; available water-holding capacity; flooding potential; limitations for septic systems; erodibility; and permeability. The soil database makes it possible to analyze and display various important watershed characteristics. The data becomes available to counties and areas throughout the United States and its territories as it is completed and verified.

A summary of some important properties of the three soils datasets is given in Table 10.9.

The attributes for the spatial data are included in a relational database. That is, the tables can be "joined" to the spatial data within the GIS, and includes values listed in Table 10.10.

The use of these data is not restricted by the federal government, and they are available at www.usda.nrcs.gov; however, the interpretation by organizations, agencies, units of government, or others is the responsibility of that party.

10.6.18 Wetlands

The National Wetlands Inventory Maps are available in digital format from the www.nwi.fws.gov/downloads.htm Web site. These, too, are based on the USGS 7½-minute quadrangles and are in the UTM 27 or 83 coordinate system. It should be noted that these wetland maps do have the polygons identified by wetland type. The accuracy of the wetland is plus or minus 50 feet. The wetlands maps were developed from digital satellite images and should not be construed as all-encompassing; in other words, there may be wetlands

TABLE 10.9 Summary of Important Properties of the Three Soils Datasets

	SSURGO	STATSGO	NATSGO
Initial Scale	1:12,000–1:63,360	1:250,000	1:750,000
Present Day Scale	1:12,000–1:24,000	1:250,000	1:750,000
Minimum Area Delineated		1,544 acres	major land resource area (MLRA) and land resource region (LRR) boundaries
Base Map	Orthophotos	USGS 1:250,000	1970 Census Bureau state and county digital database
Horizontal Datum			
Reference System	NAD83	NAD27	NAD27
Source	GRS80*	Clark 1866 Spheroid	State Generalized Soil Maps
Projection	UTM	Albers Equal Area	Albers Equal Area
Units:	Meters	Meters	Meters
Equivalent Soil-Mapping Unit	Soil types	Associations	State Generalized Soils

*Geodetic Reference System of 1980

TABLE 10.10   SSURGO and STATSGO Attributes

	SSURGO	STATSGO
Codes	X	X
Comp	X	X
Compyld	X	X
Forest	X	X
Helclass	X	
Hydcomp	X	
Inclusn	X	
Interp	X	X
Layer	X	X
Mapunit	X	X
Mucoacre	X	
Muyld	X	
Plantcom	X	X
Plantnm	X	X
Rangenm	X	
Rsprod	X	X
Ssacoac	X	
SSarea	X	
Taxclass	X	X
Windbrk	X	X
Wlhabit	X	X
Woodland	X	X
Woodmgmt	X	X
Yldunits	X	

that were not picked up through this process, so field identification should be conducted. A typical NWI GIS data layer can be found in Figure 10.28.

10.6.19   Hypsography

Hypsography, digital topographic data, shows the earth's features in relief: the depiction of the contours and elevations of the natural features on the surface of the land. Topographic information can be comprised of either digital raster graphics (DRGs), which are scanned, georeferenced images of the USGS topographic quadrangles previously described, or digital elevation models (DEMs), which are grid cell-based data of elevation cells (also discussed previously in this chapter). Digital line graphs (DLGs), which contain the planimetric data from USGS quadrangles, typically do not include the contours. Because there is a variety of digital topographic data and a number of descriptive terms that can be confusing, even to the experienced GIS user, the National Oceanic and Atmospheric Administration, National Data Center,

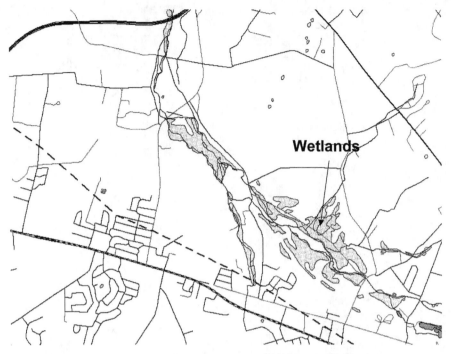

Figure 10.28   National Wetlands Inventory (NWI) digital spatial data.

National Geographic Data Committee (NGDC), has described and defined these terms and data (NOAA, 2003), which are listed here in Table 10.11.

Note the term *model* as used in DEM. Be aware that this word is used by some to mean "data" to re-create topographic elevations. Thus, care should be taken when utilizing a model to ensure that the data is detailed enough to accurately portray real-life circumstances. Again, DEMs are described in more detail earlier in the chapter.

10.6.20   Floodplains

The Federal Emergency Management Agency (FEMA), through the National Flood Insurance program (NFIP), produces Flood Insurance Study (FIS) floodplain maps and digital data for use in the NFIP and for management purposes. FEMA has now made its FIS maps available through the Digital Post Office; however, these are scanned images of the maps. (A typical map is shown in Figure 10.29.) FEMA distributes its data through the Map Service Center's (MSC's) World Wide Web site, including Digital Flood Insurance Rate Maps (DFIRMs), Flood Insurance Rate Maps (FIRMs), Flood Insurance Study reports (FIS reports), Digital Q3 flood data, Community Status Book,

TABLE 10.11   Description of Digital Topographic Data

Digital Topographic Data (DTD): Occasionally used as a generic term to describe digital topographic data.

Digital Elevation Data (DED): Same as above.

Digital Terrain Data (DTD): Same as above.

Digital Terrain Model (DTM): Same as above.

Digital Terrain Tape (DTT): Old term for early DTD produced by the U.S. Army Map Service and reformatted for distribution by the U.S. Geological Survey.

Digital Bathymetric Data (DBD): Generic term for depths in water bodies (rivers, lakes, seas, oceans).

Digital Relief Data (DRD): Generic term for elevations above and below water levels; that is, a generic term for wet and dry elevations/bathymetry.

Digital Relief Models (DRM): Same as above.

Digital Elevation Model (DEM): The most common generic acronym for digital elevation data/models. Often associated with the U.S. Geological Survey's data (and file format[s]).

Digital Terrain Elevation Data (DTED): Generally considered the specific name for digital terrain data produced by the U.S. National Imagery and Mapping Agency and its precursors (primarily the Defense Mapping Agency, and before that, the Army Map Service).

Triangulated Irregular Networks (TINs): Three-dimensional representations of surface topography. A TIN is a series of triangles or a surface representation derived from irregularly spaced sample points and breakline features. The TIN dataset includes topological relationships between points and their neighboring triangles. Each sample point has an X,Y coordinate and a surface, or Z-value. These points are connected by edges to form a set of nonoverlapping triangles used to represent the surface. TIN is also called an *irregular triangular mesh* or an *irregular triangular surface model* (ESRI, 2003).

Flood Map Status Information Service (FMSIS), Letters of Map Change (LOMC), and NFIP Insurance Manuals.

FEMA has two digital FIRM products, the Digital Flood Insurance Rate Map (DFIRM) and the digital Q3 (Quality Level 3) Flood Data. The Digital Q3 data are developed by scanning the existing FIRM hard copy and vectorizing a thematic overlay of flood risks. Vector Q3 Flood Data files contain only certain features from the existing FIRM hard copy and can be ordered from www.msc.fema.gov/q3flooda.shtml.

The Q3 Flood Data files are distributed by FEMA in three formats: DLG, ARC/INFO, and MapInfo. These formats may be accessed directly by the following GIS software packages:

- ARC/INFO
- ArcView
- ArcCAD

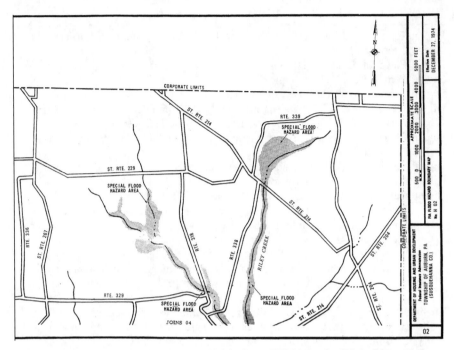

Figure 10.29   Scanned FIS map, available from FEMA at the Digital Post Office.

- MapInfo
- GENAMAP

Certain GIS software packages, such as MicroStation, provide utilities to convert DLG files into their own proprietary format. The MicroStation MGE GIS Translator module provides a utility for converting DLG files to MicroStation format (design files). Other CADD and GIS software packages, such as AutoCAD, may require the Q3 Flood Data to be translated using a third-party utility.

The DLG vector data (Q3-DLGs and DFIRM-DLGs) are in the UTM projection; the Q3 Flood Data in ARC/INFO and MapInfo formats are in geographic projection and coordinate systems; and the raster FIRMs are not georeferenced. A typical Q3 data overlain on a DOQ is shown in Figure 10.30.

10.6.21   Ecological Data

10.6.21.1   National Water Quality Assessment (NAWQA). USGS administers the National Water Quality Assessment (NAWQA) Program, which is responsible for collecting data on fish communities for 960 stream sites in

Figure 10.30   FEMA Q3 floodplain data overlaid on a DOQ.

more than 50 major river basins across the nation, which includes data on quality of water and chemical concentrations in aquatic tissue. Between 1993 and 2002, more than 1,900 fish community samples were collected; they are stored in the NAWQA Data Warehouse. This data can be downloaded from http://water.usgs.gov/nawqa/data. USGS fish community samples document the presence of all fish species and their relative abundances within designated stream reaches. A map of the watersheds for which NAWQA data is collected can be found in Figure 10.31.

10.6.21.2 GAP Analysis Program (GAP). Species diversity is fundamental to maintaining the natural ecological balance in nature. Species extinction legislation in the past has been reactive, such as the Endangered Species Act (ESA), which protects a species when it comes into danger of becoming extinct. The more proactive procedure for maintaining species diversity is to spatially identify common species' habitat and preserve those species and their habitat while they are common.

The Gap Analysis Program (GAP) takes a *biogeographic* approach to planning and managing for biodiversity. It is orchestrated by the U.S Geological Survey Biological Resources Division to identify "gaps," or missing areas that have not been assigned as areas for concern for conservation; in other

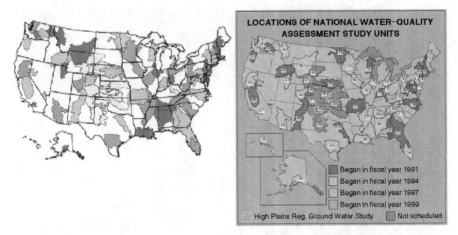

Figure 10.31 Map of NAWQA study unit (USGS, 2003).

words, it determines to what extent native animal and plant species are being protected (USGS, 2003g). The goal of the GAP program is to determine which plant and animal communities are not protected through the current network of conservation lands, based upon current land ownership. The slogan of the program is to "Keep common species common." *Common species* are those that are not threatened with extinction. The program identifies those species and their regional habitats so that they can be placed into conservation to maintain their longevity. It is acknowledged that the habitat species ranges will overlap. The program strives to protect regions rich in habitat to protect the animal species that inhabit them.

A typical application of GAP data can be found in the Iowa Rivers Information System (IRIS) project. Biological, chemical, and physical properties of streams (stream temperature, average stream flow, sinuosity, and riparian land cover) are being recorded by stream reach into the NHD stream network database (which includes attributes on stream length, stream name, route, Strahler stream order, and type) as "attributes" (IRIS, 2003). One is then able to compare relationships between fish communities, the in-stream and riparian habitat that supports them, and larger landscape-scale factors that may influence fish communities.

Aquatic GAP attribute data may include gradient, substrate, current, turbidity, physical structure, flow, temperature, aquatic vegetation, and stream type for steams and rivers. At the time of this publication, the states listed in Table 10.12 had available GAP data; data from South Carolina, Florida, and Kansas are in progress (USGS, 2003g).

Most animal species utilize a particular habitat or type of vegetation. Their range, and also vegetation type, is dependent on soils, topography, elevation, and wetlands, so these are important GIS layers to aid in the analysis. Therefore, by following the GAP process described here, range distribution can be

TABLE 10.12   States with Available GAP Analysis
Data

Arizona	Arkansas	California
Colorado	Idaho	Louisiana
Maine	Montana	Nevada
New Mexico	New York	Oregon
Pennsylvania	Utah	Virginia
Washington	West Virginia	Wyoming

predicted. The GAP process is dependent on reliable available data; in addition, field or land ownership data may be required to conclude the analysis. It general, however, the process comprises the following steps (USGS, 2003g):

1. Map the land cover types and plant community distribution.
2. Map the distribution of invertebrate and vertebrate species.
3. Overlay the previous two maps to predict species distribution.
4. Develop stewardship maps from any land ownership data (tax parcels, state or federal forest land, etc.) to determine land conservation status.
5. Assign a status ranking of 1, 2, 3, or 4 to indicate how lands are being managed. Status 1 and 2 lands are managed with conservation in mind; status 3 lands afford protection to threatened and endangered species for most of their area. They also allow extractive uses such as logging, mining, and grazing. Most private lands are considered to be status 4, unless there is a conservation easement on them.
6. Overlay the stewardship layer with the combined land cover and species distribution layer to determine if the species are adequately protected through land ownership or preservation.

GAP data is available by state and can be downloaded from www.gap. uidaho.edu/Projects/FTP.htm. A sample of GAP data can be found in Figure 10.32.

10.6.22   BASINS

Better Assessment Science Integrating Point and Nonpoint Sources, BASINS for short, is the EPA's computer assessment module that runs through ArcView. RF3 files may be retrieved by HUC from the BASINS Web site at www.epa.gov/waterscience/ftp/basins/gis_data/huc/. In addition to the RF3 files, the BASINS CD, which can be ordered by region at www.epa.gov/OST/ BASINS/, contains the base cartographic data shown in Table 10.13 (U.S. EPA, 2001). It also includes the following environmental background data shown in Table 10.14.

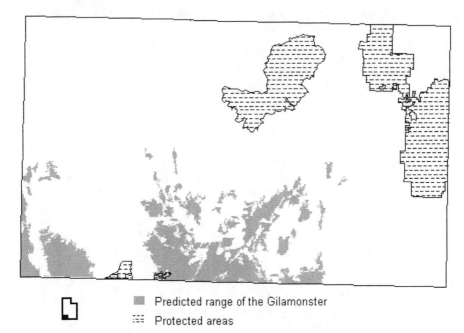

■ Predicted range of the Gilamonster

⋮⋮⋮ Protected areas

Figure 10.32 Sample GAP analysis data for the Gila Monster in southwest Utah. Notice only a very small area of its range is protected. (*Source:* www.gap.uidaho.edu/ About/Overview/IntroductionToGAShort/Protection_Afforded.htm, 12/27/03.)

TABLE 10.13 BASINS Base Cartographic Data*

Data Product	Source	Description
Hydrologic Unit Boundaries	U.S. Geological Survey (USGS)	Nationally consistent delineations of the hydrographic boundaries associated with major U.S. river basins
Major Roads	Federal Highway Administration	Interstate and state highway nework
Populated Place Locations	USGS	Location and names of populated locations
Urbanized Areas	Bureau of the Census	Delineations of major urbanized areas used in 1990 Census
State and County Boundaries	USGS	Administrtive boundaries
EPA Regions	USGS	Administrative boundaries

*From BASINS User's Manual.

TABLE 10.14 BASINS Environmental Background Data

BASINS Data Product	Source	Description
Ecoregions Level III	U.S. Environmental Protection Agency (USEPA)	Ecoregions and associated delineations
National Water Quality Assessment (NAWQA) Study Unit Boundaries	USGS	Delineation of study areas
1996 Clean Water Needs Survey	USEPA	Results of the wastewater control needs assessment by state
State Soil and Geographic (STATSGO) Database	U.S. Department of Agriculture Natural Resources Conservation Service (USDA-NRCS)	Soils information including soil component data and soils
Managed Area Database	University of California, Santa Barbara	Data layer including federal and Native American lands
Reach File Version 1 (RF1)	USEPA	Provides stream network for major rivers and supports development of stream routing for modeling purposes (1:500k)
Reach File Version 3 (RF3) Alpha	USEPA	Alpha version of Reach File 3; provides a detailed stream network and supports development of stream routing for modeling purposes (1:100K)
National Hydrography Dataset	USGS	Spatial dataset based upon the USGS DLG and the USEPA RF3, that is more refined and expanded. Contains information about surface water features which are combined to form reaches (surface water drainage network), facilitating in routing for modeling purposes (1:100K)
Digital Elevation Model (DEM)	USGS	Topographic relief mapping; supports watershed delineations and modeling

TABLE 10.14 (*Continued*)

BASINS Data Product	Source	Description
Land Use and Land Cover	USGS	Boundaries associated with land use classifications including Anderson Level 1 and Level 2
National Inventory of Dams	U.S. Army Corps of Engineers and the Federal Emergency Management Agency	This dataset provides a locational map of 75,187 dams in the contiguous United States. This database shows the age/description of the dam, number of people living downstream, and some inspection information along with some locational information.

*From BASINS User's Manual.

The DEMs in the BASINS dataset are the USGS 30-meter DEMs, but 10-meter DEMs are now available for much of the country from the USGS Web site, as described earlier.

BASINS also contains environmental monitoring data developed from existing national water quality databases, as displayed in Table 10.15.

BASINS also include data on point source discharges, as shown in Table 10.16.

Applicable BASINS data should be brought into the GIS project and analyzed. Trends or monitoring results may be explained by comparing the data obtained with the spatial location of various data layers. For instance, if PCBs are found in stream sampling results below a RCRIS site, one would have the first clue to begin the search for the source of the PCBs. One must keep in mind that BASINS is a national dataset, which is reflected in the scale of the source and product data. There may be additional or more spatially accurate state, regional, or local data that should be obtained for the watershed assessment.

10.6.23 Other Sources of Data

It is impossible to list all the data available for each individual watershed. One should complete a data-needs analysis in the early stages of a watershed assessment to determine exactly which layers are required. The various local, regional, state, and federal agencies should then be contacted for data availability before new data is created.

TABLE 10.15   BASINS Environmental Monitoring Data

BASINS Data Product	Source	Description
Water Quality Monitoring Stations and Data Summaries	USEPA	Statistical summaries of water quality monitoring for physical and chemical-related parameters; parameter-specific statistics computed by station for 5-year intervals from 1970 to 1994 and 3-year interval from 1995 to 1997
Bacteria Monitoring Stations and Data Summaries	USEPA	Statistical summaries of bacteria monitoring; parameter-specific statistics computed by station for 5-year intervals from 1970 to 1994 and 3-year interval from 1995 to 1997
Water Quality Stations and Observation Data	USEPA	Observation-level water quality monitoring data for selected locations and parameters
National Sediment Inventory (NSI) Stations and Database	USEPA	Sediment chemistry, tissue residue, and benthic abundance monitoring data for freshwater and coastal sediments
Listing of Fish and Wildlife Advisories	USEPA	State reporting of locations with advisories for fishing, including type of impairment
Gage Sites	USGS	Inventory of surface water gaging statin data including 7Q10 low and monthly mean stream flow
Weather Station Sites	National Oceanic and Atmospheric Administration (NOAA)	Location of selected first-order NOAA weather stations
Drinking Water Supply (DWS) Sites	USEPA	Location of public water supplies, their intakes, and sources of surface water supply
Watershed Data Stations and Database	NOAA	Location of selected meteorologic stations and associated monitoring information used to support modeling
Classified Shellfish Areas	NOAA	Location and extent of shellfish closure areas

TABLE 10.16 BASINS Environmental Monitoring Data

BASINS Data Product	Source	Description
Permit Compliance System (PCS) Sites and Computed Annual Loadings	USEPA	NPDES permit-holding facility information; contains parameter-specific loadings to surface waters computed using the EPA Effluent Decision Support System (EDSS) for 1990–1999
Industrial Facilities Discharge (IFD) Sites	USEPA	Facility information on industrial point source dischargers to surface waters
Toxic Release Inventory (TRI) Sites and Pollutant Release Data	USEPA	Facility information for 1987–1995 TRI public data, contains Y/N flags for each facility indicating media-specific reported releases
Superfund National Priority List Site	USEPA	Location of Superfund National Priority List sites from CERCLIS (Coprehensive Environmental Response, Compensation and Liability Information System)
Resource Conservation and Recovery Information System (RCRIS) Sites	USEPA	Location of transfer, storage, and disposal facilities for solid and hazardous waste
Minerals Availability System/Mineral Industry Location System (MAS/MILS)	U.S. Bureau of Mines	Location and characteristics of mining sites

*From BASINS User's Manual.

10.7 SUMMARY

The GIS is an invaluable tool for watershed analysis and assessment. Data is the key to being able to analyze the watershed. As time goes on, more and more data will become available; however, its source and accuracy, as explained in the metadata, should be verified before use. Data comes in all forms, projections, and accuracies, so the user must be aware of the problems that could arise from improper use. The true advantages in using a GIS comes from being able to overlay the various layers to determine patterns, trends, and interactions. In short, it provides a superior modeling advantage.

REFERENCES

Anderson, J.R., E.E. Hardy, J.T. Roach, and R.E. Witmer. "A Land Use and Land Cover Classification System for Use with Remote Sensor Data," Professional Paper 964 (Reston, VA: U.S. Geological Survey), 1976, p. 28.

Anderson, James R., Ernest E. Hardy, and John T. Roach. "A Land-Use Classification System for Use With Remote-Sensor Data," U.S. Geological Survey Circular 671, 1972.

Bailey, Robert G. "Description of the Ecoregions of the United States, March, 2nd Ed. U.S. Department of Agriculture," Forest Service, Miscellaneous publication No. 1391, March 1995.

Belward, A.S., ed. "The IGBP-DIS global 1 km Land Cover Data Set (DISCover)-Proposal and Implementation Plans: IGBP-DIS," Working Paper No. 13, Toulouse, France, 1996.

Cacchione, David A, David Drake, George B. Tate, Joanne Thede-Ferreira, and Rick Viall. "Transport of Particulate Matter in San Francisco Bay," U.S. Geological Survey, http://sfbay.wr.usgs.gov/access/DaveC/DC-Home.html, 2003g.

Chavez, P.S., Jr., and D. MacKinnon. "Automatic Detection of Vegetation Changes in the Southwestern United States Using Remotely Sensed Images," *Photogrammetric Engineering and Remote Sensing,* vol. 60, no. 5, 1994, pp. 571–583.

Clawson, Marion, and Charles L. Stewart. *Land Use Information. A Critical Survey of U.S. Statistics Including Possibilities for Greater Uniformity* (Baltimore, MD, The John Hopkins Press for Resources for the Future, Inc.), 1965.

Dana, Peter. "The Geographer's Craft Project, Map Projections," Department of Geography, The University of Colorado at Boulder, http://www.colorado.edu/geography/gcraft/notes/mapproj/mapproj.html, 1999.

DeBarry, Paul A. "GIS Applications in Storm Water Management and Nonpoint Source Pollution," Proceedings of the ASCE National Conference on Hydraulic Engineering, Nashville, TN, 1991.

DeBarry, Paul A., R.G. Quimpo, J. Garbrecht, T. Evans, L. Garcis, L. Johnson, J. Jorgeson, V. Krysanova, G. Leavesley, D. Maidment, E. James Nelson, F. Ogden, F. Olivera, T. Seybert, W. Sloan, D. Burrows, E. Engman, R. Binger, B. Evans, F. Theurer. "GIS Modules and Distributed Models of the Watershed," Report from the Task Committee on GIS Modules and Distributed Models of the Watershed, ASCE, Reston, VA, 1999.

Dewald, Thomas, Robert Greenspun, Lee Manning, Ann Montalbano, and Phillip Taylor. Storet Reach File 1 User's Guide ("STORET.HELP.REACH.RETRIEVL") (Washington, DC: U.S. EPA, Office of Water, Office of Wetlands, Oceans, and Watersheds), February 11, 1985.

Dewald, Thomas G., and Mark V. Olsen. "The EPA Reach File: A National Spatial Data Resource" (Washington, DC: U.S. Environmental Protection Agency, Office of Wetlands, Oceans, and Watersheds), May 1994; www.epa.gov/waters/doc/rfnsdr.html.

Dickinson, R.E., A. Henderson-Sellers, P.J. Kennedy, and M.F. Wilson. "Biosphere-Atmosphere Transfer Scheme (BATS) for the NCAR Community Climate Model": NCAR Technical Note NCAR/TN275+STR, Boulder, CO, 1986.

Digital Globe, www.digitalglobe.com/products/spot.shtml, accessed 12/27/2003.

Eidenshink, J.C., and J.L. Faundeen. "1-Km Avhrr Global Land Dataset: First Stages In Implementation" (Sioux Falls, SD: EROS Data Center), 2003; http://edcdaac.usgs.gov/1KM/paper.html#defn1.

Environmental Systems Research Institute (ESRI). ArcGIS Help, Definitions, ArcGIS, 2003.

Federal Geographic Data Committee. "Content Standards for Digital Geospatial Metadata," June 1994.

Fenneman, N.M., and D.W. Johnson. Physical Divisions of the United States (map) (Washington, DC: USGS), 1946.

Hydrologic Engineering Center (HEC). *HEC-RAS River Analysis System User's Manual* (CPD-68) (Davis, CA: Hydrologic Engineering Center), 1997.

Iowa Rivers Information System (IRIS), "GAP Application of the Month," www.gap.uidaho.edu/applications/Application_of_the_month_APR.htm (May 2003).

Kennedy, Melita, and Steve Kopp. *Understanding Map Projections* (Redlands, CA: ESRI Press), 2000.

Loelkes, G.L. Jr., G.E. Howard, Jr., E.L. Schwertz, Jr., P.D. Lampert, and S.W. Miller. "Land Use/Land Cover and Environmental Photointerpretation Keys" (Reston, VA: U.S. Geological Survey), 1983.

MacDonald, Andrew. *Building a Geodatabase* (Redlands, CA: ESRI Press), 2001.

Mitchell, W.B., S.C. Guptill, K.E. Anderson, R.G. Fegeas, and C.A. Hallam. "GIRAS—A Geographic Information and Analysis System for Handling Land Use and Land Cover Data," Professional Paper 1059 (Reston, VA: U.S. Geological Survey), 1977.

Natural Resources Conservation Service (NRCS) home page, www.usda.nrcs.gov, accessed 12/27/03.

National Oceanic and Atmospheric Administration, National Data Center, National Geographic Data Committee (NGDC), Digital Terrain Data: Scientific Discussion, www.ngdc.noaa.gov/seg/topo/disc (March 2003).

Olson, J.S. "Global Ecosystem Framework-Definitions: USGS EROS Data Center Internal Report, Sioux Falls, SD, 1994a.

———. Global Ecosystem Framework-Translation Strategy: USGS EROS Data Center Internal Report, Sioux Falls, SD, 1994b.

Omernik, J.M. "Ecoregions of the Conterminous United States." Map (scale 1:7,500.000). Annals of the Association of American Geographers 77(1):118–125, 1987.

Omernik, J.M. "Ecoregions: A Framework for Environmental Management." In Biological Assessment and Criteria: Tools for Water Resources Planning and Decision Making. W. Davis and T. Simon (eds.) (Chelsea, MI: Lewis Publishers), 1995.

Quenzer, Ann M., and David R. Maidment. "Constituent Loads and Water Quality in the Corpus Christi Bay System," Proceedings from Water Resources Engineering '98, ASCE, Nashville, TN, 1998.

Reese, Geoffery, and Paul A. DeBarry. "The GIS Landscape for Wellhead Protection," Proceedings of the ASCE National Hydraulics Conference, San Francisco, CA, 1993.

Running, S.W., T.R. Loveland, L.L. Pierce, R.R. Nemani, and E.R. Hunt. "A Remote Sensing-Based Vegetation Classification Logic for Global Land Cover Analysis," *Remote Sensing of Environment,* vol. 51 (1995), pp. 3,948–3,952.

Sellers, P.J., Y. Mintz, Y.C. Sud, and A. Dalcher. "A Simple Biosphere Model (SiB) for Use within General Circulation Models," *Journal of Atmospheric Science,* vol. 43 (1986), pp. 505–531.

Sellers, P.J., D.A. Randall, G.J. Collatz, J.A. Berry, C.B. Field, D.A. Dazlich, C. Zhang, G.D. Collelo, and L. Bounoua. "A Revised Land Surface Parameterization (SiB2) for Atmospheric GCMs: Part I-Model Formulation," *Journal of Climate,* vol. 9 (1996), pp. 676–705.

Spot Image. http://www.spotimage.fr/html/_167_.php, accessed December 27, 2003.

U.S. Department of Agriculture (USDA) NRCS. Soil Survey Geographic (SSURGO) Data Base—Data Use Information, Miscellaneous Publication 1527, January, 1995.

U.S. Department of the Interior, U.S. Geological Survey. Map Projections. Reston, VA, http://mac.usgs.gov/mac/isb/pubs/Map Projections/projections.html, December 15, 2000.

———. State Soil Survey Geographic (STATSGO) Data Base—Data Use Information, Miscellaneous Publication 1492, September 1995.

United States Environmental Protection Agency. U.S. EPA Reach File Version 1.0 (RF1) for the Conterminous United States (CONUS), 1996.

———. BASINS (Better Assessment Science Integrating Point and Nonpoint Sources), Version 3.0, EPA-823-C-01-003, United States Environmental Protection Agency, Office of Water, June 2001.

U.S. Fish and Wildlife Service (USFW) home page, www.nwi.us, accessed 12/27/03.

U.S. Geological Survey (USGS) home page, www.usgs.gov, accessed 12/27/03.

———. "Land Use and Land Cover and Associated Maps Factsheet" (Reston, VA: U.S. Geological Survey), 1991.

———. "Land Use and Land Cover Digital Data from 1:250,000 and 1:100,000-Scale Maps," *Data User's Guide 4* (Reston, VA: U.S. Geological Survey), 1986.

———. Geological "The National Map—The Nation's Topographic Map for the 21st Century," USGS, http://nationalmap.usgs.gov/, May 14, 2003.

———. High Resolution Land Use and Land Cover Mapping, USGS, November 1999.

———. Physical divisions of the United States, Metadata, http://water.usgs.gov/GIS /metadata/usgswrd/physio.html, 2003c.

———. Physical divisions of the United States, Water Resources NSDI Node, Retrieval for spatial data set Physiographic divisions of the conterminous U.S., http:// water.usgs.gov/lookup/getspatial?physio, 2003d.

———. Rocky Mountain Mapping Center—home page.

———. National Hydrography Dataset, http://nhd.usgs.gov/, 2003b.

———. "Gap Analysis Program (GAP)," http://www.gap.uidaho.edu/about/what_is_ gap_analysis.htm, 2003g.

———. Geographic Names Information System (GNIS), Reston, VA, 2003a http:// geonames.usgs.gov/index.html.

———. Global Land Cover Characteristics Data Base, http://edcdaac.usgs.gov/glcc/ globdoc2_0.html, 2003h.

————. http://edcwww.cr.usgs.gov/products/map/dlg.html.

————. Effects of Urbanization on Stream Ecosystems, Fact Sheet 042-02, May 2002.

Zeiler, Michael. *Modeling Our World: The ESRI Guide to Geodatabase Design* (Redlands, CA: ESRI Press), 1999.

ADDITIONAL WEB PAGE REFERENCES

http://sfbay.wr.usgs.gov (USGS San Francisco Bay Area Web site)

http://sfbay.wr.usgs.gov/access/SFB_Biblio.html (list of reference publications)

http://edcdaac.usgs.gov/samples/sf.html (San Francisco Bay Area Landsat 7 Urban Scene)

http://water.usgs.gov/nawqa (USGS National Water-Quality Assessment (NAWQA) Program

CHAPTER 11

PRECIPITATION AND STREAM FLOW DATA

11.0 INTRODUCTION

Long-term rainfall and stream flow data collection is essential to efforts to preserve our water resources, as these data help us to understand the hydrology, water budget, flooding, base flow, and hydrologic regime of a watershed. The main source of rainfall data in the United States is the National Atmospheric and Oceanic Administration National Weather Service (NOAA-NWS), and the primary agency responsible for the stream gage data is the U.S. Geological Survey (USGS).

11.1 PRECIPITATION MEASUREMENTS

Precipitation data are critical in watershed management to determine the hydrologic budget from which management measures can be established. Precipitation is a very important factor in being able to reliably model a watershed hydrologically and to obtain a truly calibrated hydrologic model. To calibrate a model, the model parameters must be established whereby the simulated hydrograph best represents an observed hydrograph for a variety of actual rainfall events. Therefore, reliable rainfall data are extremely important in calibrating a computer model, and various factors should be evaluated in choosing rainfall stations and storm events for that purpose. To that end, precipitation measurements should be recorded not only on a daily basis but also, ideally, on an hourly or even more frequent basis to note the intensity of a rain that occurs. Snowmelt also should be noted, as well as frozen ground conditions and temperature, so that the lag time from the snowfall can be understood. Local stations should be used, as rainfall data varies considerably during summer storms.

Rain gages can be classified as *recording* or *nonrecording* or total stations. Nonrecording (total rainfall) stations are those that record daily total rainfall amounts measured from midnight to midnight. Recording stations obtain rainfall at a specified increment, typically 15-minute or one-hour increments.

11.1.1 Nonrecording Stations

The least expensive equipment to collect rainfall data are the *backyard rainfall gages* that can be purchased at any hardware store. These are, of course, also the least accurate of the nonrecording stations, hence data from them should be utilized with caution; nevertheless, they may supplement National Weather Service data. Gages should be located for easy access where they can be read and emptied daily.

11.1.2 Recording Stations

Various types of rainfall recording gages are available, as summarized here. The most appropriate choice for the watershed assessment in question is typically dictated by cost or project budget. One of the most common gages is the *tipping bucket gage,* which is an accurate, cost-effective instrument that encodes precipitation increments. It is usually the input sensor for a data logger, counter, or recorder, and comes in unheated or heated versions (where equivalent rainfall values of snow measurement are required). However, the heated tipping bucket has not performed well for the National Weather Service (NWS) (Wnek, 1994).

For tipping bucket gages, precipitation is collected by a "funnel" with either a precise 8- or 12-inch diameter, which then drains to the tipping bucket assembly, as shown in Figure 11.1. The bucket assembly is comprised of two "buckets," or small cups, that collect the rainwater; it is set on a pivotal point and works like a seesaw. When it rains, the upper bucket collects the water until such time that a specified design increment of rainfall (e.g., 0.1 inch) weighs the bucket and the bucket tips, emptying the contents of the filled bucket; thereafter, the process begins again. Each tip empties one side of the seesaw and positions the other bucket under the funnel. The contents of the tipped bucket empty out the bottom of the container. The tipping movement activates a sensor that sends a signal to the attached instrument.

11.1.3 Rain Gage Locations

The exposure of a rain gage is very important for obtaining accurate measurements. The best location is where the gage is uniformly protected in all directions, such as in an opening in a grove of trees. The height of the pro-

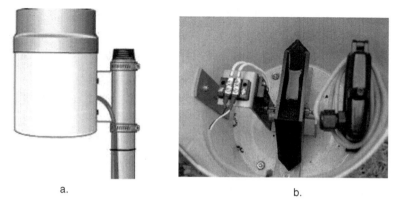

a.   b.

Figure 11.1 Tipping bucket gage. (Copyright © Campbell Scientific, Inc., Logan, Utah, USA [www.campbellsci.com]. All rights reserved.)

tection should not exceed twice its distance from the gage. As a general rule, the windier the gage location, the greater the precipitation error. Gages should not be located close to isolated obstructions such as trees and buildings, which may deflect precipitation due to erratic turbulence. Neither should gages be located in wide-open spaces or on elevated sites, such as the tops of buildings, because of wind and resulting turbulence problems.

Rainfall station locations should be closely evaluated in relation to the watershed in question. Influences such as mountain ridges, valleys, and prevailing wind direction should all be accounted for when trying to determine whether the rainfall station precipitation patterns would accurately reflect the rainfall distribution over the watershed. The more rainfall gages located either within or in close proximity to the watershed, the more reliable the data will tend to be. However, having several rain gages within the watershed is typically a luxury that is not available. Also, the number of rainfall gages to install for a new watershed assessment will be dependent on the size of the watershed.

11.1.4   Utilization of Rainfall Data

Availability of accurate data for particular rain gages in relation to the watershed is an important consideration. The best rainstorms to utilize to calibrate storm event hydrologic models that employ the unit hydrograph method are those that generate at least one inch of runoff and have a pattern that represents an S-curve distribution, as shown in Figure 11.2. Therefore, it is a good idea to develop a unit hydrograph for the particular watershed in question, first to see which peak flow would represent one inch of runoff. Storms producing this peak flow can then be chosen from the stream gage data, and the rainfall data for this event collected.

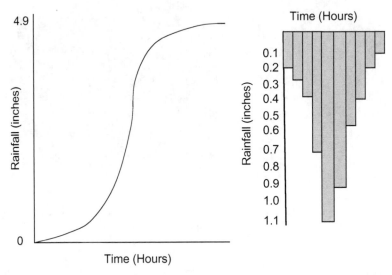

Figure 11.2   Typical "S-shaped" rainfall pattern, which works best for single-event model calibration.

Once the storm events are chosen, existing rainfall data from both recording and nonrecording gages can be obtained from the National Weather Service. Unfortunately, these data are not as readily available over the Web as are USGS stream gage data; moreover, there is a charge to download the data. Weather data can be obtained from the NOAA's National Climatic Data Center (NCDC) Web site at: www4.ncdc.noaa.gov/cgi-win/wwcgi. dll?wwAW~MP#MR.

Hourly rainfall datasets for use in hydrologic modeling can be obtained in ASCII or HTML format. A typical ASCII data file can be found in Table 11.1, and the HTML format can be found in Table 11.2 (NCDC, 2003). The ASCII format can be transferred directly into Excel for analysis and manipulation.

Most rigorous hydrologic models allow input of actual storms for model calibration and verification. All storms should be tabulated and evaluated prior to being utilized for calibration of the model. In addition, several storms should be chosen and set aside for using during the verification runs. Once the storms are tabulated, they should be developed into hyetographs so that rainfall station distribution and patterns in relation to each other and the watershed can be evaluated, and regional/physiographic effects determined.

For example, a watershed with rainfall stations on the north and south sides should have similar rainfall distribution patterns to accurately model what is occurring on the watershed. If they had completely different distribution patterns, the average computed or utilized distribution might not be representative of what is occurring over the watershed, thus averaging the rainfall would

TABLE 11.1   Typical Hourly Rainfall Data in ASCII Format

```
Month:,09/2002
Station Name:, ASHEVILLE RGNL AIRPORT,NC (AVL)
DT,Hr1, Hr2, Hr3, Hr4, Hr5, Hr6, Hr7, Hr8, Hr9, Hr10, Hr11, Hr12, Hr13, Hr14, Hr15, Hr16, Hr17, Hr18, Hr19, Hr20, Hr21, Hr22, Hr23, Hr24
1,----,--T ,--,--,--,--,--,--,--,--,--,--,--,--,--,--,--,--,--,--,--,--,--,--,
2,----,--,--,--,--,--,--,--,--,--,--,--,--,--,--,--,--,--,--,--,--,--,--,--,
3,----,--,--,--,--,--,T ,--,--,--,--,--,--,--,--,--,--,--,--,--,--,--,--,--,
4,----,--,--,--,--,--,--,--,--,--,--,--,--,--,--,--,--,--,--,--,--,--,--,--,
5,----,--,--,--,--,--,--,--,--,--,--,--,--,--,--,--,--,--,--,--,--,--,--,--,
6,----,--,--,--,.01,--,--,--,--,--,--,--,--,--,--,--,--,--,--,--,--,--,--,--,
7,----,--,--,--,--,--,--,--,--,--,--,--,--,--,--,--,--,--,--,--,--,--,--,--,
8,----,--,--,--,--,--,--,--,--,--,--,--,--,--,--,--,--,--,--,--,--,--,--,--,
9,----,--,--,--,--,--,--,--,--,--,--,--,--,--,--,--,--,--,--,--,--,--,--,--,
10,----,--,--,--,--,--,--,--,--,--,--,--,--,--,--,--,--,--,--,--,--,--,--,--,
11,----,--,--,--,--,--,--,--,--,--,--,--,--,--,--,--,--,--,--,--,--,--,--,--,
12,----,--,--,--,--,--,--,--,--,--,--,--,--,--,--,--,--,--,--,--,--,--,--,--,
13,----,--,--,--,--,--,--,--,--,--,.01,T ,--,--,--,--,--,--,--,--,--,.01,.05,
14,----,T ,.02T ,T ,----,.01,T ,T ,.05,.24,.15,.02,T ,T ,.44,.04,.05,.06,.05,.14
15,.05,.02,.03,.06,.01,.02,.02,T ,.01,T ,.02,.09,.06,.02,.04,.05,.01,T ,.01,T ,----,--,--,
16,----,--,--,--,--,--,--,--,--,--,--,--,--,--,--,--,--,--,--,--,--,--,--,--,
17,----,--,--,--,--,--,--,--,--,T ,.01,----,--,.01,T ,T ,.01,--,
18,----,T ,----,--,--,T ,T ,T ,----,--,--,--,--,--,--,--,--,--,
19,----,--,--,--,--,--,T ,T ,T ,T ,T ,----,--,--,--,--,--,--,--,
20,.01,----,T ,T ,----,T ,T ,T ,.01,T ,--,T ,.03,.26,.21,.06,----,
21,----,T ,.03,.06,.04,.19,.07,T ,--,.04,.04,.01,.04,.03,.01,T ,.01,T ,--,T ,T ,.02
22,.01,----,T ,.03,.02,.09,.11,.05,.19,.03,----,.04,T ,.02,.05,.08,.02,--,T ,T ,--,
23,----,--,--,--,--,--,--,--,--,--,--,--,--,--,--,--,--,--,--,--,--,
24,----,--,--,--,--,--,--,--,--,--,--,--,--,--,--,--,--,--,--,--,--,
25,----,T ,T ,.02,.01,.03,.1,.05,.06,.06,.03,.02,.01,T ,.01,.01,.01,T ,.01,.03,.16,----,
26,----,--,--,--,.02,.03,.02,.01,.08,.02,.01,.03,T ,--,T ,.02,.02,.01,.01,.12,.14,.09
27,.04,.03,.02,T ,.01,.02,.04,.04,.06,.01,.07,----,T ,T ,--,--,--,--,--,--,
28,----,--,--,--,--,--,--,--,--,--,--,--,--,--,--,--,--,--,--,--,--,
29,----,--,--,--,--,--,--,--,--,--,--,--,--,--,--,--,--,--,--,--,--,
30,----,--,--,--,--,--,--,--,--,--,--,--,--,--,--,--,--,--,--,--,--,
```

TABLE 11.2 Typical Hourly Rainfall Data in HTML Format

U.S. Department of Commerce
National Oceanic & Atmospheric Administration

Local Climatological Data,
Unedited Hourly Precipitation Table
Asheville Rgnl Airport, Nc (Avl)
(09/2002)

National Climatic Data Center
Federal Building
151 Patton Avenue
Asheville, North Carolina 28801

DT	A.M. -1-	-2-	-3-	-4-	-5-	-6-	-7-	-8-	-9-	-10-	-11-	-12-	-DT-	P.M. -1-	-2-	-3-	-4-	-5-	-6-	-7-	-8-	-9-	-10-	-11-	-12-	-DT-
1	—	—	T	—	—	—	—	—	—	—	—	—	1	—	—	—	—	—	—	—	—	—	—	—	—	1
2	—	—	—	—	—	—	—	—	—	—	—	—	2	—	—	—	—	—	—	—	—	—	—	—	—	2
3	—	—	—	—	—	—	—	—	—	—	—	—	3	—	—	—	—	T	—	—	—	—	—	—	—	3
4	—	—	—	—	—	—	—	—	—	—	—	—	4	—	—	—	—	—	—	—	—	—	—	—	—	4
5	—	—	—	—	—	—	—	—	—	—	—	—	5	—	—	—	—	—	—	—	—	—	—	—	—	5
6	—	—	—	—	—	—	.01	—	—	—	—	—	6	—	—	—	—	—	—	—	—	—	—	—	—	6
7	—	—	—	—	—	—	—	—	—	—	—	—	7	—	—	—	—	—	—	—	—	—	—	—	—	7
8	—	—	—	—	—	—	—	—	—	—	—	—	8	—	—	—	—	—	—	—	—	—	—	—	—	8
9	—	—	—	—	—	—	—	—	—	—	—	—	9	—	—	—	—	—	—	—	—	—	—	—	—	9
10	—	—	—	—	—	—	—	—	—	—	—	—	10	—	—	—	—	—	—	—	—	—	—	—	—	10
11	—	—	—	—	—	—	—	—	—	—	—	—	11	—	—	—	—	—	—	—	—	—	—	.05	—	11
12	—	—	—	—	—	—	—	—	—	—	—	—	12	—	T	—	—	—	—	—	—	—	.01	.05	—	12
13	—	—	—	—	—	—	—	—	—	—	—	—	13	.01	—	—	—	—	—	—	—	—	.01	—	.14	13
14	—	T	.02	T	T	—	.02	T	—	.01	T	T	14	.05	.24	.15	.02	T	T	.44	.04	.05	.06	.05	—	14
15	.05	.02	.03	.06	.01	.02	.02	T	.01	T	.02	.09	15	.06	.02	.04	.05	.01	T	.01	T	—	—	—	—	15
16	—	—	—	—	—	—	—	—	—	—	—	—	16	—	—	—	—	—	—	—	—	—	—	—	—	16
17	—	—	—	—	—	—	—	—	—	—	—	—	17	T	—	T	.01	—	—	.01	T	T	.01	—	—	17
18	—	—	—	—	T	—	—	—	—	—	—	T	18	T	T	T	T	—	—	—	—	—	—	—	—	18
19	—	—	—	—	—	—	—	—	—	—	—	—	19	T	.01	T	T	—	—	—	—	—	—	—	—	19
20	.01	—	—	T	T	.04	.19	.07	T	—	T	T	20	T	.04	.03	.01	T	.01	T	.03	.26	.21	.06	—	20
21	.01	—	T	.03	.06	.09	.11	.05	.19	.03	.04	.04	21	.01	.02	.05	.08	.02	—	T	T	—	T	T	.02	21
22	—	—	T	.03	.02	.09	.11	.05	.19	.03	—	.04	22	T	.02	—	—	—	—	T	T	—	T	—	—	22
23	—	—	—	—	—	—	—	—	—	.03	.07	—	23	—	—	—	—	—	—	—	—	—	—	—	—	23
24	—	—	—	—	—	—	—	—	—	—	—	—	24	—	—	—	—	—	—	—	—	—	—	—	—	24
25	—	—	T	T	.02	.01	.03	.1	.05	.06	.06	.03	25	.02	.01	T	.01	.01	.01	T	.01	.03	.16	.14	—	25
26	—	.03	T	T	.01	.02	.02	.03	.02	.01	.08	.02	26	.01	.03	T	—	T	.02	.02	.01	.01	.12	—	.09	26
27	.04	.03	.02	T	.01	.02	.04	.04	.06	.01	.07	—	27	T	T	—	—	—	—	—	—	—	—	—	—	27
28	—	—	—	—	—	—	—	—	—	—	—	—	28	—	—	—	—	—	—	—	—	—	—	—	—	28
29	—	—	—	—	—	—	—	—	—	—	—	—	29	—	—	—	—	—	—	—	—	—	—	—	—	29
30	—	—	—	—	—	—	—	—	—	—	—	—	30	—	—	—	—	—	—	—	—	—	—	—	—	30

282

lose the storm pattern. Many models have a routine to prevent this from happening. In addition, storm totals should be compared for consistency.

11.1.5 Existing versus New Rainfall Data

As indicated earlier, rainfall events that produce at least one inch of runoff and a hydrograph with a single peak are best for use in calibrating and verifying a hydrologic model. Watershed studies are typically not of long enough duration to enable accessing new data for several such storms. Therefore, if no preexisting data are available, and the study is not of sufficient duration to obtain reliable storm events for calibration, new data generation is typically not appropriate. That said, under certain circumstances, such as to confirm a water budget or supplement surrounding rain gage data, it may be beneficial to set up new stations and obtain rainfall data for a new study. Nationally, rainfall data are collected by the National Oceanic and Atmospheric Administration (NOAA) National Weather Service (NWS) at a series of rainfall stations. A typical weather station is shown in Figure 11.3.

Rainfall frequency atlases are available from the NCDC and provide frequency-duration rainfall amounts for the eastern and western United States, samples of which are shown in Figures 11.4 and 11.5, respectively (Hershfield, 1961).

The NCDC also makes it possible to generate graphics of annual precipitation trends by state regions, as shown in Figure 11.6. This capability allows the user to determine droughts and patterns affecting stream base flow.

Figure 11.3 Typical NWS weather station, Baltimore/Washington International Airport (NCDC, 2003).

Figure 11.4 One-year, 24-hour rainfall totals for eastern United States.

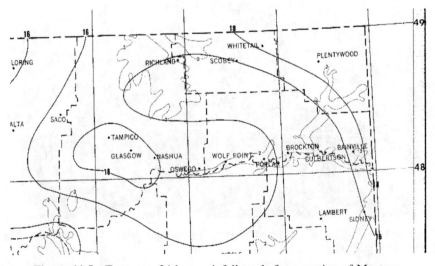

Figure 11.5 Two-year, 24-hour rainfall totals for a portion of Montana.

Precipitation (Inches)

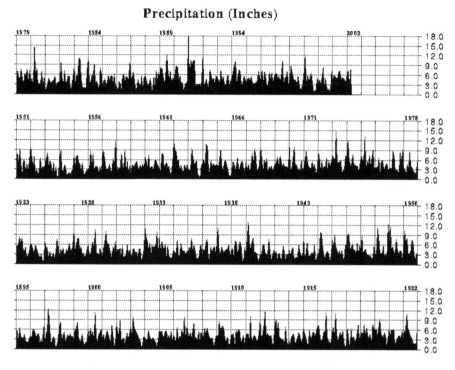

Alabama—Division 01: 1895–2002 (Monthly Averages)

Figure 11.6 Annualized precipitation for Region 1 in Alabama. (*Source:* www.ncdc.
noaa.gov/oa/climate/onlineprod/drought/xmgrg3.html, acccessed May 12, 2003.)

Note

The NCDC makes available much more data than is possible to mention
here. For more information on this weather-related data, go to the NCDC
Web site, http://www.ncdc.noaa.gov/oa/ncdc.html.

11.1.6 Flood Forecasting

Several programs obtain rainfall and stream flow data for the prevention of
flood damage. Two of these, IFLOWS and ALERT, are discussed in this
section.

11.1.6.1 IFLOWS. The National Oceanic and Atmospheric Administration
(NOAA) National Weather Service (NWS) maintains the *Integrated Flood
Observation and Warning System* (*IFLOWS*). The primary goal of IFLOWS,
which was initiated as part of the National Flash Flood Program Development
Plan in 1978, is to prevent loss of life and property from flash floods. IFLOWS

has expanded to include the Automated Flood Warning Systems (AFWS), which is a network of numerous local flood-warning systems, connected and integrated to share information. As of 2003, approximately 250 computers and 1,500 sensors in 12 states were connected (National Weather Service, 2003). IFLOWS rainfall data are available online at www.afws.net/ for the states shown in Figure 11.7.

IFLOWS rainfall data are displayed for the 15- and 30-minute and 1-, 3-, 6-, 12-, and 24-hour durations, a sample of which is shown in Table 11.3. A typical IFLOWS satellite image of rain clouds is shown in Figure 11.8.

11.1.6.2 ALERT. The National Weather Service has also developed the *Automated Local Evaluation Real-Time* (*ALERT*) program, which uses remote sensors to collect and transmit environmental data to a central computer in real time—which means data are transmitted as fast as they are collected. The U.S. Army Corps of Engineers (USACE), Bureau of Reclamation, United States Geologic Survey (USGS), and various state and local agencies have found these data to be useful for flood forecasting (ALERT, 2003).

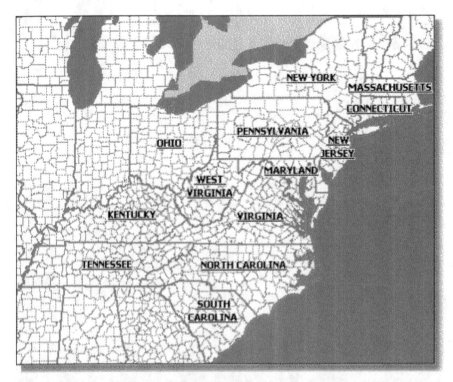

Figure 11.7 States for which IFLOWS rainfall data are available. (*Source:* www. afws.net/, accessed May 14, 2003.)

TABLE 11.3 **Typical IFLOWS Rainfall Data**

Last Updated: Wednesday May 08, 2002 at 00:15:31 AM EDT

	ID	15 MIN	30 MIN	1 HR	3 HR	6 HR	12 HR	24 HR
		Data for *Campbell* County, Tennessee						
Cross Mountain	1533	0.00	0.00	0.00	0.00	0.00	0.08	0.08
La Follette	1534	0.00	0.00	0.00	0.08	0.08	0.08	0.08
Walnut Mountain	1535	0.00	0.00	0.00	0.00	0.00	0.00	0.00
Jellico	1536	0.00	0.00	0.04	0.16	0.16	0.20	0.20
		Data for *Carter* County, Tennessee						
Holston	1604	0.00	0.00	0.00	0.00	0.00	0.00	0.00
Buladeen	1605	0.00	0.00	0.00	0.00	0.00	0.20	0.20
Roan Mountain	1606	0.00	0.00	0.04	0.04	0.04	0.04	0.04
Ripshin Lake	1607	0.00	0.00	0.00	0.00	0.00	0.00	0.00
Roan Mtn St Park	1608	0.00	0.00	0.00	0.00	0.00	0.00	0.00
Hampton Vfd	1609	0.00	0.04	0.28	0.44	0.44	0.64	0.64
Dennis Cove	1610	0.00	0.04	0.28	0.48	0.48	0.56	0.56

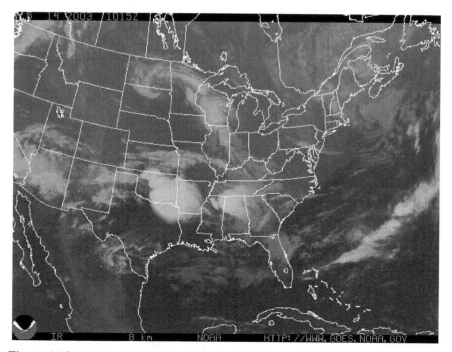

Figure 11.8 Typical IFLOWS satellite image of rain clouds. (*Source:* http://iwin. nws.noaa.gov/iwin/images/ecir.jpg, May 14, 2003.)

11.1.7 NEXRAD

Next-generation radar (NEXRAD) data are measures of reflectance of the radar from raindrops. It has therefore been correlated to rainfall amounts (Smith and Krajewski, 1993). Many universities and agencies are currently evaluating the impact of radar-rainfall estimation errors on watershed runoff predictions (DeBarry, et al, 1999). Although still being refined, NEXRAD should play an important role in the future for watershed hydrologic analyses. A typical NEXRAD image can be found in Figure 11.9, and one for hurricane Fran is shown in Figure 11.10.

NEXRAD data, which come from the Weather Surveillance Radar 88 Doppler (WSR-88D) and 158 operational NEXRAD radar systems, as shown in Figure 11.11, are managed by NOAA's Radar Operations Center (ROC) in Norman, Oklahoma (NOAA, 2003). NEXRAD provides real-time data for flood forecasting and land and water resources management.

11.2 STREAM FLOW MEASUREMENTS

The United States Geologic Survey (USGS) is the primary agency responsible for collecting and maintaining water resource data as it relates to water budg-

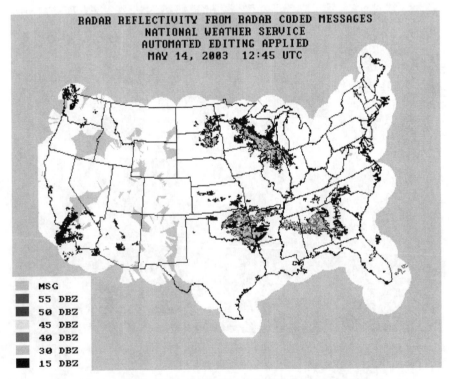

Figure 11.9 Typical NEXRAD image (NCDC, 2003).

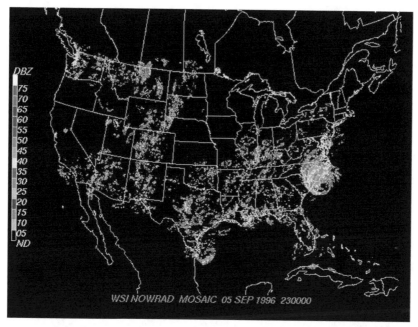

Figure 11.10 Typical NEXRAD image of Hurricane Fran, September 5, 1996 (NCDC, 2003).

ets. Collection of data is on a *water year* basis, which begins October 1 and ends September 30. The reason the water year begins in October is that the change in precipitation patterns occurs about this time of year in many regions, and accumulated snowfall records would not be split into two years.

Stream gaging is important because streams comprise the only portion of the hydrologic cycle during which the water is confined to a specific area that can be measured. Gages can be classified as nonrecording and recording (or continuous) gages.

There are many natural and man-made components of the hydrologic budget with which we are now familiar. In addition, new techniques are becoming available, such as land application of treated sewage effluent, to return water back into the aquifer and better conserve water within the hydrologic unit. In some cases, however, new techniques can complicate the water budget because users are unfamiliar with their influences. Therefore, accurate and timely tracking of water budgets should be correlated with biological as well as stream flow and groundwater monitoring data to understand the impacts resulting from disturbance of the natural balance of the hydrologic system.

11.2.1 Nonrecording Gages

Nonrecording gages do not record continuous data, but rely on systematic observations of the stage of the gage. *Stage* is the height of the water level

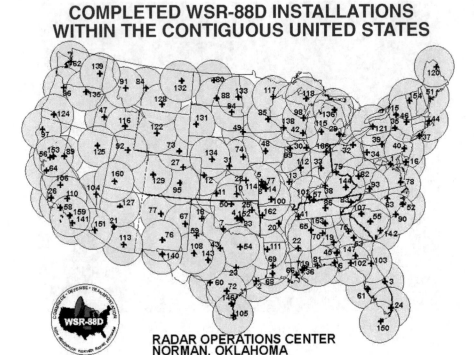

Figure 11.11 NEXRAD WSR-88D installation locations (NOAA, 2003).

in the stream in relation to a specific *datum,* that is, a known elevation. The most common elevation reference system is known as the North American Vertical Datum of 1988 (NAVD88), which describes a USGS datum above mean sea level. Although an arbitrary datum, or a datum above a benchmark with some an arbitrary elevation, may be utilized, it is recommended to use the USGS datum so that the data can be tied into various other studies, such as Flood Insurance Studies.

Nonrecording gages are not recommended by USGS, but may find their place in a watershed study due to their low cost of installation and data gathering. Nonrecording gages are often used as reference gages to verify recording gage data. Watershed associations typically utilize staff gages' data; however, data integrity should be verified. Nonrecording gages include (Buchanan and Somers, 1978):

Staff gage. A typical nonrecording gage would be a staff gage, or a marked rod placed in the area of flow of a stream. Staff gages are usually found on or along bridges in a location where the structure itself does not influence the water level of the stream. Data gathering relies on actual visitation to the gage, to record the stream stage.

Wire-weight gage. A wire-weight gage consists of a drum wound with a cable that has a weight attached at the end. The drum has a calibrated disc, and the values of stage are obtained by placing the weight at the water surface. Wire-weight gages are also affixed to bridges where the stream level would not be influenced by the bridge structure at the point of contact.

Float-tape gage. The float-tape gage works on the same principle as the wire-weight gage; however, the wire is replaced with a graduated steel tape and the weight is replaced by a float. The stage measurements are read from the tape when the float is at the water surface elevation.

Chain gage. The chain gage is a cantilevered arm that lowers a chain with a weight on the end to the water surface over a pulley. Chain gages are typically used where access is limited. The gage height is read from the mounted gage at the location of the gage marker.

Crest-stage gage. The crest-stage gage, as the name implies, measures only the peak stage from a runoff event. There are various models of the crest-stage gage, but the most common consists of a hollow pipe with a top and bottom cap, a cork within the pipe, and a staff fixed to a datum reference. The bottom cap, or the part that will be submerged, has a number of holes in it to equalize hydrostatic pressure. There is a hole in the top cap to relieve air pressure. As water rises, the cork rises. As the water falls, the cork adheres to the surface of the pipe, thus recording the peak stage.

Indirect methods. Circumstances often arise where stage or discharge measurements cannot be obtained for the desired period of record. Examples may include a power outage, storm flows too high to access the gage, destruction or vandalism of the gage, and so on. In these circumstances, discharge may have to be determined through indirect methods, which typically involve measuring the high-water marks or elevations after the flood on bridges, trees, or walls, then estimating channel flow by using Manning's equation; and calculating flows based on known stage/discharge over dams, or flow through culverts and bridges. If the channel was not stable during the flood, a field survey of the channel cross section and longitudinal slope would also be required. When utilizing indirect methods, their reliability should be verified, if possible, and limitations noted (Buchanan and Somers, 1978).

11.2.2 Recording Gages

The general objective of stream gaging is to obtain a continuous record of stage and stream flow (discharge) data, thus the term *recording* or *continuous gage*. These data can then be utilized to determine *low flow, base flow,* and *storm flow* conditions. A continuous record of stage is obtained by installing instruments that sense and record the water-surface elevation in the stream.

The following subsections summarize the most common continuous stream gages (DeBarry, 1993; Buchanan and Somers, 1984).

11.2.2.1 *Current Meter.* Discharge measurements are typically made using a current meter and involve dividing a stream cross section into equal intervals and determining the mean velocity, as well as the area of each interval, as shown in Figure 11.12. Because shallow streams restrict the use of current meters, a pygmy current meter may be utilized for these depths.

The discharge is measured manually each time a reading is required. Current meters work on the principle that flow is equal to the area of flow times its velocity, or:

$$Q = AV \tag{11.1}$$

Where:

    Q = Discharge or flow (cubic feet per second, cfs)
    A = Cross-sectional area of flow (square feet, sf)
    V = Velocity of flow (feet per second, fps)

However, channels do not have consistent velocities across a cross section, and velocities are not consistent within an interval and vary with depth, as shown in Figure 11.13. Velocity meters measure velocity at a point. Therefore, a method to obtain the mean velocity of an interval is required. Several methods to obtain mean velocity have been recognized by USGS (Buchanan and Somers, 1984) as listed here:

    Develop a vertical-velocity curve. Velocities are taken at 0.1-foot increments, and a curve is developed, as shown in Figure 11.14. The mean velocity is calculated by obtaining the area under the curve and the

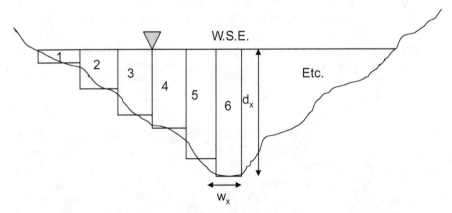

Figure 11.12   Velocity area sections (modified from Buchanan and Somers, 1984).

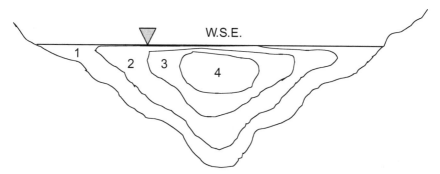

Figure 11.13 Typical variation in velocity within a channel; will vary within each channel section. Units are in feet per second (modified from Buchanan and Somers, 1984).

ordinate (x) axis, and dividing by the length of the ordinate axis. This is the most accurate—as well as most time-consuming—method.

Two-point method. Measurements are made at the 0.2 and 0.8 (or 20 and 80 percent) depth below the surface; the average value is used as the mean velocity. This approach, which has been found to be very accurate, has long been used by USGS.

Six-tenths-depth method. The velocity measurement is made at 60 percent of the depth below the surface. USGS utilizes this method depending

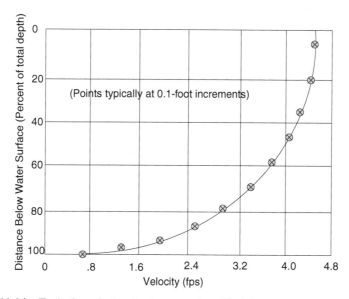

Figure 11.14 Typical vertical-velocity curve (modified from Buchanan and Somers, 1984).

on total depth and type of meter used. As a rule of thumb, use this method when: the total depth is between 0.3 feet and 2.5 feet, when surface ice precludes the 0.2 depth measurement, when bottom conditions preclude the 0.8 measurement, or when a quick measurement is needed due to fluctuating stage.

Two-tenths-depth method. The velocity is measured at 0.2 of the total depth, and a coefficient (typically 0.88) is applied to obtain the mean velocity. This method is not recommended unless specific site data on the coefficient are available.

Three-point method. The three-point method is a combination of the two-point and the six-tenths methods and includes obtaining velocity measurements at 0.2, 0.6, and 0.8 of the total depth. The mean is obtained by averaging the 0.2 and 0.8 velocities, then averaging this average with the 0.6 depth value. Mean velocities are obtained for the intervals as shown in Figure 11.12, then flow for the individual sections are computed using equation 11.2.

$$q_x = d_x \times w_x \times v_x \tag{11.2}$$

Where:

q_x = Discharge through partial section x
d_x = Depth of water at location x
w_x = Width at location x
v_x = Velocity at location x

Obtaining mean velocities is also useful in estimating flood-wave velocity. It has been found that flood-wave velocities are generally 1.3 times faster than the mean velocity (Buchanan and Somers, 1984). Travel time between reaches can therefore be expressed through the following relationship:

$$T = (L/1.3 \ v') \div 60 \ \text{sec/min} \tag{11.3}$$

Where:

T = Travel time of the flood wave (min)
L = Length of study reach (ft)
v' = Mean measured velocity (fps)

The total flow from a cross section of channel can then be computed by the sum of individual sections, as shown in Equation 11.4.

$$Q = \Sigma \ (A, v) \tag{11.4}$$

Where:

A = Cross-sectional area of flow (sf)
V = Velocity (fps)

11.2.2.2 Weir. A *weir* is a fixed cross section built across a river to regulate or measure the flow of water, divert it, or change or regulate its level. There are many types of weirs and the *Handbook of Hydraulics* (Brater and King, 1996) and other references should be consulted to determine the most appropriate type to apply.

Weir installation for flow measurement for watershed analyses includes construction of a bulkhead across the stream, with, usually, a V-notch weir in the center. The flow rate through the weir can be calculated based on flow depth using a standard equation; however, calibration utilizing a current meter or transducer is recommended. Weirs can be utilized in conjunction with a staff gage located on the weir plate or a continuous recording device (float or bubble gage).

The general equation to determine flow over a sharp-crested 90-degree V-notch weir is:

$$Q = CH^{5/2} \tag{11.5}$$

Where:

Q = Discharge (cfs)
C = Weir coefficient or coefficient of discharge
H = Static head or depth over the weir

A typical value of C for a sharp-crested 90-degree V-notch weir is 2.47; however, it has been found that coefficients can range from about 2.47 to about 2.57, depending on H.

The equation to predict flow over broad crested weirs is:

$$Q = CLH^{3/2} \tag{11.6}$$

Where:

Q = Discharge (cfs)
C = Weir coefficient or coefficient of discharge (dimensionless)
L = Length (width) of the weir (ft)
H = Static head or depth over the weir (ft)

A typical value of C is between 2.52 and 2.91, with values varying by head, slope of downward face, and length of weir, as shown in Table 11.4 (from Brater and King, 1996). However, their book, the *Handbook of Hydraulics,* should be consulted to obtain the specific value for the situation at hand.

TABLE 11.4 Coefficients of Discharge

Slope of Crest	Length of Weir	Head in Feet						
		0.1	0.2	0.3	0.4	0.5	0.6	0.7
12:1	3.0	2.58	2.87	· 2.57	2.60	2.84	2.81	2.70
18:1	3.0	2.92	2.92	2.53	2.60	2.80	2.74	2.62
18:1	10.0	2.52	2.68	2.73	2.80	2.90	2.80	2.68

The exact location of the installation of a weir is not critical; nevertheless, the more uniform the stream flow cross section, the better the calibration that can be obtained. An exact location for weir placement should be determined in the design phase of the study.

11.2.2.3 Parshall Flume. Parshall flumes are useful for measuring discharge when the depths are shallow and velocities are low. The flume has a converging section, a throat, and a diverging section, as shown in Figure 11.15. The floor of the converging or upstream section is level, both longitudinally and transversely, when properly in place. The floor of the throat section slopes downward and the floor of the diverging or downstream section slopes upward.

The flume may be operated as a free-flow, single-head measuring device, or under submerged-flow conditions where two heads are measured. The head in the converging section and the head near the downstream end of the throat section are read on staff gages or in stilling wells. Both gages have their datum at the elevation of the floor of the converging section. Note that a Parshall flume is typically economically infeasible for small streams.

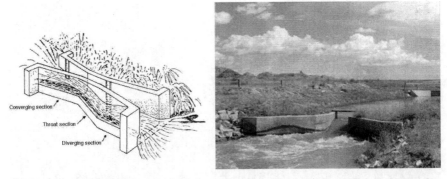

Converging section
Throat section
Diverging section

Figure 11.15 Typical Parshall flume—4-foot, short-form Parshall flume, discharging 62 cfs under free-flow conditions. Scour protection is needed for this much drop. (*Source:* U.S. Bureau of Reclamation and USDA; www.usbr.gov/pmts/hydraulics_lab/ pubs/wmm/index.htm, accesses June 12, 2003.)

11.2.2.4 Float Gage. Stage is typically measured by a float located in a stilling well that is connected to the stream by intake pipes, which maintain the stream's static head in the stilling well (see Figure 11.16). The invert of the stilling well must be below the current of the stream; its top must be above the maximum anticipated stage. The float stilling well installation is more reliable than that of the bubble gage, discussed next. The vertical accuracy of the float gage is plus or minus 0.01 feet.

11.2.2.5 Pressure Transducer. A pressure transducer measures the pressure produced by the column of water above the sensing element. The pressure transducer converts the pressure to mechanical movement of a shaft, which can be connected to a recorder and/or a telemetry device. There are two types of pressure transducers: the bubbler type and the sealed system. The *bubble-gage sensor* consists of an orifice attached to the streambed, through which gas is expelled, and a measuring device, which measures the resistance to the release of the gas produced by the column of water above it. The *sealed pressure transducer* also measures the weight of the column of water, but does not release a gas to do so. (NWS, 2003)

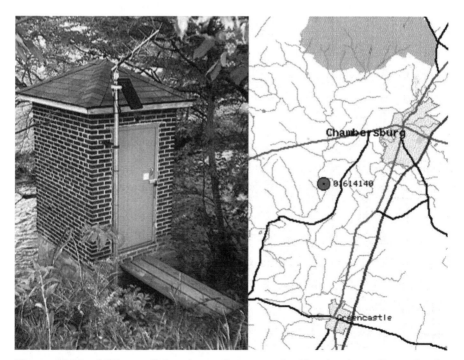

Figure 11.16 Stilling well housing a float gage, in Chambersburg, Pennsylvania, along with USGS Web-based location map.

11.2.2.6 Manometer. Manometers function on the same principle as the bubbler gage. They contain mercury, and thus have been mostly replaced by bubbler gages that are more environmentally safe (NWS, 2003).

For continuous recording of stream flow records, either a float or bubble gage stage recorder is required. In order to properly calibrate lower flows, a V-notch or other type of weir should be installed. For low and storm flows, a rating curve must be developed utilizing a current meter or hydraulic equations. There are two main types of water stage data recorders that may be activated by the float or bubble gage system: the *digital-tape* and the *slip-chart* water-level recorders. The digital-tape stage recorder punches coded stream stage values on paper tape at preselected time intervals. The slip-chart recorder produces a graphic record of the rise and fall of the stream stage over time. Both recorders can operate unattended for 60 to 90 days.

11.2.3 Gage Location

The proper location of the gage is one that takes advantage of the best available condition for stage and discharge measurements and for developing discharge *ratings*. Stream gages are typically positioned where there is a stable, well-defined channel so that a *rating curve* can be developed. A rating curve can be a simple plot of stage versus discharge for that particular stream cross section, or a function of several relations such as stage, slope, rate of change of stage, or other variables. A typical rating curve is found in Figure 11.17.

Factors to consider when locating a stream gage include (Buchanan and Somers, 1984):

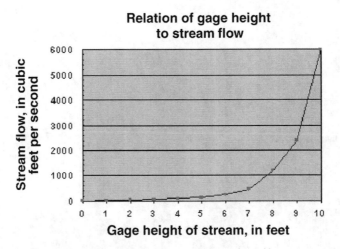

Figure 11.17   Rating curve for a stream gage. (*Source:* http://ga.water.usgs.gov/edu/gageflow.html, accessed June 12, 2003.)

- Stable stream channels, such as a rock riffle or falls
- Stream cross section where capacity is reliably computed
- Ability to measure low flow and storm flow
- Proper placement of the stage gage in relation to the channel that controls the stage
- Avoidance of movable beds, such as sand bottoms
- Avoidance of obstructions
- Absence of downstream backwater influence from obstructions, bridges, or tributaries
- Avoidance of *unsteady flow*
- Avoidance of streams where base flow or side channels may bypass the measured section
- Ability to construct the required structure
- Protection from vandalism
- Easy general access
- Access to electrical power (many gages now are solar-powered)

Fixed cross sections such as weirs, pipes, or channels are also preferred locations for gages due to the stability of the cross section and the well-defined and -tested equations to predict flow in these sections.

11.2.4 Existing versus New Stream Flow Data

A reliable source of stream flow data is the United States Geologic Survey (USGS). USGS began measuring stream flows in 1889 for special studies relating to irrigation of public lands. USGS has measured approximately 16,000 locations within the United States and has an extensive database, with 7,292 stations in operation as of 1994, and more than 400,000 station years of data (Carter and Davidson, 1977; Wahl, et al., 1995). This data is stored in the USGS National Water Information System (NWIS), which provides an invaluable source of information for watershed studies; it is available on the World Wide Web at http://waterdata.usgs.gov/nwis.

Even though the USGS has an extensive national stream flow database, at times, local data are lacking or there is an insufficient period of record to accurately determine the water budget or confidently calibrate a hydrologic model. The longer the duration of stream flow data, the more reliable the data will be, particularly in determining long-term trends. Funding should be guaranteed for USGS to continue this valuable effort.

USGS is strengthening its stream data system with the USGS National Streamflow Information Program (NSIP). NSIP addresses data in five categories, as follows (USGS, 2003):

- Stream gaging networks
- Floods and droughts
- Regional and national stream flow assessments
- Improved data delivery
- Investment in methods development and research

The gaging networks are continuously operated to fulfill the following five federal water resource goals (USGS, 2003):

Interstate and international waters. Interstate compacts, court decrees, and international treaties mandate long-term, accurate, and unbiased stream gaging by the USGS at state-line crossings, compact points, and international boundaries.

Stream flow forecasts. Real-time stage and discharge data are required to support flood forecasting by the National Weather Service across the country.

River basin outflows. Resource managers need to account for the contribution of water from each of the nation's 350 major river basins to the next downstream basin, estuary, ocean, or the Great Lakes.

Sentinel watersheds. A network of stream gaging stations is needed to describe the ever-changing status as it varies in response to changes in climate, land use, and water use in 800 watersheds across the country that are relatively unaffected by flow regulation or diversion and that typify major ecoregions and river basins.

Water quality. Stream gaging stations are needed to provide the stream flow information in support of the three national USGS water-quality networks: one that covers the nation's largest rivers; the second for intermediate-sized rivers; and the third for small, pristine watersheds.

USGS has now made its data available online, including real-time data. Real-time data typically are recorded at 15- to 60-minute intervals; they are stored at the gaging station, then transmitted to the regional USGS offices via satellite, telephone, and/or radio every one to four hours, depending on the data-relay technique used. The data are relayed through the *Geostationary Operational Environmental Satellite* (*GOES*) system and are processed automatically in near real-time. Data from real-time sites are available for public viewing on the Internet within minutes of arrival.

Note

USGS makes it clear that all real-time data are provisional and subject to revision, meaning the data have received little or no review and are subject to change.

Data are available by stage, stream flow, or both, and can be found at the USGS home page at http://waterdata.usgs.gov/nwis/sw. A recent data delivery system is Web-based, to provide the entire storehouse of USGS hydrologic data to the public. The system, called *NWIS-Web,* was first deployed in the fall of 2000. It delivers historical and current streamflow data in a variety of formats and graphical presentations. One of the benefits is that it gives the customer flexibility to create varied forms of output. Data can be retrieved by major river basin, county, or hydrologic unit code (HUC). An example of real-time data for the Snake River at Nyssa River, Oregon, can be found in Figure 11.18.

The NSIP plan goes a step further, and provides tools to perform statistical analyses for selected sites through online GIS interfaces and existing statistical models to compute streamflow charactersitics (USGS, March, 2002).

11.2.5 Stream Data for Hydrologic Modeling

Once obtained from USGS, stream gage data can be evaluated for the peak annual stream flow. The data are easy to review to determine which storms should be further analyzed; they can then be used to perform a statistical frequency analysis or to obtain equivalent design storm flows. Table 11.5 (USGS, 2004) shows a summary of peak daily flows stored in ASCII format, and Table 11.6 shows the same data in HTML format.

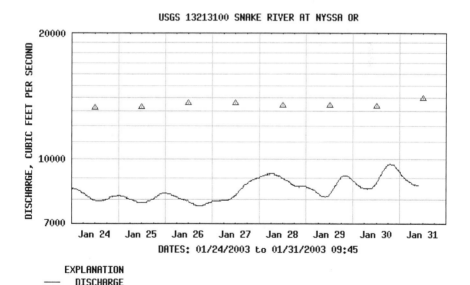

Figure 11.18 Typical plot of real-time and mean daily stream flow for the Snake River near Nyssa, Oregon. (*Source:* http://waterdata.usgs.gov/nwis/sw, accessed June 12, 2003.)

TABLE 11.5 Summary of Annual Peak Flows for Bear Creek, near Muskegon, Michigan, USGS Gage in ASCII Format

```
# U.S. Geological Survey
# National Water Information System
# Retrieved: 2003-01-31 14:58:54 EST
#
# ---------WARNING---------
# The data you have obtained from this automated
# U.S. Geological Survey database have not received
# Director's approval and as such are provisional
# and subject to revision. The data are released
# on the condition that neither the USGS nor the
# United States Government may be held liable for
# any damages resulting from its use.
#
# This file contains the annual peak streamflow data.
#
# This information includes the following fields:
#
# agency_cd  Agency Code
# site_no  USGS station number
# peak_dt  format YYYY-MM-DD
# peak_va  Annual peak streamflow value in cfs
# peak_cd  Peak Discharge-Qualification codes (see explanation below)
# gage_ht  Gage height for the associated peak streamflow in feet
# gage_ht_cd  Gage height qualification codes
# year_last_pk  Peak streamflow reported is the highest since this year
# ag_dt  Date of maximum gage-height for water year (if not concurrent with peak)
# ag_tm  Time of maximum gage-height for water year (if not concurrent with peak)
# ag_gage_ht  maximum Gage height for water year in feet (if not concurrent with peak)
# ag_gage_ht_cd  maximum Gage height code
#
```

302

```
# Sites in this file include:
# USGS 04122100 BEAR CREEK NEAR MUSKEGON, MI
#
# Peak Streamflow-Qualification Codes (peak_cd):
# 1 ... Discharge is a Maximum Daily Average
# 2 ... Discharge is an Estimate
# 3 ... Discharge affected by Dam Failure
# 4 ... Discharge less than indicated value,
#         which is Minimum Recordable Discharge at this site
# 5 ... Discharge affected to unknown degree by
#         Regulation or Diversion
# 6 ... Discharge affected by Regulation or Diversion
# 7 ... Discharge is an Historic Peak
# 8 ... Discharge actually greater than indicated value
# 9 ... Discharge due to Snowmelt, Hurricane,
#         Ice-Jam or Debris Dam breakup
# A ... Year of occurrence is unknown or not exact
# B ... Month or Day of occurrence is unknown or not exact
# C ... All or part of the record affected by Urbanization,
#         Mining, Agricultural changes, Channelization, or other
# D ... Base Discharge changed during this year
# E ... Only Annual Maximum Peak available for this year
#
# Gage height qualification codes (gage_ht_cd,ag_gage_ht_cd):
# 1 ... Gage height affected by backwater
# 2 ... Gage height not the maximum for the year
# 3 ... Gage height at different site and(or) datum
# 4 ... Gage height below minimum recordable elevation
# 5 ... Gage height is an estimate
# 6 ... Gage datum changed during this year
#
#
```

303

TABLE 11.5 (*Continued*)

| agency_cd | site_no | peak_dt | peak_tm | peak_va | peak_cd | gage_ht | gage_ht_cd | year_last_pk | ag_dt | ag_tm | ag_gage_ht | ag_gage_ht_cd |
5s	15s	10d	6s	8s	27s	8s	13s	4s	10d	6s	8s	11s
USGS	04122100	1966-02-10	274		8.43							
USGS	04122100	1967-04-22	198		7.95							
USGS	04122100	1968-02-01	193		7.92							
USGS	04122100	1969-06-26	179		7.81							
USGS	04122100	1970-03-04	177		7.79							
USGS	04122100	1971-02-27	204		7.99							
USGS	04122100	1972-08-26	131		7.29							
USGS	04122100	1972-12-31	262		8.36							
USGS	04122100	1974-05-17	500		9.58							
USGS	04122100	1975-08-31	428		9.46							
USGS	04122100	1976-03-05	930		11.00							
USGS	04122100	1977-03-13	102		6.69							
USGS	04122100	1978-08-19	233		15.11							
USGS	04122100	1979-04-26	151		14.41	2	1979-03-05		15.45	1		
USGS	04122100	1980-08-20	468		16.29							
USGS	04122100	1981-02-19	275		15.32							
USGS	04122100	1981-10-01	410		16.05							
USGS	04122100	1982-12-03	203		14.98							
USGS	04122100	1984-02-14	260		15.23							
USGS	04122100	1985-02-25	186		14.70							
USGS	04122100	1986-09-30	299		15.50							
USGS	04122100	1986-10-04	184		14.70							
USGS	04122100	1988-01-31	148		14.38							

Agency	Station	Date	Value	Flag	Value2	Flag2	Date2	Value3	Flag3
USGS	04122100	1988-11-16	123		14.10				
USGS	04122100	1990-05-10	190		14.58				
USGS	04122100	1991-04-15	257		15.15				
USGS	04122100	1991-11-30	208		14.80				
USGS	04122100	1993-04-20	170		14.49				
USGS	04122100	1994-02-21	75.0	1,2	16.19	2	1994-02-20	16.61	1
USGS	04122100	1995-01-14	145		14.26				
USGS	04122100	1996-06-18	91.0		13.48				
USGS	04122100	1997-02-22	246		14.92				
USGS	04122100	1998-04-01	94		13.55				
USGS	04122100	1999-04-23	172		14.51				
USGS	04122100	2000-04-21	145		14.26				
USGS	04122100	2001-04-12	244		15.00	2	2001-02-09	15.55	1

TABLE 11.6 Typical Table of Peak Annual Flows for the Period of Record for USGS 04122100 Bear Creek, near Muskegon in HTML Format

		Output formats
Muskegon County, Michigan		**Output formats**
Hydrologic Unit Code 04060102		*Table*
Latitude 43°17′19″, Longitude 86°13′22″ NAD27		*Graph*
Drainage area 16.70 square miles		*Tab-separated file*
Gage datum 590.00 feet above sea level NGVD29		*WATSTORE formatted file*
		Reselect output format

Water Year	Date	Gage Height (feet)	Stream-flow (cfs)	Water Year	Date	Gage Height (feet)	Stream-flow (cfs)
1966	Feb. 10, 1966	8.43	274	1984	Feb. 14, 1984	15.23	260
1967	Apr. 22, 1967	7.95	198	1985	Feb. 25, 1985	14.70	186
1968	Feb. 1, 1968	7.92	193	1986	Sep. 30, 1986	15.50	299
1969	Jun. 26, 1969	7.81	179	1987	Oct. 4, 1986	14.70	184
1970	Mar. 4, 1970	7.79	177	1988	Jan. 31, 1988	14.38	148
1971	Feb. 27, 1971	7.99	204	1989	Nov. 16, 1988	14.10	123
1972	Aug. 26, 1972	7.29	131	1990	May 10, 1990	14.58	190
1973	Dec. 31, 1972	8.36	262	1991	Apr. 15, 1991	15.15	257
1974	May 17, 1974	9.58	500	1992	Nov. 30, 1991	14.80	208
1975	Aug. 31, 1975	9.46	428	1993	Apr. 20, 1993	14.49	170
1976	Mar. 5, 1976	11.00	930	1994	Feb. 21, 1994	16.19	75.0[1,2]
1977	Mar. 13, 1977	6.69	102	1995	Jan. 14, 1995	14.26	145
1978	Aug. 19, 1978	15.11	233	1996	Jun. 18, 1996	13.48	91.0
1979	Apr. 26, 1979	14.41	151	1997	Feb. 22, 1997	14.92	246
1980	Aug. 20, 1980	16.29	468	1998	Apr. 1, 1998	13.55	94
1981	Feb. 19, 1981	15.32	275	1999	Apr. 23, 1999	14.51	172
1982	Oct. 1, 1981	16.05	410	2000	Apr. 21, 2000	14.26	145
1983	Dec. 3, 1982	14.98	203	2001	Apr. 12, 2001	15.00	244

Peak stream flow qualification codes: 1, discharge is a maximum daily average; 2, discharge is an estimate.

These peak flows can then be imported into statistical frequency programs, such as Miniex or PeakFQ, the methodologies of which are discussed in Chapter 12. In addition to developing statistical frequency analyses on the annual flood peaks, analyzing the data allows the user to determine the dates of the most significant storms, of which actual hydrographs may be obtained for use in hydrologic model calibration. For instance, Figure 11.19 displays the daily mean stream flow for the period of record for the stream gage at Wilkes-Barre, Pennsylvania. Obviously, the June 23 peak flow value is of interest (Hurricane Agnes), so more detailed data on the hydrograph would be obtained.

Caution is necessary when obtaining actual measured data. Preferably, for a small watershed, gather data on a 15-minute basis, and for a large watershed,

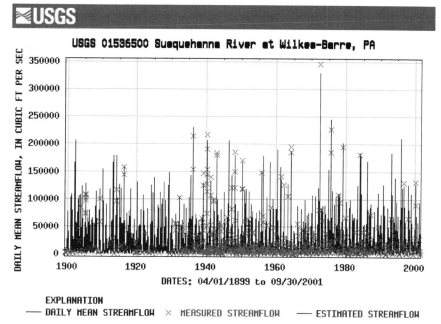

Figure 11.19 Mean daily flow values for the stream gage in Wilkes-Barre, Pennsylvania, June 23 to 26, 1972.

on a one-hour basis; and do not confuse this data with mean daily flow data. The mean daily flow data "smooths" the hydrographs, so the peaks and actual shape of the hydrographs may not be represented, as shown in Figure 11.20. Figure 11.20 shows the full mean daily stream flow hydrograph for the flood of Hurricane Agnes June 23 to 26, 1972, in Wilkes-Barre, Pennsylvania.

Discretion should be utilized when choosing storm events to calibrate a model so that the frozen ground, snowmelt, and so on do not influence the flood peak, as these would not be represented in the rainfall data.

11.3 BASE FLOW SEPARATION

Determination of the water budget for a watershed, or a specific site (as described in Chapter 3), is becoming an increasingly important aspect of managing watersheds. Watershed water budgets need to be determined in order to develop a program or management plan to maintain stream base flow and aquifer recharge or the natural hydrologic regime, and to manage groundwater and surface withdrawals. Standards need to be developed so that, as development increases, the goal of maintaining the natural hydrologic regime

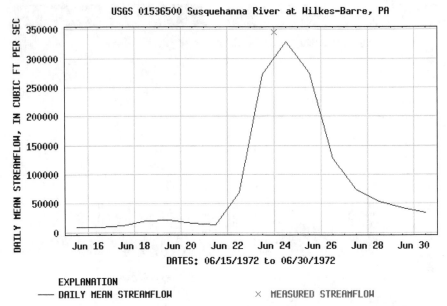

Figure 11.20 Hydrograph of Hurricane Agnes in Wilkes-Barre, Pennsylvania, June 23 to 26, 1972. Notice the mean daily flow peak misses the actual measured peak (x) in both peak flow and time.

is considered. Although ideal, this goal is almost impossible to achieve due to the fact that the evapotranspiration (ET) component of the water budget equation (discussed in Chapter 3) is almost impossible to replicate when a site is converted from forest to commercial. This evapotranspiration volume is the flow that gets passed downstream after a site becomes developed and is a major cause of streambank erosion problems. What to do with this ET volume is a challenge to the engineering community. Nevertheless, infiltrating, at a minimum, the amount of rainwater that infiltrated under natural conditions should be obtained. (Stormwater management will be discussed in further in Chapters 18 and 19.)

Rainfall and total runoff data are, more often than not, readily available. In contrast, the separation of the rainfall that runs off the land (surface or overland runoff) and the amount that recharges the groundwater system (essentially, equal to stream base flow or groundwater discharge) are not easy factors to determine, particularly without stream gage data. If the user is lucky, USGS will have performed a base flow separation on a gage located in the watershed being studied. If not, the study professional will need to perform a base flow separation.

There are several manual methods to separate base flow from surface runoff, two of which are shown in Figure 11.21 (Viessman et al., 1977). The

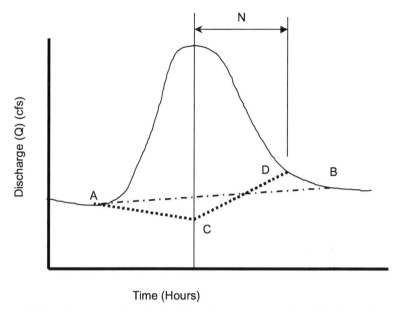

Figure 11.21 Two manual methods of base flow separation (modified from Viessman, et al., 1977).

first method connects the point at which the surface runoff or hydrograph begins (A) to the flow near which the hydrograph began (B), to form a near-horizontal line. The second method continues the base flow slope until the hydrograph peaks (C), then increases to a point equal to N (Point D):

$$N = A^{0.2} \tag{11.7}$$

Where:

N = Time in days
A = Drainage area in square miles

This accounts for the lag in time in groundwater flow in relation to surface flow.

Manual methods are, unfortunately, subjective and difficult to replicate. USGS has, therefore, developed several computer programs, including the recession index (RECESS), Rorabaugh Method (RORA), hydrograph separation (HYSEP), and stream flow partitioning (PART) programs, that allow base flow separation (Rutledge, 2003). One of those, PART, uses stream flow partitioning to estimate a daily record of groundwater discharge under the stream flow record. More specifically, it, and HYSEP, use the fixed-interval, sliding-

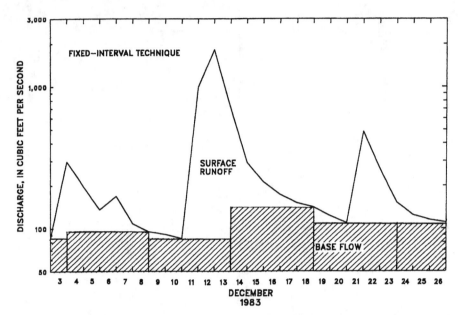

Figure 11.22 Hydrograph analysis using the fixed-interval technique (White and Sloto, 1990).

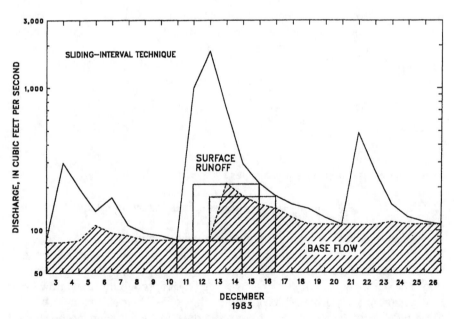

Figure 11.23 Hydrograph analysis using the sliding-interval technique (White and Sloto, 1990).

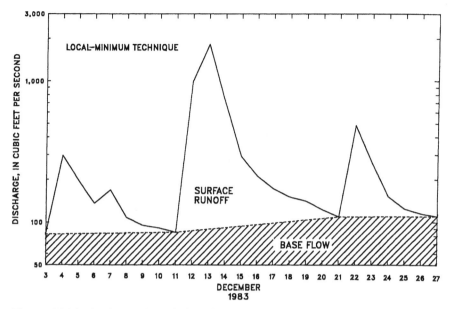

Figure 11.24 Hydrograph analysis using the local-minimum technique (White and Sloto, 1990).

interval, and local-minimum techniques developed by W.A. Pettyjohn and Roger Henning (1979), as shown conceptually in Figures 11.22 through 11.24.

The algorithms systematically draw connecting lines between low points of the stream flow hydrograph to define the base flow hydrograph (White and Sloto, 1990). The advantage of these computer programs is that they allow efficient analysis of a large number of records, which then enables determination of various stream and base flow statistics, such as mean annual base flow and annual minimum base flows for recurrence intervals.

TABLE 11.7 Base Flows at Selected Recurrence Intervals

	Sample Creek Near Sample	
Recurrence Interval (Year)	Local-Minimum Method (mgd/mi^2)*	Fixed-Interval Method (mgd/mi^2)*
2	0.572	0.638
5	0.456	0.510
10	0.426	0.441
25	0.383	0.417
50	0.342	0.385
100	0.320	0.353

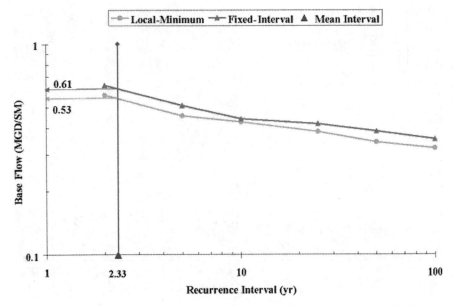

Figure 11.25 Typical plot of base flow and determination of mean annual (2.33 year) base flow.

Utilizing the USGS computer-assisted empirical methods to separate the groundwater and surface-runoff components for the selected stream flow stations, the base flow for the 2-, 5-, 10-, 25-, 50-, and 100-year recurrence intervals can determined, as shown in Table 11.7.

The data in Table 11.7 can be plotted as shown in Figure 11.25; or a regression equation can be developed to then obtain mean annual base flow (recurrent interval equal to 2.33 years). For the example in Figure 11.25, the mean annual base flow is 0.57 mgd/mi^2. Application of this for on-site stormwater management will be discussed in Chapter 19.

11.4 SUMMARY

Rainfall and stream flow data are essential for analyzing a watershed. Usually, many years of record are required to perform any type of statistical analysis to determine trends on the data, meaning that installing gages just for a study will not likely provide statistically significant results. That said, there may be some circumstance, such as when determining a lake's water budget, where short-term rainfall and stream flow data could be utilized. Stream gages are invaluable for understanding our water resources, and national funding should be increased to increase the number of gages.

REFERENCES

ALERT. Automated Local Evaluation in Real Time Web page www.alertsystems. org/, accessed December 21, 2003.

American Society of Civil Engineers (ASCE). *Hydrology Handbook,* 2nd ed. (New York: ASCE), 1996.

Brater, Ernest F., and H.W. King. *Handbook of Hydraulics,* 7th ed. (New York: McGraw-Hill Book Company), March 1, 1996.

Buchanan, Thomas J., and William P. Somers. "Techniques of Water-Resources Investigations of the United States Geological Survey," Chapter A7, Book 3, *Stage Measurements at Gaging Stations* (Washington, DC: USGS, United States Printing Office), 1978.

———. "Techniques of Water-Resources Investigations of the United States Geological Survey," Chapter A8, Book 3, *Discharge Measurements at Gaging Stations* (Washington, DC: USGS, United States Printing Office), 1984.

Carter, R.W., and Jacob Davidson. "Techniques of Water-Resources Investigations of the United States Geological Survey," Chapter A6, Book 3, *General Procedure for Gaging Streams* (Washington, DC: USGS, United States Printing Office), 1977.

DeBarry, Paul A. "Stream Gage Alternative Analysis for Swiftwater Creek," Swiftwater, PA, unpublished report, May 25, 1993.

DeBarry, Paul A., R.G. Quimpo, J. Garbrecht, T. Evans, L. Garcis, L. Johnson, J. Jorgeson, V. Krysanova, G. Leavesley, D. Maidment, E. James Nelson, F. Ogden, F. Olivera, T. Seybert, W. Sloan, D. Burrows, E. Engman, R. Binger, B. Evans, F. Theurer. *GIS Modules and Distributed Models of the Watershed* (Reston, VA: American Society of Civil Engineers), 1999.

Hershfield, D.M. "Rainfall Frequency Atlas of the United States for Durations from 30 Minutes to 24 Hours and Return Periods from 1 to 100 Years," Technical Paper 40 (Washington, DC: Weather Bureau, U.S. Department of Commerce), 1961.

Hulsing, Harry. "Techniques of Water-Resources Investigations of the United States Geological Survey," Chapter A5, Book 3, *Measurement of Peak Discharge at Dams by Indirect Methods* (Washington, DC: USGS, United States Printing Office), 1967.

Hydrology Subcommittee, Interagency Advisory Committee on Water Data. "Guidelines for Determining Flood Flow Frequency," Bulletin 17B, March 1982.

Jennings, M.E., W.O. Thomas, Jr., and H.C. Riggs. "Nationwide Summary of U.S. Geological Survey Regional Regression Equations for Estimating Magnitude and Frequency of Floods for Ungaged Sites, 1993," U.S. Geological Survey Water-Resources Investigations Report 94-4002, 1994.

Matthai, H.E. "Techniques of Water-Resources Investigations of the United States Geological Survey," Chapter A3, Book 3, *Measurement of Peak Discharge at Width Contractions by Indirect Methods* (Washington, DC: USGS, United States Printing Office), 1967.

National Climatic Data Center (NCDC). "NEXRAD National Mosaic Reflectivity Images," www.ncdc.noaa.gov/cgi-n/wwcgi.dll?WWNEXRAD~Images2, accessed May 14, 2003.

National Oceanic and Atmospheric Administration (NOAA). "NEXRAD WSR-88D," Radar Operations Center, www.roc.noaa.gov/, accessed May, 14, 2003.

National Weather Service (NWS). Integrated Flood Observation and Warning System (IFLOWS) www.afws.net/, accessed May 2003.

———. *Service Hydrologist Reference Manual,* www.nws.noaa.gov/om/hod/, accessed July 2003.

Pettyjohn, W.A., and Roger Henning. "Preliminary Estimate of Ground-water Recharge Rates, Related Streamflow, and Water Quality in Ohio," Ohio State University Water Resources Center Project Completion Report 552, 1979.

Ries III, K.G., and M.Y. Crouse. "The National Flood Frequency Program, Version 3: A Computer Program for Estimating Magnitude and Frequency of Floods for Ungaged Sites," Water-Resources Investigations Report 02-4168, U.S. Geological Survey, Reston, VA, 2002.

Rutledge, A.T. "Computer Programs for Describing the Recession of Ground-Water Discharge and for Estimating Mean Ground-Water Recharge and Discharge from Streamflow Records—Update," U.S. Geological Survey Water Resources Investigations Report 98-4148, December 2, available on-line at http://water.usgs.gov/pubs/wri/wri984148/, accessed Dec 21, 2003.

Smith, J.A., and W.F. Krajewski. "A Modeling Study of Rainfall Rate-Reflectivity Relationships," *Water Resources Research,* vol. 29, no. 8, pp. 2505–2514 (1993).

U.S. Bureau of Reclamation and USDA. *Water Measurement Manual: A Water Resources Technical Publication—A Guide to Effective Water Measurement Practices for Better Water Management,* 3rd ed. (Washington, DC: U.S. Government Printing Office), 1997; www.usbr.gov/pmts/hydraulics_lab/pubs/wmm/index.htm.

U.S. Geological Survey, Five Components of the National Streamflow Information Program, The National Streamflow Information Program, http://water.usgs.gov/nsip/components.html, last update: 15:42:21 Tue 12 Mar 2002, accessed Jan. 14, 2004.

U.S. Geological Survey. Surface Water for USA: Peak Streamflow, http://waterdata.usgs.gov/nwis/peak?, retrieved on 2004-01-14 19:15:33 EST.

U.S. Water Resources Council. "A Uniform Technique for Determining Flood Flow Frequencies," Bulletin 15, December 1967.

Viessman, Jr., Warren, John W. Knapp, Gary L. Lewis, and Terence E. Harbaugh. *Introduction to Hydrology,* 2nd ed. (New York: Harper and Row Publishers), 1977.

Wahl, Kenneth L., Wilbert O. Thomas, Jr., and Robert M. Hirsch. "Stream-Gaging Program of the U.S. Geological Survey," U.S. Geologic Survey Circular 1123, Reston, VA, 1995.

White, Kirk E., and Ronald A. Sloto. "Base-Flow-Frequency Characteristics of Selected Pennsylvania Streams," U.S. Geological Survey, Water Resources Investigations Report 90-4160, Harrisburg, PA, 1990.

Winter, T.C., J.W. Harvey, O.L. Franke, and W.M. Alley. Ground Water and Surface Water as a Single Resource, U.S. Geological Survey Circular 1139, U.S. Government Printing Office, Denver, CO, 1998.

Wnek, R.A. "1993–1994 Frise Liquid Precipitation Accumulation Sensor Winter Test Report," September 1994, from The Tipping Bucket Rain Gauge, National Weather Service Web page http://www.nws.noaa.gov/asos/tipbuck.htm, accessed June 12, 2003.

CHAPTER 12

WATERSHED HYDROLOGY
AND MODELING

12.0 INTRODUCTION

Predicting runoff and, in turn, stream flows, requires essentially three parameters: drainage area, a time of concentration, and an infiltration/runoff parameter typically based upon soils and land use. Each will be described in more detail in this chapter. There are many methods to compute these parameters, hence a text on hydrology such as *The Hydrology Handbook* (ASCE, 1996) or *Handbook of Hydrology* (Maidment, 1992) should be consulted to compare the various methods. Many hydrologic methods have been incorporated into computer models to aid in the massive computations required to model a watershed's runoff processes. One of the first hydrologic watershed simulation models was the Stanford Watershed Model (Crawford and Linsley, 1966), which later became the Hydrologic Simulation Program—FORTRAN, or HSPF (Johanson et al., 1980).

Of the many methods, the two main ones used to compute runoff are the traditional *lumped parameter method* and, since the advent of GIS, the *distributed model*. A distributed model is defined as "a set of algorithms that performs hydrologic/hydraulic modeling by considering subunits of the watershed under study" (DeBarry et al., 1999). These subunits are typically grid cells from a digital elevation model. Lumped parameter models utilize "lumped," or average, variables that generalize rainfall-runoff relationships such as the NRCS curve number.

The two most commonly used lumped parameter methods, the SCS (NRCS) *Curve Number Method* and the *Rational Method,* will be discussed in this chapter. Lumped parameter methods typically utilize a standard synthetic design storm applied to the entire watershed and average *antecedent moisture conditions (AMC)*, also referred to as *antecedent runoff conditions (ARC)*. Antecedent runoff conditions comprise an index of runoff potential for a storm event, which varies from storm to storm and is based upon the number of days since the last storm. NRCS developed three AMCs, I, II, and III, based upon the five-day antecedent rainfall. For design storm analysis, AMC II, for median conditions, should be utilized (SCS, 1982). SCS's three AMCs are described in Table 12.1 (USDA SCS, 1985).

12.1 DRAINAGE AREA

One factor that influences the peak and volume of flow at a point of interest (POI) is the contributing drainage area. The drainage area is the land mass from which runoff will contribute to a point of interest in a watershed measured in acres, hectares, square miles, etc. The point of interest may be the downstream confluence point or outlet of the stream or river, a stream gage, or, in the case for developing watershed-wide stormwater management plans, the downstream location of each of the many individual subwatersheds comprising the watershed being studied. In land development, locations for points of interest where flows are required may be where the size of culverts, bridges, swales, best management practices, or detention basins are required. The term "drainage area" is often used interchangeably with the terms "basin," "watershed," "subbasin," "subwatershed," or "subarea" but should refer to the measured area of the contributing land mass. The drainage area is a required input parameter in the majority of hydrologic models and regression equations.

12.2 TIME-OF-CONCENTRATION AND WATERSHED LAG

To aid in hydrograph generation, one must look at the general process of overland runoff from storm events and visualize the process of stream formation. There are many ways to compute a watershed's time-of-concentration, including the SCS segmental approach (USDA SCS, 1986) and others. Kibler and colleagues (1982) compared the results of the Kirpich, SCS Lag, Kine-

TABLE 12.1 Seasonal Rainfall Limits for AMC

| | Total Five-Day Antecedent Rainfall (in.) | |
AMC	Dormant Season	Growing Season
I	<0.5	<1.4
II	0.5–1.1	1.4–2.1
III	>1.1	>2.1

matic Wave, SCS Average Velocity Chart, and Kerby Equation, among others. The USDA Natural Resources Conservation Service's segmental approach classifies runoff into *overland* or *sheet flow, shallow concentrated flow,* and *channel flow,* and has derived equations to predict the travel time for each, which are described in the Technical Release 55 (TR-55) publication (USDA SCS, 1986). (Note: NRCS is continually conducting research on time-of-concentration methodologies and revising this methodology. Obtain the latest information from NRCS.)

Sheet flow is flow over surface planes and begins the runoff process. Sheet flow can be determined from a form of Manning's equation (Equation 12.1):

$$T_{t1} = 0.007 \ (nL)^{0.8} \div ((P2)^{0.5} \ s^{0.4}) \tag{12.1}$$

Where:

T_{t1} = Travel time (minutes)
n = Manning's roughness coefficient
L = Flow length (ft)
$P2$ = Two-year, 24 hour rainfall (in.)
s = Slope of the hydraulic grade line (land slope, (ft/ft)).

NRCS does not recommend utilizing a value of more than 150 feet for the overland flow length (L). There are various sources of data for Manning's n, values with the original sources, such as hydraulics handbooks, displaying n values for channel flow. Sources that display n values specifically for overland flow should be obtained for Equation 12.1. State Department of Transportation Drainage handbooks may be a good source of Manning's n values. Typical n values for sheet flow can be found in Table 12.2 (U.S. Army Corps of Engineers 1987; USDA SCS TR-55, 1986). Note that n values are significantly different for sheet flow versus channel flow.

Although TR-55 shows n values of up to 0.8 for dense woods, an n value of greater than 0.4 is not recommended for overland flow.

Shallow concentrated flow occurs when sheet flow accumulates. This velocity will vary with path length. The velocity equations as published in the 1986 version of TR-55 are found in equations 12.2 and 12.3; however, NRCS is considering revising the equations (Aron, 2003).

$$v = 16.1345 \ (s)^{0.5} \ \text{for unpaved areas} \tag{12.2}$$

$$v = 20.3282 \ (s)^{0.5} \ \text{for paved areas} \tag{12.3}$$

Where:

v = Velocity (fps)
s = Watercourse slope (hydraulic grade line) (ft/ft)

TABLE 12.2 Manning's Roughness Coefficient (n) Values for Overland/Sheet Flow

Surface Description	Range of Manning's n Values
Dense Growth	0.4–0.5
Pasture	0.3–0.4
Lawns	0.2–0.3
Bluegrass Sod	0.2–0.5
Short Grass Prairie	0.1–0.2
Sparse Vegetation	0.05–0.13
Bare Clay: Loam Soil (eroded)	0.01–0.03
Concrete/Asphalt:	
Very shallow depths (less than 1/4 inch)	0.10–0.15
Small depths (1/4 inch to several inches)	0.05–0.10
Fallow (no residue)	0.05
Cultivated Soils	
Residue Cover ≤20%	0.06
Residue Cover >20%	0.17
Grass	
Dense Grasses	0.24
Bermuda Grass	0.41
Range (natural)	0.13
Woods (light underbrush)	0.40

The travel time is the ratio of flow length to velocity and can be determined by equation 12.4:

$$T_{t2} = L \div 60v \qquad (12.4)$$

Where:

T_t = Travel time (minutes)
L = Flow length (feet)
v = Velocity (feet per second)
60 = Conversion from seconds to minutes

Channel flow can be computed using Manning's equation 12.5:

$$v = (1.486 \, r^{2/3} \, s^{1/2}) \div n \qquad (12.5)$$

Where:

v = Average velocity (ft/s)
r = Hydraulic radius (ft), equal to the cross-sectional area/wetted perimeter (a/wp)

a = Cross-sectional area of flow (sf)
s = Watercourse slope (hydraulic grade line) (ft/ft)
n = Manning's roughness coefficient

Details on the equations and Manning's roughness coefficients can be found in the references cited in the preceding discussion. Typical values for channel flow can be found in Table 12.3. These principles are also important to understand how the runoff process begins the fluvial formation process.

The travel time for channel flow (T_{t3}) would be determined utilizing equation 12.4. The time-of-concentration would therefore be:

$$Tc = T_{t1} + T_{t2} + T_{t3} \tag{12.6}$$

Knowing the flow and cross-sectional area of flow, velocity can also be determined from equation 12.7.

$$v = Q \div A \tag{12.7}$$

Where:

v = Velocity (feet per second)
Q = Flow (cubic feet per second)
A = Cross-sectional area of flow (square feet)

The *lag time* of a watershed is defined as the time from center of mass of the *rainfall excess* to the time of peak of the hydrograph (Dunne and Leopold, 1978), and can be measured in a variety of ways, including plotting rainfall runoff accumulation curves (Dunne and Leopold, 1978), Snyder's method (Chow, 1964), or through the time-of-concentration of the watershed. Rainfall excess is the amount of rainfall that does not infiltrate, or equates to *direct surface runoff*. Lag time is related to the time-of-concentration by equation 12.8 (U.S. Army Corps of Engineers, 1987):

TABLE 12.3 Manning's Roughness Coefficient (n) Values for Channel Flow

Reach Description	Manning's n
Natural stream: clean, straight; no rifts or pools	0.03
Natural stream: clean, winding; some pools or shoals	0.04
Natural stream: winding, pools, shoals; stony with some weeds	0.05
Natural stream: sluggish deep pools and weeds	0.07
Natural stream or swale: very weedy or with timber underbrush	0.10
Concrete pipe, culvert, or channel	0.012
Corrugated metal pipe	0.012–0.027*

*Depending upon type, coating, and diameter.

$$\text{Lag} = 0.6 \ Tc \tag{12.8}$$

The Snyder Lag is determined by empirical equation 12.9.

$$t_L = C_t \ (L \ L_{ca})^{0.3} \tag{12.9}$$

Where:

t_L = Basin lag time in hours
L = Length of the longest watercourse from the point of interest to the boundary of the watershed (mi.)
L_{ca} = Length along the longest watercourse from the point of interest to a point opposite and perpendicular to the centroid of the watershed
C_t = A regional coefficient related to basin slope

Equation 12.10 explains the relation of C_t to slope for the north and middle Atlantic states (ASCE, 1996). Ranges of C_t are between 0.4 and 8, and should be obtained from regional specific studies if available. Generalized values are found in Table 12.4.

$$C_t = 0.6 \div S^{1/2} \tag{12.10}$$

Where:

S = Basin slope

There are several other methods to compute basin lag, which should be consulted before choosing a method.

12.3 RUNOFF

Surface runoff, also termed *direct runoff* or *precipitation excess,* in terms of hydrology or stormwater management, refers to the amount of precipitation

TABLE 12.4   Snyder Regional Coefficient (C_t) Values

Region	C_t Values
Mountainous Appalachian Highlands	2
Southern California	0.4
Sierra Nevada	0.7–1.0
Eastern Gulf of Mexico	8

that is not lost to abstractions (evapotranspiration (ET), depression storage, infiltration). Two methods to compute runoff from storm events are described in this section: the Rational Method and the SCS (NRCS) Runoff Curve Number Method, sometimes referred to as the Soil Cover Complex Method.

12.3.1 Rational Method

One of the most widely used methods for determining peak flows for design storms is the Rational Method, as shown in Equation 12.11. Watershed management begins with managing runoff from new development sites, so this method is included here for reference; it is a tried-and-true method for obtaining ballpark flows for small drainage areas. The Rational Method is applicable to drainage areas of 200 acres and less, although this value varies by reference and is utilized primarily for on-site stormwater system design.

$$Q = C\,i\,A \tag{12.11}$$

Where:

Q = Runoff (cfs)
i = Rainfall intensity (from regional I-D-F curve) (in./hr)
A = Drainage area (acres) of the subwatershed
c = Dimensionless runoff coefficient

The rainfall intensity (i) is for a duration equal to the time-of-concentration of the subwatershed. This theory comes from the observation that the maximum discharge from a subwatershed will occur when the entire subwatershed is contributing to the point of interest. At first glance, it appears that the units do not match in the rational equation. However, the next example shows that the units do work out:

A (acres) $\times$ i (in./hr) $\times$ 1 ft/12 in. $\times$ 43,560 sf/acre $\times$ 1 hr/3600 sec

$\times$ c = Q 1.0083 (cfs)

The 1.0083 value is rounded to 1. The c-coefficient should be obtained from regional tables if available and varies widely by source. Typical values from one source are shown in Table 12.5 (from the New Jersey Department of Environmental Protection, *Technical Manual for Stream Encroachment*, August, 1984). Many sources do not distinguish the c-coefficients by hydrologic soil group.

There are many other sources of Rational C values, most of which provide values much lower than those in Table 12.5. Although it may appear that utilizing higher C values, which provides higher peak flows, is conservative,

TABLE 12.5 Rational Runoff Coefficients (AMC II)

Land Use Description	Hydrologic Soil Group			
	A	B	C	D
Cultivated land: Without conservation treatment	.49	.67	.81	.88
With conservation treatment	.27	.43	.61	.67
Pasture or range land: Poor conditions	.38	.63	.78	.84
Good conditions	—*	.25	.51	.65
Meadow: Good conditions	—*	—*	.44	.61
Wood or forest land: Thin stand, poor cover, no mulch	—*	.34	.59	.70
Good cover	—*	—*	.45	.59
Open spaces, lawns, parks, golf courses, cemeteries				
Good conditions: grass cover on 75% or more of the area	—*	.25	.51	.65
Fair conditions: grass cover on 50% to 75% of the area	—*	.45	.63	.74
Commercial and business areas (85% impervious)	.84	.90	.93	.96
Industrial districts (72% impervious)	.67	.81	.88	.92
Residential:				
Average lot size Average % Impervious				
1/8 acre or less 65	.59	.76	.86	.90
1/4 acre 38	.25	.49	.67	.78
1/3 acre 30	—*	.49	.67	.78
1/2 acre 25	—*	.45	.65	.76
1 acre 20	—*	.41	.63	.74
Paved parking lots, roofs, driveways, and so on	.99	.99	.99	.99
Streets and roads:				
Paved with curbs and storm sewers	.99	.99	.99	.99
Gravel	.57	.76	.84	.88
Dirt	.49	.69	.80	.84

Values are based on SCS definitions and are average values derived by an advisory committee for the manual.
*Values indicated by a dash should be determined by the design engineer based on site characteristics.

in actuality, when computing predevelopment runoff rates, this higher flow allows a higher flow to be discharged from a detention basin, which could cause problems, therefore the opposite is sometimes true.

It is important to point out that the Rational Method computes peak flow. Several attempts have been made to construct hydrographs from rational peaks, but this practice is not recommended except for minor drainage systems. When constructing hydrographs from rational peaks, the appropriate sources should be referenced to ensure proper application. There are also other modifications to the Rational Method, including the *Modified Rational Method*.

12.3.2 NRCS (SCS) Curve Number (Soil Cover Complex) Method

The NRCS runoff equation is universally accepted to predict stormwater runoff from precipitation events:

$$Q = \frac{(P - 0.2S)^2}{(P + 0.8S)} \tag{12.12}$$

and

$$S = 1000 \div CN - 10 \tag{12.13}$$

Where:

Q = Runoff (inches)
P = Rainfall (inches)
S = Potential maximum retention after runoff begins (inches)
CN = NRCS Curve Number

The equation can also be written:

$$Q = \frac{(P - I_a)^2}{(P - I_a) + S} \tag{12.14}$$

Where:

I_a = initial abstraction = $0.2S$

The initial abstraction (I_a) is all losses before runoff begins, including water retained in depressions, water intercepted by vegetation, evaporation, and infiltration. The NRCS method determines the amount of runoff based on the NRCS Curve Number, as shown in Table 12.6. The values of CN in Table 12.6a–d are based upon an ARC of II, or median conditions.

12.3.3 Unit Hydrograph

The *unit hydrograph* is defined as "a hydrograph of *direct runoff* resulting from a unit depth of *effective rainfall* uniformly distributed over the basin area during a specified period of time known as the unit duration" (ASCE, 1996). In the United States, the unit depth is typically one inch, and the unit duration typically 24 hours. Once developed for a watershed, unit hydrographs are useful tools for constructing actual hydrographs from actual storm events. To perform this, the direct runoff from a storm must be determined. Direct runoff is the effective rainfall or precipitation excess that runs off the surface

TABLE 12.6a Runoff Curve Numbers for Urban Areas[1] (Modified From NRCS SCS TR-55)

Cover description		Curve numbers for hydrologic soil group			
Cover type and hydrologic condition	Average percent impervious area[2]	A	B	C	D
Fully developed urban areas (vegetation established)					
Open space (lawns, parks, golf courses, cemeteries, etc.)[3]:					
Poor condition (grass cover <50%)		68	79	86	89
Fair condition (grass cover 50% to 75%)		49	69	79	84
Good condition (grass cover >75%)		39	61	74	80
Impervious areas:					
Paved parking lots, roofs, driveways, etc. (excluding right-of-way)		98	98	98	98
Streets and roads:					
Paved; curbs and storm sewers (excluding right-of-way)		98	98	98	98
Paved; open ditches (including right-of-way)		83	89	92	93
Gravel (including right-of-way)		76	85	89	91
Dirt (including right-of-way)		72	82	87	89
Western desert urban areas:					
Natural desert landscaping (pervious areas only)[4]		63	77	85	88
Artificial desert landscaping (impervious weed barrier, desert shrub with 1- to 2-inch sand or gravel mulch and basin borders)		96	96	96	96
Urban districts:					
Commercial and business	85	89	92	94	95
Industrial	72	81	88	91	93
Residential districts by average lot size:					
1/8 acre or less (town houses)	65	77	85	90	92
1/4 acre	38	61	75	83	87
1/3 acre	30	57	72	81	86
1/2 acre	25	54	70	80	85
1 acre	20	51	68	79	84
2 acres	12	46	65	77	82
Developing urban areas					
Newly graded areas (pervious areas only, no vegetation)[5]		77	86	91	94
Idle lands (CN's are determined using cover types similar to those in table 2-2c).					

TABLE 12.6a *(Continued)*

[1]Average runoff condition, and $I_a = 0.2S$.

[2]The average percent impervious area shown was used to develop the composite CN's. Other assumptions are as follows: impervious areas are directly connected to the drainage system, impervious areas have a CN of 98, and pervious areas are considered equivalent to open space in good hydrologic condition. CN's for other combinations of conditions may be computed using figure 2-3 or 2-4.

[3]CN's shown are equivalent to those of pasture. Composite CN's may be computed for other combinations of open space cover type.

[4]Composite CN's for natural desert landscaping should be computed using figures 2-3 or 2-4 based on the impervious area percentage (CN = 98) and the pervious area CN. The pervious area CN's are assumed equivalent to desert shrub in poor hydrologic condition.

[5]Composite CN's to use for the design of temporary measures during grading and construction should be computed using figure 2-3 or 2-4 based on the degree of development (impervious area percentage) and the CN's for the newly graded pervious areas.

and directly into the stream during a storm event. Effective rainfall is the total precipitation minus the hydrologic abstractions (depression storage, infiltration, evapotranspiration). It is the rainfall that is responsible for the large storm flows.

Unit hydrographs can be developed for any watershed from any storm by subtracting the base flow from the total runoff, thus producing the direct runoff. Dividing each direct runoff ordinate by the storm total precipitation excess will provide a unit hydrograph. The precipitation excess can be determined by dividing the sum of the direct runoff ordinates by the total drainage area. Development of a unit hydrograph for a 10-square-mile watershed can be found in Table 12.7.

Once a unit hydrograph is developed for a watershed, a hydrograph from a particular storm can be determined based on the precipitation excess determined by expanding Table 12.7, as shown in Table 12.8.

The precipitation excess and resultant unit hydrograph and storm hydrograph can be found in Figure 12.1.

12.4 ROUTING

Routing is the transposition of a flood wave as it moves through a river or stream reach or reservoir. This subject is covered in detail in the recommended hydrologic engineering texts, and so will only be summarized here. As flows expand into the floodplain, floodplain storage has the effect of temporarily storing water, thus attenuating flow and increasing the flood wave travel time through that reach. This is depicted in Figure 12.2, which shows the importance of avoiding the filling of floodplains and the disadvantages of levees and floodwalls in flood peak attenuation.

Routing methods include *hydraulic routing* based upon the *St. Venant Equations* and *hydrologic methods* based upon the *continuity equation* and

TABLE 12.6b Runoff Curve Numbers for Cultivated Agricultural Lands[1] (Modified From NRCS SCS TR-55)

Cover description			Curve numbers for hydrologic soil group			
Cover type	Treatment[2]	Hydrologic condition[3]	A	B	C	D
Fallow	Bare soil	—	77	86	91	94
	Crop residue cover (CR)	Poor	76	85	90	93
		Good	74	83	88	90
Row crops	Straight row (SR)	Poor	72	81	88	91
		Good	67	78	85	89
	SR + CR	Poor	71	80	87	90
		Good	64	75	82	85
	Contoured (C)	Poor	70	79	84	88
		Good	65	75	82	86
	C + CR	Poor	69	78	83	87
		Good	64	74	81	85
	Contoured & terraced (C&T)	Poor	66	74	80	82
		Good	62	71	78	81
	C&T + CR	Poor	65	73	79	81
		Good	61	70	77	80
Small grain	SR	Poor	65	76	84	88
		Good	63	75	83	87
	SR + CR	Poor	64	75	83	86
		Good	60	72	80	84
	C	Poor	63	74	82	85
		Good	61	73	81	84
	C + CR	Poor	62	73	81	84
		Good	60	72	80	83
	C&T	Poor	61	72	79	82
		Good	59	70	78	81
	C&T + CR	Poor	60	71	78	81
		Good	58	69	77	80
Close-seeded or broadcast legumes or rotation meadow	SR	Poor	66	77	85	89
		Good	58	72	81	85
	C	Poor	64	75	83	85
		Good	55	69	78	83
	C&T	Poor	63	73	80	83
		Good	51	67	76	80

[1]Average runoff condition, and $I_a = 0.2S$.
[2]Crop residue cover applies only if residue is on at least 5% of the surface throughout the year.
[3]Hydraulic condition is based on combination factors that affect infiltration and runoff, including (a) density and canopy of vegetative areas, (b) amount of year-round cover, (c) amount of grass or close-seeded legumes, (d) percent of residue cover on the land surface (good ≥20%), and (e) degree of surface roughness.
Poor: Factors impair infiltration and tend to increase runoff.
Good: Factors encourage average and better-than-average infiltration and tend to decrease runoff.

**TABLE 12.6c   Runoff Curve Numbers for Other Agricultural Lands[1]
(Modified From NRCS SCS TR-55)**

Cover description		Curve numbers for hydrologic soil group			
Cover type	Hydrologic condition	A	B	C	D
Pasture, grassland, or range—continuous forage for grazing.[2]	Poor	68	79	86	89
	Fair	49	69	79	84
	Good	39	61	74	80
Meadow—continuous grass, protected from grazing and generally mowed for hay.	—	30	58	71	78
Brush—brush-weed-grass mixture with brush the major element.[3]	Poor	48	67	77	83
	Fair	35	56	70	77
	Good	30[4]	48	65	73
Woods—grass combination (orchard or tree farm).[5]	Poor	57	73	82	86
	Fair	43	65	76	82
	Good	32	58	72	79
Woods.[6]	Poor	45	66	77	83
	Fair	36	60	73	79
	Good	30[4]	55	70	77
Farmsteads—buildings, lanes, driveways, and surrounding lots.	—	59	74	82	86

[1]Average runoff condition, and $I_a = 0.2S$.
[2]*Poor:*   <50% ground cover or heavily grazed with no mulch.
Fair: 50 to 75% ground cover and not heavily grazed.
Good: >75% ground cover and lightly or only occasionally grazed.
[3]*Poor:*   <50% ground cover.
Fair: 50 to 75% ground cover.
Good: >75% ground cover.
[4]Actual curve number is less than 30; use CN = 30 for runoff computations.
[5]CN's shown were computed for areas with 50% woods and 50% grass (pasture) cover. Other combinations of conditions may be computed from the CN's for woods and pasture.
[6]*Poor:*   Forest litter, small trees, and brush are destroyed by heavy grazing or regular burning.
Fair: Woods are grazed but not burned, and some forest litter covers the soil.
Good: Woods are protected from grazing, and litter and brush adequately cover the soil.

simplified forms of the *energy equation*. These equations and methods can be obtained from standard hydraulic and hydrologic engineering texts.

12.5   MODELING PROCESS

Before a watershed is modeled, the goals and objectives determined in the assessment process will determine what and how much data is required, which

TABLE 12.6*d* **Runoff Curve Numbers for Arid and Semiarid Rangelands**[1] **(Modified From NRCS SCS TR-55)**

Cover description		Curve numbers for hydrologic soil group			
Cover type	Hydrologic condition[2]	A[3]	B	C	D
Herbaceous—mixture of grass, weeds, and low-growing brush, with brush the minor element.	Poor		80	87	93
	Fair		71	81	89
	Good		62	74	85
Oak-aspen—mountain brush mixture of oak brush, aspen, mountain mahogany, bitter brush, maple, and other brush.	Poor		66	74	79
	Fair		48	57	63
	Good		30	41	48
Pinyon-juniper—pinyon, juniper, or both; grass understory.	Poor		75	85	89
	Fair		58	73	80
	Good		41	61	71
Sagebrush with grass understory.	Poor		67	80	85
	Fair		51	63	70
	Good		35	47	55
Desert shrub—major plants include saltbush, greasewood, creosotebush, blackbrush, bursage, palo verde, mesquite, and cactus.	Poor	63	77	85	88
	Fair	55	72	81	86
	Good	49	68	79	84

[1]Average runoff condition, and $I_a = 0.2S$. For range in humid regions, use table 2-2c.
[2]Poor: <30% ground cover (litter, grass, and brush overstory).
 Fair: 30 to 70% ground cover.
 Good: >70% ground cover.
[3]Curve numbers for group A have been developed only for desert shrub.

computer model to use, and what the product of the modeling will be. In modeling, the watershed must be subdivided into subwatersheds. The subdivision process should consider location of obstructions, known flooding, drainage or erosion problems, dams, and tributary confluences. The most downstream point of each of these areas is considered a "point of interest," in which increased runoff must be analyzed for its potential impact. These subwatersheds may then become management areas.

The ultimate goal for selecting the key points of interest is to provide overall watershed stormwater runoff control through effective control of individual subarea storm runoff. Thus, comprehensive control of stormwater runoff for the entire watershed can be achieved through stormwater management in each subwatershed.

12.6   SENSITIVITY ANALYSIS

Before calibrating the model, a parameter sensitivity analysis should be performed to become familiar with how each parameter affects flow peak, vol-

TABLE 12.7   Development of a Unit Hydrograph

Drainage Area (DA) (SM) = 10
Precipitation Excess (Pe)* = Direct RO Sum Total ÷ DA = 1.88

Column	A Time (Hours)	B Q (cfs)	C Base Flow (cfs)	D Direct Runoff (cfs)	E UHG Ordinate** (cfs/in)
	1	1,000	1,000	0	0
	2	1,200	980	220	117
	3	1,400	960	440	234
	4	1,700	980	720	382
	5	2,300	990	1,310	695
	6	3,000	1,000	2,000	1,061
	7	2,800	1,010	1,790	950
	8	2,500	1,020	1,480	785
	9	2,200	1,030	1,170	621
	10	1,900	1,020	880	467
	11	1,700	1,020	680	361
	12	1,500	1,010	490	260
	13	1,400	1,010	390	207
	14	1,300	1,010	290	154
	15	1,200	1,000	200	106
	16	1,100	1,000	100	53
	17	1,000	1,000	0	0
	18	1,000	1,000	0	0
				12,160	

*Pe = 12,160 cfs ÷ 10 sm × 12 in./ft ÷ (1 mi ÷ 5,280 ft) (1 mi ÷ 5,280 ft) (3,600 sec ÷ hr) = 1.88 in.
**UHG: Unit Hydrograph Ordinate = DA ÷ Pe

ume, and timing (DeBarry, 1988). Parameters such as flow lengths, the way watershed and channel slopes are measured, routing parameters, and land cover methodology should be compared. The magnitude of the resultant hydrograph peak, volumes, and timing should be noted with each parameter change. The model should be run with a high, middle, and low parameter value within the parameter range. For those parameters that are highly sensitive—that is, they produce a significant change in the resultant hydrograph—the model should be rerun with smaller increments. Only one parameter should be changed at a time to best evaluate each parameter's sensitivity. Those parameters that produce runoff should be evaluated separately from those parameters that affect flood routing, as both can be used as calibration parameters. A hypothetical example of a sensitivity analysis is shown in Table 12.9. Initial conditions values are a SCS curve number of 75 and a lag of 5.6 hours.

From analyzing the values in Table 12.9, one can see that the curve number (CN) does not affect the time to peak. The lag affects both the time to peak

TABLE 12.8 Development of a Storm Hydrograph from a Unit Hydrograph

Drainage Area (DA) (SM) = 10
Precipitation Excess (Pe)[1] = Direct RO Sum Total ÷ DA = 1.88

Column	A Time (hrs)	B Q (cfs)	C Base Flow (cfs)	D Direct Runoff (cfs)	E UHG Ordinate[2] (cfs/in.)	F Precipitation Excess		L g*e	M h*e	N i*e	O j*e	P Total Discharge[3] (cfs)
	1	1,000	1,000	0	0	0		0	0	0	0	1,000
	2	1,200	980	220	117	0.2		23	0	0	0	1,003
	3	1,400	960	440	234	0.6		47	70	0	0	1,077
	4	1,700	980	720	382	1.1		76	140	128	0	1,325
	5	2,300	990	1,310	695	0.8		139	229	257	93	1,709
	6	3,000	1,000	2,000	1,061	0	g	212	417	420	187	2,237
	7	2,800	1,010	1,790	950	0	h	190	637	765	306	2,907
	8	2,500	1,020	1,480	785		i	157	570	1,168	556	3,471
	9	2,200	1,030	1,170	621		j	124	471	1,045	849	3,520
	10	1,900	1,020	880	467		k	93	373	864	760	3,110
	11	1,700	1,020	680	361			72	280	683	628	2,684
	12	1,500	1,010	490	260			52	217	514	497	2,289
	13	1,400	1,010	390	207			41	156	397	374	1,978
	14	1,300	1,010	290	154			31	124	286	289	1,740
	15	1,200	1,010	200	106			21	92	228	208	1,549
	16	1,100	1,000	100	53			11	64	169	166	1,409
	17	1,000	1,000	0	0			0	32	117	123	1,272
	18	1,000	1,000	0	0			0	0	58	85	1,143
				12,160								

[1]Pe = 12,160 cfs ÷ 10 sm × 12 in./ft ÷ (1 mi ÷ 5,280 ft) (1 mi ÷ 5,280 ft) (3,600 sec ÷ hr) = 1.88 in.
[2]UHG Unit Hydrograph Ordinate = DA ÷ Pe
[3]Total Discharge = Baseflow + columns L, M, N, and O

330

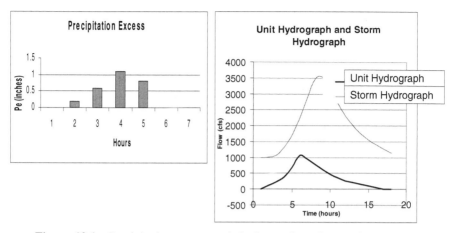

Figure 12.1   Precipitation excess, unit hydrograph, and storm hydrograph.

and peak flow and volume. This will aid in calibration of the model by allowing the user to choose the right parameters to calibrate with.

12.7   GIS

As stated earlier in this book, GIS is a powerful spatial analysis and database management tool that allows unique and very specific watershed assessment functions. The physical parameters discussed in previous chapters can be brought into the GIS, overlaid, and evaluated, and trends determined. Early GIS applications included the development of computer model parameters, such as NRCS curve numbers, drainage area, and others (DeBarry, 1990). GIS has since advanced to allow distributed and time series modeling and three-dimensional analysis. Many applications have been developed to integrate GIS with the hydrologic modeling procedure discussed previously, and which will be discussed in greater detail in Chapter 17.

12.8   MODEL CALIBRATION PROCESS

In order to hydrologically model a watershed with confidence and reliability, the chosen computer model should be calibrated against field data and actual storm events if available, and if not available, statistical frequency analyses or regression methods (DeBarry, 1988). Calibration is the process of fine-tuning the model by adjusting model input parameters within their tolerance to accurately simulate observed conditions, and is an integral component of any modeling process. The method that provides the most reliable results is to calibrate against actual rainfall and stream gage data, if accurate data is

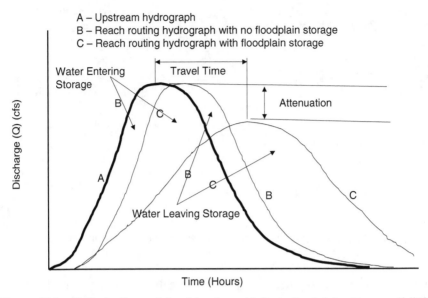

Figure 12.2   Typical effects of flood routing with large floodplain storage and little floodplain storage potential.

available; however, reliable watershed-specific data may not be available for each watershed.

Calibration can be performed utilizing a variety of measures, including statistical frequency analyses and regional regression methods. These methods allow only peak flow calibration. The most accurate data available should always be used to calibrate the model. Models should be calibrated utilizing the methods in the order established in Table 12.10.

Each of these methods is described in more detail in the following subsections.

TABLE 12.9   Hypothetical Sensitivity Analysis for Watershed Parameters

		Results		
Parameter	Value	Time to Peak (hrs)	Peak Flow (cfs)	Hydrograph Volume (ac-ft)
Runoff parameter (i.e., CN)	98	6	12,000	560,000
Runoff parameter (i.e., CN)	75	6	9,000	450,000
Runoff parameter (i.e., CN)	65	6	7,500	250,000
Lag (hrs)	4.3	5	10,000	390,000
Lag (hrs)	6.3	6.7	8,500	420,000

TABLE 12.10 Advantages and Disadvantages of Calibration Data

Priority	Data	Advantages	Disadvantages
1	Actual (recorded) rainfall hyetographs and stream flow hydrographs	Can calibrate peak, timing, and volume.	Data not always available; time-consuming.
2	Statistical Frequency Analysis	Based on actual data.	Calibrates peaks only; does not calibrate volumes or time.
3	Regional Regression Methods	Quick	Watershed may not fit "Regional Trend." Calibrates peaks only; does not calibrate volumes or time.

12.8.1 Calibration against Actual Storm Events

12.8.1.1 Rainfall Data. Rainfall events recorded in hourly or 15-minute increments can be obtained from the National Weather Service (NWS), as can daily total data for rain gages. Rainfall data may also be available from a local university or television station.

Ideally, those storms that produce at least one inch of runoff (rule of thumb: two or more inches of rainfall) in a 24-hour period with a typical bell-shaped distribution or a distribution similar to the SCS type curve of the region where the study is being performed should be chosen for calibration. This is not, however, realistically obtainable; nevertheless, this rainfall pattern should be kept in mind when choosing the storms. Single peaking storms—those that typically represent a unit hydrograph from the SCS distribution—are also ideal for calibrating and verifying the model. Therefore, the stream gage data should be compared with the rainfall data before finalizing the storms to use for calibration.

Data should be analyzed for completeness and reliability. Winter storms should be omitted from further analysis to avoid obtaining misleading results influenced by frozen ground or snowmelt. In addition, thunderstorms that have a high variation in regional rainfall distribution and amounts should be analyzed closely for consistency and usability.

Utilization of total station data should be based upon three factors:

- Proximity of gaging station to the watershed
- Consistency of the data throughout the gaging station period of record
- Consistency of the data with other stations for the particular storms in question

12.8.1.2 Stream Data. Stream gage data should be obtained from the United States Geologic Survey (USGS) or other reliable source. One can review the annual peak series and choose approximately two dozen storms to calibrate against. Of the two dozen storms, approximately six should be set aside for verification; the other six will be thrown out for one reason or another, as discussed in Section 12.8.1.1. It is unusual to find a gaging station on the stream at the location at which it is required for the analysis at hand.

Other available data may include staff gage information recorded by FEMA, USGS, the Army Corps of Engineers, or a local agency or group. Typically, the staff gage data will give peak storm flow values, which can aid in model calibration. The USGS data are based on stream stage, therefore, a rating curve is required from USGS to convert the stage data to stream discharge. It is important to ensure that the correct rating curve is applied to the stage data.

The model should be run; then a hydrograph should be developed and graphically compared to the observed hydrograph. The simulated hydrograph peak, volume, and time most closely matches the observed hydrograph when the Nash-Sutcliffe value nears 1 and the root mean square (RMS) errors are minimized, as shown in Figure 12.3. The Nash-Sutcliffe coefficient (R^2) is given in equation 12.15:

$$R^2 = 1 - \frac{\sum\limits_{Sim=1}^{n} (q_{sim} - q_{obs})^2}{\sum\limits_{Sim=1}^{n} (q_{sim} - q_{mean})^2} \tag{12.15}$$

Where:

R^2 = Nash-Sutcliffe coefficient
q_{sim} = Simulated discharge at time I
q_{obs} = Observed discharge at time I
q_{mean} = Mean observed discharge for the event

An R^2 value of 1 represents a perfect fit. The RMS error equation 12.16 is:

$$RMS = ((1 \div n) \{n \text{ SUM } i = 1\} (q_{sim}, i - q_{obs}, i)^2)^{1/2} \tag{12.16}$$

Where:

RMS = Root mean square error
q_{sim} = Simulated discharge at time i
q_{obs} = Observed discharge at time i

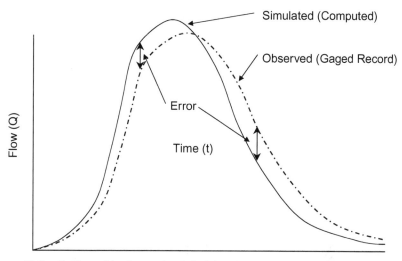

Figure 12.3 Calibrated hydrograph minimizing the root mean square errors (modified from U.S. Army Corps of Engineers, 1987).

An RMS value of 0 represents a perfect set of data. Some hydrologic computer models such as HEC-1 (U.S. Army Corps of Engineers, 1987) and HEC-HMS (U.S. Army Corps of Engineers, 2003) have optimization routines that minimize the RMS error to obtain a calibrated hydrograph. HEC-HMS allows many options for optimization. Naturally, the initial calibrated parameters for one storm will have the same value as the initial calibration run of another storm in the calibration step. The parameter values that provide the lowest total RMS values for all calibration storms should be the preverification final calibrated parameters. Once these parameter values are calibrated, the verification runs should be done, and any final minor adjustments made to the parameters.

The deviation of the runoff volumes, D_v, can be computed for annual, seasonal, or event runoff as shown in equation 12.17:

$$D_v[\%] = \frac{V_R - V_R'}{V_R} \cdot 100 \tag{12.17}$$

Where:

D_v = Deviation of runoff volume
V_R = Measured yearly, seasonal, or event runoff volume
V_R' = Simulated yearly, seasonal, or event runoff volume

12.8.2 Statistical Frequency Analysis

A *frequency analysis* is a probability analysis of annual maximum storms, used to determine flow, stage, or precipitation frequency of occurrence. It is used to determine the *return period (Tr)*, also referred to as the *recurrence interval,* of a given flood peak, or conversely, the magnitude of flow of a given return period. The return period is the average time interval between actual occurrences of a hydrologic event of an equal or greater magnitude (ASCE, 1996). In performing a frequency analysis on storm peaks, the peak annual storm peaks, or *annual series,* is obtained based on water year. The frequency analysis is then performed on these annual peaks.

One method to compute return periods is through the *plotting position* method. Several equations are available to plot a cumulative distribution of the return period (Tr):

$$\text{Weibull: } Tr = (n + 1) \div m \tag{12.17}$$

$$\text{Cunnane: } Tr = (n + 0.2) \div (m - 0.4) \tag{12.18}$$

$$\text{Hazen: } Tr = 2n \div (2m - 1) \tag{12.19}$$

Where:

n = Number of years of record
m = Rank of event (m = 1 for the largest)

The general process would be as follows:

1. Compile the annual series from n-year historical period at a single stream gage station.
2. Rank in descending order and assign a plotting position (estimate of Tr).
3. Select probability (log-log) paper and plot Q versus Tr.
4. Fit a straight line through the plotted points to obtain the least square of errors.

An example of flow data and the return period (Tr) computed for equations 12.17, 12.18, and 12.19 can be found in Table 12.11.

These can then be plotted on a semilogarithmic scale, as shown in Figure 12.4, and flows for defined return periods obtained directly from the graph. Theoretically, one should only use the data for a return period determined from the number of years of record. In other words, predicting a storm with a 1 percent chance of occurrence in any one year (100-year flood) with only 11 years of record may provide unreliable results. That said, the period of record may also influence the results. For instance, for a watershed in the

TABLE 12.11   Weibull, Cunnane, and Hazen Plotting Positions

Years of record (n) = 11				
Rank (m)	Flow	Weibull Tr (yrs)	Cunnane Tr (yrs)	Hazen Tr (yrs)
7	2,090	1.71	1.70	1.69
2	3,720	6.00	7.00	7.33
6	2,190	2.00	2.00	2.00
3	3,710	4.00	4.31	4.40
5	2,190	2.40	2.43	2.44
10	1,250	1.20	1.17	1.16
11	1,180	1.09	1.06	1.05
8	2,000	1.50	1.47	1.47
1	3,760	12.00	18.67	22.00
4	2,890	3.00	3.11	3.14
9	1,820	1.33	1.30	1.29

northeastern United States, relying on only five years of record for the years 1998 to 2002, when a severe drought occurred with no major rainfall events, would produce arbitrarily low flow values for the five-year return period. As can be seen from Figure 12.4, the small number of data points leads to a higher variability in accuracy and a log-log plot would provide a straighter line.

Another method to correlate return period with peak flows is a *frequency factor* method, or fitting a theoretical distribution to an annual series. There are several frequency factor methods, including the Gumbel, Log Gumbel, and Log Normal. In order to provide consistency in flood frequency prediction, the U.S. Water Resources Council published Bulletin 15, "A Uniform Technique for Determining Flood Flow Frequencies," in December 1967, which utilized the Log Pearson Type II distribution. Bulletin 15 was updated by the Hydrology Subcommittee, Interagency Advisory Committee on Water

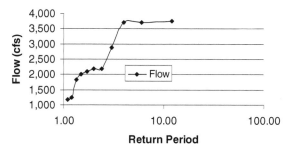

Figure 12.4   Plot of frequency analyses.

Data (1982) in Bulletin 17B, which utilizes the Log Pearson Type III methodology. The recommended technique for fitting a Log Pearson Type III distribution to observed annual peaks is to compute the logarithms base 10 of the discharge (Q) (Hydrology Subcommittee, 1982):

$$\text{Log } Q_{Tr} = \bar{y} + KS_y \qquad (12.20)$$

Where:

Tr = Return period
y = Logarithm of annual peak flow
$\bar{y}$ = Mean logarithm
Sy = Standard deviation
K = $f(T_r, Cs_y)$ from Table 12.12 and tables in Bulletin 17B (expanded version)
Cs_y = Skew coefficient (from Figure 12.5)
n = Number of years of record

The mean logarithm ($\bar{y}$), standard deviation (Sy), and skew coefficient (Cs_y) can each be further defined as:

$$\bar{y} = 1/n \; \Sigma \log Q = \Sigma \, y \div n \qquad (12.21)$$

$$Sy = [1 \div (n-1) \; \Sigma \; (\log Q^{-y})^2]^{1/2} = [(\Sigma \; (y - \bar{y})^2) \div (n-1)]^{0.5} \quad (12.22)$$

or

$$Sy = \sigma \log Q = \sqrt{\text{variance}} \qquad (12.23)$$

and

$$Cs_y = (n \; \Sigma \; (y - \bar{y})^3) \div ((n-1)(n-2)Sy^3) \qquad (12.24)$$

(Note: the skew coefficient Cs_y is referred to as "G" in Bulletin 17B)

or:

$$Cs_y = \frac{n \times \displaystyle\sum_{i}^{n} (\log Q - \text{avg}(\log Q))^3}{(n-1)(n-2)(\sigma \log Q)^3} \qquad (12.25)$$

TABLE 12.12 K Values for Pearson Type III Distribution

	Recurrence Interval in Years										
	200	100	50	25	10	5	2	1.2500	1.1111	1.0526	1.0101
	Exceedance Probability										
Skew Coeff. G	.05	.01	.02	.04	.10	.20	.50	.80	.90	.95	.99
	Positive Skew										
3.0	4.970	4.051	3.152	2.278	1.180	0.420	−0.396	−0.636	−0.660	−0.665	−0.667
2.9	4.909	4.013	3.134	2.277	1.195	0.440	−0.390	−0.651	−0.681	−0.688	−0.690
2.8	4.847	3.973	3.114	2.275	1.210	0.460	−0.384	−0.666	−0.702	−0.711	−0.714
2.7	4.783	3.932	3.093	2.272	1.224	0.479	−0.376	−0.681	−0.724	−0.736	−0.740
2.6	4.718	3.889	3.071	2.267	1.238	0.499	−0.368	−0.696	−0.747	−0.762	−0.769
2.5	4.652	3.845	3.048	2.262	1.250	0.518	−0.360	−0.711	−0.771	−0.790	−0.799
2.4	4.584	3.800	3.023	2.256	1.262	0.537	−0.351	−0.725	−0.795	−0.819	−0.832
2.3	4.515	3.753	2.997	2.248	1.274	0.555	−0.341	−0.739	−0.819	−0.850	−0.867
2.2	4.444	3.705	2.970	2.240	1.284	0.574	−0.330	−0.752	−0.844	−0.882	−0.905
2.1	4.372	3.656	2.942	2.230	1.294	0.592	−0.319	−0.765	−0.869	−0.914	−0.946
2.0	4.298	3.605	2.912	2.219	1.302	0.609	−0.307	−0.777	−0.895	−0.949	−0.990
1.9	4.223	3.553	2.881	2.207	1.310	0.627	−0.294	−0.788	−0.920	−0.984	−1.037
1.8	4.147	3.499	2.848	2.193	1.218	0.643	−0.282	−0.799	−0.945	−1.020	−1.087
1.7	4.069	3.444	2.815	2.179	1.324	0.660	−0.268	−0.808	−0.970	−1.056	−1.140
1.6	3.990	3.388	2.780	2.163	1.329	0.675	−0.254	−0.817	−0.994	−1.093	−1.197
1.5	3.910	3.330	2.743	2.146	1.333	0.690	−0.240	−0.825	−1.018	−1.131	−1.256
1.4	3.828	3.271	2.706	2.128	1.337	0.705	−0.225	−0.832	−1.041	−1.168	−1.318
1.3	3.745	3.211	2.666	2.108	1.339	0.719	−0.210	−0.838	−1.064	−1.206	−1.383
1.2	3.661	3.149	2.626	2.087	1.340	0.732	−0.195	−0.844	−1.086	−1.243	−1.449
1.1	3.575	3.087	2.585	2.066	1.341	0.745	−0.180	−0.848	−1.107	−1.280	−1.518

TABLE 12.12 (Continued)

	Recurrence Interval in Years										
	200	100	50	25	10	5	2	1.2500	1.1111	1.0526	1.0101
	Exceedance Probability										
Skew Coeff. G	.05	.01	.02	.04	.10	.20	.50	.80	.90	.95	.99
1.0	3.489	3.022	2.542	2.043	1.340	0.758	−0.164	−0.852	−1.128	−1.317	−1.588
.9	3.401	2.957	2.498	2.018	1.339	0.769	−0.148	−0.854	−1.147	−1.353	−1.660
.8	3.312	2.891	2.453	1.993	1.336	0.780	−0.132	−0.856	−1.166	−1.388	−1.733
.7	3.223	2.824	2.407	1.967	1.333	0.790	−0.116	−0.857	−1.183	−1.423	−1.806
.6	3.132	2.755	2.359	1.939	1.328	0.800	−0.099	−0.857	−1.200	−1.458	−1.880
.5	3.041	2.686	2.311	1.910	1.323	0.808	−0.083	−0.856	−1.216	−1.491	−1.955
.4	2.949	2.615	2.261	1.880	1.317	0.816	−0.066	−0.855	−1.231	−1.524	−2.029
.3	2.856	2.544	2.211	1.849	1.309	0.824	−0.050	−0.853	−1.245	−1.555	−2.104
.2	2.763	2.472	2.159	1.818	1.301	0.830	−0.033	−0.850	−1.258	−1.586	−2.178
.1	2.670	2.400	2.107	1.785	1.292	0.836	−0.017	−0.846	−1.270	−1.616	−2.252
.0	2.576	2.326	2.054	1.751	1.282	0.842	0	−0.842	−1.282	−1.645	−2.326
Negative Skew											
−.1	2.482	2.252	2.000	1.716	1.270	0.846	0.017	−0.836	−1.292	−1.673	−2.400
−.2	2.388	2.178	1.945	1.680	1.258	0.850	0.033	−0.830	−1.301	−1.700	−2.472
−.3	2.294	2.104	1.890	1.643	1.245	0.853	0.050	−0.824	−1.309	−1.726	−2.544
−.4	2.201	2.029	1.834	1.606	1.231	0.855	0.066	−0.816	−1.317	−1.750	−2.615
−.5	2.108	1.955	1.777	1.567	1.216	0.856	0.083	−0.808	−1.323	−1.774	−2.686
−.6	2.016	1.880	1.720	1.528	1.200	0.857	0.099	−0.800	−1.328	−1.797	−2.755
−.7	1.926	1.806	1.663	1.488	1.183	0.857	0.116	−0.790	−1.333	−1.819	−2.824
−.8	1.837	1.733	1.606	1.448	1.166	0.856	0.132	−0.780	−1.336	−1.839	−2.891

-.9	-2.957	-1.858	-1.339	-0.769	0.148	0.854	1.147	1.407	1.549	1.660	1.749
-1.0	-3.022	-1.877	-1.340	-0.758	0.164	0.852	1.128	1.366	1.492	1.588	1.664
-1.1	-3.087	-1.894	-1.341	-0.745	0.180	0.848	1.107	1.324	1.435	1.518	1.581
-1.2	-3.149	-1.910	-1.340	-0.732	0.195	0.844	1.086	1.282	1.379	1.449	1.501
-1.3	-3.211	-1.925	-1.339	-0.719	0.210	0.838	1.064	1.240	1.324	1.383	1.424
-1.4	-3.271	-1.938	-1.337	-0.705	0.225	0.832	1.041	1.198	1.270	1.318	1.351
-1.5	-3.330	-1.951	-1.333	-0.690	0.240	0.825	1.018	1.157	1.217	1.256	1.282
-1.6	-3.388	-1.962	-1.329	-0.675	0.254	0.817	0.994	1.116	1.166	1.197	1.216
-1.7	-3.444	-1.972	-1.324	-0.660	0.268	0.808	0.970	1.075	1.116	1.140	1.155
-1.8	-3.499	-1.981	-1.318	-0.643	0.282	0.799	0.945	1.035	1.069	1.087	1.097
-1.9	-3.553	-1.989	-1.310	-0.627	0.294	0.788	0.920	0.996	1.023	1.037	1.044
-2.0	-3.605	-1.996	-1.302	-0.609	0.307	0.777	0.895	0.959	0.980	0.990	0.995
-2.1	-3.656	-2.001	-1.294	-0.592	0.319	0.765	0.869	0.923	0.939	0.946	0.949
-2.2	-3.705	-2.006	-1.284	-0.574	0.330	0.752	0.844	0.888	0.900	0.905	0.907
-2.3	-3.753	-2.009	-1.274	-0.555	0.341	0.739	0.819	0.855	0.864	0.867	0.869
-2.4	-3.800	-2.011	-1.262	-0.537	0.351	0.725	0.795	0.823	0.830	0.832	0.833
-2.5	-3.845	-2.012	-1.250	-0.518	0.360	0.711	0.771	0.793	0.798	0.799	0.800
-2.6	-3.889	-2.013	-1.238	-0.499	0.368	0.696	0.747	0.764	0.768	0.769	0.769
-2.7	-3.932	-2.012	-1.224	-0.479	0.376	0.681	0.724	0.738	0.740	0.740	0.741
-2.8	-3.973	-2.010	-1.210	-0.460	0.384	0.666	0.702	0.712	0.714	0.714	0.714
-2.9	-4.013	-2.007	-1.195	-0.440	0.390	0.651	0.681	0.683	0.689	0.690	0.690
-3.0	-4.051	-2.003	-1.180	-0.420	0.396	0.636	0.660	0.666	0.666	0.667	0.667

Source: Leonard et al., 1992.
After Water Resources Council, Bulletin No. 15 (13).

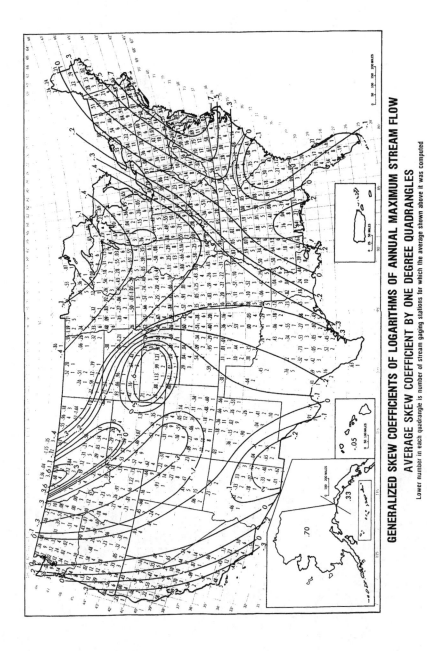

GENERALIZED SKEW COEFFICIENTS OF LOGARITHMS OF ANNUAL MAXIMUM STREAM FLOW

AVERAGE SKEW COEFFICIENT BY ONE DEGREE QUADRANGLES

Lower number in each quadrangle is number of stream gaging stations for which the average shown above it was computed

Figure 12.5 Generalized skew coefficients of logarithms of annual maximum stream flow (Hydrology Subcommittee, Bulletin 17B, 1982).

342

The variance can be defined as:

$$\text{Variance} = \frac{\sum_{i}^{n} (\log Q - \text{avg}(\log Q))^2}{n - 1} \qquad (12.26)$$

The process has been developed in Bulletin 17B and can be seen through the following central Oklahoma example:

Tr = Return Period: = 100
$\bar{y}$ = $1/n \sum \log Q$: y = 2436 cfs log y = 3.359
Sy = standard deviation = $[1/(n - 1) \sum (\log Q^{-y})^2]^{1/2}$: Sy = 950 cfs, log Sy = 0.177
K = $f(T_r, Cs_y)$ from tables: K_{100} = 2.165 from Table 12.12
Cs_y = skew coefficient: -0.208 from Figure 12.5
Log Q_{100} = 3.359 + 2.165 (0.177) = 3.742
Q_{100} = 5,523 cfs

Regional skew coefficients (Cs_r) should be utilized if there are fewer than 25 years of record. A weighted skew coefficient (Cs_w) should be used between 25 and 100 years of record, and a station skew (Cs_s) should be utilized if there are more than 100 years of record. Cs_r = can be obtained from Figure 12.5 (from Bulletin 17B).

$$Cs_w = (n - 25) \div 75 \times C_{\text{station}} + (1 - (n - 25 \div 75)) C_{\text{region}} \qquad (12.27)$$

$$Cs_s = (1 + 6 \div n)\, Cs_1 \qquad (12.28)$$

$$Cs_1 = n \div ((n - 1)(n - 2)) \times (\sum (x_i - \bar{x})^3 \div S_x^3) \qquad (12.29)$$

When performing frequency analyses, one should consult Bulletin 17B for advice on outliers, incomplete records, and so on.

The United States Geologic Survey (USGS, 2002) has developed the computer program PEAKFQ, a flood frequency analysis based upon Bulletin 17B, published by the Interagency Advisory Committee on Water Data in 1982. It is a sibling of the WATSTORE program developed by W.H. Kirby in 1981. WATSTORE, an acronym for Water Data Storage and Retrieval System, has been overseen by USGS since 1981 until recently, and consists of several files in which water data are grouped and stored by common characteristics and data-collection frequencies. It has recently been replaced by the USGS National Water Information System (NWIS), which is available on the World Wide Web at http://waterdata.usgs.gov/nwis. Files are maintained for the storage of:

- Surface-water, quality-of-water, and groundwater data measured daily or more frequently

- Annual peak values and peaks above a base flow for stream flow stations
- Chemical analyses for surface- and groundwater sites
- Geologic and inventory data for groundwater sites
- Water use summary data

Daily values, water-quality, and peak-flow files are generally used in retrieving data from NWIS. In addition, NWIS maintains an index file of sites for which data are stored in the system. NWIS data files with a station header, station identification, and peak-flow format can be read directly from PEAKFQ using the Input/ASCII option.

PEAKFQ uses the *method of moments* to fit the Pearson Type III distribution to the logarithms of annual flood peaks. The method of moments equates sample moments to parameter estimates. Method of moments computations are easy to implement, but have the disadvantage that they are often not available. The skew may be from the Bulletin 17B skew map or a user-developed generalized skew for a region, computed from the data or weighted between the generalized skew and computed station skew. Adjustments can be made for high and low outliers and historic information. PEAKFQ is available in DOS for Windows applications and Sun UNIX platforms; it can be downloaded from http://water.usgs.gov/software/surface_water.html. The input and output screens for PEAKFQ are shown in Figures 12.6 and 12.7, respectively.

The output provides for a statistical frequency analysis, as shown in Figure 12.8. Official versions of U.S. Geological Survey water-resources analysis software, including PEAKFQ, are available for electronic retrieval via the World Wide Web from http://water.usgs.gov/software/.

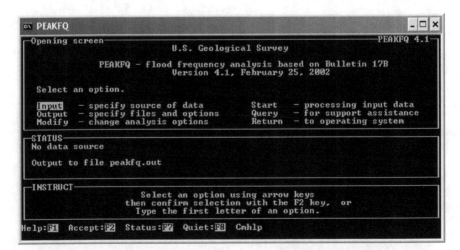

Figure 12.6 PEAKFQ input screen. (Thomas, et al., 1998).

ANNUAL FREQUENCY CURVE PARAMETERS -- LOG-PEARSON TYPE III

| | FLOOD BASE | | LOGARITHMIC | | |
| | EXCEEDANCE | | | STANDARD | |
	DISCHARGE	PROBABILITY	MEAN	DEVIATION	SKEW
SYSTEMATIC RECORD	0.0	1.0000	3.3684	0.2456	0.730
BULL.17B ESTIMATE	0.0	1.0000	3.3684	0.2456	0.668

ANNUAL FREQUENCY CURVE -- DISCHARGES AT SELECTED EXCEEDANCE PROBABILITIES

| ANNUAL EXCEEDANCE PROBABILITY | BULL.17B ESTIMATE | SYSTEMATIC RECORD | 'EXPECTED PROBABILITY' ESTIMATE | 95-PCT CONFIDENCE LIMITS FOR BULL.17B ESTIMATES | |
				LOWER	UPPER
0.9950	773.1	797.9	717.3	510.6	1019.0
0.9900	829.6	851.4	778.2	558.6	1083.0
0.9500	1038.0	1050.0	999.3	740.6	1313.0
0.9000	1192.0	1199.0	1163.0	880.1	1485.0
0.8000	1438.0	1439.0	1418.0	1106.0	1760.0
0.5000	2194.0	2181.0	2194.0	1796.0	2660.0
0.2000	3657.0	3645.0	3743.0	2996.0	4722.0
0.1000	4959.0	4966.0	5208.0	3946.0	6848.0
0.0400	7067.0	7134.0	7778.0	5358.0	10700.0
0.0200	9031.0	9181.0	10430.0	6589.0	14660.0
0.0100	11390.0	11660.0	13920.0	7995.0	19780.0
0.0050	14220.0	14670.0	18620.0	9605.0	26370.0
0.0020	18830.0	19650.0	27510.0	12100.0	38030.0

EMPIRICAL FREQUENCY CURVES -- WEIBULL PLOTTING POSITIONS

WATER YEAR	RANKED DISCHARGE	SYSTEMATIC RECORD	BULL.17B ESTIMATE
1955	8800.0	0.0400	0.0400
1956	8280.0	0.0800	0.0800
1961	4340.0	0.1200	0.1200
1968	3630.0	0.1600	0.1600
1953	3220.0	0.2000	0.2000
1952	3170.0	0.2400	0.2400
1962	3060.0	0.2800	0.2800
1949	3020.0	0.3200	0.3200
1948	2970.0	0.3600	0.3600
1958	2500.0	0.4000	0.4000
1951	2490.0	0.4400	0.4400
1945	2290.0	0.4800	0.4800
1947	2220.0	0.5200	0.5200
1960	2140.0	0.5600	0.5600
1959	1960.0	0.6000	0.6000
1963	1780.0	0.6400	0.6400
1954	1760.0	0.6800	0.6800
1967	1580.0	0.7200	0.7200
1946	1470.0	0.7600	0.7600
1964	1380.0	0.8000	0.8000
1957	1310.0	0.8400	0.8400
1950	1210.0	0.8800	0.8800
1966	1040.0	0.9200	0.9200
1965	980.0	0.9600	0.9600

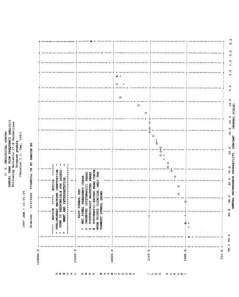

Figure 12.7 PEAKFQ output screen (Thomas et al., 1998).

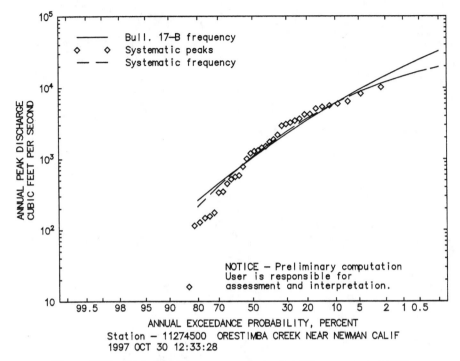

Figure 12.8 Statistical frequency analysis plot (Thomas et al., 1998).

In addition to PEAKFQ, there are several other computer programs to determine extreme value floods. Table 12.13 displays a summary output of a flood frequency analysis using the Penn State Mini Extreme Value Analysis (MiniExvan) program (Aron, 1987), which computes flows using not only the Log Pearson analysis, but also the Gumbel, Log Gumbel, Log Normal, and Log Pearson with regional skew methods. The variability in flows should be noted and carefully selected to calibrate the model. For instance, the Gumbel typically underestimates the higher frequency (1.11-year) flows, whereas the Log Gumbel typically overestimates the lower frequency (100-year) flows.

Naturally, stream gages rarely exist at the exact location where flows are required in the watershed being studied. Therefore, flows must be developed at ungaged locations, whether that point of interest is another location within the watershed or in a hydrologically similar watershed. A hydrologically similar watershed would be one with similar geology, land use, climate, and so on. One approach to test similarity is correlation of base flow measurements at the site of interest to daily means at a nearby gage.

If there is a gage in the watershed, this can be accomplished through the ratio of drainage area and flows of the gage to the site in question. The ratio is raised to a power to account for regional variability, as shown in equation 12.30.

TABLE 12.13 Summary Output of a Flood Frequency Analysis Using the Penn State Mini Extreme Value Analysis (MiniExvan) Program

Ogden Creek at Carrol, Kentucky	Gage No	01470000
Drainage Area:	162 square miles	
Water Years 1936–2004		

Return Period	Gumbel	Log Gumbel	Log Pearson	Log Normal	Log Pearson with Regional Skew (Skew = +0.47)
1.11	964	2,237	2,200	2,116	2,184
1.25	2,016	2,645	2,671	2,693	2,673
2.00	4,487	3,919	4,069	4,271	4,101
2.33	5,108	4,326	4,485	4,706	4,521
5	7,811	6,651	6,639	6,774	6,665
10	10,012	9,440	8,825	8,620	8,800
25	12,794	14,694	12,224	11,145	12,054
50	14,857	20,403	15,281	13,158	14,928
100	16,905	28,263	18,837	15,272	18,218

*Results based on 68 years of record.

$$Q_{\text{site}} = (DA_{\text{site}} \div DA_{\text{gage}})^x \times Q_{\text{gage}} \qquad (12.30)$$

Where:

Q_{site} = Flow required at the point of interest
DA_{site} = Drainage area at the point of interest
DA_{gage} = Drainage area at the gage
Q_{gage} = Flow determined for a particular recurrence interval at the gage
X = Power to adjust flows for regional effects

The x value should be obtained from the regional or state reports and is typically in the range of 0.4 to 1.1. It will vary by state, region, and return period. If x is not defined for a state or region, the value of 1 should be used (USGS, 2002), but with caution.

Assuming the region had an x value defined as 0.83, the flows calculated at the gage in Table 12.13 could be transposed to another point of interest in the watershed with a drainage area of 45 square miles using equation 12.30, as shown in Table 12.14.

12.8.3 Regional Regression Methods

Another means to predict stream flow is through *regression methods,* which are equations that have been regionally developed to determine discharges with exceedance probabilities, means, standard deviations, and so on from

TABLE 12.14   Transposition of Flows from Gage to Point of Interest

		Ogden Creek at River Road Drainage Area: Water Years 1936–2004*		45 square miles $x = 0.83$	
Return Period	Gumbel	Log Gumbel	Log Pearson	Log Normal	Log Pearson with Regional Skew (+0.47)
1.11	333	773	760	731	754
1.25	696	913	923	930	923
2.00	1,550	1,353	1,405	1,475	1,416
2.33	1,764	1,494	1,549	1,625	1,561
5	2,698	2,297	2,293	2,339	2,302
10	3,458	3,260	3,048	2,977	3,039
25	4,418	5,075	4,222	3,849	4,163
50	5,131	7,046	5,277	4,544	5,155
100	5,838	9,761	6,505	5,274	6,292

*Results based on 68 years of record.

regionally derived parameters and patterns. The general form of a simple linear regional regression equation is given in equation 12.31:

$$Y = a + bX \tag{12.31}$$

Where:

Y = Dependent variable
a = Regression constant
b = Regression coefficient
X = Independent variable

The independent variables may be slope of the main stream, drainage area, watershed width-to-length ratio, percent impervious area, and so on. These parameters can easily be obtained from available maps or by using the GIS. For instance, several watersheds within a physiographic region may exhibit the same flow patterns when accounting for their parameter differences. Obtaining the gage data for these particular watersheds and correlating them provides an equation that can then predict flows on ungaged or adjacent watersheds, or in a location in the same watershed or on the same stream where a gage is not located. The general form of a multiple linear regional regression equation would be:

$$Q_x = a + b_1(w \div l)^z + b_2 \log (s) - i^y \tag{12.32}$$

Where the parameters might be defined as follows:

Q_x = Flow from a flood with a $1 \div x$ percent chance of occurrence

a = Drainage area

b_1 and b_2 = Coefficients determined for the equation

$(w \div l)^z$ = Watershed width-to-length ratio raised to a power determined for the equation

s = Slope of main stem of stream from watershed divide to outlet

i^y = Percent impervious area raised to a power determined for the equation

Equation 12.32 is just an example of the general format. Federal, state (environmental protection, dam safety, departments of transportation), regional, and local agencies should be contacted to ask whether there are any regional regression methods applicable to the watershed being studied.

Many regional regression equations have been developed by the U.S. Army Corps of Engineers, USGS, state DOTs, and others. USGS compiled the many regional regression equations for the United States and Puerto Rico, and in 1994 released the National Flood Frequency (NFF) computer program (Jennings, et al.). An updated version, 3.0, with new equations was released in 2002 (Ries, 2002). It contains 2,065 regression equations and 289 flood regions nationwide. The software and documentation are available at http://water.gov/software/nff.html.

NFF provides equations for weighting flows for stream gaging stations and ungaged sites, and improves on equation 12.32 by including in the weighting the flow obtained from regression equations. It also incorporates state or regional regression equations, with each reference hyperlinked to its Web site. For instance, in the sample given here, the equations in NFF are obtained from the report "Analysis of the Magnitude and Frequency of Floods in Colorado," by J.E. Vaill (2000). Through the Web site, all the state Water-Resources Investigation Reports that contain the regression equations contained in NFF can be accessed for estimating magnitude and frequency of peak flows in gaged and ungaged, urban and rural watersheds.

NFF Version 3.0 allows (Ries, 2002):

- Estimation of flood frequencies for ungaged rural and urban basins
- Creation of hydrographs
- Creation and manipulation of flood frequency curves
- Improved estimates of flood flows frequency at gages through weighted estimates at the gages using regression equations
- Improved estimates of flood flows frequency at ungaged sites through weighted estimates from the gages based on ratio of drainage area (raised to a power)

The input screen for NFF for a sample rural watershed in the mountain region of Colorado can be seen in Figure 12.9. Note that, for this region of Colorado, the regression equation has only two variables, *drainage area* and

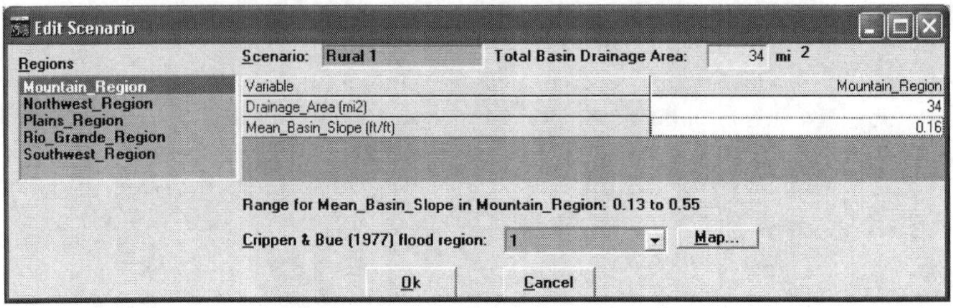

Figure 12.9 NFF input screen.

mean basin slope, as just described. The actual equations are shown in Table 12.15 (Vaill, 2000).

It is important to know how the mean basin slope is defined in the equations being utilized, as it may vary with application. Sample NFF output can be found in Figures 12.10, 12.11, and 12.12.

To develop the hydrograph, the lag time for the watershed is required, as shown in Figure 12.13.

Crippen and Bue (1977) developed 17 flood frequency regions and curves for the conterminous United States. Peak flood flows from thousands of observation sites were studied to provide a guide for estimating potential maximum flood flows. Data were selected from 883 sites with drainage areas of less than 10,000 square miles (25,900 square kilometers) and were grouped into regional sets. Outstanding floods for each region were plotted on graphs showing selected peak discharges versus drainage areas, and envelope curves

TABLE 12.15 **Regional Flood-Frequency Equations, Mountain Region Colorado**

Recurrent Interval, in Years	Regression Equation	Standard Error of the Model, in Percent	Average Standard Error of Prediction, in Percent
Mountain region			
2	$Q = 11.0 \, (A)^{0.663} \, (S + 1.0)^{3.465}$	58	52
5	$Q = 17.9 \, (A)^{0.677} \, (S + 1.0)^{2.739}$	48	47
10	$Q = 23.0 \, (A)^{0.685} \, (S + 1.0)^{2.364}$	44	45
25	$Q = 29.4 \, (A)^{0.695} \, (S + 1.0)^{2.004}$	41	44
50	$Q = 34.5 \, (A)^{0.700} \, (S + 1.0)^{1.768}$	41	44
100	$Q = 39.5 \, (A)^{0.706} \, (S + 1.0)^{1.577}$	42	44
200	$Q = 44.6 \, (A)^{0.710} \, (S + 1.0)^{1.408}$	44	45
500	$Q = 51.5 \, (A)^{0.715} \, (S + 1.0)^{1.209}$	48	47

[Q, discharge, in cubic feet per second; A, drainage area, in square miles; P, mean annual precipitation, in inches; S, mean drainage-basin slope, in foot per foot]

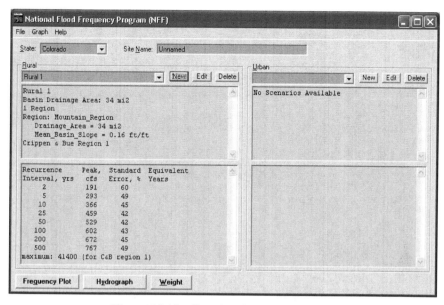

Figure 12.10   Sample NFF tabular output.

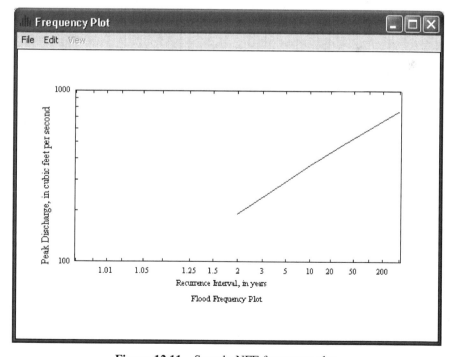

Figure 12.11   Sample NFF frequency plot.

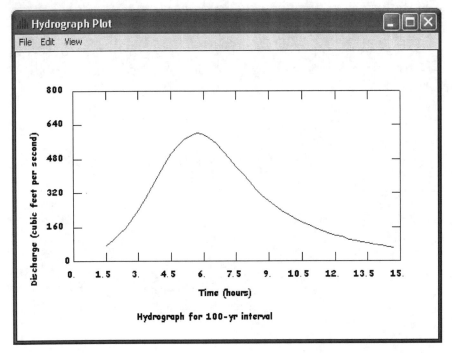

Figure 12.12 Sample NFF hydrograph.

were computed that offer reasonable limits for estimates of maximum floods. The curves indicate that floods may occur that are two to three times greater than those known for most streams.

Although Crippen and Bue performed this work in 1977, and the peak flows used in the analysis are currently being reexamined, one significant result of this study was the development of the flood regions map, which can be compared to physiographic provinces, ecoregions, and so on when performing watershed assessments. In developing the maps, Crippen and Bue

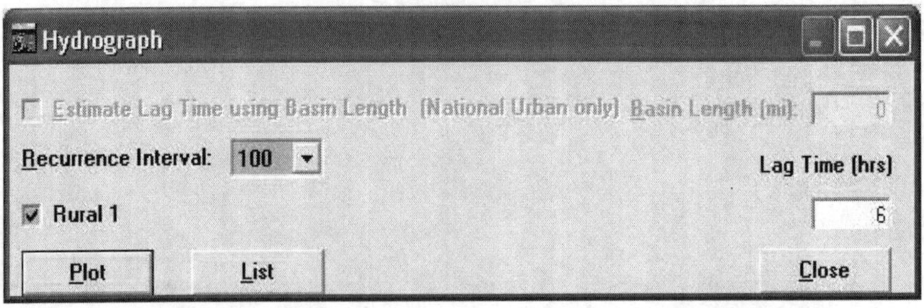

Figure 12.13 Required watershed lag time menu.

utilized a combination of the physiographic provinces (Fenneman, 1931, 1938) and variations in rainfall intensity (U.S. Weather Bureau, 1961) as the initial basis for the flood regions. They then further refined the maps to reflect hydrologic and flood data and differences, separating or accumulating like areas. The Crippen and Bue Flood Region Map is reproduced in Figure 12.14. The full report can be found at http://choctaw.er.usgs.gov/new_web/reports /other_reports/flood_frequency/. To keep updated on the National Flood Frequency Bulletin, visit the National Flood Frequency Web site at www. floodmaps.fema.gov/listserv/nf_may02.htm.

12.8.4 Calibration Summary

Various watersheds react differently to higher-frequency versus lower-frequency storms, depending mostly on available floodplain storage. This will affect the calibration procedure. For instance, channels with little floodplain storage may have little attenuation in the routed hydrograph. These calibrated routing parameters may, therefore, be the same for the lower- and higher-frequency storms. Flows should be computed utilizing the frequency analysis and regression methods for the range of storm frequencies—that is, 1- through

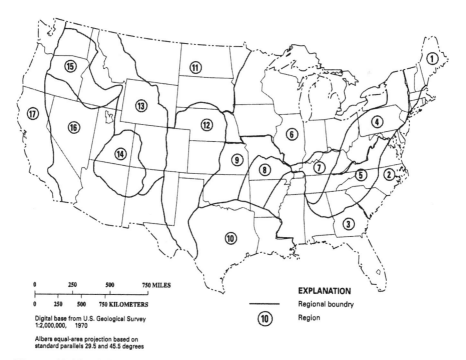

Figure 12.14 Crippen and Bue flood region map (Thomas and Kirby, 2002, after Crippen and Bue, 1977).

100-year storms—and compared to design storm runs from the computer model.

In contrast, a watershed with expansive floodplains may have significantly different calibrated routing parameter values for the 2- versus the 100-year storms. Therefore, calibration should be performed utilizing several methodologies, starting with the most accurate rainfall and stream flow data for several design storms and several points of interest or subwatersheds, to ensure the most reliable model.

One should calibrate the model based upon 6 to 12 storms, and reserve at least six storms for verification. Calibrating against actual storms provides the most accurate data, flood peak and volume prediction, and verification of watershed timing. The statistical frequency analysis allows calibration against actual peaks, but does not include a volume or timing comparison. The regression methods allow comparison against flood peaks for design storms, but also do not allow for volume or timing comparison. The statistical frequency analysis and the regional regression analyses can often be utilized in conjunction with the calibration and verification process, even if actual rainfall and stream gage data are available. A generalized calibration and verification flowchart can be found in Figure 12.15 (DeBarry, 1992). A typical observed versus simulated hydrograph for a particular storm is found in Figure 12.16.

Calibration of gage data provides a model calibrated to the gage. The point of interest for which flows are required does not often fall on the gage. In addition, the gage watershed may contain nonuniform subwatersheds. Tributaries may react very differently to rainfall events due to varying topographic, physiographic, hydrologic, and anthropogenic properties. Dams are a hydrologic property that would significantly attenuate flood flow peaks, hence should be included in the hydrologic model. It is therefore advisable to not only calibrate to the gage, but to several points of interest (POI) in the watershed. These points of interest (which are explained in the regional watershed management section of this book, Chapter 18) would include the upstream and downstream locations of tributary confluences, obstructions, problem areas, dams, and stream gage locations. The watershed should then be subdivided into subwatersheds (or subareas) so that flows can be obtained at these points of interest. The points of interest should then also be utilized as model calibration points to calibrate the model generated flows against the flows obtained by the methods discussed in this chapter. For example, flows can be transposed to the various locations in the watershed using the ratio of watershed drainage area methodology described in equation 12.30.

Other factors influencing calibration points are portions of a watershed underlain by limestone, sharp changes in channel or watershed slope, changes in land use, or other watershed physical characteristics that would cause a significant change in the rainfall/runoff process. Figure 12.17 shows the difference in stream flow patterns in a watershed underlain by limestone. Many of the streams appear to flow underground in the limestone area. This has a major hydrologic impact on that portion of the watershed, and shows why

GENERALIZED CALIBRATION/VERIFICATION FLOW CHART

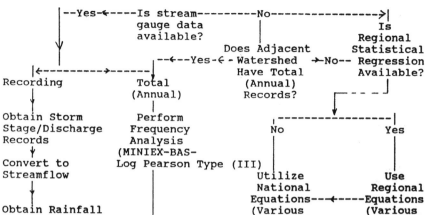

Figure 12.15 Generalized calibration and verification flowchart (DeBarry, 1993).

portions of the watershed need to be calibrated separately based on topographic, physiographic, hydrologic, and anthropogenic differences.

12.9 VERIFICATION

Verification is the process of ensuring that the calibrated model represents the true hydrologic conditions of the watershed. Verification methods will depend on the data used for calibration, whether actual rainstorm events or statistical/regression methods are used. If actual rainfall and stream flow data are available, verification would include running the calibrated model on storm events that were not used in the calibration process to determine whether the calibrated model will predict accurate hydrographs. The verification should be conducted by comparing the results with all the available methods previously described.

If not needed in the calibration process, an extreme-value statistical analysis should be performed for each gaging station for which data are available,

Calibrated Run

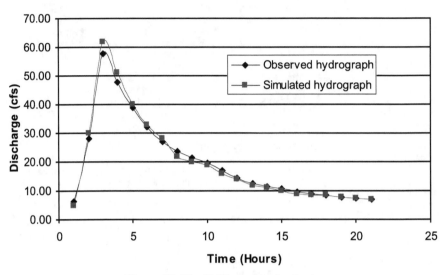

Figure 12.16 Calibrated hydrograph.

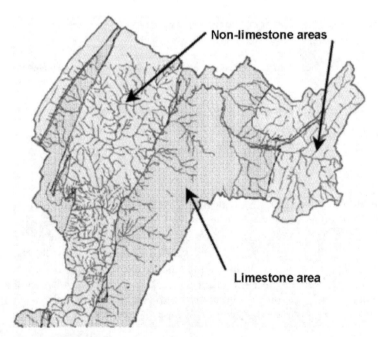

Figure 12.17 The difference in stream flow patterns in an area underlain by limestone versus nonlimestone geology. (Source: Paul DeBarry courtesy of Borton-Lawson Engineering, 2003.)

and a Log Pearson Type III determined for several storm frequencies. The calibrated hydrologic model should also be run for various storm frequencies. The peak flow values from the hydrologic model run and the statistical analysis method can then be plotted on log-log paper for comparison purposes. As a double-check, peak flow values obtained using a statistical regression approach should also be plotted. This should be performed for each point of interest in the watershed.

Peak flows can also be checked against peak flows developed as part of the Federal Flood Insurance Program and tabulated in Flood Insurance Studies (FIS). FIS flows are typically developed using statistical frequency analyses, where gage data are available, or regional regression methods, when flow data are not available. The methodology used to develop the FIS flows should be obtained.

12.10 CALIBRATED MODEL RUNS

Using a model that has not been calibrated can result in inaccurate results, which in turn may lead to facilities that are not designed properly. Stormwater facilities may be over- or underdesigned, leading to improper functioning or failure. Once the model is calibrated and verified, the watershed can then be modeled to determine the hydrologic response for the 1-, 1.5-, 2-, 2.33-, 5-, 10-, 25-, 50-, and 100-year storm events for the 1-, 6-, 12-, and 24-hour storms.

The modeling process addresses:

- Peak discharge values at various locations along the stream and its tributaries
- Time to peak for the above discharge
- Runoff contributions of individual subareas at selected downstream locations
- Flow values contained in the channel and overflow values
- Overall watershed timing

Additional model runs should be made for the purpose of assessing alternative stormwater management approaches, or what-if scenarios. This may involve an evaluation of individual development detention versus a "regional" facility. The *release rate percentage,* described in Chapter 18, can also be evaluated, in addition to the possibility of waiving the release rate percentage concept in favor of a downstream facility controlling multiple subbasins.

12.11 SUMMARY

A basic understanding of watershed hydrology and some of the methodologies utilized to evaluate it is required for watershed assessments. All watershed plans should incorporate a stormwater management plan component that can only be developed after detailed hydrologic and hydraulic modeling as will be discussed in Chapters 18 and 19. The watershed should be managed based on watershed management districts. To have a reliable watershed model, it is essential to do model calibration and verification against actual data, statistical frequency analyses, and regression equations.

REFERENCES

Aron, Gert. "Mini-Exvan PC Program—Extreme Value Analysis on Annual Series of Flood or Storm Events Manual," The Pennsylvania State University, University Park, PA, 1987.

Aron, Gert. Personal conversation, July 12, 2003.

ASCE. *Hydrology Handbook,* 2nd ed. (New York: American Society of Civil Engineers), 1996.

Brater, Ernest F., and H.W. King. *Handbook of Hydraulics,* 7th ed. (New York: McGraw-Hill Book Company), 1976.

Buchanan, Thomas J., and William P. Somers. "Techniques of Water-Resources Investigations of the United States Geologic Survey," Chapter A7, Book 3, *Stage Measurements at Gaging Stations* (Washington, DC: USGS, United States Printing Office), 1978.

———. "Techniques of Water-Resources Investigations of the United States Geologic Survey," Chapter A8, Book 3, *Discharge Measurements at Gaging Stations* (Washington, DC: USGS, United States Printing Office), 1984.

Carter, R.W., and Jacob Davidson. "Techniques of Water-Resources Investigations of the United States Geologic Survey," Chapter A6, Book 3, *General Procedure for Gaging Streams* (Washington, DC: USGS, United States Printing Office), 1977.

Chow, Ven Te. *Open Channel Hydraulics* (New York: McGraw-Hill Book Company), 1964.

Crawford, N.H., and R.K. Linsley Jr. "Digital Simulation in Hydrology: Stanford Watershed Model IV," Department of Civil Engineering, Stanford University, Tech Report No. 39, July 1966.

Crippen, J.R., and C.D. Bue. "Maximum Floodflows in the Conterminous United States," U.S. Geological Survey Water-Supply Paper 1887, 1977.

DeBarry, Paul A. "Stream Gage Alternative Analysis for Swiftwater Creek." Unpublished report, May 25, 1993.

DeBarry, Paul A. "Regional Stormwater Management," ASCE Short Course, 1992.

DeBarry, Paul A., and J. Carrington. "Computer Watersheds" in Civil Engineering, American Society of Civil Engineers (A.S.C.E.), Washington, DC, June 1990.

DeBarry, P.A. "Model Calibration and Parameter Sensitivity" in Watershed Modeling, Floodplain and Stormwater Management Symposium, Penn State University, University Park, PA, 1988.

DeBarry, Paul A., R.G. Quimpo, J. Garbrecht, T. Evans, L. Garcis, L. Johnson, J. Jorgeson, V. Krysanova, G. Leavesley, D. Maidment, E. James Nelson, F. Ogden,

F. Olivera, T. Seybert, W. Sloan, D. Burrows, E. Engman, R. Binger, B. Evans, F. Theurer. "GIS Modules and Distributed Models of the Watershed," Report from the Task Committee on GIS Modules and Distributed Models of the Watershed, ASCE, Reston, VA, 1999.

Dunne, T., and Luna L. Leopold. *Water in Environmental Planning* (New York: W.H. Freeman), 1978.

Federal Highway Administration, U.S. Department of Transportation. Hydraulic Design of Highway Culverts, Hydraulic Design Series No. 5, Publication No. FHWA-NHI-01-020, September 2001.

Fenneman, N.M. *Physiography of Eastern United States* (New York: McGraw-Hill Book Co.), 1938.

———. *Physiography of Western United States* (New York: McGraw-Hill Book Co.), 1931.

Hydrology Subcommittee, Interagency Advisory Committee on Water Data. "Guidelines for Determining Flood Flow Frequency," Bulletin No. 17B, March 1982.

Jennings, M.E., W.O. Thomas, Jr., and H.C. Riggs. Nationwide Summary of U.S. Geological Survey Regional Regression Equations for Estimating Magnitude and Frequency of Floods for Ungaged Sites, 1993: U.S. Geological Survey Water-Resources Investigations Report 94—4002, 1994.

Johanson, F.A., J.C. Imhoff, and H.H. Davis Jr. "Users Manual for the Hydrologic Simulation Program—FORTRAN (HSPF)." Report EPA 600/9-80-015, Environmental Protection Agency, Athens, GA, April 1980.

Kibler, David F., Gert Aron, Karen Riley, Gilbert Osei-Kwadwo, and Elizabeth L. White. "Recommended Hydrologic Procedures for Computing Urban Runoff from Small Developing Watersheds in Pennsylvania," Research Report to the Bureau of Dams and Waterway Management, Division of Storm Water Management, Department of Environmental Resources, Commonwealth of Pennsylvania, Penn State University, University Park, PA, January 1982.

Kirby, W.H. "Annual Flood Frequency Analysis Using U.S. Water Resources Council Guidelines (program J407)": U.S. Geological Survey Open-File Report 79-1336-I, WATSTORE User's Guide, vol. 4, ch. I, sec. C, 1981.

Leonard, S., G. Staidl, J. Fogg, K. Gebhardt, W. Hagenbuck, and D. Prichard. "Riparian Area Management," Technical Reference 1737-7 (Denver, CO: U.S. Department of the Interior, Bureau of Land Management), 1992.

Maidment, David R. (Ed.). *Handbook of Hydrology* (New York: McGraw-Hill, Inc.), 1992.

New Jersey Department of Environmental Protection (NJDEP), Flood Plain Management—Technical Manual for Stream Encroachment, NJDEP, Division of Water Resources, Bureau of Flood Plain Management, August 1984.

Ries, K.G. III, and M.Y. Crouse. "The National Flood Frequency Program, Version 3: A Computer Program for Estimating Magnitude and Frequency of Floods for Ungaged Sites," Water-Resources Investigations Report 02-4168, U.S. Geological Survey, Reston, VA, 2002.

Thomas, W.O., Jr., A.M. Lumb, K.M. Flynn, and W.H. Kirby. *User's Manual for Program PEAKFQ,* Annual Flood Frequency Analysis Using Bulletin 17B Guidelines, 1998.

Thomas, W.O., Jr., and W.H. Kirby. "Estimation of Extreme Floods," in Ries, K.G., III, and Crouse, M.Y., "The National Flood Frequency Program, Version 3: A Computer Program for Estimating Magnitude and Frequency of Floods for Ungaged Sites," U.S. Geological Survey Water-Resources Investigations Report 02-4168, 2002.

U.S. Army Corps of Engineers. *HEC-1—Flood Hydrograph Package Users Manual* (Davis, CA: Hydrologic Engineering Center), March 1987.

———. *Hydrologic Modeling System—HMS User's Manual* (Davis, CA: Hydrologic Engineering Center), 2003.

USDA Soil Conservation Service (SCS). "Computer Program for Project Formulation Hydrology, Technical Release 20 (TR-20)," Washington, DC, 1982.

———. "Urban Hydrology for Small Watersheds," Technical Release 55 (TR-55), 2nd ed., Washington, DC, June 1986.

———. *National Engineering Handbook,* Section 4: "Hydrology," Washington, DC, March 1985.

USGS. PEAKFQ Computer Program, February 25, 2002.

U.S. Water Resources Council. "A Uniform Technique for Determining Flood Flow Frequencies," Bulletin 15, December 1967.

U.S. Weather Bureau. "Rainfall Frequency Atlas of the United States for Durations from 30 Minutes to 24 Hours and Return Periods from 1 to 100 Years," Technical Paper 40 (TP 40), prepared by Davd, M. Hershfield, U.S. Department of Commerce, Government Printing Office, Washington, DC, 61 pp, 1961.

Vaill, J.E. "Analysis of the Magnitude and Frequency of Floods in Colorado," USGS Water-Resources Investigations Report 99-4190 (Denver, CO: U.S. Geological Survey, Branch of Information Services), 2000.

Wahl, Kenneth L., Wilbert O. Thomas, Jr., and Robert M. Hirsch. Stream-Gaging Program of the U.S. Geological Survey, U.S. Geological Survey Circular 1123, Reston, VA, 1995.

Winter, T.C., J.W. Harvey, O.L. Franke, and W.M. Alley. "Ground Water and Surface Water, a Single Resource," Circular 1139, 2003.

CHAPTER 13

STREAM AND RIVER MORPHOLOGIC ASSESSMENT

"The patterns of rivers are naturally developed to provide for the dissipation of the kinetic energy of moving water, and the transportation of sediment."
— Dave Rosgen in "Applied River Morphology," 1996

13.0 INTRODUCTION

Fluvial geomorphology (FGM) is the study of why a stream has formed, based on its watershed characteristics, such as climate, topography, vegetation, and geology. The father of fluvial geomorphology has to be Luna P. Leopold, much of whose work was done in the 1960s, and who laid the framework for watershed/stream relationships. More recently, FGM has been popularized by the publication in 1996 of a groundbreaking book, *Applied River Morphology,* by David Rosgen. Accelerated stream bank erosion caused by human-induced changes in the watershed increases sediment load and inhibits aquatic life. To bring the channel back into "equilibrium," the hydraulic and geomorphic processes should be understood and natural channel design techniques employed. Natural channel restoration prevents stream bank erosion and provides habitat for aquatic life.

Although past structural failures demonstrated that channelization, levees, dikes, and rip-rap were not the answer to biologic and hydraulic stream restoration, there was no well-defined "manual" describing how to apply the principles developed by Leopold. Rosgen's book, as well as others (Harrelson et al., 1994; FISRWG, 1998), provides a comprehensive collection of principles, processes, practices, and illustrations for natural stream channel evaluation and design. The intent of this chapter is to provide readers with an overview of the essential information presented by Leopold, Rosgen, and others, so that they may better understand FGM processes and concepts. To that end, these and other quality publications will be summarized here.

Note

Readers are recommended to read the publications summarized in this chapter, in conjunction with the proper education and training in hydraulics before attempting FGM classification or stream restoration.

13.1   AGE CLASSIFICATIONS

Rivers and streams can be classified as *youthful, mature,* and *old,* and knowing the age of a river or stream will aid the practitioner in making decisions on its predicted characteristics. But the distinctions between age classifications are not straightforward, as the boundaries are diffuse between the classes, making a precise definition of each class somewhat subjective; hence, various alternatives are accepted by different practitioners. The youthful, mature, and old classifications generally equate to the headwaters, transfer, and depositional zone, respectively, described later in this chapter.

Youthful streams are characteristically very irregular and V-shaped, due to vertical cutting. They are typically developed by accumulated overland flow in fractured nonerosive and erosive materials in mountains, thus the composition of the bed materials. The river channels have widened through lateral bank cutting, and slope gradients are reduced in mature channels. The slope and energy of the mature stream are just sufficient to transport the material to it, thus causing a *streambed* with a *graded condition.* Meanders and narrow floodplains are indicative of mature channels. Old rivers have valleys or floodplains of great width with low relief. The meanders have expanded their *sinuosity,* but do not extend beyond the river valley. Natural levees have formed along the stream banks; beyond them are swamps. Often, in old rivers, the channel will break out of the natural levee and then flow parallel to the main channel until finding the path of least resistance in the bank and reentering the channel (Drye et al., 2003). Due to their relatively flat topography, old channel floodplains are often encroached upon and developed on by humans, the problems of which will be discussed in Chapter 20.

13.2   STREAM FORM (PATTERN)

Rivers can also be classified as *straight, braided,* or *meandering,* as shown in Figure 13.1. Over time, streams may tend to form *meanders,* which then form *chutes,* then *oxbows,* and, eventually, *oxbow lakes.* This could also occur over a short time span in alluvial soils of floodplains, where the vegetation

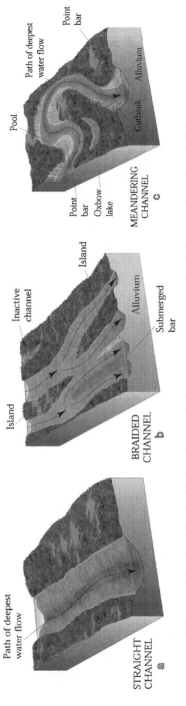

Figure 13.1 Straight, braided, and meandering streams with associated features (Murck and Skinner, 1998).

has been removed and the banks are unstable and where there is a high sediment load due to human influence. Knowing these classifications helps the practitioner decipher what the channel's current equilibrium condition should be, and thus makes it possible to predict its behavior.

Where terrain is flat enough to allow a stream to move sideways, which is typically where alluvial deposition has occurred, the stream's constant undercutting of the outside curve bank forms a cycle of enlarged curves and loops, termed meanders. During the Pleistocene era, when a stream was flowing on new geologically uplifted sedimentary rock strata, the meandering stream, pushing toward equilibrium, "rejuvenated" and cut through the uplifted strata, resulting in steep-walled canyons, with the meanders following essentially the same pattern as the original stream, a term called *incised meanders*. As meanders enlarge, the extreme curvature of the loops may take a shortcut through the "neck" of the loop, forming an extremely curved meander or oxbow, as shown in Figure 13.2. The water may flow in both channels for a period of time until the velocity distribution allows sediment to accumulate at the natural channel bank location of the short-cut channel, essentially cutting off the oxbow from the channel flow and forming an oxbow lake.

Where sediment in the channel is greater than the normal flow is capable of carrying, a series of sandbars and low islands form. This is typical in semiarid regions, and results in streams with a twisted ropelike pattern, known as *braided streams*. This deposition is often washed downstream during periods of high flow, re-forming braided streams as the flows and velocities resend, again forming a series of sandbars. This process occurs at the outlet of the stream or river, where velocities slow due to very flat slopes and possible backwater effects from lakes or oceans, and forms a *delta*.

13.3 STREAM FACTORS

In order to determine the classification of the streams, the FGM specialist must determine their physical characteristics. These characteristics can be grouped into *dimension* factors, *profile* factors, and *pattern* factors, each of which is defined in the following subsections.

13.3.1 Dimension Factors: Lateral Stream (Cross-Sectional) Definitions

This section discusses physical factors or actual physical dimensions of the channels. In general, rivers can be divided into two classifications, those with floodplains and those without. Floodplains are not always associated strictly with *overbank* flow; they are also the result of lateral movement of the river channel across the plain through geologic time. The increase flows and velocities of urbanization often make a stream *downcut* itself deep enough into

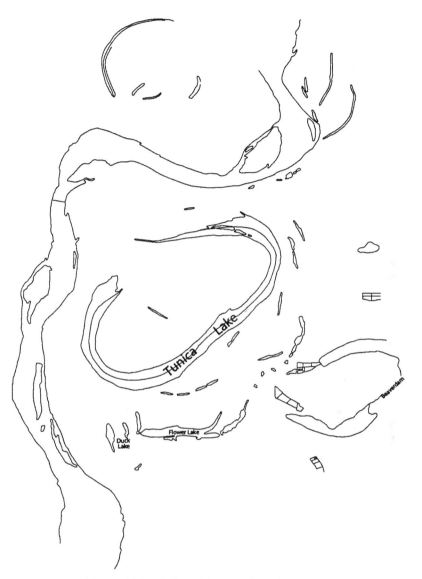

Figure 13.2 Oxbow lake on Mississippi River.

the underlying strata to leave the existing floodplain high above the modern-day flood levels. These high-and-dry floodplains are termed *terraces*. This phenomenon also occurs naturally.

A typical stream and stream valley is shown in Figure 13.3. Many of the parameters shown in this figure are utilized by geomorphologists, floodplain managers, biologists, and others for their individual intended purposes, re-

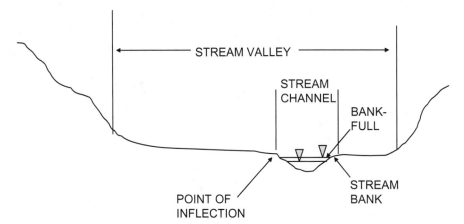

Figure 13.3 Typical cross-section of channel and valley.

sulting in sometimes-conflicting definitions; therefore the goal here is to attempt to standardize the terms. A stream *channel* can be classified as a channel that flows at least part of the year. This would not include natural or man-made channels that flow only during rainstorm events. The channel is defined by the channel or stream *banks*. When evaluating the physical appearance of streams, the stream banks are the point of inflection of the smallest radius of the cross-section curvature, as shown in Figure 13.3. This may be a different elevation on the left and right bank; however, the lowest bank dictates the *bankfull* capacity or discharge. When determining the left and right bank, engineers typically refer to them as one looks downstream; thus, looking downstream, the left bank would be on one's left. Biologists, in contrast, look at the system through the eyes of the fish that are facing upstream, and therefore have the reverse orientation. The point is, practitioners should compare notes to maintain one consistent orientation for the study.

Bankfull typically refers to the stage or elevation of the top of a stream banks, just before spilling onto the floodplains. *Channel bank* and *bankfull discharge (flow)*, or capacity, are terms defined in many different ways, depending on application and discipline. Bankfull discharge would therefore be the flow at which the stage or elevations of the banks would be obtained. Literature has shown that this is typically the storm with a recurrence interval of about 1.5 years (Dunne and Leopold, 1978; Rosgen, 1996). It is this storm frequency that tends to dictate the morphological characteristics of the stream through sediment transportation over time. This 1.5-year recurrence interval is a typical or average value, and many factors will influence this, including urbanization, climate, and others. For instance, one study (Dunne and Leopold, 1978) found the bankfull discharge to be approximately 390 cfs, while the 1.5-year discharge was 600 cfs. Because streams in similar physiographic regions support similar classifications, Dunne and Leopold (1978) have shown that bankfull discharge (as well as average width, depth, and cross-sectional area) is directly related to watershed drainage area in similar physiographic areas, often referred to as a *regional curve,* as shown in Figure 13.4.

Dunne and Leopold have also determined that bankfull discharge can be generally defined by the equation:

$$Q_F = cD_A^n \qquad (13.1)$$

Where:

Q_F = the bankfull discharge (cfs)
c = A coefficient of climate and frequency of flood
D_A = Drainage area (square miles)
n = An exponent less than 1

Typical values of n are 0.7 to 0.8. The average bankfull discharge can then be determined from equation 13.1; however, this factor should be field-verified. This relationship can be extremely useful in a variety of applications, from FGM definition to hydraulic routing of channel flows.

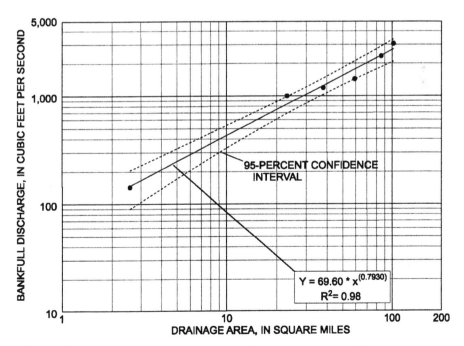

Figure 13.4 Regional curve representing relation between bankfull discharge and drainage area (White, 2001).

If, however, stream gage data is available, the most accurate stream bank discharge would come from field measurements compared against the gage data frequency analysis, as explained in Chapter 19. Not to be confused with bankfull discharge is the *effective discharge,* or the flow that provides for maximum sediment transport over time.

The channel *width* is one of the factors that various disciplines define differently. For instance, using a typical trapezoidal channel configuration, the width may mean the width of the channel at the channel invert or the width of the channel at the channel bank, as shown in Figure 13.5. In hydraulic studies, such as for floodplain analysis or bridge design, width is typically the width of flow for a specified discharge, that is, the discharge for the 100-year flood, and therefore should be differentiated from channel width. Again, it's important to be aware of the definition being used when referencing past studies and/or research data. And for future use, one should clearly distinguish between top width or bottom width.

For his part, Rosgen (1996) refers to the width as "measured at the bankfull stage," often referred to as the *top width.* The bottom width is defined as the width of the channel at the inflection point of the bank and the bed. According to the Federal Interagency Stream Restoration Working Group (FISRWG, 1998), the *thalweg* is the path of deepest flow, or the deepest part of the channel. FISRWG shows the thalweg as having a width, whereas other practitioners assume it's a line, running up the longitudinal profile of the stream. Hydraulic modeling, such as with HEC-RAS, utilizes the centerline of the channel to develop the longitudinal profile, which may or may not be the thalweg. Data required to determine where the thalweg is located is usually not required for hydraulic modeling. One should be cautious in utilizing referenced graphs, such as regional curves and data, as different sources define width differently and, as just stated, the distinction must be made between top width or bottom width.

The *mean depth* typically refers to the mean depth of flow at bankfull capacity, or simply the mean depth of flow when the stream is flowing at

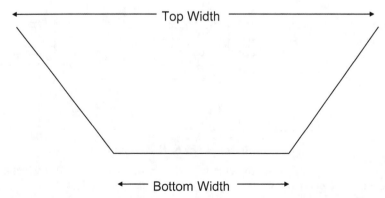

Figure 13.5 Width of the channel defined as the bottom width and the top width.

bankfull stage. Mean depth should not be confused with *normal depth of flow* (y_o) used in open channel hydraulics. Normal depth of flow is defined as the depth of flow during *uniform* (or *laminar*) *flow* conditions. Flow can be classified as *uniform* or *nonuniform.* Nonuniform flow is otherwise know as *turbulent flow.* Uniform flow, in the case of open channels, occurs when the water slope and cross section remain constant over a certain reach of the channel. The depth of flow will remain constant if the flow is in a uniform state. (A stream *reach* is another term that is utilized inconsistently and may be defined differently by hydraulic modelers and biologists; it is defined in the next section.)

The vertical containment of a stream and the degree to which it is incised in the valley floors is its *entrenchment* (Kellerhalls and Bray, 1971). The entrenchment ratio is the width of the geomorphologic *flood-prone area* to the surface width of the bankfull channel (Rosgen, 1996). Rosgen defines the flood-prone area as the active floodplain and the low terrace; the flood-prone area width is measured at the elevation that corresponds to twice the maximum depth of the bankfull channel as taken from the established bankfull stage (it should not be confused with the Federal Flood Insurance Program defined *100-year floodplain,* shown in Figure 20.1; see Chapter 20).

Another factor used to help define channel characteristics is the *width-to-depth ratio* (W/D), which is an index value that indicates the channel cross-section shape. It is defined by Rosgen as the ratio of the bankfull surface width to the mean depth of the bankfull channel. The W/D ratio is important to understand the available energy within a channel, which translates to its capability to transport sediment.

13.3.2 Profile Factors: Longitudinal Stream Definitions

The *stream corridor* can be considered on the watershed scale and would be equivalent to a highway right-of-way, which contains a specific pavement type, shoulder type, and land use up to the edge of the right-of-way. The stream corridor includes the stream channel and floodplain and includes both man-made and natural features such as land use, wetlands, habitat, and so on. Two terms similar in meaning to that of *corridor* are the stream *reach* and *element.* Reach, another term that is defined in many ways, generally refers to a segment of a stream channel; however, depending on the application, it may need to be further refined. A reach may be a stretch of the stream with a specific slope and, in turn, velocity; a *riffle* or *pool;* a segment between two tributary *confluences;* or a segment with biological similarities.

Riffles are the shallow portions of the stream that typically have a bed of larger rocks. Pools are the deeper portions that typically have sand or sand mixed with a few cobbles (Dunne and Leopold, 1978). A typical riffle and pool can be seen in Figure 13.6. The shallower depth of the riffles, and in turn area of flow, causes increased velocities of the flowing water that carries the smaller particles with it, according to the relationship given in equation 13.2, allowing it to settle and depositing it in the pools as velocities slow,

Figure 13.6   Typical riffle (foreground) and pool (background).

according to a derivation of Darcy's Law, which is discussed in Chapter 12, equation 12.7.

$$V = Q \div A \tag{13.2}$$

Where:

   V = Velocity (fps)
   Q = Flow (cfs)
   A = Area of flow (sf)

Rosgen (1996) refers to a reach as a segment of channel with common physical characteristics based on changes in geology and tributary influence, and is utilized to establish reference dimensions, profile, and shape. In an FGM assessment a reach is defined based on a transition to a different stream type (correlating to the section of stream channel that is unstable). USGS (White, 2001) defines a stream reach as "a section of stream extending between 10 and 20 bankfull widths in length"; however, for small streams, this results in relatively small reach lengths.

The term reach is also often used in hydrologic and hydraulic modeling, which requires data in a specific format. Often a reach is defined as the segment of stream between tributaries; however, in hydrologic or hydraulic modeling, the reach should have a consistent *flow regime,* that is, consistent flow, slope, channel dimensions, and so on. For instance, in the HEC-RAS model (U.S. Army Corps of Engineers, 2002), the reach length is defined as the distance between model cross sections. These parameters should be consistent in a reach in hydrologic and hydraulic modeling so that velocities and travel times are accurately represented.

River reaches have been digitally developed and combined into *geometric networks* as part of the National Hydrography Dataset (USGS, 2003) and incorporated into the Arc Hydro Model (Maidment, 2002). In the digital model, a network is a series of reaches, or "edges," called *flowlines,* topologically connected at junctions or nodes. *Topologically connected* means that one digital line knows where it stands in relation to the digital junction and the next line. These lines also have a *from node* and a *to node,* allowing for flow direction modeling.

Though the Arc Hydro Model is a breakthrough in the digital modeling of our nation's waterways, one point to keep in mind in regard to this model is the transient nature of stream flow lines, which may shift over time. On a national scale, this should be insignificant as it affects hydrologic modeling; however, when performing more localized analyses, the length change from shifting should be addressed. In unstable systems being analyzed as part of the development of a comprehensive watershed restoration plan, what defines a reach constantly changes—in short, reach designation is something that relates to a snapshot in time. Often, in the time it takes to complete the restoration plan and obtain implementation money, reach designations have changed and instabilities have migrated upstream or downstream.

In the past, the term *element* has often been used interchangeably with the term *reach,* but more recently it has been defined as a particular hydrologic parameter and may refer to a reach, subwatershed (subarea), or reservoir (U.S. Army Corps of Engineers, 2002).

Related to stream order and FGM classification is the classification of a stream as it transverses the watershed into three zones based on location in the watershed and sediment qualities. The *headwaters zone,* which can be equated to headwater streams, has characteristically steep streams, which are still forming through erosion processes. Runoff begins here, and this is the principal area from which dissolved solids and sediments are derived.

In the second zone, the *transfer zone,* or the *medium streams,* the channel widens as more tributaries are added; the gradient decreases and the steams receive the sediment from the headwater streams. Inflow of water roughly equals outflow, and assuming stability, influxes of dissolved solids roughly equal fluxes of dissolved solids out.

The final zone, the *depositional zone,* somewhat similar to the lower *river classification,* is characterized by wider and flatter channel slopes, generally large floodplains, sediment deposition, and meanders. These areas are characterized by *alluvial-fan, alluvial-plain, estuarine, deltaic,* and *coastal* environments. The headwater, transfer, and depositional zones, and the general relationship to stored alluvium deposits, discharge, velocity, mean channel depth and width, and bed material grain size are shown in Figure 13.7 (FISRWG, 1998). Chapter 18 explains how this corresponds to stormwater management planning.

When a stream is disturbed, channelized, leveed, and so on, in a particular location, the effects of the change are felt both upstream and downstream, as the channel tries to readjust back to equilibrium conditions. Often, therefore, it is relevant to construct a *stream profile,* or the longitudinal plot of elevation change versus horizontal distance, as shown in Figure 13.8. This will allow

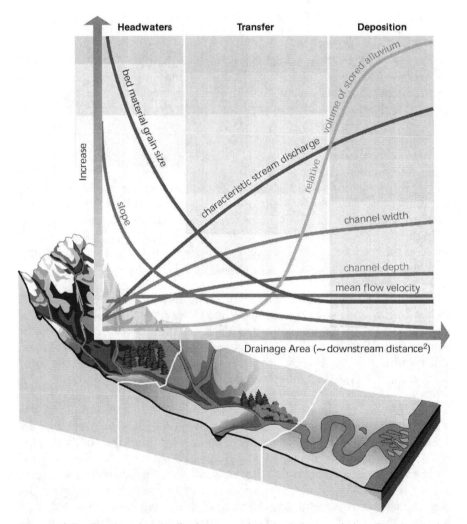

Figure 13.7 The three longitudinal zones and channel characteristics. (Reprinted with permission of Federal Interagency Stream Restoration Working Group, 1998.)

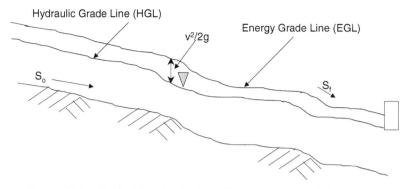

Figure 13.8 Typical longitudinal profile of a stream defining slopes.

a quick indication of where reaches may be defined, based on slope of the stream, and can indicate the classification of a stream segment as described. The slope of the stream channel invert, (S_o), or the bottom of the channel, may or may not be the slope of the flowing water surface or hydraulic grade line. Rosgen's classification system uses water surface slope measured from the same relative position on a riffle or pool. This is because stream invert slope varies considerably due to bottom facets. Both these slopes should not be confused with the friction slope (S_f), which is the slope of the energy grade line. The energy grade line is the total energy exerted by the flowing water and is equal to the depth of flow (or hydraulic grade line) plus the velocity head $(v^2/2g)$, where:

v = Velocity (fps)
g = Acceleration due to gravity (ft/sec^2)

The channel slope, water slope, and friction slope may all be different due to influential factors, such as backwater from bridges or abrupt changes in channel slope; however, under normal depth conditions for a stabilized reach, they would be the same.

13.3.3 Pattern Factors: Stream and Channel Form

The amount of curvature of a stream is known as its *sinuosity* and is computed by dividing the centerline of the channel reach by the length of the valley centerline, as shown in Figure 13.9. It has been defined by Rosgen as the stream length/valley length or valley slope/channel slope. The stream length is measured by the stream's thalweg, which is the path of deepest flow. Rosgen (1996) classifies channels with a sinuosity of greater than 1.3 as meandering, whereas the U.S. Department of Transportation (1975) considers channels with a sinuosity of less than 1.5 as straight.

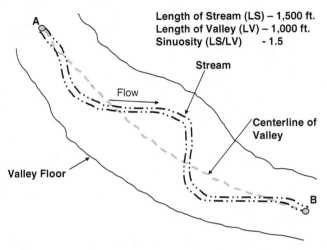

Figure 13.9 Stream sinuosity: the higher the sinuosity number, the more meander the stream has.

Other factors that help define stream patterns are the *meander wavelength,* the *belt width,* and the *meander width ratio,* which is the meander belt width/ bankfull width that defines the degree of lateral channel containment and the radius of curvature.

13.4 STREAM CLASSIFICATIONS

As mentioned, streams reach equilibrium conditions based on influencing factors such as the climate and physiographic conditions that typify its location. Therefore, streams under similar conditions can be classified into categories; streams with the same natural conditions should have the same physical factors or patterns. Thus, when analyzing a stream that does not have the "natural" stream characteristics, it may be regarded as "under stress," and the goal would be to restore the stream to its classification conditions.

In developing a watershed management plan, an engineer or hydrologist who specializes in FGM typically is part of the team. The FGM expert needs to assess and define the physical characteristics of the stream in order to evaluate its condition. The FGM expert will therefore first classify the stream based on its physical conditions. Assigning a stream to a particular class makes it possible to assess the potential for natural recovery, sediment supply, stream sensitivity to disturbance, channel response to changes in flow regime, and aquatic habitat, and to recommend a restoration methodology. *Reference reaches,* stream reaches that match the classification or category in which they fall, will then be located and defined. Reference reaches are areas of the stream that have no degradation or are stable for a significantly long stretch, and hence can provide design ratios for the restoration of the degraded stream sections. Essentially, these are the stream reaches that serve as a template for the design of study stream restoration project.

Rosgen has developed a broad classification system for major stream types, utilizing a series of nine alpha characters (Aa+, A, B, C, D, DA, E, F, and G) based on morphological characteristics such as entrenchment, slope, width/depth ratio, and sinuosity, as shown in Figure 13.10 and described in Table 13.1 (Rosgen, 1996). The slope in Table 13.1 refers to the slope of the water surface averaged for 20 to 30 channel widths. According to Rosgen, Level I is a broad-level delineation performed from USGS 7½-minute topographic quadrangle maps. It delineates streams based mainly on slope and sinuosity, introducing error due to the accuracy of the maps as well as the lack of width-to-depth and entrenchment ratios from a field survey. However, it provides some general data that can be used for further stream classification when needed.

Each of the nine categories (Aa+ through G) is further classified into one of six subcategories based on dominant bed material, ranging from bedrock to silt/clay (subcategories 1 through 6), as shown in Figure 13.11. The resulting letter/number classification provides a great deal of information on a stream based on data collected from a large sample size of streams with the same Rosgen classification. This information can then be used in a multitude of applications, including engineering, fish habitat enhancement, stream res-

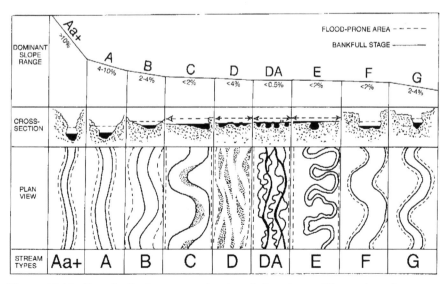

Figure 13.10 Longitudinal, cross-sectional, and plan views of Rosgen's major stream types (Rosgen, 1996).

TABLE 13.1 General Stream Type Descriptions and Delineative Criteria for Broad-level Classification (Level I)

Stream Type	General Description	Entrenchment Ratio	W/D Ratio	Sinuosity	Slope	Landform/Soils/Features
Aa+	Very steep, deeply entrenched, debris transport, torrent streams.	<1.4	<12	1.0 to 1.1	>.10	Very high relief. Erosional, bedrock or depositional features; debris flow potential. Deeply entrenched streams. Vertical steps with deep scour pools; waterfalls.
A	Steep, entrenched, cascading, step/pool streams. High energy/debris transport associated with depositional soils. Very stable if bedrock or boulder dominated channel.	<1.4	<12	1.0 to 1.2	.04 to .10	High relief. Erosional or depositional and bedrock forms. Entrenched and confined streams with cascading reaches. Frequently spaced, deep pools in associated step/pool bed morphology.
B	Moderately entrenched, moderate gradient, riffle dominated channel, with infrequently spaced pools. Very stable plan and profile. Stable banks.	1.4 to 2.2	>12	>1.2	.02 to .039	Moderate relief, colluvial deposition, and/or structural. Moderate entrenchment and W/D ratio. Narrow, gently sloping valleys. Rapids predominate w/scour pools.
C	Low gradient, meandering, point-bar, riffle/pool, alluvial channels with broad, well defined floodplains.	>2.2	>12	>1.4	<.02	Broad valleys w/terraces, in association with floodplains, alluvial soils. Slightly entrenched with well-defined meandering channels. Riffle/pool bed morphology.
D	Braided channel with longitudinal and transverse bars. Very wide channel with eroding banks.	n/a	>40	n/a	<4	Broad valleys with alluvium, steeper fans. Glacial debris and depositional features. Active lateral adjustment, w/abundance of sediment supply. Convergence/divergence bed features, aggradational processes, high bedload and bank erosion.

DA	Anastomosing (multiple channels) narrow and deep with extensive, well vegetated floodplains and associated wetlands. Very gentle relief with highly variable sinuosities and width/depth ratios. Very stable streambanks.	>2.2	Highly variable	Highly variable	<.005	Broad, low-gradient valleys with fine alluvium and/or lacustrine soils. Anastomosed (multiple channel) geologic control creating fine deposition w/well-vegetated bars that are laterally stable with broad wetland floodplains. Very low bedload, high wash load sediment.
E	Low gradient, meandering riffle/pool stream with low width/depth ratio and little deposition. Very efficient and stable. High meander width ratio.	>2.2	<12	>1.5	<.02	Broad valley/meadows. Alluvial materials with floodplains. HIghly sinuous with stable, well-vegetated banks. Riffle/pool morphology with very low width/depth ratios.
F	Entrenched meandering riffle/pool channel on low gradients with high width/depth ratio.	<1.4	>12	>1.4	<.02	Entrenched in highly weathered material. Gentle gradients, with a high width/depth ratio. Meandering, laterally unstable with high bank erosion rates. Riffle/pool morphology.
G	Entrenched "gully" step/pool and low width/depth ratio on moderate gradients.	<1.4	<12	>1.2	.02 to .039	Gullies, step/pool morphology w/ moderate slopes and low width/depth ratio. Narrow valleys, or deeply incised in alluvial or colluvial materials, i.e., fans or deltas. Unstable, with grade control problems with high bank erosion rates.

Source: Rosgen, 1996.

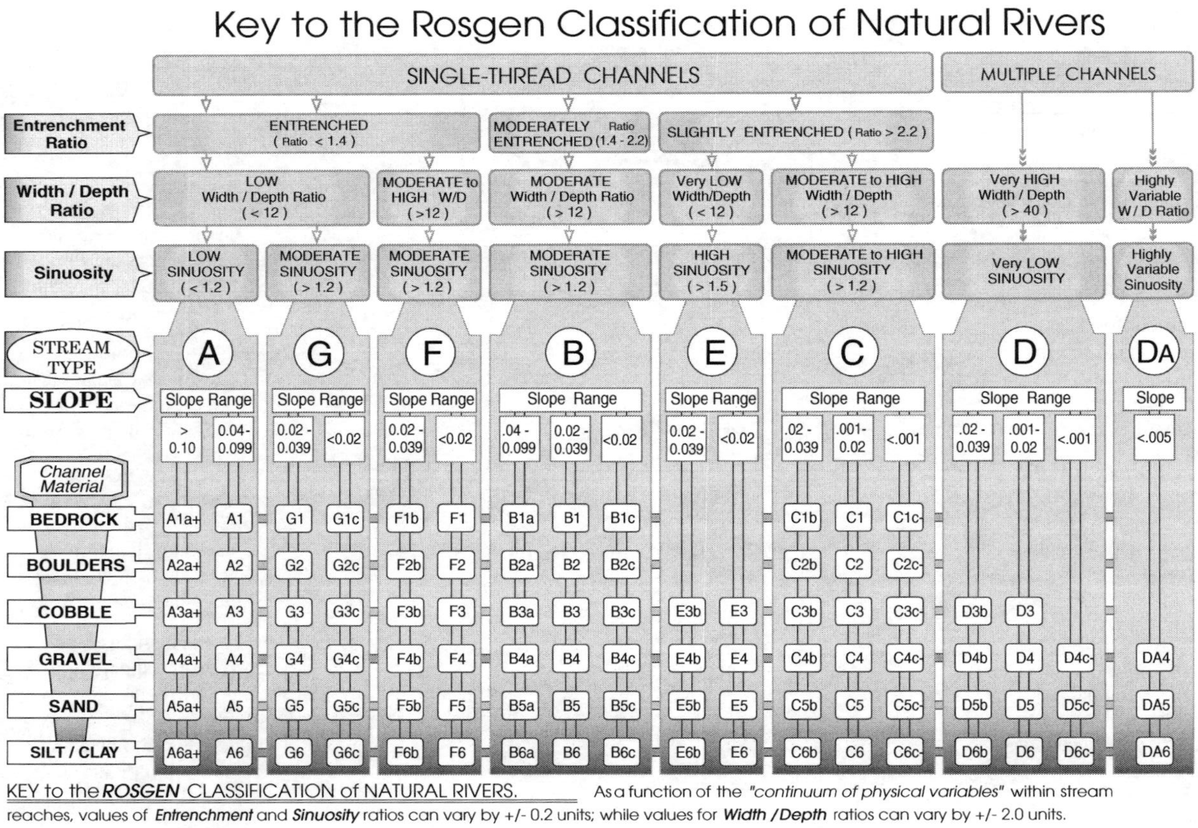

Figure 13.11 Rosgen's classification key for natural rivers (Rosgen, 1996).

KEY to the **ROSGEN** CLASSIFICATION of NATURAL RIVERS. As a function of the *"continuum of physical variables"* within stream reaches, values of *Entrenchment* and *Sinuosity* ratios can vary by +/- 0.2 units; while values for *Width / Depth* ratios can vary by +/- 2.0 units.

toration, and water resource applications. A very detailed description of the stream types, methodology for classifications, and fluvial geomorphologic principles can be found in Rosgen's book, *Applied River Morphology* (1996).

13.5 PERFORMING A FLUVIAL GEOMORPHOLOGICAL ASSESSMENT

Achieving natural stream stability plays an important role in minimizing stream bank erosion and resultant sediment pollution and, in turn, water quality and aquatic habitat preservation. Natural stream stability is achieved by allowing the stream to develop stable dimensions (stream bankfull width, width/depth ratio, and capacity), profile, and pattern so that the stream system neither degrades nor aggrades. These dimensions for stability can be mathematically determined using the Rosgen method of fluvial geomorphological (FGM) assessment. Once the stream is categorized and instability problems identified, effective and sustainable stream restoration measures can be recommended to bring the stream back into a stable condition through proper targeted stormwater management and recommended restoration measures.

Preservation of stream geomorphology is an important goal in the identification of stream reaches that require restoration, and aids in developing a sustainable flood protection program; hence it should be integrated with a comprehensive stormwater management and stormwater quality management program. Living resources and water quality considerations must be part of the evaluation of alternative runoff control strategies, such as stormwater release rate requirements from new development, as they have long-term implications for morphologically sensitive areas. Such an evaluation would include:

- Relating the extent of stream bank erosion, sedimentation, and downstream water quality problems to changes in stormwater flows (both volume and peak).
- Considering living resource protection through aquatic habitat preservation.
- Identifying changes in channel configuration in response to changes in stormwater runoff that might contribute to flooding problems in the future as the stream strives to reach a new equilibrium.
- Ensuring adequate protection of sewerage infrastructure.
- Protecting living resources through aquatic habitat preservation.
- Recommending effective and sustainable stream restoration measures.

13.5.1 Four Levels of Assessment

Assessing stream stability requires an FGM assessment and baseline determination. This frequently is performed based on the Rosgen classification methodology (Rosgen, 1996). To properly characterize the watershed, it is recommended that measurement of geomorphological parameters and physical and hydraulic relationships be performed at both Level I and Level II. This will address some of the root causes of stream bank erosion and sedimentation, habitat loss, and water quality impairments, and can be performed in conjunction with a stormwater analysis and modeling results and assessment of flooding impacts. The assessment will provide critical information for use in identifying and understanding existing and future problems and in devising an effective framework for stormwater management that will protect any future stream restoration efforts.

13.5.1.1 Level I: Desktop Survey. A Level I FGM assessment of the watershed should be performed based on Rosgen classification methodology (Rosgen, 1996). This is desktop delineation of the stream using generalized major stream types A through G based on available topographic information, geological maps, soils maps, and aerial photographs. The purpose of this inventory is to provide an initial framework for organizing and targeting subsequent field assessments of targeted or important reaches where problems are known to occur or are anticipated to occur.

Specific drainage areas for selected stream reaches within the watershed should be calculated where needed. Using *regional curve data* developed for the area of interest, ranges of hydraulic geometry relationships based on the bankfull discharge should be estimated.

Stream reaches should initially be classified by stream type (Level II) based on objective comparisons of land forms, soils, slope, and channel patterns. Field verification will be required where stream types change or where distinct variations in conditions are observed.

13.5.1.2 Level II: Reach Stream Survey. Field teams should be sent out to walk along the designated lengths of each stream and tributary and estimate by observation the following parameters:

Channel Morphology

Bankfull elevation	Sinuosity range
Bankfull width	Channel slope range
Entrenchment ratio range	Channel materials (pebble count)
Width/depth ratio range	Meander pattern

In order to assure that the field teams produce consistent results, at the beginning of the study a modified Wolman Pebble Count should be performed on riffle bars or reaches, and special note should be made of their location (Wolman, 1954). To perform the count, 100 rocks chosen at random and the b axis of the particles (intermediate axis) are measured and recorded. Rocks (particles) are tallied by using the pebble count size classes given in Table 13.2 and then plotted by size class and frequency.

TABLE 13.2 Pebble Count Size Classes (Source: USDA NRCS, 2004)

Inches	Size Class	Millimeters	Inches	Size Class	Millimeters
—	Silt/Clay	<0.062	2.5–3.8	Small Cobble	45–64
—	Very Fine Sand	0.062–0.125	3.8–5.0	Small Cobble	90–128
—	Fine Sand	0.125–.025	2.5–3.8	Large Cobble	128–180
—	Medium Sand	0.25–0.50	7.6–10	Large Cobble	180–256
—	Coarse Sand	0.50–1.0	10–15	Small Boulder	256–362
0.04–0.08	Very Coarse Sand	1.0–2.0	15–20	Small Boulder	362–512
0.08–0.16	Very Fine Gravel	2–4	20–40	Medium Boulder	512–1024
0.16–0.24	Fine Gravel	4–5.7	40–80	Large Boulder	1024–2048
0.24–0.31	Fine Gravel	5.7–8	80–160	Very Large Boulder	2048–4096
0.31–0.47	Medium Gravel	8–11.3	—	Bedrock	—
0.47–0.63	Medium Gravel	11.3–16			
0.63–0.94	Coarse Gravel	16–22.6			
0.94–1.26	Coarse Gravel	22.6–32			
1.26–1.9	Very Coarse Gravel	32–45			
1.9–2.5	Very Coarse Gravel	45–64			

Cross sections are taken utilizing the traditional survey equipment or GPS. One cross-sectional survey (by rod, measuring tape, and level) is obtained at a representative *crossover* (i.e., riffle) location that includes stream invert, maximum depth, bankfull depth, and flood-prone level (enough stations to determine whether the entrenchment is greater than 2.2 [Rosgen, 1996]). The stations within the cross section are taken to pick up significant slope changes and should in no case be greater than two feet apart. An Excel spreadsheet program can be used for entering and plotting the data and cross section to scale. Each cross section is marked in the field with labeled flagging, located approximately with GPS, and indicated on an area map.

Photographs are taken at strategic points throughout the surveyed portions of the streams and coded for future reference. In addition, any obvious erosion or stream blockages are noted and placed in the GIS.

A Rosgen Level II Reach Field Form and a Watershed Data Summary Sheet (Rosgen, 1996) are completed for each reach for the parameters observed. The result is a *measured reach,* that is, a stream classification Level II morphological description of the stream reaches for which Level II data have been collected.

The distribution of reaches to be measured is determined from the surveyed reach evaluation, the assessment of where problems are occurring, and the importance attached to the stream segment. Typically, a single assessment reach should not be more than about 1,000 feet in length. Commonly, an average of five measured reaches per stream mile is anticipated.

Part of the classification includes estimating the bankfull discharge. Bankfull depth is determined through field visits, and bankfull stage is calibrated to known stream flows from the results of the hydrologic and hydraulic analysis. This is done using the existing USGS discharge gage information on the creek, if available, or nearby creek, and transposed to the site in question based on drainage area. The difference between the water surface elevation and the bankfull elevation is carried back to the gage (as long as stream type does not change). Then the elevation difference is added to the water surface elevation at the gage to determine the bankfull stage elevation relative to the gage staff. Next, the elevation difference is added to the water surface elevation at the gage to determine the bankfull stage elevation relative to the gage staff. Once the stage-discharge relationship has been established, the recurrence interval for the bankfull stage can be calculated, as can the hydraulic geometry data for width, depth, velocity, and cross-sectional area versus stream discharge.

The FGM assessment can be utilized not only to determine reaches requiring restoration, but also to support the development of geomorphologically sensitive stormwater management requirements (storage, release rates, etc.).

13.5.1.3 *Level III: Stream Assessment and Departure from Potential.*
As described by Rosgen (1996), a level III assessment involves analyzing the stream condition in relation to its stream potential, or "best channel condi-

tion," and its departure from normal conditions. It looks at riparian vegetation, deposition pattern, debris occurrence, meander pattern, and several other channel stability and flow regime characteristics. Rosgen should be consulted for this assessment

13.5.1.4 Level IV: Field Data Verification. Level IV involves field sediment and stream flow measurements and stability and bank erosion rate trends. Again, Rosgen (1996) is the recommended reading.

13.6 SUMMARY

Assessing the streams' morphology, reference reaches, and problem areas is an important step in the watershed assessment. Stream bank erosion is a major source of sediment pollution in the streams. Determining where stream banks are eroding, and applying restoration techniques, can enhance fish habitat and improve water quality. Having a watershed-wide stormwater management plan in place, to prevent flows from increasing in the future due to new development, is essential before undertaking any restoration activities. Many restoration efforts have been literally washed away as the upstream watershed urbanizes and flows increase.

REFERENCES

Drye, Jocelyn, Heidi Glaesel, and Jimmie Agnew. "New River," Elon University, www.elon.edu/geo/new_river.htm, accessed September 22, 2003.

Dunne, T., and Luna L. Leopold. *Water in Environmental Planning* (New York: W.H. Freeman), 1978.

FISRWG. "Stream Corridor Restoration: Principles, Processes, and Practices," Federal Interagency Stream Restoration Working Group (FISRWG), GPO Item 0120-A; SuDocs No. A 57.6/2:EN 3/PT.653, 1998.

Harrelson, Cheryl C., C.L. Rawlins, and John P. Potyondy. "Stream Channel Reference Sites: An Illustrated Guide to Field Technique," General Technical Report RM-245, USDA Forest Service, April 1994.

Kellerhalls, R., and D.I. Bray. "Sampling Procedures for Coarse Fluvial Sediments," ASCE Journal of the Hydraulics Division, v. 97, no. HY8, p. 1165–1179, 1971.

Keystone Stream Team. "Natural Stream Channel Design in Pennsylvania," revised March 2003.

Leopold, L.B., M.G. Molman, and J.P. Miller. *Fluvial Processes in Geomorphology* (San Francisco, CA: Freeman), 1964.

Maidment, David R., ed. *Arc Hydro: GIS for Water Resources* (Redlands, CA: ESRI Press), 2002.

Murck, Barbara W., and Brian J. Skinner. *Geology Today* (New York: John Wiley & Sons, Inc.), 1998.

Rosgen, David. *Applied River Morphology* (Fort Collins, CO: Wildland Hydrology), 1996.

Strahler, Arthur N., and Alan H. Strahler. *Elements of Physical Geography* (New York: John Wiley & Sons, Inc.), 1976.

U.S. Army Corps of Engineers, Hydrologic Engineering Center. Hydrologic Modeling System (HEC-HMS) User's Manual, 2002.

U.S. Department of Agriculture (USDA), Natural Resources Conservation Service (NRCS). Regional Hydraulic Geometry Curves—NWMC Procedure, National Water Management Center, http://wmc.ar.nrcs.usda.gov/technical/HHSWR/Geomorphic/procedure.html, accessed January 16, 2004.

U.S. Department of Transportation, Federal Highway Administration. "Highways in the River Environment: Hydraulic and Environmental Design Considerations, Training and Design Manual," prepared by Civil Enginering Department, Engineering Research Center, Colorado State University, Fort Collins, CO, May 1975.

USDA—Natural Resources Conservation Service (NRCS)—formerly SCS—Urban Hydrology for Small Watersheds, Technical Release 55, June 1986.

USGS, National Hydrography Dataset, http://nhd.usgs.gov/, 2003b.

White, Kirk E. "Regional Curve Development and Selection of a Reference Reach in the Non-Urban, Lowland Sections of the Piedmont Physiographic Province, Pennsylvania and Maryland," Water-Resources Investigations Report 01-4146, USGS, New Cumberland, PA, 2001.

Wolman, M.G. "A method of sampling coarse river-bed material," Trans. Am. Geophys. Union, v. 35, p. 951–956, 1954.

CHAPTER 14

ECOLOGY: HABITAT CLASSIFICATION AND ASSESSMENT

14.0 INTRODUCTION

In analyzing a watershed and the effect of physical changes in the watershed, it is important first to understand the habitat and ecosystems of the receiving water bodies. Understanding these factors, in conjunction with each other, water quality, fluvial geomorphology, hydraulics, sediment, and related characteristics, aids the scientist in determining what impacts may be occurring in the watershed and its water bodies.

14.1 LONGITUDINAL CHANGES IN STREAM ECOSYSTEMS

As they move from the headwaters to the mature valleys, stream channels experience physical and geomorphological changes, such as stream slope, width, size, flow, meander, and temperature. Stream biota and ecosystems adapt and change along with these geomorphological changes. Ecosystems can be grouped into homogeneous classes of streams and rivers to provide a geographic framework for more efficient aquatic resource identification, classification, and management. As discussed in Chapter 5, streams can be classified regionally by physiographic province, ecosystem province, stream order, and so on. These classification parameters are not necessarily interrelated. Because geomorphic processes dictate stream types, it has been found that placing streams and stream habitats in a geographic, spatial hierarchy is an efficient aid in characterizing streams (Frissell et al. 1986). Frissell and colleagues proposed five systems, or spatial levels, in order of increasing detail: order, stream, segment, reach, pool/riffle, and microhabitat. The U.S. Geological Survey National Water-Quality Assessment (NAWQA) modified this approach by eliminating the term *system,* equating the term *stream segment* to *basin,* and combining the pool/riffle into the stream reach spatial scale. The resulting four classifications of the NAWQA system are basin, stream segment, stream reach, and microhabitat, as shown in Figure 14.1.

A popular way of defining how the aquatic community and stream ecosystem changes with longitudinal stream characteristics is the River Continuum Concept (RCC), a model to aid in determining the biotic community based on the size of the stream itself. Invertebrates (microbes, collectors, grazers, predators), particulate matter (sediment, organic matter), vegetation, and fish all change as one progresses from upstream to downstream, (WOW, 2003) as shown in Figure 14.2 and Table 14.1 (adapted from Ward, 1992).

Another method of stream habitat and species distribution classification is the *patch dynamics model.* This model is based on the theory that species distribution is not determined by the location of a stream reach, but rather as a random occurrence (Townsend, 1989). In actuality, both the patch dynamics model and the RCC concept are valid. On a short-term and local level, patchy or random variation of aquatic communities will occur; however, longitudinal patterns emerge over larger scales and longer time periods.

14.2 ECOSYSTEM CLASSIFICATIONS

Ecoregion is the general term for areas that have similar physical geography, geology, soils, vegetation, land use, wildlife, and climate. Ecoregions denote

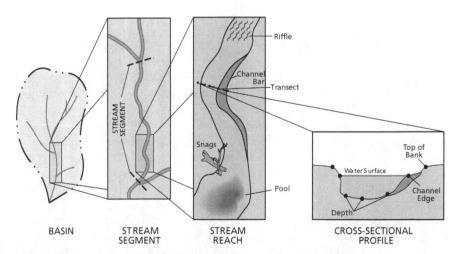

Figure 14.1 Spatial hierarchy of basin, stream segment, stream reach, and cross-sectional profile (Lewis and Turtora, 1998, after Frissell et al., 1986).

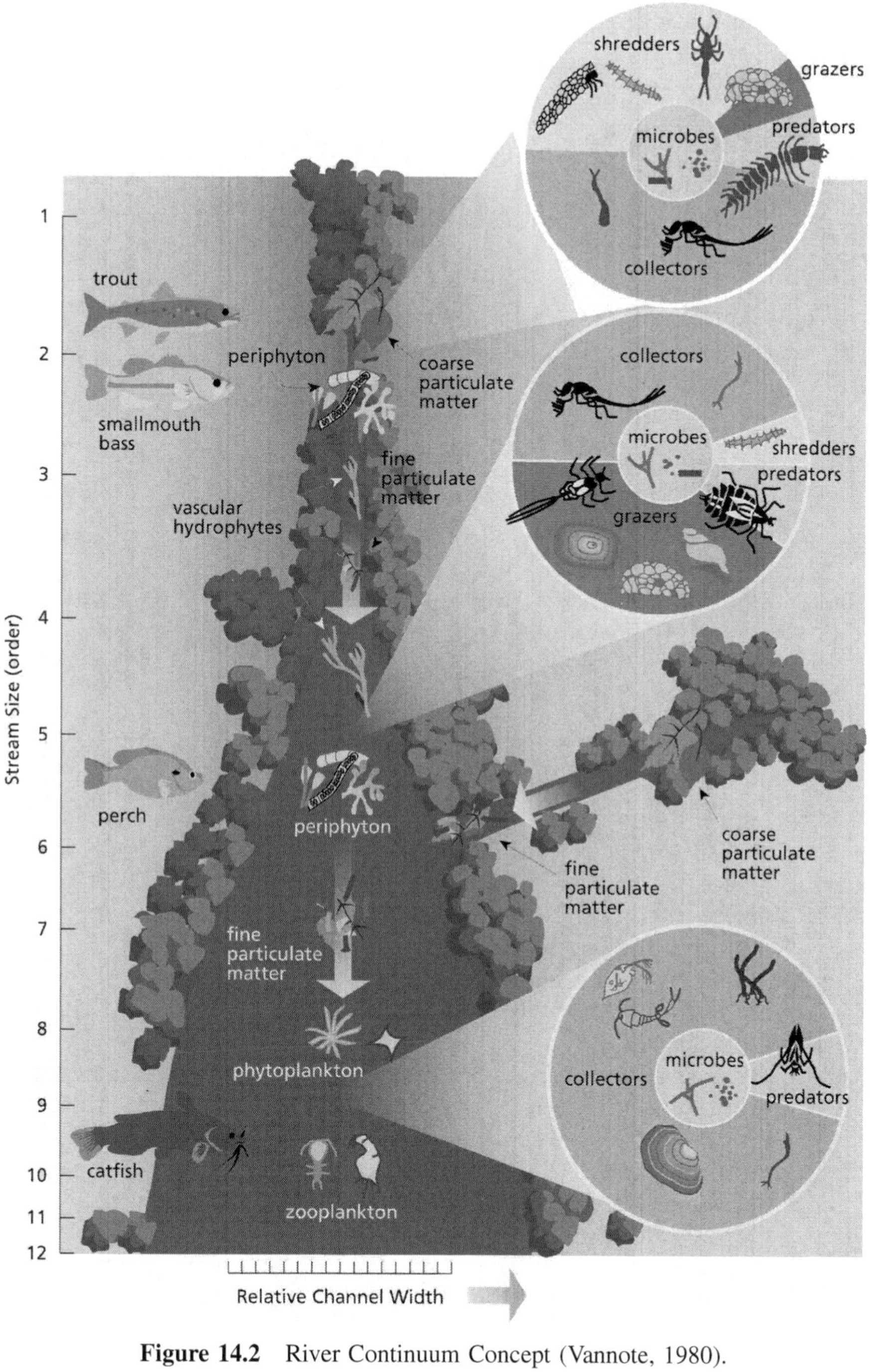

Figure 14.2 River Continuum Concept (Vannote, 1980).

TABLE 14.1   Commonly Observed Changes Associated with River Continuum Concept (Adapted from Ward, 1992.)

Simplified RCC	Upper Reaches	Middle Reaches	Lower Reaches
Steam order	1–3	3–6	6 and above
Substrate	coarse	sand gravel	fine
Current	fast		slow
Oxygen	saturated		periodic deficits
Sunlight	low	high	low
Temperature	max. <20°C, fairly constant	highly variable	max. >20°C, variable
Particulate matter	coarse		fine
Nutrients	low	high	low
Dominant energy source	leaf filter	primary producers	transport devitus
Dominant primary producers	primary producers rare	attached	plankton
Dominant invertebrates	shredders, collectors	grazers (scrapers), collectors	collectors
Fish habitat and food preferences	cool water, swift current insects	fish and insects	slow current plankton, bottom matter
Biological diversity	low	high	low

areas within ecosystems where the type, quality, and quantity of environmental resources are generally similar. The U.S. Forest Service originally developed a system for classifying ecoregions in 1978 as described in section 14.2.1 (Bailey, 1995). The U.S. Environmental Protection Agency (EPA) began working with this system, but eventually developed its own.

Ecoregions are useful in watershed assessment and management, for, as with other classification systems, they establish a "benchmark" for evaluating the ecosystem and watershed parameters. For example, valleys generally have deep soils and warm temperatures and are relatively flat compared to mountains, which have steep slopes, shallow soils, and cooler climates. Mountain streams naturally have a different slope, sinuosity, width-to-depth ratio and water quality than valley streams (Woods et al., 1996, 1999).

Ecoregions also aid in evaluating a stream's response to natural and anthropogenic (human) disturbances, and determining which management solutions may be applicable. For instance, some watersheds in a northeastern ecoregion that contain soils without pH buffering capacity may exhibit deteriorating stream water quality due to acid rain. Ecoregions typically have similar land uses due to topographic and soil similarities. Therefore, ecoregions also aid in evaluating the effects of land uses, such as nonpoint source pollution from forestry or agriculture, or acid mine drainage from mining disturbances, in the watershed.

14.2.1 U.S. Forest Service System: National Hierarchical Framework of Ecological Units

The U.S. Forest Service ecoregion classification system was originally published in 1978 to provide a general description of the ecosystem geography of the nation as shown on the map titled "Ecoregions of the United States," which was printed in 1976 and reprinted in 1980 by the Forest Service, Washington, DC, as Miscellaneous Publication 1391. The technique of mapping ecoregions was subsequently expanded to include the rest of North America (Bailey and Cushwa, 1981) and the world (Bailey, 1989).

More recently, the Ecological Classification and Mapping Task Team— made up of representatives from the U.S Forest Service, the USDA Natural Resources Conservation Service, and The Nature Conservancy—refined the data in a publication titled "Ecoregions and Subregions of the United States" (ECOMAP, 1993). This publication provided a standardized method for classifying, mapping, and describing ecological units at various geographic planning and analysis scales. The ecoregions were adapted for use in ecosystem management to illustrate interrelationships. This geographic definition of ecological units is being used by federal agencies and others for a variety of broad planning and assessment efforts, including the National Interagency Ecoregion-Based Ecological Assessments. These assessments present information at the subregion scale on a wide range of environmental, biological, and cultural characteristics of ecosystems that influence the energy, moisture, and nutrient gradients that regulate the structure and function of ecosystems. These characteristics include geomorphology, lithology and stratigraphy, soil taxa, potential natural vegetation, fauna, climate, surface-water characteristics, disturbance regimes, land use, and cultural ecology. The task team also developed the National Hierarchical Framework of Ecological Units, which breaks ecosystems into these categories: Domain, Division, Province, and Sections, Subsections, Land Type Association, and Land Type Phase. These are delineated in Table 14.2 (Bailey, 1994).

Ecosystem provinces are large-scale land classifications based on landform, climate, vegetation, soils, and fauna. The ecosystem provinces for the United States can be found in Figure 14.3.

A typical hierarchy of the Domain, Division, Province, and Section for the Dry Domain of the Great Plains and Rockies of the central United States is shown in the next paragraph (U.S. Forest Service, 2003). A prefix of "M" is used to indicate mountainous provinces. The complete listing for the entire United States can be found at www.fs.fed.us/land/pubs/ecoregions/toc.html.

TABLE 14.2   U.S. Forest Service National Hierarchical Framework of Ecological Units

Planning and Analysis Scale	Ecological Units	Purpose, Objectives, and General Use	General Size Range
Global	Domain	Broad applicability for modeling and sampling, strategic planning and assessment, and international planning	Millions to tens of thousands of square miles
Continental	Division		
Regional	Province		
Subregion	Section Subsection	Forest, area-wide planning and watershed analysis	Thousands to hundreds of square miles
Landscape	Land type Association	Forest, area-wide planning and watershed analysis	Thousands to hundreds of acres
Land Unit	Land type Association Landtype Phase	Project and management area planning and analysis	Hundreds to less than 10 acres

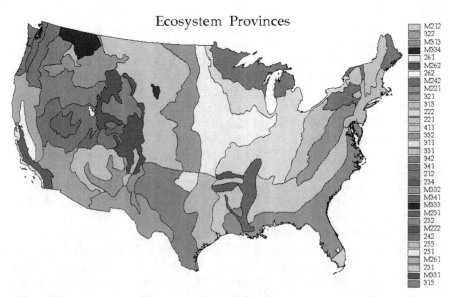

Ecosystem Provinces

Figure 14.3   Ecosystem provinces of the United States. (*Source:* "Description of the Ecoregions of the United States," Robert G. Bailey et al., 1994.)

300 Dry Domain
  330 Temperate Steppe Division
    331 Great Plains-Palouse Dry Steppe Province
    332 Great Plains Steppe Province
  M330 Temperate Steppe Division—Mountain Provinces
  M332 Middle Rocky Mountain Steppe—Coniferous Forest—Alpine Meadow Province
      Section M332A—Idaho Batholith
      Section M332B—Bitterroot Valley
      Section M332C—Rocky Mountain Front
      Section M332D—Belt Mountains
      Section M332E—Beaverhead Mountains
      Section M332F—Challis Volcanics
      Section M332G—Blue Mountains

The principal map unit design of the National Hierarchical Framework of Ecological Units and map scale and polygon size of ecological units can be found in Tables 14.3 and 14.4 (Cleland et al., 1997), respectively.

14.2.2 U.S. EPA System

James Omernik (1987) originally identified 76 ecoregions in the conterminous United States for the U.S. EPA by using information from small-scale (1:3,168,000 scale) maps, which were published at even a smaller scale (1:7,500,000). Development of ecoregions is based on the premise that ecological regions can be classified through patterns of biotic and abiotic phenomena that reflect ecosystem quality and integrity differences (Wiken, 1986; Omernik, 1987, 1995).

This approach, which relied on the combination and relative importance of characteristics that explain ecosystem regionality, and may vary from one place to another and from one hierarchical level to another, was modeled after the approach used by Environment Canada (Wiken, 1986). In describing ecoregionalization in Canada, Wiken stated:

Ecological land classification is a process of delineating and classifying ecologically distinctive areas of the earth's surface. Each area can be viewed as a discrete system which has resulted from the mesh and interplay of the geologic, landform, soil, vegetative, climatic, wildlife, water, and human factors which may be present. The dominance of any one or a number of these factors varies with the given ecological land unit. This holistic approach to land classification can be applied incrementally on a scale-related basis from very site-specific ecosystems to very broad ecosystems.

The U.S. EPA has developed a hierarchical classification system based on the level of detail or scale of the criteria. The ecoregions for North America are defined in Table 14.5 (U.S. EPA, 2003).

TABLE 14.3 Principal Map Unit Design Criteria of Ecological Units

Ecological Unit	Principal Map Unit Design Criteria
Domain	Broad climatic zones or groups (e.g., dry, humid, tropical)
Division	Regional climatic types (Koppen, 1931; Trewatha, 1968)
	Vegetational affinities (e.g., prairie or forest)
	Soil order
Province	Dominant potential natural vegetation (Kuchler, 1964)
	Highlands or mountains with complex vertical climate-vegetation-soil zonation
Section	Geomorphic province, geologic age, stratigraphy, lithology
	Regional climatic data
	Phases of soil orders, suborders, or great groups
	Potential natural vegetation
	Potential natural communities (PNC)
Subsection	Geomorphic process, surficial geology, lithology
	Phases of soil orders, suborders, or great groups
	Subregional climatic data
	PNC—formation or series
Land Type Association	Geomorphic process, geologic formation, surficial geology, and elevation
	Phases of soil subgroups, families, or series
	Local climate
	PNC—series, subseries, plant associations
Land Type	Landform and topography (elevation, aspect, slope gradient, and position)
	Phases of soil subgroups, families, or series
	Rock type, geomorphic process
	PNC—plant associations
Land Type Phase	Phases of soil subfamilies or series
	Landform and slope position
	PNC—plant associations or phases

Note: The criteria listed are broad categories of environmental and landscape components. The actual classes of components chosen for designing map units depend on conditions and relative importance of factors within respective geographic areas.

TABLE 14.4 Map Scale and Polygon Size of Ecological Units

Ecological Unit	Map Scale Range	General Polygon Size
Domain	1 : 30,000,000 or smaller	1,000,000s of square miles
Division	1 : 30,000,000 to 1 : 7,500,000	100,000s of square miles
Province	1 : 15,000,000 to 1 : 5,000,000	10,000s of square miles
Section	1 : 7,500,000 to 1 : 3,500,000	1,000s of square miles
Subsection	1 : 3,500,000 to 1 : 250,000	10s to low 1,000s of square miles
Land Type association	1 : 250,000 to 1 : 60,000	1,000s to 10,000s of acres
Land Type	1 : 60,000 to 1 : 24,000	100s to 1,000s of acres
Land Type phase	1 : 24,000 or larger	<100 acres

TABLE 14.5 Ecoregion Levels

Level	Description	Number of Classes
I	Most coarse	15
II		52
III		98
IV	Most refined	Incomplete

The exact number of ecological regions at each hierarchical level is still changing slightly as the framework undergoes development at the international, national, and local levels. Level IV classification has been completed only for portions of the United States. A description of a typical Level III ecoregion is as follows (U.S. EPA, 2003):

BLUE RIDGE MOUNTAINS (66)*

The Blue Ridge Mountains extend from southern Pennsylvania to northern Georgia, varying from narrow ridges to hilly plateaus to more massive mountainous areas with high peaks. The mostly forested slopes, high-gradient, cool, clear streams, and rugged terrain occur on a mix of igneous, metamorphic, and sedimentary geology. Annual precipitation of over 200 centimeters can occur on the well-exposed high peaks of the Great Smoky Mountains that reach over 1830 meters. The southern Blue Ridge is one of the richest centers of biodiversity in the eastern U.S. It is one of the most floristically diverse ecoregions, and includes Appalachian oak forests, northern hardwoods, and southeastern spruce-fir forests. Shrub, grass, and heath balds, hemlock, cove hardwoods, and oak-pine communities are also significant.

14.3 ECOREGIONS AND WATERSHED MANAGEMENT

Aquatic systems are integrated with the watershed's terrestrial features, such as soils, geology, and topography, and with climatic conditions. These physical characteristics affect water chemistry, geomorphology, and hydrology. The national hierarchy and U.S. Forest Service or U.S. EPA ecoregion approach enable a scientist to evaluate on a preliminary basis the linkages between terrestrial and aquatic ecosystems and watershed management. Aquatic resource systems in steep slope regions will exhibit different features from those of aquatic systems in flat terrain pocketed with lakes and ponds. Aquatic ecosystems within an ecoregion should exhibit similar hydrology, habitat, and organisms. The U.S. General Accounting Office (GAO) report on ecosystem management, titled "Ecosystem Management: Additional Actions Needed to Adequately Test a Promising Approach," emphasized the need to manage natural resources based on ecosystem boundaries, rather than political or ad-

* (66)—refers to Level III ecoregion identification number.

ministrative boundaries (U.S. GAO, 1994; Omernik, 1997). Omernik states, "Although ecoregions and watersheds are intended for different purposes, they can be complementary." What he means is that ecosystem management focuses on managing the biology and habitat of the area, and ecosystems can be classified based on regional similarities. Watersheds may be managed for stormwater management, flood control, water budget preservation, water yield, or sustainable base flow, which relates to ecosystem sustainability. There, the two entities should be utilized in conjunction with each other, as will be described later in this text, to achieve ecosystem management goals on a watershed level approach. In other words, the sustainability of the ecosystem should be considered in managing the land and water resources. Analyzing natural and anthropogenic features in the watershed can aid in explaining variances of the ecosystem from that which is expected based on the classification. Variances in the ecosystem may aid in determining impacts on the watershed.

14.4 RAPID BIOASSESSMENT PROTOCOL (RBP)

The Rapid Bioassessment Protocol (RBP) is a biological assessment methodology designed to provide basic aquatic life data for water quality management purposes. By assessing the aquatic life, the relative health of the water body and the presence of any stressors may be determined. The RBP promotes an integrated assessment, comparing biological measures, water quality (chemistry), and habitat (physical structure, flow regime) with empirically defined reference conditions. Although reference conditions could be established from historical data, extrapolation, or modeling, they are best established from actual systematic monitoring of reference sites that represent the range of variation in minimally disturbed conditions (Gibson et al., 1996). The RBP is a scientific, cost-effective method that can be used to characterize regional biotic reference conditions and severity of water resource impairment, help identify the source of impairment, evaluate the effectiveness of restoration activities, and aid in problem screening, site ranking, and long-term trend monitoring. Bioassessment protocols typically refer to the Strahler method of stream orders discussed in Chapter 5.

The U.S. EPA's RBP manual, *Rapid Bioassessment Protocols for Use in Wadeable Streams and Rivers, Periphyton, Macroinvertebrates, and Fish* (1999) provides a comprehensive look at the use of this protocol. As the title implies, it was developed for three aquatic assemblages: periphyton, benthic macroinvertebrates, and fish. The EPA's version of this protocol is the most commonly used today, and is a compilation of methods previously employed by several state agencies (Ohio Environmental Protection Agency, Florida Department of Environmental Protection, Delaware Department of Natural Resources and Environmental Control, Massachusetts Department of Environmental Protection, Kentucky Department of Environmental Protection, and

Montana Department of Environmental Quality). It can be used alone or to complement other methods.

Before conducting an expensive RBP, one should consult the U.S. Geological Survey's National Water-Quality Assessment Program (NAWQA) database (Moulton, et al., 2002). The NAWQA study is collecting data on algal, invertebrate, and fish communities. Information from these ecological studies, together with chemical and physical data, provide an integrated assessment of water quality at local, regional, and national scales. Analysis and interpretation of water quality data at these various geographic scales require accurate and consistent application of sampling protocols and sample-processing procedures. The USGS NAWQA studies follow the report, "Revised Protocols for Sampling Algal, Invertebrate, and Fish Communities as Part of the National Water-Quality Assessment Program," USGS Open-File Report 02-150, (Moulton et al., 2002). This report, which is available at http://water.usgs.gov/nawqa/protocols/OFR02-150/index.html, provides more information on fish-sampling protocols.

14.5 SPECIES DIVERSITY INVENTORY

Species diversity is an indicator of good water quality. Each species plays a role in survival of the others, all of whose interactions with their environment sustain their life-support system. When species are lost, the entire ecosystem suffers. Once the number of species of invertebrates has been collected for the RBP, the species diversity can be determined using the Shannon Index of Species Diversity formula, which states:

$$H = \sum_{i=1}^{s} (p_i)(\log p_i) \tag{14.1}$$

Where:

H = diversity index
s = Number of species (or orders)
i = Species (or order) number
p_i = Proportion of individuals of the total sample belonging to the ith species (or order).

The higher the H value, the higher the diversity of the sample.

14.6 AQUATIC HABITAT

Acquatic habitat can be defined as the "quality of the instream and riparian habitat that influences the structure and function of the aquatic community in a stream" (U.S. EPA, 1999). In order to assess aquatic habitat, physical characteristics of the water body and surrounding land must be ascertained. Habitat plays an important part in the long-term viability of the organisms that it supports. Habitat structure, or the actual physical features that provide hiding places, and substrate for living and reproducing must be maintained.

To characterize the physical conditions of a stream, the following information should be obtained.

- General land use
- Stream origin and type
- Riparian vegetation features
- Instream parameters

Instream parameters include:

- Stream depth, width, length
- Stream flow
- Sinuosity
- Meander ratio

14.7 CONDUCTING AN AQUATIC SURVEY

Most water bodies (streams, lakes, estuaries, wetland aquifers) have been designated for specific use(s) by the state's water quality agency, including:

- Aquatic life
- Fish for consumption
- Drinking water supply
- Swimming or other high-contact recreation
- Boating and other minimal-contact recreation
- Agriculture (irrigation/livestock)
- Industrial

Before undertaking a watershed monitoring program, the monitoring program should be defined and should consider the scale of the study, specific goals for the monitoring efforts, what watershed indicators will be monitored, data quality objectives, quality assurance/quality control procedures, methods, station locations, and responsible parties.

Sampling locations that will provide data representative of the stream reach or impoundment and that are spatially distributed so that future changes in biology and chemistry can be detected should be chosen. Factors to consider in determining sample locations include physiographic conditions, watershed land uses, known point source discharges, nonpoint source problem areas, and hydrology and volume of the water body. Data on water chemistry (nutrients, pH, dissolved oxygen), qualitative or semiquantitative biological data, and physical data must be collected. Physical data would include measurements/observations of stream substrate, aquatic habitat, hydrologic/bathymetric conditions, depth, flow, channel dimensions, and temperature. Collecting the physical data is important in order to obtain a reference for the conditions sampled at that time. If future changes in water biology or chemistry are noted, their relationship to the physical data can be compared, thus making it possible to determine whether they are related.

For *flowing water,* the stream is divided into homogeneous reaches based on physiographic and demographic similarities. Sampling should be performed in each stream reach during low flow conditions, the discharge should be measured, and instream and riparian conditions should be recorded at the time of sampling. Chemical *grab* samples, or those collected at a particular point in time, are collected at midchannel and middepth, unless hydrologic, stream width, stream flow, or observed biological conditions indicate stratification of water. Composite (mixed) or depth-integrated samples may be required if the water is stratified. Biological samples should be collected across a transect or throughout a large portion of the water body to ensure inclusion of all available habitat.

For *impounded water bodies,* sampling should be performed during the peak growing season or summer stratification period. Chemical water samples are recommended to be collected at depths of 1 meter below the surface and 1 meter above the bottom, utilizing a minimum of two sampling locations. Three locations are preferred, one near the inlet beyond any influences of inflow, one in the centroid or deepest part of the impoundment, and one toward the outlet. Biological samples should be collected throughout the water column. Plankton should be collected throughout the *photic* zone, whereas chlorophyll should be collected midway in that zone. Specific conductance, pH, dissolved oxygen, and temperature should be collected at 1-meter intervals throughout the water column. Transparency can be measured from the surface with a Secchi disc, a round disk with alternating black-and-white wedges, which may eventually disappear as it is lowered.

Typical physical, chemical, and biological parameters to sample are displayed in Tables 14.6 and 14.7 for flowing and impounded waters, respectively. Naturally, these parameters may vary by state, region, or water body.

This type of data is also used as part of a state's water quality standards review process, to assess current *attainment* or potential attainability of designated fish and aquatic life uses. In flowing water bodies, this assessment involves the comparison of the portion of the water body under study with a *reference segment* or *reach* on the same or a similar water body. The reference segment provides baseline data on physical, chemical, and biological characteristics of the water body, which define its potential uses; the same data from the study segment yield information on the current attainment of designated water uses. In impounded water bodies, this assessment involves analysis of specific selected area(s) in order to define existing or potential uses (PaDEP, 2002).

TABLE 14.6 Use Attainability Surveys for Flowing Water Bodies

Physical Evaluations	Chemical Evaluations	Biological Evaluations
Instream characteristics	Field analyses	Biological inventory
Riparian characteristics	Laboratory analyses	Biological conditions
Instream cover	Temperature	Fishes
Bank condition	Alkalinity	Macroinvertebrates
Epifaunal substrate	Dissolved oxygen	Indices:
Vegetative protection	Hardness	Benthic macroinvertebrates
Embeddedness/pool substrate	pH	Taxa richness
Grazing or other disruptive pressure	Total dissolved solids (TDS)	Periphyton/algae/slimes
Velocity and depth regime/pool variability	Specific conductance	Macrophytes
Riparian vegetative zone width	Total suspended solids (TSS)	Percent dominant taxon
Channel alteration	Total residual chlorine (TRC)	Fecal coliform organisms
Sediment deposition	Chloride (Cl)	Modified percent mayflies
Weather	Sulfate (SO_4)	
Frequency of riffles/channel sinuosity	Laboratory analyses	
Air temperature	pH	
Channel flow status	Phosphorus, total (P)	
Cloud cover	Total organic carbon (TOC)	
Preceding/prevailing winds	Nitrate nitrogen (NO_3-N)	
	Biochemical oxygen demand (BOD_5 and BOD_{20})	
	Nitrite nitrogen (NO_2-N)	
	Ammonia nitrogen (NH_3-N)	

Source: PaDEP, 2002.

TABLE 14.7 Use Attainability Surveys for Impounded Water Bodies

Physical Evaluations	Chemical Evaluations	Biological Evaluations
Drainage basin characteristics	*Field Analyses* * (*at 1-meter intervals through water column*)	Fishes
Hydrology	*Temperature	Macrophytes
Size	*Dissolved oxygen	Benthic macroinvertebrates
Inflows	*pH	Plankton
Geology	*Specific conductance	
Outflows	Secchi disc	
Soils	*Laboratory Analyses (1 meter below surface and 1 meter from bottom)*	
Mean retention time	Silica (dissolved)	
Land use (percent urban, suburban,	Total phosphate (as P)	
agricultural, forest, lake)	Calcium (dissolved)	
Weather	Total soluble phosphate (as P)	
Air temperature	Magnesium (dissolved)	
Cloud cover	Total dissolved solids (TDS)	
Preceding/prevailing winds	Sodium (dissolved)	
	Total hardness (as $CaCO_3$)	
	Potassium (dissolved)	
	Total alkalinity (as $CaCO_3$)	
	Sulfate (dissolved)	
	Total organic carbon	
	Chloride (dissolved)	
	pH	
	Iron (dissolved)	
	Specific conductance	
	Manganese (dissolved)	
	Dissolved oxygen	
	Nitrate nitrogen	
	Temperature	
	Ammonia nitrogen	
	Chlorophyll a (surface only)	
	Nitrite nitrogen	

Source: PaDEP, 2002.

399

A typical biological *antidegradation survey protocol* would follow the following procedure:

A. Using USGS 1:24,000 topographic maps:
1. Determine the basin location and stream order.
2. Determine tentative sampling locations based on:
   - Downstream end of study area
   - Main stem every 2 to 3 stream miles
   - Mouth of major tributaries
   - Upstream and downstream of impoundments
   - Population centers
   - Known sources of nonpoint and point source pollution
   - Changes in land use

   Actual frequency of locations will depend on noticeable changes in stream flow, instream habitat structure, riparian habitat, land use, and land cover. Minor tributaries may be targeted. Chemical samples should be collected on the minor tributaries even if biological sampling is not conducted.
3. Determine final sampling site locations within a stream reach based on riffle/run habitat in high to moderate gradient streams, or pool/glide habitat in low gradient streams. These habitats provide the most productive macroinvertebrate conditions.

14.8   AQUATIC ORGANISM ASSESSMENT

A *dip net* is placed at the bottom of the stream just downstream of an area of swift water while the substrate just upstream of the net is sufficiently disturbed to dislodge the benthic organisms. The dislodged organisms will flow into the net. The net is used to obtain a "kick" sample of organisms that is representative of the organisms present. Rocks within the disturbed area should be overturned and the attached organisms collected and included in the sample.

14.9   SUMMARY

Maintaining the ecological integrity of a watershed, or the various unique areas of the watershed, should be a goal of every watershed assessment plan. Evaluating the ecoregions aids in determining what aquatic and terrestrial conditions to expect in an area. Performing a rapid bioassessment protocol, aquatic survey, and water chemical analysis determines the condition and/or stresses on a particular stream reach.

REFERENCES

Bailey, R.G. "Description of the Ecoregions of the United States," Miscellaneous Publication 1391 (Washington, DC: U.S. Department of Agriculture), 1980, 2nd Edition, Revised and Enlarged March 1995.

————. Explanatory Supplement to Ecoregions Map of the Continents. Environmental Conservation 16:307–309. With separate map at 1:30,000,000, 1989.

————. "Delineation of Ecosystem Regions," *Environmental Management,* vol. 7, pp. 365–373, 1983.

Bailey, R.G., and C.T. Cushwa. *Ecoregions of North America* [map]. (FWS/OBS-81/29.) Washington, DC: U.S. Fish and Wildlife Service. 1:12,000,000, 1981.

Bailey, R.G., P.E. Avers, T. King, and W. H. McNab. *Ecoregions and Subregions of the United States* [map]. Washington, DC: USDA Forest Service. 1:7,500,000. With supplementary table of map unit descriptions, compiled and edited by W.H. McNab and R.G. Bailey. http://www.fs.fed.us/land/ecosysmgmt/pages/ecoreg1home.html, 1994.

Center for Educational Technologies and the COTF/Classroom of the Future, Exploring the Environment, Wheeling Jesuit University, A NASA funded project, www.cotf.edu/ete/modules/waterq3/WQglossary.html#watershed. Accessed July 25, 2003.

Cleland, David T., Peter E. Avers, W. Henry McNab, Mark E. Jensen, Robert G. Bailey, Thomas King, and Walter E. Russell. "National Hierarchical Framework of Ecological Units." In *Ecosystem Management: Applications for Sustainable Forest and Wildlife Resources,* Mark S. Boyce and Alan Haney (Eds.) (New Haven: Yale University Press), pp. 181–200, 1997.

ECOMAP. *National Hierarchical Framework of Ecological Units.* Washington, DC: U.S. Forest Service, 1993.

Frissell, C.A., W.J. Liss, C.E. Warren, and M.D. Hurley. "A Hierarchical Framework for Stream Habitat Classification: Viewing Streams in a Watershed Context," *Environmental Management,* vol. 10, pp. 199–214, 1986.

Gibson, G.R., M.T. Barbour, J.B. Stribling, J. Gerritsen, and J.R. Karr. Biological Criteria: Technical Guidance for Streams and Small Rivers, revised edition (Washington, DC: U.S. Environmental Protection Agency, Office of Water, EPA 822-B-96-001), 1996.

Koppen, J.M. *Grundriss der Klimakunde* (Berlin: Walter de Grayter), 1931.

Kuchler, A.W. "Potential Natural Vegetation of the Conterminous United States," American Geographic Society Special Publication 36, 116 p., 1964.

Lewis Lori J., and Michael Turtora. "Stream Habitat Characteristics at Selected Sites in the Georgia-Florida Coastal Plain," USGS Water-Resources Investigations Report 98-4013, Tallahassee, FL, 23 p., 1998.

Moulton, Stephen R., II, Jonathan G. Kennen, Robert M. Goldstein, and Julie A. Hambrook. "Revised Protocols for Sampling Algal, Invertebrate, and Fish Communities as Part of the National Water-Quality Assessment Program," USGS Open-File Report 02-150, U.S. Geological Survey, Reston, VA, 2002.

Omernik, J.M. "Ecoregions of the Conterminous United States," *Annals of the Association of American Geographers,* vol. 77, no. 1, pp. 118–25, 1987.

————. "Ecoregions: A Spatial Framework for Environmental Management." In: *Biological Assessment and Criteria: Tools for Water Resource Planning and Decision Making,* W. Davis and T. Simon (Eds.) (Boca Raton, FL: Lewis Publishers), pp. 49–62, 1995.

Omernik, James M., and Robert G. Bailey. "Distinguishing between Watersheds and Ecoregions," *Journal of the American Water Resources Association,* vol. 33, no. 5, pp. 935–94, October 1997.

Pennsylvania Department of Environmental Protection (PaDEP). "Aquatic Life—Use Attainable Studies for Flowing and Impounded Water Bodies," (Harrisburg, PA: PaDEP), April 18, 2002.

Townsend, C.R. "The Patch Dynamic Concept of Stream Community Ecology," *Journal of the North American Benthological Society,* vol. 8, 1989, pp. 36–50.

Trewartha, G.T. *An Introduction to Climate 4th ed.* (New York, NY: McGraw-Hill), 1968.

U.S. EPA. "Level III Ecoregions," Corvallis, OR: U.S. EPA, Western Ecology Division, ftp://ftp.epa.gov/wed/ecoregions/us/useco_desc.doc, May 30, 2003, accessed July 25, 2003.

U.S. EPA. "Rapid Bioassessment Protocols for Use in Wadeable Streams and Rivers, Periphyton, Macroinvertebrates, and Fish," 2nd ed., EPA 841-B-99-002, July 1999.

U.S. Forest Service, "Ecological Subregions of the United States," www.fs.fed.us/land/pubs/ecoregions/toc.html, accessed March 22, 2003.

U.S. General Accounting Office (GAO). "Ecosystem Management: Additional Actions Needed to Adequately Test a Promising Approach," GAO/RCED-94-111 U.S. General Accounting Office, Washington, DC, 1994.

U.S. Geological Survey. "Methods for Sampling Fish Communities as a Part of the National Water-Quality Assessment Program," U.S. Geological Survey, Open-File Report 93-104; Raleigh, NC, available at http://water.usgs.gov/nawqa/protocols/OFR-93-408/habit1.html, 1993.

Vannote, R.L., G.W. Minshall, K.W. Cummins, J.R. Sedell, and C.E. Cushing. "The River Continuum Concept," *Canadian Journal of Fisheries and Aquatic Sciences,* 37, pp. 130–137, 1980.

Ward, J. V. *Aquatic insect ecology 1. Biology and habitat* (New York, NY: John Wiley and Sons), 1992.

Wiken, E.B. (compiler). "Terrestrial Ecozones of Canada," *Ecological Land Classification,* Series No. 19 (Hull, PQ: Environment Canada), 1986.

Woods, Alan J., James M. Omernik, and Douglas D. Brown. "Level III and IV Ecoregions of Pennsylvania and the Blue Ridge Mountains, the Ridge and Valley, and the Central Appalachians of Virginia, West Virginia, and Maryland," U.S. Environmental Protection Agency National Health and Environmental Effects Research Laboratory, Corvallis, Oregon 97333, June 1996 (Addenda May 1999). http://www.dcnr.state.pa.us/wrcf/contents.htm. http://www.epa.gov/ceisweb1/ceishome/atlas/maiaatlas/mid_atlantic_ecoregions.html.

WOW. "Water on the Web—Monitoring Minnesota Lakes on the Internet and Training Water Science Technicians for the Future—A National On-line Curriculum Using Advanced Technologies and Real-Time Data," University of Minnesota-Duluth, Duluth, MN 55812, http://waterontheweb.org, accessed April 28, 2003.

## CHAPTER 15

NONPOINT SOURCE POLLUTION LOAD ASSESSMENT

15.0 INTRODUCTION

Quantification of pollutant loads is important in determining the overall health of water bodies and loads from the surrounding watershed, and in evaluating the total maximum daily load (TMDL) that a water body can receive without being degraded. Pollutant loads can be determined in one of several ways: field measurement and calibration, simple equations or spreadsheets, the mass balance approach, or simulation models.

Pollutant loads are typically described in terms of either mass (pounds/ acre/year, pounds/day or tons/acre/year) or concentration (mg/l or ppm). The pollutant load occurs either at a point in time (grab sample) or during a storm event (event mean concentration, or EMC). Each will be discussed in this chapter.

15.1 EXISTING POLLUTANT LOADING DATA (NURP)

One source of nonpoint source loading data is the National Urban Runoff Program (NURP). The NURP study was conducted by the Environmental Protection Agency between 1978 and 1983 to determine pollutant loads from urban runoff (U.S. EPA, 1983). Samples were collected from 81 sites in order to establish median event mean concentrations (EMCs) for the various urban land use categories. The results can be found in Table 15.1 (U.S. EPA, 1983b; 1999).

By analyzing the NURP data, a relationship has been established between sediment levels and the size of the urban watershed. This relationship can be

TABLE 15.1 Median Event Mean Concentrations for Urban Land Uses (USEPA, 1999)

Pollutant	Units	Residential		Mixed		Commercial		Open/Nonurban	
		Median	COV	Median	COV	Median	COV	Median	COV
BOD	mg/l	10	0.41	7.8	0.52	9.3	0.31	—	—
COD	mg/l	73	0.55	65	0.58	57	0.39	40	0.78
TSS	mg/l	101	0.96	67	1.14	69	0.85	70	2.92
Total Lead	μg/l	144	0.75	114	1.35	104	0.68	30	1.52
Total Copper	μg/l	33	0.99	27	1.32	29	0.81	—	—
Total Zinc	μg/l	135	0.84	154	0.78	226	1.07	195	0.66
Total Kjeldahl Nitrogen	μg/l	1,900	0.73	1288	0.50	1179	0.43	965	1.00
Nitrate + Nitrite	μg/l	736	0.83	558	0.67	572	0.48	543	0.91
Total Phosphorus	μg/l	383	0.69	263	0.75	201	0.67	121	1.66
Soluble Phosphorus	μg/l	143	0.46	56	0.75	80	0.71	26	2.11

COV: Coefficient of variation

found in Figure 15.1 (Schueler, 1987) and can be used as a guide for sediment pollution prediction.

15.2 FIELD DATA COLLECTION AND MONITORING

Measuring pollutants is very reliable if properly controlled, but the process also can be very expensive, depending on the accuracy of the methodology utilized, the quality assurance/quality control (QA/QC) procedures followed, the number of sampling points, the constituents being sampled, and the duration of the sampling period. Cost is typically also directly related to watershed size. In order to determine the extent of the sampling program, the cost of the proposed sampling and analysis programs must be balanced with the required accuracy of the data.

Before establishing a monitoring program, available water quality data should first be obtained. The sources of all existing data should be checked to determine their quality and reliability. Federal or state agency data collection usually follows strict QA/QC procedures and is typically very reliable. In contrast, the QA/QC procedures followed by volunteer groups should be reviewed carefully before including their data in the study. The goals and objectives of the watershed assessment and long-term use compatibility determination should be kept in mind when developing a water quality moni-

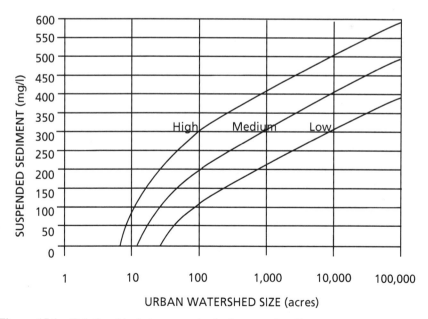

Figure 15.1 Relationship between watershed area and sediment event mean concentration (EMC) (Schueler, 1987).

toring program. This, along with watershed characteristics (land use, geology, regional location) and the computer model(s) that will be utilized, will also dictate which water quality parameters to collect. Water quality data are typically used as input for water quality models such as the Storm Water Management Model (SWMM), the Illinois Urban Drainage Area Simulator (ILLUDAS), the Enhanced Stream Water Quality Model (Qual2E), the Better Assessment Science Integrating Point and Nonpoint Sources (BASINS) Model, and the Virginia Stormwater Model (VAST). In addition, the model(s) should be closely examined to determine the correct pollutant form to sample and analyze; for example, each of the various forms of nitrogen (TKN, total nitrogen, nitrate, ammonia) has a different decay function. Water quality samples can be obtained manually or with the use of automatic samplers, and data collection should be coordinated with rainfall and stream flow data collection.

Table 8.6 in Chapter 8, which lists the parameters typically associated with various land cover types and various pollutants of concern in impoundments and streams, can be consulted to determine a preliminary or initial range of pollutants to test within the watershed. To minimize costs, those parameters that do not appear in concentrations of concern in the initial sampling can be dropped from the monitoring program. However, a reevaluation should be performed if conditions change in the watershed.

Water quality parameter laboratory testing can often be expensive; in fact, the costs for testing many parameters can consume a large portion of a watershed assessment budget. Therefore, it is best to make a preliminary assessment of parameters to test for based on the parameters associated with the land use or industry in the area, then prioritize which parameters to test for. And, as noted previously, though grab sampling often involves a chance that a pollutant will be found, it does have its applications—for example, base flow ambient water quality, as long as flow is obtained at the time of the sampling. That said, much grant funding has been wasted by sampling programs whereby a grab sample is collected irrespective of flow, days since last rain, and so on; or there could be a polluter upstream making discharges at night, and the day grab sample would miss it. Thus, the results do not "tell a story." Storm event sampling, in contrast, allows the plot of the pollutograph, comparison to hydrograph and runoff volume, and subsequent loading from the watershed.

Table 15.2 (Brodhead Watershed Association, 2003) summarizes a number of water quality parameters that could be tested for—whether they may be baseline, storm event, lake, or sediment samples—and the method, minimum detection limits (MDL), holding time, type of sample, and preservative.

Sampling locations should be strategically located based on up-gradient land uses, industry, discharges, just-upstream-of-tributary confluences, predetermined management areas, and so on.

Pollutant concentrations are usually expressed in terms of milligrams per liter; however, this represents a concentration at one point in time and does

TABLE 15.2 Water Quality Sampling Parameters

Parameter	Method	MDL	Holding	Sample	Bottle	Pres.
Baseline Monitoring						
Water level	Staff guage	n/a	n/a	Field	n/a	n/a
Water flow	Field sensor	n/a	n/a	Field	n/a	n/a
Temperature	Field sensor	n/a	n/a	Field	n/a	n/a
pH	Field sensor	n/a	n/a	Field	n/a	n/a
Conductivity	Field sensor	n/a	n/a	Field	n/a	n/a
Dissolved oxygen	Field sensor	n/a	n/a	Field	n/a	n/a
Total dissolved solids	Field sensor	n/a	n/a	Field	n/a	n/a
pH	EPA 150.1	n/a	ASAP	Grab	1-L-P	n/a
Nitrite	SM 4500 NO2B	0.005 mg/l	48 hours	Grab	1-L-P	n/a
Nitrate	SM 4500 NO3D	0.1 mg/l	48 hours	Grab	1-L-P	n/a
Total phosphorus	EPA 365.2	0.02 mg/l	28 days	Grab	1-L-P	H2SO4
Total suspended solids	EPA 160.2	1.0 mg/l	7 days	Grab	1-L-P	n/a
Fecal coliform	SM 9222B	1 CFU/100 ml	6 hours	Grab	1-120-P	Na2S2O3
Alkalinity	SM 2320B	1.0 mg/l	7 days	Grab	1-L-P	n/a
Total hardness	EPA 130.2	1.0 mg/l	28 days	Grab	1-L-P	n/a
Aluminum	EPA 202.1	0.1 mg/l	6 months	Grab	1-L-P	HNO3
Antimony	EPA 204.1	0.1 mg/l	6 months	Grab	1-L-P	HNO3
Arsenic	EPA 206.2	0.005 mg/l	6 months	Grab	1-L-P	HNO3
Barium	EPA 208.1	0.1 mg/l	6 months	Grab	1-L-P	HNO3
Beryllium	EPA 210.1	0.02 mg/l	6 months	Grab	1-L-P	HNO3
Cadmium	EPA 213.1	0.02 mg/l	6 months	Grab	1-L-P	HNO3
Calcium	EPA 215.1	0.02 mg/l	6 months	Grab	1-L-P	HNO3
Chromium	EPA 218.1	0.02 mg/l	6 months	Grab	1-L-P	HNO3
Cobalt	EPA 219.1	0.05 mg/l	6 months	Grab	1-L-P	HNO3
Copper	EPA 220.1	0.02 mg/l	6 months	Grab	1-L-P	HNO3
Iron	EPA 236.1	0.03 mg/l	6 months	Grab	1-L-P	HNO3
Lead	EPA 239.1	0.10 mg/l	6 months	Grab	1-L-P	HNO3

TABLE 15.2 (*Continued*)

Parameter	Method	MDL	Holding	Sample	Bottle	Pres.
Mercury	EPA 245.1	0.0005 mg/l	28 days	Grab	1-L-P	HNO3
Magnesium	EPA 242.1	0.02 mg/l	6 months	Grab	1-L-P	HNO3
Manganese	EPA 243.1	0.04 mg/l	6 months	Grab	1-L-P	HNO3
Nickel	EPA 249.1	0.05 mg/l	6 months	Grab	1-L-P	HNO3
Selenium	EPA 270.1	0.005 mg/l	6 months	Grab	1-L-P	HNO3
Silver	EPA 272.1	0.02 mg/l	6 months	Grab	1-L-P	HNO3
Thallium	EPA 279.1	0.1 mg/l	6 months	Grab	1-L-P	HNO3
Vanadium	EPA 286.1	2.0 mg/l	6 months	Grab	1-L-P	HNO3
Zinc	EPA 289.1	0.02 mg/l	6 months	Grab	1-L-P	HNO3
Potassium	EPA 258.1	0.02 mg/l	6 months	Grab	1-L-P	HNO3
Sodium	EPA 273.1	0.02 mg/l	6 months	Grab	1-L-P	HNO3
Storm Events						
Water level	Staff gage	n/a	n/a	Field	n/a	n/a
Water flow	Field sensor	n/a	n/a	Field	n/a	n/a
Temperature	Field sensor	n/a	n/a	Field	n/a	n/a
pH	Field sensor	n/a	n/a	Field	n/a	n/a
Conductivity	Field sensor	n/a	n/a	Field	n/a	n/a
Dissolved oxygen	Field sensor	n/a	n/a	Field	n/a	n/a
pH	EPA 150.1	n/a	ASAP	Composite	1-L-P	n/a
Nitrite	SM 4500 NO2B	0.005 mg/l	48 hours	Composite	1-L-P	n/a
Nitrate	SM 4500 NO3D	0.1 mg/l	48 hours	Composite	1-L-P	n/a
Total phosphorus	EPA 365.2	0.02 mg/l	28 days	Composite	1-L-P	H2SO4
Total suspended solids	EPA 160.2	1.0 mg/l	7 days	Composite	1-L-P	n/a
Fecal coliform	SM 9222B	1 CFU/100 ml	6 hours	Grab	1-120-P	Na2S2O3
Oil & grease	EPA 1664	1.6 mg/l	28 days	Grab	1-L-G	H2SO4
Like Samples						
Temperature	Field sensor	n/a	n/a	Field	n/a	n/a

Parameter	Method	Detection limit	Holding time	Sample type	Container	Preservative
pH	Field sensor	n/a	n/a	Field	n/a	n/a
Conductivity	Field sensor	n/a	n/a	Field	n/a	n/a
Dissolved oxygen	Field sensor	n/a	n/a	Field	n/a	n/a
Specific conductance	Field sensor	n/a	n/a	Field	n/a	n/a
Alkalinity	SM 2320B	1.0 mg/l	7 days	Grab	1-L-P	n/a
Total suspended solids	EPA 160.2	1.0 mg/l	7 days	Grab	1-L-P	n/a
Total phosphorus	EPA 365.2	0.02 mg/l	28 days	Grab	1-L-P	H2SO4
Chlorophyll -a				Grab	1-L-P	
Hardness	EPA 130.2	1.0 mg/l	28 days	Grab	1-L-P	n/a
Dissolved reactive phosphorus	EPA 365.3	0.02 mg/l	28 days	Grab	1-L-P	n/a
Ammonia nitrogen	SM 4500 NH3E	0.1 mg/l	28 days	Grab	1-L-P	H2SO4
Nitrate nitrogen	SM 4500 NO3D	0.1 mg/l	48 hours	Grab	1-L-P	n/a
Nitrite nitrogen	SM 4500 NO2B	0.005 mg/l	48 hours	Grab	1-L-P	n/a
Total Kjeldahl nitrogen	SM 4500-NorgB	0.3 mg/l	28 days	Grab	1-L-P	H2SO4
Sediment Samples						
PCBs	SW-846(8020)	0.5 mg/kg	7 days	Grab	4oz-G	n/a
Aluminum	SW-846(6010)	3.31 mg/kg	28 days	Grab	4oz-G	n/a
Cadmium	SW-846(6010)	0.091 mg/kg	28 days	Grab	4oz-G	n/a
Copper	SW-846(6010)	0.16 mg/kg	28 days	Grab	4oz-G	n/a
Iron	SW-846(6010)	4.77 mg/kg	28 days	Grab	4oz-G	n/a
Lead	SW-846(6010)	0.94 mg/kg	28 days	Grab	4oz-G	n/a
Manganese	SW-846(6010)	0.07 mg/kg	28 days	Grab	4oz-G	n/a
Mercury	SW-846(7471A)	0.0105 mg/kg	28 days	Grab	4oz-G	n/a
Zinc	SW-846(6010)	0.36 mg/kg	28 days	Grab	4oz-G	n/a
Total phosphorus	EPA 365.1	10 mg/kg	28 days	Grab	4oz-G	n/a
Total organic carbon	EPA 415.1	60 mg/kg	28 days	Grab	4oz-G	n/a
Grain size and distribution	ASTM D-422	1 or 75 um	n/a	Grab	4oz-G	n/a
Digestion for sed. samples	various	n/a	n/a	n/a	n/a	n/a

little to explain the true conditions of water quality. The Event Mean Concentration (EMC) methodology was, therefore, established to quantify water quality sampling data. The EMC is computed using the half-volume method and is equal to a total storm event load divided by the total storm event runoff volume, as shown in Figure 15.2.

Flow and concentrations of a particular water quality parameter are determined at incremental stages, as shown in the following equations. Concentrations of the pollutant constituents are determined for each sample. The volumes under the curve, $V1$ through $V6$, are the incremental volumes of runoff leading up to the time the sample was taken. The volumes calculated for each concentration are, therefore, labeled $C1$ through $C7$. Using Simpson's Rule as stated in the following equations, the average volumes of concentration ($C1$–$C7$) for each increment would therefore be calculated as:

$$C1 = V1 + (0.5 \times V2) \tag{15.1}$$

$$C2 = (0.5 \times V2) + (0.5 \times V3) \tag{15.2}$$

$$C3 = (0.5 \times V3) + (0.5 \times V4) \tag{15.3}$$

$$C7 = (0.5 \times V6) + (0.5 \times V7) \tag{15.4}$$

The total load for an event is, therefore, the summation of the concentration of each sample multiplied by the volume assigned to that concentration:

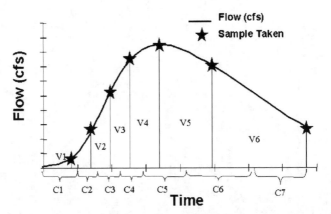

Figure 15.2   Calculation of EMC by the half-volume method (DeBarry, 1996).

$$\text{Total Load} = (V1 \times C1) + (V2 \times C2) + \ldots (V7 \times C7) \quad (15.5)$$

Total annual load can be determined from the EMC by obtaining one year's worth of storm event data.

Because watershed management recommendations are typically needed within the short time span of a study, it is not feasible to undertake a comprehensive nonpoint source (NPS) pollution monitoring study prior to developing the watershed plan. Further, since completion of the EPA's Nationwide Urban Runoff Program, NURP (U.S. EPA, 1983a), there is a general consensus that local monitoring studies may no longer be required to characterize NPS pollution loadings due to the extreme expense of laboratory costs. Results utilizing loading values and available in the literature have been found to be in agreement with those determined through expensive local monitoring programs.

It is recommended that a long-term monitoring program be established to track changes in the watershed over time, or the changes in water quality due to changes in, for instance, land use. The program should be established before monitoring begins, to ensure consistency throughout the process.

15.3 TOTAL MAXIMUM DAILY LOAD (TMDL)

The *total maximum daily load,* or *TMDL,* is a geographically based approach to preparing load and waste load allocations for sources of stress that might impact water body integrity (U.S. EPA, 1994). It is the sum of the individual waste load allocations for point, nonpoint, and natural sources with a margin of safety expressed in terms of mass per time, toxicity, or other unit of measure, and quantifies the pollutant loading capacity of a water body (stream, river, or lake) for a particular pollutant. Thus it provides a quantitative method for allocating loadings from pollutant sources (U.S. EPA, 1994). A TMDL is a tool used to achieve *water quality standards (WQS)* via a mathematical analysis of the percent reduction of a pollutant from a particular source needed to meet the concentration of that pollutant specified in the WQS.

The goal of the Clean Water Act, passed in 1972, is to achieve established water quality criteria at least 99 percent of the time. The Clean Water Act assigned states the task of, one, identifying surface waters or portions thereof that required TMDL development; two, prioritizing them; and three, developing TMDLs for these waters. Waters requiring TMDL development are those water bodies or stream reaches that the state identifies as failing or not expected to meet the WQS after applying technology-based controls. The states were also tasked to establish *water quality-based effluent limitations (WQBELs)* on point source discharges, to be incorporated into the TMDL equation. States must also develop *waste load allocations (WLAs),* or the portion of the surface water's loading capacity allocated to existing and future *point* source discharges. WLAs serve as the basis for determining the WQBELs for point source discharges relating to the *National Pollutant Discharge Elimination System* (*NPDES*) permitting and monitoring. States must also designate *load allocations* (*LAs*), which are the portions of the surface water's loading capacity assigned or allocated to existing and future *nonpoint* sources and natural quality. LAs are the basis for development of nonpoint source restoration plans.

In developing TMDLs and WQBELs, states consider the following relevant design factors:

- Water quality criteria duration
- Flow duration and frequency
- Natural seasonal variability in water temperature
- Natural variability of pH and hardness
- Pollution contributions from other sources
- Assumption that all pollutants are conservative
- Physical characteristics of the watershed
- Reserve factors
- Factors of safety

Treating all pollutants as conservative means that they are assumed not to decay, transform, or degrade for analysis purposes. If evidence is available to the contrary, the actual data may be utilized for that "unconservative" pollutant. Reserve factors are a portion of the effluent flow held to provide for projected future waste loads. The factor of safety is a margin that takes into account uncertainties in the relationships between effluent limitations and water quality, including inaccuracies in water quality models. As can be seen from the design factors listed, TMDL development is a complex and interrelated process. Mathematical models integrated with the GIS, such as the ArcView Generalized Watershed Loading Function Model (AVGWLF) (Evans et al., 2001), are useful tools to aid in this process.

The waste load allocation procedure is as follows.

WLAs, LAs and effluent limitations are assigned or allocated to individual pollutant sources based on the more stringent of the following:

- The pollutant loading that will achieve the water quality protection criteria specified in the applicable state code.
- The pollutant loading required by applicable state nutrient discharge requirements. Nutrient regulations may require:
  - Land disposal of wastewater
  - Additional phosphorus limitations
  - Additional and more specific TMDL requirements for surface runoff into lakes

- The pollutant loading required by applicable state and thermal discharge requirements. Thermal regulations may require:
  - No change in surface water temperature greater than 2 degrees Fahrenheit during a one-hour period.
  - The allowable heat content of the discharge will be limited to an evaluation conducted in accordance with Section 316(a) of the Federal Clean Water Act (33 U.S.C.A Section 1326(a)) or applicable state code.
  - Additional and more specific TMDL requirements for surface runoff into lakes.
- The pollutant loading authorized to be discharged under applicable technology-based requirements.

Water quality design criteria for point source discharges are based on applicable steady state design flows, which may vary based on the parameter being modeled. Typical steady state design flows are shown in Table 15.3.

The Q_{7-10} is the actual or estimated lowest seven-consecutive-day average flow that occurs once in 10 years for a stream with unregulated flow, or the estimated minimum flow for a stream with regulated flow. The Q_{30-10} is the estimated 30-day average minimum flow for a stream with regulated flow. The harmonic mean flow is the flow that is determined by taking the reciprocal of the arithmetic mean of daily flow values.

15.4 POLLUTANT LOADING EQUATIONS AND METHODS

The ultimate goal of nonpoint source pollution assessment is the identification of critical loadings so that best management practices (BMPs) can be incorporated into stormwater controls in those areas where it is critical to reduce NPS loads. This is particularly important where streams and reservoirs in the watershed serve as municipal water supplies, and to meet the more stringent treatment requirements for drinking water supplies imposed by the 1986 Federal Safe Drinking Water Act. In addition, if streams within the watershed

TABLE 15.3 Recommended Steady State Design Flows and Water Quality Criteria

Water Quality Criteria	Steady State Design Flows
Ammonia-nitrogen (NH_3-N)	Q_{30-10}
Fish and aquatic life, except NH_3-N	Q_{7-10}
Threshold human health	Q_{7-10}
Nonthreshold human health (Carcinogens)	Harmonic mean flow

are classified as "high quality waters" or "exceptional value waters," it is best to determine where pollutants are originating in order to properly manage land use and resultant pollutant loads from NPS runoff. Uncontrolled future development within the watershed could jeopardize water supplies, impoundment quality, and coldwater fisheries.

Pollutant loads or export can also be predicted by utilizing one of several simple methods. These methods assume that all runoff has a constant concentration of pollutant. The parameters typically analyzed are suspended solids, total phosphorus, total nitrogen, BOD_5, and fecal coliforms.

15.4.1 Schueler's Simple Method

A method, commonly known as the "Simple Method," was developed by T.R. Schueler (1987) to determine pollutant export loads from development sites. Even if one does not wish to utilize the equations, Schueler's report, "Controlling Urban Runoff," is an excellent source of information on urban runoff and best management practices. The method is applicable to less than 1 square mile and utilizes an empirical equation to estimate total mass pollutant loads from a development site, as shown in equation 15.6.

$$L = [(P)\ (Pj)\ (Rv)/12]\ (C)\ (A)\ (2.72) \tag{15.6}$$

Where:

L = Pollutant export in pounds

P = Rainfall depth (inches) over the desired time interval. The value of P will vary with duration of the load estimate desired. Values can be obtained from regional or local standard rainfall-duration-frequency curves, if available, or TP-40, if standard curves are not available. If a specific design storm load is required, then the local rainfall data for this storm can be utilized. The average annual rainfall amount can be utilized if an annual load is desired.

Pj = Factor that corrects P for storms that produce no runoff. Pj is used for the fraction of annual or seasonal rainfall in the calculation period that will not produce storm runoff. Schueler determined that about 10 percent of the storms in the Washington, DC, area do not produce runoff; therefore, the Pj value would be 0.9. Individual storms would have a Pj value of 1.

Rv = Runoff coefficient, which expresses the fraction of rainfall that is converted to runoff. Rv depends on the nature of soils, topography, and land cover, and is a measure of the site's response to rainfall events. Rv can be determined by equations 15.7 and 15.8.

C = Flow-weighted mean concentration of the pollutant (mg/l). This value generally represents the pollutant levels from stabilized urban areas. It is the flow-weighted mean concentration of the pollutant (mg/l) obtained from the NURP study, and can be obtained from Table 15.4 (from Schueler, 1987) for the Washington, DC, area. Localized data should be used to better reflect local conditions. Schueler's publication outlines a procedure to obtain localized C values. NPS pollutant concentrations are reasonable and reliable for nonpoint source pollution decisions on a site-planning level when computed using a methodology such as the Washington, DC, Council of Governments Simple Method. However, to indicate areas of potential critical nonpoint source pollution loadings on a macro, or watershedwide, scale, the mass balance approach described later in this chapter may be more appropriate.

A = The site's drainage area in acres. Appendix A, Section 6 of Schueler's "Controlling Urban Runoff" defines a methodology to account for base flow in drainage areas greater than 640 acres. 12 and 2.72 = unit conversions.

$$Rv = r/p \tag{15.7}$$

Or

$$Rv = 0.05 + 0.009\ (I) \tag{15.8}$$

Where:

Rv = Runoff coefficient
r = Storm runoff (inches)
p = Storm rainfall (inches)
I = Percent imperviousness area of the site

15.4.2 Mass Balance Approach

Monitoring and modeling methods to determine nonpoint source pollutant loadings can be very accurate, but are often very time-consuming and expensive. If the data to properly calibrate a model are not available, the accuracy of the model could be in question. In addition, with so many variables in a watershed, the model's results may give a false sense of security as to their reliability. For instance, if there were one construction site in the watershed contributing sediment to the receiving waters, and it was not documented, the model might be calibrated against exaggerated sediment levels. Also, land activities taking place in the watershed are dynamic, and the models would have to incorporate changes to them to accurately represent the situation. The

TABLE 15.4 Urban C-Values for Use with the Simple Method

Pollutant	New Suburban NURP Sites (Washington, DC)	Older Urban Areas (Baltimore)	Central Business District (Washington, DC)	National NURP Study Average	Hardwood Forest (North Virginia)	National Urban Highway Runoff
Phosphorus						
Total	0.26	1.08	—	0.46	0.15	—
Ortho	0.12	0.26	1.01	—	0.02	—
Soluble	0.16	—	—	0.16	0.04	0.59
Organic	0.10	0.82	—	0.13	0.11	—
Nitrogen						
Total	2.00	13.6	2.17	3.31	0.78	—
Nitrate	0.48	8.9	0.84	0.96	0.17	—
Ammonia	0.26	1.1	—	—	0.07	—
Organic	1.25	—	—	—	0.54	—
TKN	1.51	7.2	1.49	2.35	0.61	2.72
COD	35.6	163.0	—	90.8	>40.0	124.0
BOD (5-day)	5.1	—	36.0	11.9	—	—
Metals						
Zinc	0.037	0.397	0.250	0.176	—	0.380
Lead	0.018	0.389	0.370	0.180	—	0.350
Copper	—	0.105	—	0.047	—	—

actual "trapping" efficiency of erosion and sediment pollution control measures is extremely hard to achieve, and hence adds another layer of uncertainty to the results of computer models.

Given that results are often unreliable, coupled with the expanse of the watershed area and the expense of modeling, an educated estimate of pollutant loads may be suitable for the intended purpose. The *mass balance approach* is a method that applies national, regional, or locally derived annual pollutant loads (typically found in units of pounds/acre/year), from the various land uses in the area of contribution to the point of interest. Naturally, the more site-specific the data, the more accurate the results of the analysis. An extensive literature search should be performed to obtain unit aerial pollutant loadings (lbs/ac/yr) by land use for each of the pollutant factors to be analyzed. The land uses can be classified as in USDA's National Engineering Handbook (NEH-4) (USDA, 1985), which is duplicated in TR-55 (USDA, 1986) for consistency with hydrologic modeling. The results of the literature search should yield a table of nonpoint source pollutant loadings (lbs/ac/yr) for various land uses and soil types. These results can then be compared and analyzed for consistency, variability, and regionality. Nonpoint pollution loadings tend to be governed by the amount of impervious area associated with each land use. Loading rates or land use export coefficients tend to be fairly consistent but do vary slightly by region. These factors were considered when developing the loading rates. Typical pollutant loading data results, compiled from NURP data (U.S.. EPA, 1983) and then refined or regionalized with local data from a nearby EPA Clean Lakes Program Study (Lake Wallenpaupack Watershed Management District, 1983) can be found in Table 15.5 (DeBarry, 1991).

Composite loadings can then be calculated as shown in the following example in Table 15.6 for total nitrogen.

The unit watershed loading would, therefore, be the total loading divided by the area, or 704 pounds/year/200 acres equals 3.52 pounds/acre/year. This process could also be utilized to predict the total maximum daily load (TMDL) for a watershed. The results could then be displayed as shown in Table 15.7.

This process was incorporated into the GIS as early as 1991 (DeBarry, 1991). A 172-square-mile watershed was divided into 206 subwatersheds. The pollutant loading data was included in the GIS attribute data of the individual land uses, and by cross-referencing (overlaying) the hydrologic soil groups and the subwatersheds, pollutant loads by subwatershed were determined in the GIS. Pollutant loads were then computed for each subwatershed and the results could be displayed graphically. This then highlighted the troubled subwatersheds and provided a tool for prioritizing and better managing the land use within these subwatersheds.

Land use loading rates were used to obtain average annual and total pollutant loads for each of the 206 subareas in the watershed for existing and future conditions (see Figure 15.3). The results were then plotted graphically

TABLE 15.5   Typical Pollutant Loading Values for a Watershed in Northeast Pennsylvania

	Nonpoint Source Pollution Loading Factors							
	Total: Nitrogen (lbs/ac/yr) Hydrologic Soil Group				Total: Phosphorus (lbs/ac/yr) Hydrologic Soil Group			
Land Use	A	B	C	D	A	B	C	D
Open Space	0.60	0.60	0.60	0.60	0.08	0.08	0.08	0.08
Meadow	0.06	0.60	0.60	0.60	0.08	0.08	0.08	0.08
Newly Graded	—	—	—	—	3.10	4.70	5.60	6.60
Forest	0.60	0.60	0.60	0.60	0.08	0.08	0.08	0.08
Commercial	13.20	13.20	13.20	13.20	1.60	1.60	1.60	1.60
Industrial	11.20	11.30	11.20	11.00	1.30	1.35	1.35	1.30
Residential								
1/8 acre or less	11.70	12.90	13.20	12.00	1.60	1.60	1.70	1.75
1/4–1/3 acre	7.90	8.80	8.80	8.60	0.90	1.10	1.10	1.10
1/2–1 acre	6.40	7.35	7.30	7.20	0.70	0.90	0.95	0.95
2–4 acre	3.30	3.80	3.55	3.55	0.24	0.33	0.33	0.33
Smooth Surface	11.60	11.60	11.60	11.60	2.00	2.00	2.00	2.00

Land Use	TSS (lbs/ac/yr)	BOD (lbs/ac/yr)	Fecal Coliforms (mpn/ac/yr)
Open Space	3	2.1	14.22
Meadow	3	2.1	0.07
Newly Graded	59,600	40	100
Forest	140	2.1	0.07
Commercial	754	33.9	0.82
Industrial	1504	29	2.64
Residential			
1/8 acre or less	952	24.6	32.9
1/4–1/3 acre	900	15.1	20
1/2–1 acre	851	11.4	1.65
2–4 acre	800	7.7	1.0
Smooth Surface	400	39.2	40.0

based on the pollutant loading rate criteria. Several pollutant parameters were plotted on individual maps to show the critical areas with the highest pollutant loads for several parameters. Critical areas were identified to aid the participating municipalities in targeting priority areas for nonpoint source pollution control.

Both existing and proposed conditions were analyzed. Existing conditions show those areas that currently contribute high amounts of nonpoint source pollution. Future conditions were evaluated to indicate where problems might arise from potential development if nonpoint source pollution controls were

TABLE 15.6 Sample NPS Loading Calculations for Total Nitrogen

Land Use	Area (AC)	Percent of Total	HSG*	Unit Loading Rate**	Total
Loading§					
Residential	70	35	C	3.55	248
Forest	100	50	C	0.60	60
Commercial	30	15	C	13.20	396
Totals	**200**				**704**

*Hydrologic soil group
**Pounds per acre per year
§Pounds per year

not implemented. Future conditions were based on proposed future land use as shown in local zoning maps.

15.4.3 USDA Phosphorus Index

Regulations have now been imposed by the USDA's Natural Resources Conservation Service (NRCS) to help reduce phosphorus (P) levels from farms. As of June 1, 2002, any new cost-share contract funded by NRCS when NRCS provides technical assistance for manure storage, handling, or treatment facilities, requires the operator to implement a P-based nutrient management plan.

USDA, NRCS (1994) has developed the Phosphorus Index, a valuable tool to determine the potential risk of phosphorus movement to water bodies. The P Index is a simple 8 characteristic by 5 value category matrix that relates site characteristics with a range of value categories as shown in Table 15.8. The P Index is obtained by selecting a rating value for each site characteristic and multiplying the two values, then summing the weighted values and comparing the sum to the values at the end of Table 15.8. The Phosphorus Index uses parameters that can have an influence on phosphorus availability, retention, management, and movement. Some analytic testing of the soil and organic material is required to determine the rating levels. This soil and material analysis is considered essential as a basis for the assessment.

15.5 REVISED UNIVERSAL SOIL LOSS EQUATION (RUSLE)

The amount of soil erosion or loss from a particular site, and in turn sediment load, can be predicted using several methods or equations. Calculation of sediment yield can be useful in estimating the sediment loads to waterways. The original Universal Soil Loss Equation (USLE), developed on agricultural watersheds at the Center of Agricultural Research Service located at Purdue University (Wischmeier and Smith, 1965; Beasley, 1978), was designed to predict *soil loss* from sheet and rill erosion. It represents *detachment,* but does

TABLE 15.7 Nonpoint Source Pollutant Loading Calculations

NONPOINT SOURCE POLLUTANT LOADINGS
EXISTING CONDITIONS

SUB AREA NO.	AREA (ACRES)	TOTAL NITROGEN TOTAL (LB/YR)	TOTAL NITROGEN UNIT (LB/AC/YR)	TOTAL PHOSPHORUS TOTAL (LB/YR)	TOTAL PHOSPHORUS UNIT (LB/AC/YR)	B.O.D. TOTAL (LB/YR)	B.O.D. UNIT (LB/AC/YR)	FECAL COLIFORMS TOTAL (MPN/YR)	FECAL COLIFORMS UNIT (MPN/YR)	SUSPENDED SOLIDS TOTAL (LB/YR)	SUSPENDED SOLIDS UNIT (LB/AC/YR)
1	1807.264	1547.61	0.86	227.63	0.13	4480.62	2.48	242.35	0.13	298958.06	165.42
2	1619.675	971.8	0.6	134.74	0.08	3401.31	2.1	113.37	0.07	225561.64	139.26
3	688.681	679.68	0.99	83.25	0.12	1944.39	2.82	869.55	1.26	145258.28	210.92
4	947.789	568.67	0.6	85.99	0.09	1990.35	2.1	66.34	0.07	130057.33	137.22
5	375.042	246.97	0.66	33.94	0.09	818.06	2.18	1458.66	3.89	38334	102.21

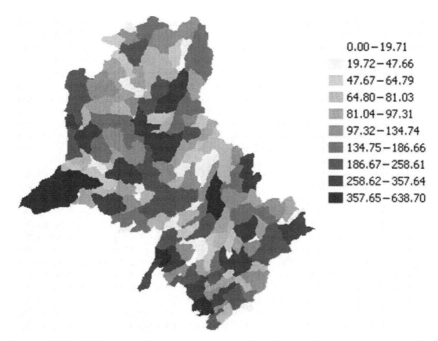

0.00−19.71
19.72−47.66
47.67−64.79
64.80−81.03
81.04−97.31
97.32−134.74
134.75−186.66
186.67−258.61
258.62−357.64
357.65−638.70

Figure 15.3 Pollutant loading map for total phosphorus (lb/ac/yr), based on land cover and hydrologic soil group by subarea and calculated using the GIS mass balance approach (DeBarry, 1991).

not predict where these particles end up; nor does it predict *sediment yield,* which is the amount of eroded soil that is delivered to a particular point of interest. One common misuse of the equation is that soil loss does not necessarily equate to field sediment yield. Eroded soil may be deposited in depressions within the field, along the toe of slopes or within other such physical features, and the equation does not account for this deposition. It was developed for sheet and rill erosion and was not originally intended to predict the effects of concentrated runoff (Dion, 2002).

RUSLE must be used specifically for its intended purpose. Applying the equation to situations for which it was not designed will result in erroneous results. For example, applying RUSLE to a complex watershed by generalizing varying slopes, land uses, and soil types would be incorrect. The watershed would need to be divided into homogeneous subwatersheds correctly representing the *LS, K,* and *C* values, as described later (Soil Conservation Society of America, 1977). Based on new data and technology for evaluating equation factors, Renard and colleagues (1996) issued a guide to support the use of the revised equation (RUSLE) for conservation planning. RUSLE was further refined and put into a computer program, version 1.06b (or latest version) (Galetovic et al., 1998), which addresses soil loss from construction

TABLE 15.8 Phosphorus Index for Assessing the Vulnerability of a Land Unit

SITE CHARACTERISTIC (weight)	PHOSPHORUS LOSS RATING (VALUE)				
	NONE (0)	LOW (1)	MEDIUM (2)	HIGH (4)	VERY HIGH (8)
SOIL EROSION (1.5)	NOT APPLICABLE	<5 TONS/AC	5–10 TONS/AC	10–15 TONS/AC	>15 TONS/AC
IRRIGATION EROSION (1.5)	NOT APPLICABLE	TAILWATER RECOVERY or QS < 6 for very erodible soils or QS < 10 for other soils	QS > 10 for erosion resistant soils	QS > 10 for erodible soils	QS > 6 for very erodible soils
RUNOFF CLASS (0.5)	NEGLIGIBLE	VERY LOW or LOW	MEDIUM	HIGH	VERY HIGH
SOIL P-TEST (1.0)	NOT APPLICABLE	LOW	MEDIUM	HIGH	EXCESSIVE
P-FERTILIZER APPLICATION RATE (0.75)	NONE APPLIED	1–30 P_2O_5 LBS/AC	31–90 P_2O_5 LBS/AC	91–150 P_2O_5 LBS/AC	>150 P_2O_5 LBS/AC
P-FERTILIZER APPLICATION METHOD (0.5)	NONE APPLIED	PLACED WITH PLANTER DEEPER THAN 2 INCHES	INCORPORATED IMMEDIATELY BEFORE CROP	INCORPORATED >3 MONTHS BEFORE CROP or SURFACE APPLIED <3 MONTHS BEFORE CROP	SURFACE APPLIED >3 MONTHS BEFORE CROP

ORGANIC P SOURCE APPLICATION RATE (1.0)	NONE APPLIED	1–30 P$_2$O$_5$ LBS/AC	31–60 P$_2$O$_5$ LBS/AC	61–90 P$_2$O$_5$ LBS/AC	>90 P$_2$O$_5$ LBS/AC
ORGANIC P SOURCE APPLICATION METHOD (1.0)	NONE	INJECTED DEEPER THAN 2 INCHES	INCORPORATED IMMEDIATELY BEFORE CROP	INCORPORATED >3 MONTHS BEFORE CROP or SURFACE APPLIED <3 MONTHS BEFORE CROP	SURFACE APPLIED TO PASTURE, or >3 MONTHS BEFORE CROP

Total of Weighted Rating Values	Site Vulnerability
<8	LOW
8–14	MEDIUM
15–32	HIGH
>32	VERY HIGH

Summation of the weighted rating value is used to determine the site vulnerability.

An on-line P-Index calculator is available at http://cattlefeeder.ab.ca/manure/pindex2.shtml and is a valuable tool for a quick phosphorus pollution potential analysis.

sites and mined and reclaimed lands. It includes practice-factor subroutines that address sediment yield. The documentation is provided in USDA Agricultural Handbook 703 and can be downloaded from www.sedlab.olemiss.edu/ rusle/document.html. A general overview of the equation parameters will be summarized here, but this handbook should be consulted when applying RUSLE. Dion (2002) provides an effective summary of RUSLE as it relates to land development design. The RUSLE is found in equation 15.9.

$$A = R \, K \, (LS) \, C \, P \tag{15.9}$$

Where:

A = Computed soil loss (lbs/acre, kg/hectare)
R = Rainfall factor
K = Soil erodibility factor
LS = Soil loss ratio
C = Cropping management factor
P = Erosion control practice factor

Each parameter is discussed more fully in the following list.

A: The computed average spatial and temporal soil loss per unit area for a period selected for R expressed in units of K, typically tons per acre per year.

R: A measure of the erosive forces from rainfall and runoff or a number of rainfall erosion index (EI) units for a particular region. The EI value was determined from research that found that soil losses from fields are directly proportional to the rainfall kinetic energy times its maximum 30-minute intensity (Beasley, 1978). R-values can be obtained from equation 15.10 or from locational average annual values of EI as displayed in the sample isoerodent map of western United States shown in Figure 15.4 (Renard et al., 1996).

$$R = 1/n \sum_{j=1}^{N} \left[\sum_{k=1}^{m} (E)(I_{30})_k \right] \tag{15.10}$$

Where:

R = Average annual rainfall erosivity
E = Total storm kinetic energy (hundreds of foot-tons per acre)
I_{30} = Maximum 30-minute rainfall intensity (in/hr)
j = Index of number of years used to produce the average
k = Index of number of storms in a year

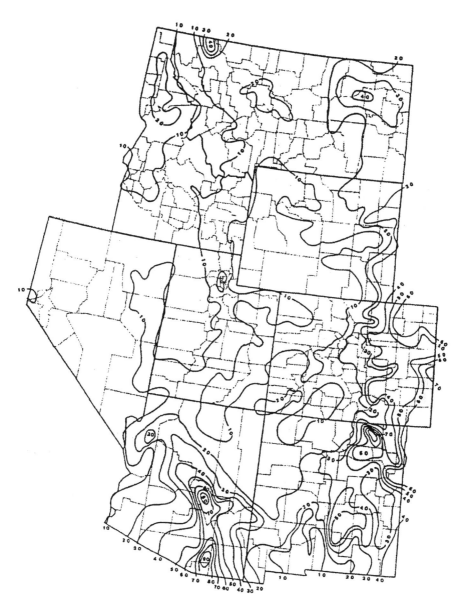

Figure 15.4   RUSLE isoerodent map of the western United States. Units are hundreds ft · tonf · in (ac · h · yr)$^{-1}$ (Renard et al., 1996).

n = Number of years used to obtain average R (minimum 22 years preferred)

m = Number of storms in each year

Or:

$$R = \left(\sum_{i=1}^{j} (EI_{30})_i \right) \div N \qquad (15.11)$$

Where:

$(EI_{30})_i$ is for storm i

j = Number of storms in an N-year period

RUSLE version 1.06b (Galetovic, et al., 1998) can utilize a stored database file for R and EI for many locations in the United States.

K: The inherent erodibility of a soil, which varies depending on soil texture, structure, permeability, chemical properties, and so on. Through experimental test plots, it has been determined that K equals the average soil loss per unit factor of R. K-values can be obtained from the STATSGO and MUIR tables (discussed in Chapter 10) and in many of the original soil surveys.

LS: Soil erosion is greatly affected by the steepness and length of slope. L and S are dimensionless length and slope factors that adjust the soil loss rate for length and steepness of slope, respectively. The point at which the length of slope is determined, or the point at which sediment deposition would occur, can be determined through the following relationship and example:

1. Add the upper slope and lower slope (10% + 6%)
2. Divide by 2 (8%).
3. Eight percent is the point at which deposition would occur at the end of the 10 percent segment length.

LS values may be obtained from Table 15.9 (Renard, 1996).

C: The cover and management factor (also called the vegetation cover factor) reflects the effect of ground cover, agricultural cropping, and soil conditions. It is the ratio of soil loss from land under specific cover conditions to that which would occur on clean-tilled continuous fallow. It reflects the effects of plants, roots, residue, soil cover, and any soil-disturbing activities. The determination of the C-value can become quite complex, therefore use of these values should be obtained from Handbook 703 (Renard, et al., 1996).

TABLE 15.9   Values for Topographic Factor, LS, for Low Ratio of Rill to Interrill Erosion[1] (Renard, 1996)

Slope (%)	Horizontal slope length (ft)																
	<3	6	9	12	15	25	50	75	100	150	200	250	300	400	600	800	1000
0.2	0.05	0.05	0.05	0.05	0.05	0.05	0.05	0.05	0.05	0.05	0.05	0.05	0.05	0.05	0.05	0.05	0.05
0.5	0.08	0.08	0.08	0.08	0.08	0.08	0.08	0.08	0.09	0.09	0.09	0.09	0.09	0.09	0.09	0.09	0.09
1.0	0.12	0.12	0.12	0.12	0.12	0.13	0.13	0.14	0.14	0.15	0.15	0.15	0.15	0.16	0.16	0.17	0.17
2.0	0.20	0.20	0.20	0.20	0.20	0.21	0.23	0.25	0.26	0.27	0.28	0.29	0.30	0.31	0.33	0.34	0.35
3.0	0.26	0.26	0.26	0.26	0.26	0.29	0.33	0.36	0.38	0.40	0.43	0.44	0.46	0.48	0.52	0.55	0.57
4.0	0.33	0.33	0.33	0.33	0.33	0.36	0.43	0.46	0.50	0.54	0.58	0.61	0.63	0.67	0.74	0.78	0.82
5.0	0.38	0.38	0.38	0.38	0.38	0.44	0.52	0.57	0.62	0.68	0.73	0.78	0.81	0.87	0.97	1.04	1.10
6.0	0.44	0.44	0.44	0.44	0.44	0.50	0.61	0.68	0.74	0.83	0.90	0.95	1.00	1.08	1.21	1.31	1.40
8.0	0.54	0.54	0.54	0.54	0.54	0.64	0.79	0.90	0.99	1.12	1.23	1.32	1.40	1.53	1.74	1.91	2.05
10.0	0.60	0.63	0.65	0.66	0.68	0.81	1.03	1.19	1.31	1.51	1.67	1.80	1.92	2.13	2.45	2.71	2.93
12.0	0.61	0.70	0.75	0.80	0.83	1.01	1.31	1.52	1.69	1.97	2.20	2.39	2.56	2.85	3.32	3.70	4.02
14.0	0.63	0.76	0.85	0.92	0.98	1.20	1.58	1.85	2.08	2.44	2.73	2.99	3.21	3.60	4.23	4.74	5.18
16.0	0.65	0.82	0.94	1.04	1.12	1.38	1.85	2.18	2.46	2.91	3.28	3.60	3.88	4.37	5.17	5.82	6.39
20.0	0.68	0.93	1.11	1.26	1.39	1.74	2.37	2.84	3.22	3.85	4.38	4.83	5.24	5.95	7.13	8.10	8.94
25.0	0.73	1.05	1.30	1.51	1.70	2.17	3.00	3.63	4.16	5.03	5.76	6.39	6.96	7.97	9.65	11.04	12.26
30.0	0.77	1.16	1.48	1.75	2.00	2.57	3.60	4.40	5.06	6.18	7.11	7.94	8.68	9.99	12.19	14.04	15.66
40.0	0.85	1.36	1.79	2.17	2.53	3.30	4.73	5.84	6.78	8.37	9.71	10.91	11.99	13.92	17.19	19.96	22.41
50.0	0.91	1.52	2.06	2.54	3.00	3.95	5.74	7.14	8.33	10.37	12.11	13.65	15.06	17.59	21.88	25.55	28.82
60.0	0.97	1.67	2.29	2.86	3.41	4.52	6.63	8.29	9.72	12.16	14.26	16.13	17.84	20.92	26.17	30.68	34.71

[1]Such as for rangeland and other consolidated soil conditions with cover (applicable to thawing soil where both interrill and rill erosion are significant).

P: The erosion or support control practice, which is used to depict the
effectiveness of various structural and nonstructural agricultural
conservation practices such as contour terracing, strip cropping,
contour plowing, and so on. It is the ratio of soil loss using a
particular conservation practice to the soil loss from up- and
downhill plowing. One of its benefits is that, by varying the
erosion control practice, the practice that best reduces soil loss
can be chosen. RUSLE allows BMPs such as gravel filters, silt
fences, berms, rock check-dams, bench terraces, and sediment
basins to influence the P-factor (Dion, 2002).

Although originally developed to predict soil loss from agricultural areas,
USLE, or the revised USLE, has been applied to urban settings and incorporated in hydrologic models such as the Generalized Watershed Loading
Function (GWLF) and SWMM to aid in predicting sediment loads. It can be
used to predict the change in erosion from future land use scenarios. It is
recommended that one become thoroughly familiar with the cited sources and
correct applications of the RUSLE before applying it in watershed assessment.

15.6 COMPUTER SIMULATION MODELS

In modeling a watershed to obtain NPS loadings, a great amount of time and
effort must be spent collecting data, calibrating the model, verifying, and
eventually comparing values to those published in the literature to justify the
model. An alternative approach is to apply the widely accepted published
loading rates to the land use map already digitized in the GIS to obtain a unit
aerial loading for each subarea in the watershed, as described previously. The
costs and benefits of developing a detailed water quality model must be
weighed against utilizing the Simple Method or mass balance approach. Many
computer simulation models have been developed over the last two decades
to aid in the process of predicting runoff and nonpoint source pollution. Such
models account for the hydrologic processes such as rainfall, infiltration, runoff, and the nonpoint source component, including pollutant accumulation and
washoff. They provide a framework for integrating data to describe land-
surface characteristics and predict processes to determine pollutant loads to
water bodies.

Model input parameter requirements typically include data on rainfall, land
use or cover (percent impervious area), and soil types. Two fundamental data
types are required for water quality models, input parameters and time-series
data, and measured discharge and water quality samples for model calibration
and verification. Input parameters and time-series data include precipitation,
drainage area, land use-land cover or percent impervious area, a runoff coefficient and watershed timing, and quality parameters such as pollutant constituent concentration, coefficient of variations, regression relationships,

buildup and washoff parameters, soils, chemical characteristics, and decay and reaction rates. Obtaining adequate water quality sampling data for proper model calibration may require an extensive monitoring network, and hence can become quite expensive.

The utilization of watershed models can sometimes be a tedious and time-consuming task because of the broad temporal and spatial scales that must be considered, as well as the large amount of data that must be collected, compiled, integrated, analyzed, and interpreted. The quality of the model's output is only as good as the accuracy of the input data.

The most commonly used water quality models include the Agricultural Nonpoint Source Pollution Model (AGNPS) (Young et al., 1986), the Areal Nonpoint Source Watershed Environment Response Simulation (ANSWERS) (Beasley and Huggins, 1981), the Chemicals, Runoff, and Erosion from Agricultural Management Systems (CREAMS) (Knisel, 1980), the Distributed Routing Rainfall Runoff Model (DR3M-QUAL) (Alley and Smith, 1982), the Hydrologic Simulation Program-Fortran (HSPF) (Johanson et al., 1980), the Storm Water Management Model (SWMM) (Metcalf and Eddy Inc. et al., 1971), and the Storage Treatment Overflow Runoff Model (STORM) (Hydrologic Engineering Center, 1977), the Illinois Urban Drainage Area Simulator (ILLUDAS), the Enhanced Stream Water Quality Model (Qual2E), the Better Assessment Science Integrating Point and Nonpoint Sources (BASINS) Model, and the Virginia Stormwater Model (VAST). An excellent overview of many of the models mentioned, in addition to others, is provided by Donigian and Huber (1991). Summaries of and links to several water quality computer simulation models can be found on the following Web sites:

USEPA Surface Water Software Page: *http://www.epa.gov/ceampubl/swater/index.htm*

USEPA Water Quality Models Page: *http://www.epa.gov/water/soft.html#qual*

USEPA Groundwater Models Page: *www.epa.gov/ada/csmos/models.html*

USGS Surface Water Software Page: *http://water.usgs.gov/software/surface_water.html*

USGS Water Quality Software Page: *http://water.usgs.gov/software/water_quality.html*

USGS Groundwater Software Page: *http://water.usgs.gov/nrp/gwsoftware*

The last decade or so of model development and enhancement has coincided with rapid advances in the development and use of geographic information system (GIS) technology, which provides an efficient means for compiling, organizing, manipulating, and analyzing data (DeBarry, 1990). But perhaps the most important advantage of GIS is its capability to graphically present spatially referenced model input and output data. Models such as

BASINS and AVGWLF (Evans, et al., 2001) are two GIS-integrated water quality models.

15.7 SUMMARY

Nonpoint source pollution management should be a goal of any watershed assessment and management plan, even if current regulations do not require it. In order to manage nonpoint source pollution, the sources and loadings from particular land uses, areas, or subwatersheds should be determined. There are many nonpoint source prediction tools, from the basic Schueler's Simple Method to the comprehensive SWMM model.

Nonpoint source pollution management may be necessary on new development in a watershed plan if associated local ordinances require it. Controlling nonpoint source pollution from existing sites is more difficult to achieve (unless it is a municipal separate storm sewer system MS4 community as discussed in Chapter 16) unless a grant can be obtained to correct an existing problem. The BMPs discussed in Chapter 19 may be applied to new and existing development to reduce nonpoint source pollution.

REFERENCES

Alley, W.M., and P.E. Smith. "Distributed Routing Rainfall-Runoff Model, Version II," USGS Open File Report 82-344, Gulf Coast Hydroscience Center, NSTL Station, MS, 1982.

Beasley, R.P. *Erosion and Sediment Pollution Control* (Ames, IA: Iowa State University Press), 1978.

Beasley, R.P. and L.F. Huggins. *ANSWERS User's Manual,* EPA-905/9-82-001, U.S. Environmental Protection Agency, Region V. Chicago, IL, 1981.

Brodhead Watershed Association. "Paradise Creek Watershed Sampling and Analysis Plan," 2003.

Cook, G. Dennis, and Robert H. Kennedy. "Water Quality Management for Reservoirs and Tailwaters," Report 1, in *Reservoir Water Quality Management Techniques,* Technical Report E-89-1, U.S. Army Corps of Engineers Waterways Experiment Station, Vicksburg, MS, January 1989.

Crawford, N.H., and R.K. Linsley Jr. "Digital Simulation in Hydrology: Stanford Watershed Model IV," Department of Civil Engineering, Stanford University, Technical Report No. 39, July 1966.

DeBarry, Paul A., and J. Carrington. "Computer Watersheds," in *Civil Engineering,* American Society of Civil Engineers (A.S.C.E.), Washington, DC, June 1990.

DeBarry, P.A. "G.I.S. Applications in Nonpoint Source Pollution Assessment," Proceedings from ASCE's 1991 National Conference on Hydraulic Engineering, Nashville, TN, 1991.

————. "Tobyhanna Creek Watershed Management Plan," Monroe County Conservation District, Monroe County Planning Commission, 1996.

Dion, T.R. *Land Development for Civil Engineers,* 2nd ed. (Hoboken, NJ: John Wiley & Sons, Inc), 2002.

Donigian, Anthony S. Jr., and Wayne C. Huber. "Modeling of Nonpoint Source Water Quality in Urban and Non-Urban Areas," U.S. Environmental Protection Agency Report EPA/600/3-91/039, Athens, GA, June 1991.

Evans, Barry M., Scott A. Sheeder, Kenneth J. Corradini, and Will S. Brown. AVGWLF User's Guide, Version 3.2 (University Park, PA: Penn State University), 2001.

Galetovic, J.R., T.J. Toy, G.R. Foster, K.G. Renard, G.A. Weesies, S.A. Schroeder, W.F. Kuenstler, G.W. Wendt, and R. Warner. *Guidelines for the Use of the Revised Universal Soil Loss Equation (RUSLE): Version 1.06 on Mined Lands, Construction Sites, and Reclaimed Lands* (Denver, CO: Office of Technology Transfer, Western Regional Coordinating Center of the Office of Surface Mining), 1998.

Hydrologic Engineering Center (HEC). *Storage, Treatment, Overflow, Runoff Model (STORM) User's Manual,* Generalized Computer Program 723-S8-L7520 (Davis, CA: U.S. Army Corps of Engineers), 1977.

Johanson, R.C., J.C. Imhoff, J.L. Kittle, and A.S. Donigian. *Hydrological Simulation Program—Fortran (HSPF): User's Manual for Release 8,* EPA-600/3-84-066 (Athens, GA: U.S. Environmental Protection Agency), 1980.

Knisel, W. (ed). CREAMS: "A Field–Scale Model for Chemicals, Runoff, and Erosion from Agricultural Management Systems," U.S. Department of Agriculture Conservation Research Report 26, 1980.

Lake Wallenpaupack Watershed Management District, Phase I Diagnostic-Feasibility Lake Study, Lake Wallenpaupack, PA, EPA Clean Lakes Program, 1983.

Marsh, D. "When a Watershed Is Gripped by Nonpoint Source Pollution," in *All Riled Up: Nonpoint Source Pollution; the Point of No Return* (Madison, WI: Wisconsin Department of Natural Resources), undated.

Metcalf and Eddy, Inc., University of Florida, and Water Resources Engineers, Inc. *Storm Water Management Model,* Volume I: Final Report, EPA Report 11024D)C0-7/71 (NTIS PB-203289) (Washington, DC: U.S. Environmental Protection Agency), 1971.

Pennsylvania Department of Environmental Protection, Pennsylvania Code, Title 25 Environmental Protection, Chapter 93, "Water Quality Standards," December 15, 2001.

Renard, K.G., G.R. Foster, G.A. Weesies, D.K. McCool, and D.C. Yoder. USDA Agricultural Research Service, Agricultural Handbook 703, *Predicting Soil Erosion by Water: A Guide to Conservation Planning with the Revised Universal Soil Loss Equation (RUSLE)* (Washington DC: U.S. Government Printing Office), 1996.

Schueler, T.R. *Controlling Urban Runoff* (Washington, DC: Metropolitan Washington Council of Governments [MWCoG]), July 1987.

Soil Conservation Society of America (SCS). "Soil Erosion: Prediction and Control," Special Publication 21, Soil Conservation Society of America, Akeny, IA, 1977.

USDA Natural Resources Conservation Service. "The Phosphorous Index—A Phosphorous Assessment Tool, Technical Note," Series No. 1901, USDA, NRCS, August 1994.

USDA Soil Conservation Service. "Water Quality Indicators Guide: Surface Water," SCS-TP-161, USDA-SCS, Washington, DC, September 1989.

——. "Urban Hydrology for Small Watersheds," Technical Release 55 (TR-55), 2nd ed., Washington, DC, June 1986.

——. *National Engineering Handbook,* Section 4, "Hydrology," Washington, DC, March 1985.

U.S. Environmental Protection Agency (U.S.EPA). BASINS (Better Assessment Science Integrating Point and Nonpoint Sources), Version 3.0 EPA-823-C-01-003, United States Environmental Protection Agency, Office of Water, Washington, DC, June 2001.

——. "Lake and Reservoir Restoration Guidance Manual," EPA 440/5-88-002, U.S Environmental Protection Agency, Washington, DC, February 1988.

——. "Results of the Nationwide Urban Runoff Program: Executive Summary," EPA 841S83101, December 1983a.

——. "Results of the Nationwide Urban Runoff Program: Volume I, Final Report," EPA 832R83112, December 1983b.

——. "Watershed Protection: TMDL Note 2, Bioassessment and TMDLs," EPA 841-K-94-005a, U.S Environmental Protection Agency, Washington, DC, 1994.

U.S. Geological Survey (USGS). "Effects of Urbanization on Stream Ecosystems," Fact Sheet 042-02, May 2002.

Wischmeier, W.H., and D.D. Smith. *Predicting Rainfall Erosion Losses from Cropland East of the Rocky Mountains,* Agricultural Handbook 282 (Washington, DC: U.S. Government Printing Office), May 1965.

——. *Predicting Rainfall Erosion Losses,* Agricultural Handbook 537 (Washington, DC: U.S. Government Printing Office), 1978.

Young, R.A., C.A. Onstad, D.D. Bosch, and W.P. Anderson. *Agricultural Nonpoint Source Pollution Model: A Watershed Analysis Tool* (Morris, MN: Agricultural Research Service, U.S. Department of Agriculture), 1986.

PART C:

WATERSHED MANAGEMENT

CHAPTER 16

AGENCY ROLES, PROGRAMS, REGULATIONS, AND POLICIES

16.0 INTRODUCTION

Watershed plans should be consistent with federal, state, and local legislation, and should provide a source of uncontested engineering and planning. All existing data sources should be explored, collected, and analyzed to prevent duplicate efforts and thereby waste funds. Standards and criteria to prevent degradation should be developed. To that end, this chapter reviews the most common grant programs, ongoing federal programs that supply data, and rules and regulations related to watershed assessment and planning.

Up to now, regulations relating to various topics of water resources protection have been developed independently. There are regulations for point source pollution, nonpoint source pollution, dams, stormwater, erosion and sedimentation control, flooding, groundwater protection, water supply protection, and wetlands protection. As presented in this text, water resources are integrated; and often, regulations geared toward one aspect of water resources management are contradictory to the goals of protecting another area of water resources. The EPA and states are beginning to take a more comprehensive watershed approach to water resources management. Perhaps, through education, future laws can have a more inclusive scope.

One should become familiar with water rights when considering a watershed plan. In the eastern portion of the United States, where water is plentiful, riparian rights govern the appropriation of water. That is, those adjacent to or who have access to surface- and groundwater have the ability to withdraw the water for their use. In the western portion of the United States, the Appropriation Doctrine governs, which generally follows the principle of "first in time is first in right." That is, users are assigned water rights that are based on the date that the water was first put to beneficial use. Senior rights owners, or those that reserved their appropriated water first, are the last to be curtailed in times of water shortages. Junior priority rights may be fully or partially curtailed in times of short water supply to ensure that senior appropriators (water right holders) receive their full appropriation (legal share).

16.1   WATER QUALITY CRITERIA AND STANDARDS

In order to maintain the sustainability of our water resources, water quality standards must be developed based on scientific data. A *standard* is a specific rule or measure established by an authority, such as a river basin commission, or by the federal or state governments. There are two main means of establishing water quality standards: *governmental stipulation* and a policy of *anti- or minimal degradation* (Tchobanoglous, 1987).

Standards set by governmental stipulation impose specific standards on industry; however, they may be relaxed if they would hinder economic growth, and this "loophole" can lead to water quality and environmental degradation. Classification of streams or reaches of stream based on existing or desired water quality is one government stipulation method utilized to quantify standards. Various classification systems are available. For example, a classification range may go from 1, meaning high water quality, to 5, for poor water quality. The elimination of water pollution focuses on pollutants found to be harmful to water quality and, in turn, on controlling the sources of these pollutants.

Standards based on anti- or minimal degradation, in contrast, are based on an estimation of a no-pollution discharge objective, set forth in the Federal Water Pollution Control Act Amendments of 1972 (Public Law 92-500). *Water quality standards (WQS)* define a designated use or uses of a water body in order to set the criteria necessary to protect those uses, thus preventing degradation through antidegradation provisions. Water quality standards are developed to protect the public health, safety, and welfare and to enhance water quality and protect biological integrity. That said, it is economically and/or technologically impossible to remove all pollutants from point or nonpoint source discharge; therefore, the objective of antidegradation (used interchangeably with nondegradation) of a receiving stream can be met through *effluent trading*. That is, by recognizing that one discharger cannot conceivably remove all its pollutants, it may be possible to make up for this deficiency elsewhere in the watershed. The objective is to maintain the water body in near-natural conditions to maintain aquatic life and biodiversity. The difficulties in nondegradation policies include the concept of "fairness" of the effluent trading policy and antidegradation in one segment of the watershed at the expense of another segment.

A combination of stipulation and nondegradation policies may be applied. Water quality standards adopted by many states are, therefore, the combination of *water uses* and the instream *water quality criteria* necessary to protect those uses. The water uses of a water body are for water supply, recreation, and propagation of fish and wildlife. The water quality criteria comprise the descriptive and numerical biological, chemical, and physical conditions of the stream that must be maintained to continually support those uses. Thus, the water quality standards determine to what degree a water body may incur degradation without jeopardizing the existing uses.

The United States has adopted a *best available treatment* (*BAT*) or *best technology in process and treatment* (*BTPT*) policy to minimize effects of discharges on receiving waters. This policy applies standards to minimize water quality degradation based on type of wastewater produced, while considering that either criterion might not be economically or technologically achievable. The pollutant discharge criteria may also be dictated by the quality of the receiving water body. Dischargers are required to apply BAT.

Before a permit can be issued, the regulatory agency must determine the following:

- Whether an applicant's proposed discharge would lower water quality.
- Whether lowering water quality is necessary because no alternative technology is available.
- Whether the discharge would support important social and economic development.

If the proposed lowering of water quality is deemed unnecessary either because of pollution prevention or the availability of alternative technology, or if the discharge will not support important social and economic development, the agency must deny the request to lower water quality. If the lowering of water quality is deemed necessary and will support important social and economic development, the agency must allow that part of the proposed lowering necessary to accommodate the development; however, the agency must not allow water quality to be lowered below the minimum level required to fully support existing and designated uses.

Water quality criteria are established in the United States by the EPA, state environmental regulatory agencies, and various regional commissions, which may have a special interest in particular bodies of water. Water quality criteria can be found in the following sources, however, the criteria are continually being updated, so the latest reference should be obtained:

U.S. EPA's "1986 Quality Criteria for Water" (often referred to as the "Goldbook")

U.S. EPA's "Guidelines for Deriving Numerical National Water Quality Criteria for the Protection of Aquatic Organisms and their Uses" (1985)

U.S. EPA's "Ambient Water Quality Criteria Development Documents,"
   and updates

Aquatic life toxicity data available in published scientific literature

Great Lakes Initiative (GLI) Clearinghouse

State codes and regulations

16.2   REGULATIONS

The purpose of *regulations* is to provide for the public health, safety, and
welfare. In essence, regulations are made to ensure controlled management
of our natural resources for a sustainable future. U.S. codes can be found at
www4.law.cornell.edu/uscode/42/index.html. This Web site provides a complete listing of code by subject, and applies not only to water resources, but
to all statutes. A summary of those most closely related to watershed management is given in the next section.

16.2.1   Federal Clean Water Act (CWA)

Perhaps the law with the most significant impact on how watershed assessments and management occur is the Federal Water Pollution Control Act,
initially passed in 1948. A major rewrite in 1972 put the law into its current
form and with these changes became known as the Clean Water Act (CWA).
The objective of the Clean Water Act (*PL 92-500; 33 U.S.C. 1251 et seq.*)
as amended in 1977, revised in 1981, and further amended in 1987, is "to
restore and maintain the chemical, physical, and biological integrity of the
Nation's waters." Therefore, it is explored in detail here.

16.2.1.1   CWA Timeline. A summary of the evolution of the Clean Water
Act is shown in Table 16.1.

   A detailed description of the most recent major changes made to the Clean
Water Act are discussed below.

1972.   The goal of the 1972 amendment to the Clean Water Act is to "prohibit
the discharge of any pollutant to waters of the United States from a point
source unless the discharge is authorized by a National Pollutant Discharge
Elimination System (NPDES) permit." The NPDES program is designed to
track point sources, such as industrial wastewater and municipal sewage, and
to require the implementation of the controls necessary to minimize the discharge of pollutants. These discharge sources were easily identified as the
cause of poor, often drastically degraded, water quality conditions. As a part
of the CWA, the EPA has the responsibility of leading efforts to achieve its
goals. The EPA can, however, delegate this responsibility to states or Native
American tribes.

TABLE 16.1   Clean Water Act and Major Amendments (Codified Generally as 33 U.S.C. 1251-1387)

Year	Act	Public Law
1948	Federal Water Pollution Control Act	P.L. 80-845 (Act of June 30, 1948)
1956	Water Pollution Control Act of 1956	P.L. 84-660 (Act of July 9, 1956)
1961	Federal Water Pollution Control Act Amendments	P.L. 87-88
1965	Water Quality Act of 1965	P.L. 89-234
1966	Clean Water Restoration Act	P.L. 89-753
1970	Water Quality Improvement Act of 1970	P.L. 91-224, Part I
1972	Federal Water Pollution Control Act Amendments (Clean Water Act)	P.L. 92-500
1977	Clean Water Act of 1977	P.L. 95-217
1981	Municipal Wastewater Treatment Construction Grants Amendments	P.L. 97-117
1987	Water Quality Act of 1987	P.L. 100-14

Source: Environmental Media Services, 2004.

This process was extremely successful in managing the pollution from point source discharges and began to improve the nation's waters. However, it became increasingly evident that much of the nation's water impairment was coming from nonpoint or diffuse sources of pollution, particularly stormwater runoff from urban, agricultural, and other land uses. Nonpoint source pollution was also preventing the CWA's goal of maintaining *designated beneficial uses* from being achieved.

16.2.1.2   1987. Congress amended the CWA with Section 402(p) to require implementation, in two phases, of a comprehensive national program for addressing stormwater discharges. The EPA, consequently, strengthened its process for regulating discharges to receiving waters. Section 319, dealing with nonpoint source pollution, was also added. This section, which required states to assess and control nonpoint sources of pollution, met with limited overall success, thanks to voluntary participation and funding through provisions of Section 319.

Phase I (55 FR 47990) of the CWA, adopted November 16, 1990, defines enforceable policy, addresses the effects of nonpoint source pollution, and requires NPDES permits for stormwater discharges from a large number of priority sources, including municipal separate storm sewer systems (MS4s) serving populations of 100,000 or more and several categories of industrial activity, as these uses were identified as the major contributors to nonpoint source pollution. For example, construction sites that disturb five or more acres of land are cited as industrial use that requires a Phase I NPDES permit. MS4s were targeted so that illicit discharges to the storm sewer systems could be detected. An *illicit discharge* is defined by federal regulations as "any discharge to an MS4 that is not composed entirely of stormwater," and is termed "illicit" because there is no treatment or processing mechanism at the end of the pipe to treat nonstormwater wastes. There are a few exceptions, including discharges from NPDES-permitted industrial sources from allowable nonstormwater discharges; for example, fire hydrant flushing.

Specifically listed under Section 402(p)(2), Phase I regulates:

- A discharge subject to an NPDES permit before February 4, 1987
- Discharges from construction activities disturbing five or more acres of land
- A discharge associated with industrial activities
- A discharge from a municipal separate storm sewer system (MS4) serving a population of 250,000 or more (large MS4)
- A discharge from a municipal separate storm sewer system serving a population of 100,000 or more but less than 250,000 (medium MS4)
- A discharge that an NPDES permitting authority determines to be contributing to a violation of a water quality standard or a significant contributor of pollutants to the waters of the United States

The NPDES permit requires gathering data on precipitation and stormwater runoff volume and quality, as well as developing a site-specific management program or *pollution control plan* (*PCP*) to reduce pollutants in runoff. Note that the details of permit requirements may vary among the states administering the NPDES program under the EPA's guidance.

The interim second phase of the stormwater program expanded the Phase I program to include discharges of stormwater from:

- Municipal separate storm sewer systems (MS4) in urbanized areas with populations between 10,000 and 100,000 per square mile (small MS4)
- Construction sites disturbing from one to five acres.

The Phase II rule allows certain sources to be excluded from the national program if a lack of impact on water quality can be demonstrated. The rule also allows other sources not automatically regulated on a national basis (fewer than 10,000 people per square mile) to be designated for inclusion based on increased likelihood of localized adverse impact on water quality on a schedule consistent with the state watershed permitting approach.

States have the responsibility to adopt a customized plan for maintaining water quality under the guidance and support of the EPA. The Phase II rule required that the NPDES permitting authority (the EPA or state governing agency) issue general NPDES permits by December 2002.

The goals of the NPDES program are achieved through the use of best management practices (BMPs), which are stormwater management measures designed to improve the quality of the stormwater outflow from the BMP. (BMPs are described in further detail in Chapter 19.) The implementation of the final Phase II rule, which replaced this interim rule, did not come into effect until December 8, 2002; it is discussed in greater detail in the December 8, 1999, segment of this timeline.

16.2.1.3 September 9, 1992. The EPA published a notice requesting information and public comment as to how to prepare regulations under CWA Section 402(p)(6). The notice specified that potential sources for coverage under the Section 402(p)(6) regulations would fall into two main categories: *municipal separate storm sewer systems (MS4s)* and *individual (commercial and residential) sources.*

16.2.1.4 August 7, 1995. The EPA promulgated a final rule, referred to as "the Interim Phase II Rule," that required facilities regulated under Phase II to apply for an NPDES permit by August 7, 2001, unless the NPDES permitting authority required a permit by an earlier date (60 FR 40230).

16.2.1.5 1996. The EPA's National Water Quality Inventory showed that 40 percent of surveyed U.S. water bodies were still impaired by pollution and did not meet National Water Quality Standards.

16.2.1.6 1998. The EPA's National Water Quality Inventory Report to Congress stated that nonpoint source pollution, which includes urban stormwater runoff, is considered the leading cause of impairment of our nation's water quality.

16.2.1.7 December 8, 1999. The "National Pollutant Discharge Elimination System (NPDES) Regulations for Revision of the Water Pollution Control Program Addressing Storm Water Discharges; Final Rule" (40 CFR Parts 9, 122, 123, and 124) was published in the *Federal Register.* The December 8, 1999 rule replaced the Interim Phase II rule, established implementation of Phase II as described previously, and extended the deadline by which certain industrial facilities owned by small MS4s had to obtain coverage under an NPDES permit from August 7, 2001, until March 10, 2003.

It had been found that effluent limitations required by Section 301 of the Clean Water Act and/or local or state regulations were, in some instances, not stringent enough to attain and maintain applicable water quality standards. Therefore, a goal was set of establishing total maximum daily loads (TMDLs) to ensure that water quality standards are attained and maintained. The regulations define a TMDL as "a quantitative assessment of the pollutants that are contributing to water quality impairment from a specific area." A TMDL specifies the amount of each particular pollutant that may be present in a water body impaired by both point sources *and* nonpoint sources and allocates allowable pollutant loads among sources. If one source cannot meet those established load requirements, another may reduce its contribution over the amount required to compensate for the area where standards cannot be met, a process termed *effluent trading*. The final rule clarified the EPA's regulatory requirements for establishing TMDLs under the Clean Water Act.

The EPA requires, under the Final Phase II Regulation of the National Pollutant Discharge Elimination System (adopted on October 28, 1999), that owners and operators of small, urbanized municipal separate storm sewer systems (MS4s) reduce the pollutant loading from regulated systems to the "maximum extent practicable" in order to protect water of the United States. The EPA has required that this be accomplished through a permitting program established by the states, or, if a state does not have the capability to handle the program, by the EPA.

The Phase II final rules state that, first, operators of MS4s are required to develop, implement, and enforce a stormwater management (SWM) program designed to reduce the amount of pollutants discharged from their site. The affected municipalities must develop a stormwater management program, if they do not already have one in place. If a political division has an established stormwater management program and is subject to the provisions of the Phase II Rule, then those provisions of the rule needed to satisfy the requirements must be added (Pennsylvania Department of Environmental Protection, 2002).

One way to implement Phase II is through the use of BMPs, which for stormwater management can be defined as recognized practices, schedules of activities, prohibited practices, maintenance procedures, and the use of pollution control devices and other means to prevent or reduce the amount of pollutant loading being discharged in stormwater runoff into water bodies of the United States.

Second, the Phase II final rules state that municipalities required to implement the MS4 program must address the following six *minimum control measures (MCMs)*:

- Public education and outreach
- Public involvement/participation
- Illicit discharge detection and elimination
- Construction site stormwater runoff control
- Postconstruction stormwater management in new development and redevelopment
- Pollution prevention/good housekeeping for municipal operations

Good housekeeping may be such measures as ensuring proper material disposal, management and recyling, not dumping wastes down inlets, pet waste collection, proper vehicle washing and maintenance, parking lot and street cleaning, use of alternative products, hazardous materials storage, material recycling, and road salt application and storage.

The following is a summary of how MS4 municipalities can implement the MCMs as compiled by the Pennsylvania Department of Environmental Protection (Pennsylvania Department of Environmental Protection, 2002).

I. Public Education and Outreach on Stormwater Impacts

A. Political division *must* implement a public education program, including distributing educational materials that:
- Describe impacts of stormwater.
- Describe steps to reduce stormwater pollution.

B. Political division *should* inform households and individuals about steps they can take, such as:
- Perform proper septic system maintenance.
- Limit use and runoff of garden chemicals.
- Restore local streams.
- Stencil storm drains.
- Protect stream banks.

C. Political division *should* direct information to targeted groups, including commercial, industrial, and institutional entities likely to cause stormwater impacts. Examples include restaurants (potential grease clogging/blocking of storm drains) and auto service facilities.

D. Political division *should* address the viewpoints and concerns of minorities and the disadvantaged, development/construction companies, businesses, educational groups, government entities, and industrial facilities.

II. Public Involvement/Participation

A. Political division *must* comply with state and local public notice requirements (adoption of stormwater management program, policies, ordinances, etc.).

B. Political division *should* involve the public in developing, implementing, and reviewing stormwater management programs, as follows:
- Reach out to and engage all economic and ethnic groups.
- Consider establishing a citizen group to participate in decision making.
- Work with volunteers.

III. Illicit Discharge Detection and Elimination (IDD&E)

A. Political division *must* develop stormwater system maps, which:
- Show location of major pipelines, outfalls, and topography.
- Show areas of concentrated activities likely to be a source of stormwater pollutants.

B. Political division *must* effectively prohibit illicit discharges into MS4 system, via:

- Passage of use ordinances, orders, and the like.
- Implementation of enforcement procedures.

C. Political division *must* implement a plan to detect illicit discharges and illegal dumping.

D. Political division *must* inform public employees, businesses, and citizens of hazards arising from illegal discharges and improper disposal of waste.

The method to determine a potential illicit discharge is through field survey of the streams and location of storm sewer outfalls. The field survey should be conducted three days after the last rainfall, at which time storm-only flow should be nonexistent. Those outfalls that do not have flow can be considered preliminarily as not having an illicit discharge. However, before making this determination, evidence at the outfall, such as stains and odors, should be used to determine if there may be sporadic illicit discharges, for example, only at night.

An IDD&E map is shown in Figure 16.1. Although the EPA did not devise a national IDD&E data collection form, several other agencies have created their own, as shown in Figures 16.2 and 16.3.

Although many political divisions are undertaking the IDD&E survey independently, it may be more advantageous to coordinate the activities on a watershed basis, as is done in Berks County, Pennsylvania. The municipalities

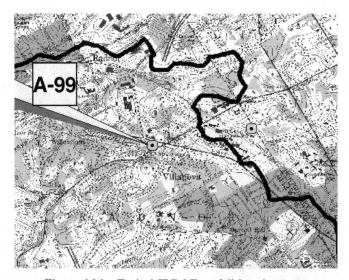

Figure 16.1 Typical IDD&E outfall location map.

ILLICIT DISCHARGE FIELD SCREENING PROGRAM
Data Collection Form

OUTFALL #: _____ Date: _____ Time: _____

TIME SINCE LAST RAIN: ≥72 hours <72 hours
QUANTITY OF LAST RAIN: ≥0.1 inches <0.1 inches
INSPECTION TEAM: _____

SITE DESCRIPTION:
LOCATION (Narrative Description): _____

STRUCTURE TYPE: OPEN CHANNEL MANHOLE OUTFALL OTHER: _____

DOMINANT WATERSHED LAND USES: INDUSTRIAL COMMERCIAL RESIDENTIAL UNKNOWN
OTHER: _____

FLOW ESTIMATION:
WAS FLOW OBSERVED? NO YES IF YES, PLEASE ANSWER a. - d. BELOW.
  a. WIDTH OF WATER SURFACE (feet): _____
  b. APPROXIMATE DEPTH OF WATER (feet): _____
  c. APPROXIMATE FLOW VELOCITY (feet per second): _____
  d. FLOW RATE (cubic feet per second) = a x b x c = _____

VISUAL OBSERVATIONS:
WAS A PHOTO TAKEN? NO YES (Roll and Photo Number: _____)

ODOR: NONE MUSTY SEWAGE ROTTEN EGGS SOUR MILK OTHER: _____

COLOR: CLEAR RED YELLOW BROWN GREEN GREY OTHER: _____

CLARITY: CLEAR CLOUDY OPAQUE

FLOATABLES: NONE OILY SHEEN GARBAGE/SEWAGE OTHER: _____

DEPOSITS/STAINS: NONE SEDIMENTS OILY OTHER: _____

VEGETATION CONDITION: NONE NORMAL EXCESSIVE GROWTH INHIBITED GROWTH

STRUCTURAL CONDITION: NORMAL CONCRETE CRACKING METAL CORROSION OTHER: _____

BIOLOGICAL: MOSQUITO LARVAE BACTERIA/ALGAE OTHER: _____

FIELD ANALYSIS:
WATER TEMP:	_____ °F / °C	CHLORINE (Total):	_____ mg/l
pH:	_____	COPPER:	_____ mg/l
PHENOL:	_____ mg/l	DETERGENTS:	_____ mg/l

WAS A LABORATORY SAMPLE COLLECTED? NO YES
(if yes attach copy of chain-of-custody record)

COMMENTS: _____
_____
_____

DATA SHEET FILLED OUT BY: (signature): _____ DATE: _____

(print name): _____

This form was modified from *Municipal Urban Runoff Program: A How-To Guide For Developing Urban Runoff Programs for Small Municipalities,* by the City of Monterey, City of Santa Cruz, California Coastal Commission, Monterey Bay National Marine Sanctuary, Association of Monterey Bay Area Governments, Woodward-Clyde, and Central Coast Regional Water Quality Control Board, 1998.

Figure 16.2 Typical IDD&E data collection form (modified from the City of Monterey, California, 1998).

Pipe and Storm Drain Mapping Survey Sheet

Surveyors: _____   Date: _____   Weather in past 3 days: _____

Municipality: _____   Name of receiving water: _____

Pipe Swale # Photo #	Storm Drain? Yes/No Not sure	Time (00:00) am/pm	Diameter (inches)	Material and condition of pipe or swale Is there a headwall or riprap at the outfall?	Slope (steep, gradual, or flat) Is the bank scoured below the pipe?	Distance from pipe/swale to stream channel (feet)	Indicate, depth of flow in pipe	Water Color/Odor (specify if floatables, algae or sediment present)	Describe access to the outlet. (include streets or landmarks, etc.) Include GPS coordinates if available.	Describe land use of upstream drainage area.	Rating (0-2)
							◯				
							◯				
							◯				
							◯				
							◯				
							◯				

Rating System:

0= No observed impairment (No dry weather flow, no solids, floatables or debris, no erosion or sediment, pipe in good repair)

1= Needs rechecking (some dry weather flow, moderate scouring or sediment deposition, some floatables or debris, odor, algae, pipe in poor condition)

2= Impairment - needs investigation (flow exhibiting odors, foam, solids, turbidity or oil sheen, considerable sediment deposition, algae or debris, pipe collapsed or crumbling, blocked catchbasins or drain)

NOTES:

Figure 16.3   Typical IDD&E data collection form (Massachusetts Riverways Program, 2002).

in the Schuylkill River watershed utilized a consistent mapping procedure and data collection forms, which were explained in a joint workshop (Figures 16.4a and b).

IV. Construction Site Stormwater Runoff Control

    A. Political division *must* develop, implement, and enforce a program to reduce nonpoint stormwater runoff from construction activities to regulated MS4s, to include:

        • Control construction sites greater than or equal to one acre.

        • Pass an ordinance that controls erosion and sedimentation.

        • Control construction site waste materials (discarded building material, concrete washout, sanitary waste, etc.).

    B. A political division's program *must* include:

        • Requirement for construction site owners or operators to implement BMPs.

        • Preconstruction review of site plans.

        • Procedures to receive and consider public input.

        • Regular inspections during construction.

        • Penalties to ensure compliance.

V. Postconstruction Stormwater Management in New Development and Redevelopment

    A. In order to maintain predevelopment runoff conditions, political division *must* develop, implement, and enforce a program to address stormwater runoff from new development and redevelopment projects, to include:

(a)   (b)

Figure 16.4a and 16.4b   Classroom and on-site IDD&E data collection training.

- Land disturbance sites greater than or equal to one acre and discharge to regulated MS4.
- Project sites that discharge to MS4.

B. A political division's program *must:*
- Include site-appropriate, cost-effective structural and nonstructural BMPs.
- Ensure long-term ownership and maintenance of BMP connected to regulated MS4s.
- Ensure that controls are in place that prevent or minimize water quality impacts.

C. A political division's program *should* include structural and nonstructural BMPs, such as:
- Locally based watershed planning.
- Measures to prevent or minimize water quality impacts.

The EPA *recommends* (for the fifth minimum control measure):
- BMPs that minimize water quality impacts.
- BMPs that maintain predevelopment runoff conditions.
- Nonstructural BMPs that emphasize management and source controls, such as:
  - Implementing policies and ordinances that protect natural resources and prevent runoff
  - Limiting growth to identified areas
  - Protecting sensitive areas such as wetlands
  - Minimizing the amount of impervious surface
  - Maintaining open space
  - Minimizing disturbance of soils and vegetation
- Structural BMPs, which may include:
  - Storage facilities (retention/detention ponds)
  - Filtration facilities (grassed swales, sand filters, filter strips)
  - Infiltration facilities (recharge basins, porous pavement)

VI. Pollution Prevention/Good Housekeeping for Municipal Operations

A. A political division must develop and implement a cost-effective infrastructure, operations, and maintenance (O&M) program to prevent or reduce pollutant runoff from municipal operations.

B. Political division *must* provide employee training in the following areas:
- Park and open space maintenance
- Fleet maintenance
- Planning

- Building management
- Stormwater system maintenance

The EPA *recommends* (for the sixth minimum control measure) that, at a minimum, the following be *considered* as components of the political division's program:

- Maintenance activity schedules and inspections to reduce floatable and other pollutants.
- Controls for reducing pollutants from streets, parking lots, yards, and solid waste operations.
- Proper disposal of waste removed from storm drains.
- Assess water quality impact of new flood control projects.
- Maximize current activities before adding new ones.

The third aspect of the Phase II final rules require that operators of MS4s identify those BMPs applicable to the site and set *measurable* goals for each BMP. At a minimum, municipal entities regulated under MS4 must:

- Specify BMPs and implement them to the "maximum extent practicable."
- Identify measurable goals for control measures.
- Develop an implementation schedule of activities or frequency of activities.
- Define the entity responsible for implementation.

The affected municipalities were required to obtain a permit by submitting a *letter of intent* (*LOI*) to the overseeing agency (state or the EPA) by March 10, 2003.

The CWA also contains language to require farms classified as Concentrated Animal Feeding Operations (CAFOs) to meet certain technology and performance requirements set by the EPA as Effluent Limitation Guidelines (ELGs) and develop nutrient management plans. Discharge from CAFOs is considered point source discharge of pollutants. There are 1.3 million livestock and poultry operations (LPOs) throughout the United States; approximately 238,000 are considered animal feeding operations (AFOs). Of these AFOs, only 15,500 are CAFOs. That means only a small portion of the LPOs are covered; however, through education and volunteer participation, the remaining LPOs should be identified in the watershed management plans.

These CAFO regulations were updated in December 2002 by court order after a Natural Resources Defense Council lawsuit found the EPA's implementation of the CAFO rules to be inadequate. The thrust of the new regulations reduces the size (in terms of animal numbers) required to qualify as a CAFO and to comply with the CAFO rules. Additional information on CAFOs can be found at http://cfpub.epa.gov/npdes/afo/cafofinalrule.cfm.

16.2.1.8   CWA Sections Related to Watershed Management. The primary goals of the CWA are the "elimination of pollutant discharges" and the maintenance of the "fishable and swimmable" quality of U.S. waters, referred to as *designated uses*. The initial phases of the act focused on the improvement of the quality of water from municipal and industrial "point" discharges, or discharges from pipes. More recent phases focus on nonpoint sources of pollution because the EPA feels that the point sources are reasonably under control. The act also contains very detailed wetlands protection measures.

The major sections of the act as they relate to watershed management are explained in greater detail below.

16.2.1.9   CWA Sections 201-209 (33 U.S.C. 1281–1289): Funding for Publicly Owned Treatment Works (POTWs). CWA Sections 201-209 provided grants for construction of POTWs. Originally, these sections awarded federal grants for the construction of wastewater treatment plants; the 1987 amendments replaced the grant program with a revolving loan fund.

16.2.1.10   CWA Section 301 (33 U.S.C. 1311): Effluent Limitations. The discharge of any pollutant into the nation's waters except for discharges in compliance with the CWA is prohibited, according to this section. Limitations are placed on existing sources, which vary according to the nature of the pollutant discharged and where the outfall is directed.

16.2.1.11   CWA Section 302 (33 U.S.C. 1312): Water-Quality-Related Effluent Limitations. Section 302 regulates point sources of pollution that interfere with the attainment or maintenance of desired water quality, by imposing more stringent effluent limitations.

16.2.1.12   CWA Section 303 (33 U.S.C. 1313): Water Quality Standards and Implementation Plans. 40 CFR Part 130, Water Quality Planning and Management Section 303, requires water quality standards (WQS) and water-quality-based regulatory controls to be required by states to protect designated uses of water bodies. Technological capability is not a consideration in setting WQS.

The states are required to establish policies for water quality planning and management, and implementation of Section 303. The Water Quality Management process from the CWA provides the authority for a "consistent national approach for maintaining, improving, and protecting water quality while allowing states to implement the most effective individual programs." States must achieve the WQS by issuing permits, or more importantly, by not issuing permits if a discharge is not in compliance with the WQS, by building POTWs, or by requiring best management practices (BMPs) through a water quality management plan. This can be achieved through determination and enforcement of total maximum daily loads (TMDLs) and implementation of individual water-quality-based effluent limitations in 130.7. States are required to submit water quality reports to the EPA Regional Administrator in accordance with 305(b) of the CWA (130.8). Final points of the regulation cover state submittals to EPA and program management.

16.2.1.13 CWA Section 304 (33 U.S.C. 1314): Information and Guidelines.

Section 304 requires the EPA to administer the National Pollutant Discharge Elimination System (NPDES) program and develop water quality criteria and guidelines for effluent limitations and pretreatment programs.

16.2.1.14 CWA Section 305(b): Water Quality Assessments and Reports.

CWA Section 305(b) establishes a process for states, the District of Columbia, territories, certain river basin commissions, and some tribes to establish ground- and surface-water monitoring programs and report information on the quality of our nation's waters to Congress through the National Water Quality Inventory. It establishes biological assessment techniques and reporting to:

- Determine the designated/beneficial and aquatic life use status of water resources.
- Evaluate the causes of degraded water resources and relative contributions of pollutant sources.
- Report the activities under way to restore water resource use.
- Determine the control of mitigation programs.
- Measure the success of watershed management programs (U.S. EPA, 1999).

16.2.1.15 CWA Section 306 (33 U.S.C. 1316): National Standards of Performance.

Section 306 specifies a list of industries that must conform to technology-based new pollution source performance standards based on demonstrated best available technology (BAT). The industries are categorized by effluent sources.

16.2.1.16 CWA Section 307 (33 U.S.C. 1317): Toxic and Pretreatment Effluent Standards.

Section 307 mandates the establishment of pretreatment standards (Part b), requires surveys of new sources of pollutants into publicly owned treatment works (POTWs) (Part c), and requires that industries discharging toxic pollutants meet effluent limits by utilizing the BAT that is economically achievable.

16.2.1.17 CWA Section 309 (33 U.S.C. 1319): Enforcement.

Section 309 authorizes administrative, civil, and criminal enforcement and compliance orders by the states.

16.2.1.18 CWA Section 311 (33 U.S.C. 1321): Oil and Hazardous Substance Liability.

Section 311 addresses the discharges of oil or hazardous substances in harmful quantities into waters and adjoining shorelines, and requires all facilities that handle, transport, and store oil to develop a spill prevention, control, and countermeasure (PCC). It also requires that liability of any spill or discharge of a harmful quantity of oil be assigned to the facilities' owners and operators and that they be reported to the National Response Center.

16.2.1.19 CWA Section 319 (33 U.S.C. 1329): Nonpoint Source Assessment and Management Programs. Section 319 deals with nonpoint source pollution and requires that states assess and control nonpoint sources of pollution. More specifically, this section requires that states identify waters that are not able to meet WQS because of nonpoint sources. The activities responsible for the pollution are to be identified, and a management plan created, to help correct the nonpoint source problem.

16.2.1.20 CWA Section 402 (33 U.S.C. 1342): National Pollutant Discharge Elimination System (NPDES). This section of the CWA has perhaps the greatest impact on business, industry, and agency oversight, as it establishes the National Pollutant Discharge Elimination System (NPDES). NPDES must be administered by the states under the EPA-delegated authority or by the EPA. This section is significant in that it translates standards into enforceable limitations and establishes permits for point source discharges of pollutants after the public hearing process is complete.

16.2.1.21 CWA Section 404 (33 U.S.C. 1344): Permits for Dredged or Fill Material. Another significant section of the CWA, Section 404, requires permits from the U.S. Army Corps of Engineers with EPA concurrence for any dredging or filling of wetlands. The Corps of Engineers was determined to be the regulating agency because it is responsible for navigable waters of the United States, and wetlands are an integral part of the hydrologic system.

40 CFR Part 230 of Section 404 (b)(1) provides guidelines and specifications for the disposal of dredged or fill material on sites to maintain the biological, chemical, and physical integrity (antidegradation) of the United States. It places the burden of proving that alternative sites or discharges are not available on the landowner.

The 1997 decision (*U.S. v. Wilson*) by the Fourth Circuit U.S. Court of Appeals overturned federal regulation and EPA's and the U.S. Army Corps of Engineers' jurisdiction to protect "isolated" wetlands. The decision not only applied to the Fourth Circuit (Maryland, Virginia, West Virginia, North Carolina, and South Carolina), but to the nation. A January 2001 decision in *Solid Waste Agency v. U.S. Army Corps of Engineers* (SWANCC decision) reiterated this ruling, stating that regulating "isolated waters" goes beyond the authority granted to the Corps under Section 4044(a) of the Clean Water Act. Isolated wetlands are wetlands that are not adjacent to navigable waters of the United States (streams, lakes or oceans, etc.) It is estimated by the EPA that 10 to 20 million acres of wetlands in the contiguous United States could lose federal protection due to these decisions.

The decisions were made because they lack a surface connection to navigable waters. However, the court failed to see that below-surface connections or connections through groundwater are the most important link between these isolated bodies and regulated waters. Therefore, Congress is considering two proposed laws, S. 473 and H.R. 962, that would provide more strength in preserving the wetlands, isolated wetlands, and watersheds of the United States. These bills are also being considered because of the known value of treating water quality, recharging groundwater (in some cases), providing habitat and flood control, and the known domino effect in preserving the public's health, safety, and welfare. The economic costs (loss of waterfowling, increased flooding, diminished water supply) must also be considered. States and local jurisdictions may adopt their own regulations to preserve wetlands and buffers surrounding them.

16.2.2   Safe Drinking Water Act (SDWA) of 1974

The Safe Drinking Water Act (SDWA) is the main federal law that ensures the quality of Americans' drinking water. Under the SDWA, the EPA sets standards for drinking water quality and oversees the states, localities, and water suppliers that implement those standards. The 1974 law had largely an after-the-fact regulatory focus. It was amended in 1996 (Safe Drinking Water Act Amendments of 1996) to transform it into an environmental statute that better provided for the sustainable use of water by our nation's public water systems and their customers. The 1996 amendments increased state flexibility, provided for more efficient investments by water systems, provided better information to consumers, and strengthened the EPA's scientific work, including the use of risk and benefit cost considerations in setting drinking water standards. The 1996 amendments also established a strong new emphasis on preventing contamination problems through *Source Water Protection* (*SWP*) and enhanced water system management. Prevention is accomplished through state-enacted prevention programs and helping water systems improve operations to avoid contamination problems. As part of the Source Water Protection program, states had to submit a program for delineating source water areas of public water systems and for assessing the susceptibility of such source waters to contamination. States make grants available for wellhead and surface source delineation, and the implementation should be considered a part of any watershed assessment and management plan.

For the SDWA to establish standards, the EPA determined maximum contaminant level goals based on health data and enforceable *maximum contaminant levels* (*MCLs*) for each chemical that is a potential health hazard.

16.2.3   The National Environmental Policy Act of 1969 (NEPA)

This discussion refers to the amended version of NEPA (P.L. 91-190, 42 U.S.C. 4321-4347, January 1, 1970, as amended by P.L. 94-52, July 3, 1975, P.L. 94-83, August 9, 1975, and P.L. 97-258, §4(b), September 13, 1982).

NEPA established a national policy for the environment, provided for the establishment of a Council on Environmental Quality (CEQ), and required Environmental Impact Statements (EIS) to assess environmental benefit/cost ratios for proposed federal projects, among other actions.

The stated purposes of the act are "To declare a national policy which will encourage productive and enjoyable harmony between man and his environment; to promote efforts which will prevent or eliminate damage to the environment and biosphere and stimulate the health and welfare of man; to enrich the understanding of the ecological systems and natural resources important to the Nation; and to establish a Council on Environmental Quality."

16.2.4 Fish and Wildlife Act of 1956 (16 U.S.C. 742-742j)

As amended, the Fish and Wildlife Act of 1956 authorizes the Secretary of the Interior to take steps "required for the development, advancement, management, conservation, and protection of fish and wildlife resources, including, but not limited to, acquisition by purchase or exchange of land and water or interests therein."

16.2.5 Fish and Wildlife Coordination Act (16 U.S.C. 661-66c), 1934

As amended, the Fish and Wildlife Coordination Act authorizes federal water resource agencies to acquire lands or interests in connection with water use projects specifically for mitigation and enhancement of fish and wildlife, and provides for management of such lands by the U.S. Fish and Wildlife Service or by state wildlife agencies.

16.2.6 Migratory Bird Conservation Act (16 U.S.C 715-715r), 1929

As amended, the Migratory Bird Conservation Act establishes a Migratory Bird Conservation Commission to approve acquisition of land, water, or land and water, recommended by the Secretary of the Interior as suitable for use by migratory birds. The act also authorizes appropriation of funds for the construction of dams, dikes, ditches, flumes, spillways, buildings, and other necessary improvements.

16.2.7 Endangered Species Act of 1973 (16 U.S.C. 1531-1543)

As amended, the Endangered Species Act of 1973 provides for the conservation of threatened and endangered species of fish, wildlife, and plants by federal action and by encouraging the establishment of state programs.

16.2.8 McCarran Amendment (43 U.S.C. 666), 1952

The McCarran Amendment (43 U.S.C. 666) of 1952 waives the sovereign immunity of the United States and permits state courts to adjudicate all federal water rights where there is a general adjudication designed to establish all rights in a watershed and the United States is properly served.

16.2.9 The National Wildlife Refuge System Improvement Act of 1997 (16 U.S.C. 668dd et seq.)

The National Wildlife Refuge System Improvement Act of 1997 amends the National Wildlife Refuge System Administration Act of 1966 to improve the overall management of the Refuge System, to include a unified system mission, to implement a new process for determining compatible uses, and to require preparation of refuge-specific comprehensive conservation plans.

16.2.10 Natural Resource Information System (NRIS)

The U.S. Forest Service Natural Resource Information System (NRIS) is designed to support field-level users of national forests with a common set of basic data and data standards, in a common computing environment, but that can be applied elsewhere. This system provides agency employees and partners and the public with access to data used for natural resource decision making.

16.2.11 Resource Conservation and Recovery Act (RCRA) (42 U.S.C. s/s 6901 et seq.), 1976

The Resource Conservation and Recovery Act (RCRA) authorized the EPA to control hazardous waste from the "cradle to grave." This includes the generation, transportation, treatment, storage, and disposal of hazardous waste. RCRA also set forth a framework for the management of nonhazardous wastes.

The 1986 amendments to RCRA enabled the EPA to address environmental problems that could result from underground tank storage of petroleum and other hazardous substances. RCRA focuses only on active and future facilities; it does not address abandoned or historical sites, which are covered by the Comprehensive Environmental Response, Compensation, and Liability Act (Superfund) (CERCLA). Coordination of these programs and projects should be included in any watershed plan.

16.2.12 Chesapeake Bay Program (CBP)

The Chesapeake Bay Program (CBP) was formed to provide coordination and support the goals of the historic Chesapeake Bay Agreement of 1983. It covers, naturally, only the land area of the Chesapeake Bay Watershed, but its goals, objectives, and procedures can be applied anywhere. On June 28, 2000, the Chesapeake Executive Council signed Chesapeake 2000, a comprehensive and far-reaching bay agreement that will guide Maryland, Pennsylvania, Virginia, the District of Columbia, the Chesapeake Bay Commission, and the U.S. Environmental Protection Agency (EPA) in their combined efforts to restore and protect the Chesapeake Bay. The Chesapeake 2000 agreement guides the Bay Program partnership efforts through the year 2010.

16.2.13 Coastal Zone Management Program (CZMP)

The Coastal Zone Management Program (CZMP) is authorized by the Coastal Zone Management Act of 1972 and is administered at the federal level by the Coastal Programs Division (CPD) within the National Oceanic and Atmospheric Administration's Office of Ocean and Coastal Resource Management (OCRM). The National Coastal Management Program is a federal-state partnership dedicated to comprehensive management of the nation's coastal resources, ensuring their protection for future generations while balancing competing national economic, cultural, and environmental interests. It covers the landmass drainage directly or indirectly into the coasts.

The term *coastal zone* refers to the coastal waters (including the lands therein and thereunder) and the adjacent shore lands (including the waters therein and thereunder), strongly influenced by each other and in proximity to the shorelines of the several coastal states, and includes islands, transitional and intertidal areas, salt marshes, wetlands, and beaches. The zone extends in Great Lakes waters, inland from the shorelines only to the extent necessary to control shore lands, the uses of which have a direct and significant impact on the coastal waters, and to control those geographical areas that are likely to be affected by or vulnerable to sea level rise.

16.2.14 National Estuary Program (NEP)

In 1987, the U.S Congress established the National Estuary Program (NEP) as part of the Clean Water Act. Its purpose is to protect and restore the health of estuaries while still supporting economic and recreational activities. The program, through the EPA, aids development of local partnerships to develop and implement Comprehensive Conservation and Management Plans (CCMPs). The CCMPs may have such goals as protecting against contaminated sediments, improving livestock grazing practices, reducing stormwater impacts, predicting wetland habitat changes, and preventing shoreline erosion.

16.2.15 Major Environmental Laws

A summary of the major environmental laws, including the ones discussed here, related to watershed management and their year of enactment are listed in Table 16.2.

TABLE 16.2 Major Environmental Laws

Date	Law
1938	Federal Food, Drug, and Cosmetic Act
1947	Federal Insecticide, Fungicide, and Rodenticide Act
1948	Federal Water Pollution Control Act (also known as the Clean Water Act)
1955	Clean Air Act
1965	Shoreline Erosion Protection Act
1965	Solid Waste Disposal Act
1969	National Environmental Policy Act
1970	Pollution Prevention Packaging Act
1970	Resource Recovery Act
1971	Lead-Based Paint Poisoning Prevention Act
1972	Coastal Zone Management Act
1972	Marine Protection, Research, and Sanctuaries Act
1972	Ocean Dumping Act
1973	Endangered Species Act
1974	Safe Drinking Water Act
1974	Shoreline Erosion Control Demonstration Act
1975	Hazardous Materials Transportation Act
1976	Resource Conservation and Recovery Act
1976	Toxic Substances Control Act
1977	Surface Mining Control and Reclamation Act
1978	Uranium Mill-Tailings Radiation Control Act
1980	Asbestos School Hazard Detection and Control Act
1980	Comprehensive Environmental Response, Compensation, and Liability Act
1982	Nuclear Waste Policy Act
1984	Asbestos School Hazard Abatement Act
1986	Asbestos Hazard Emergency Response Act
1986	Emergency Planning and Community Right to Know Act
1988	Indoor Radon Abatement Act
1988	Lead Contamination Control Act
1988	Medical Waste Tracking Act
1988	Ocean Dumping Ban Act
1988	Shore Protection Act
1990	National Environmental Education Act

16.3 STATE, REGIONAL, AND LOCAL REGULATIONS

16.3.1 State Regulations

It is ironic that, until recently, many federal and state regulations have actually forced a divergence of the disciplines and physical sciences that need to be addressed in order to obtain watershed sustainability. Many states have adopted separate water quality protection regulations, groundwater protection regulations, stormwater management regulations, and erosion and sedimentation (E&S) pollution control regulations. In addition, many states have developed separate stormwater management, E&S, and/or best management practices manuals to apply many of the principles introduced in this text, even though the stormwater runoff, erosion and sedimentation, and water quality degradation process occur in a single storm event. It is, however, encouraging to see that states such as Pennsylvania, which just developed a Comprehensive Stormwater Management Policy (PADEP, 2002), and New Jersey, which is developing a new BMP manual to include stormwater and E&S provisions (NJDEP, 2003), are beginning to encompass the holistic approach to water resources management.

This disconnect becomes obvious by citing an example of a state where septic drain fields could not be placed within 100 feet of an existing well, although the same state had no regulations on wells. That meant a well could legally be drilled within 10 feet of an existing septic drain field. Even today, grant programs are structured around a particular water resource (groundwater, stormwater, nonpoint source pollution), and applicants who take a holistic approach to water resources management are still being turned down if they expand the scope of work beyond the auspices of the grant framework.

Many manuals and regulations are available online and are being updated constantly, so it is recommended that a detailed evaluation of each state's regulations and manuals be performed before undertaking a watershed management project.

16.3.2   Regional Water Resources Management Commissions

Many quasigovernmental or multistate agencies have regulatory authority over water resources in the basin over which they have jurisdiction. Examples include the Delaware River Basin Commission, Susquehanna River Basin Commission, Ohio River Basin Commission, and Great Lakes River Basin Commission. These commissions typically have jurisdiction over maintaining water quantity and quality for drinking water supply purposes. For instance, the Delaware River watershed has many impoundments that supply New York City with drinking water through aqueducts, and the cities of Trenton, New Jersey, and Philadelphia, Pennsylvania, maintain direct intakes on the river itself. The commissions regulate water withdrawals throughout their respective watersheds via a permitting process, to ensure base flows will be sustained during periods of drought. These commissions typically employ highly qualified engineers, scientists, and planners who perform very valuable watershed, water budget, and water quality studies in conjunction with federal, state, and local agencies. They are an excellent source of data and technical resources when performing a watershed assessment and developing a watershed water resource management plan.

For example, the Delaware River Basin Commission (DRBC) requires permits for wells, or systems of wells, that withdraw more than 10,000 gallons per day averaged over any 30-day period. Its records contain permitted allocations, but not current water use records. The DRBC listing of "Most Recent Groundwater Withdrawal Data" can be used as a guide for controlling subbasin water withdrawal rates. The values are based partially on water consumption data provided by the member states' environmental protection agencies. DRBC also calculates net, instead of gross, water use and, therefore, does not represent total groundwater withdrawal.

16.3.3 Local Ordinances

In many areas, land development is controlled through local land use regulations in the form of ordinances. Most local governments have subdivision and land development (SALDO), stormwater management, and floodplain ordinances, requiring stormwater management measures to be installed on new construction or development sites. Unfortunately, many of the ordinances do not properly address a comprehensive approach to maintain sustainable water resources. Many local ordinances also are not updated frequently enough to contain water quality management or groundwater recharge provisions. Water quality control has been difficult to implement on construction or development sites, as it is more difficult to quantify than water quantity control. Moreover, many local ordinances were derived from a template that is outdated and could actually be detrimental. These ordinances have boilerplate language that is applied on a political division basis, and which does not address water resource management on a watershedwide basis. For instance, many ordinances require that the postdevelopment peak flow be reduced to the predevelopment peak flow for some specified design storm or storms. However, it has been found that placing detention basins in the lower portions of the watershed may actually release the water at the time of peak of the watershed, thus actually *increasing* flows in the stream (see Chapter 19).

Another example of ordinances causing more harm than good is one that requires detention basins to be designed to reduce the 100-year postdevelopment flow to the 100-year predevelopment flow. What this actually produces are detention basins with large-capacity outlet structures that would allow the postdevelopment flows of lesser-frequency storms to pass through the basin undetained, thereby causing flooding downstream. Local ordinances must be developed in conjunction with a master watershed assessment and plan to be consistent with the goals of the plan. It is up to the professional community to ensure that ordinances are comprehensive.

Standards and criteria should be developed following a detailed comprehensive watershed assessment and plan and be incorporated into a model ordinance. The plan should include the means to monitor, manage, and plan future growth and/or development. A typical watershed ordinance should manage water quantity, quality, recharge, and stream bank erosion flows. The key to successful implementation of such an ordinance is to involve the responsible officials who must administer the ordinance early in the process. They must be included in the watershed plan advisory meetings and encouraged to provide input throughout the process, which will help them to implement the ordinance after adoption. The other key to successful implementation is doing a thorough evaluation of the standards and criteria existing prior to incorporation of the ordinance and customizing the ordinance to a particular watershed and governing body.

16.4   FEDERAL PROGRAMS

16.4.1   United States Geologic Survey Water-Quality Assessment Program

The USGS is investigating the effects of urbanization on stream ecosystems through its National Water-Quality Assessment (NAWQA) Program. In the NAWQA program, urbanization is defined as the conversion from rural land uses to residential and commercial uses that are typical of recent, generally sprawling, urban growth patterns.

The program is intended to better define the interrelationships between urbanization, water quality, ecosystems, hydrology, and stream habitat. It will lead to development of key indicators of the effects of urbanization on stream quality. Identification of the magnitude and pattern of stream ecosystem degradation should provide engineers and other stakeholders with tools to prioritize restoration projects, along with a strategy for managing, protecting, and restoring impaired waters.

Assessments began in 1999 and 2001 in 15 metropolitan areas, as shown in Table 16.3 and Figure 16.5.

The goals of the study are extremely important in understanding the intricacies of watershed processes and will aid in establishing and strengthening protocols in the future. The study hopes to answer the following questions related to urbanization of watersheds (USGS, 2003):

- What is the magnitude and pattern of response in stream hydrology, water chemistry, and biological communities?

TABLE 16.3   NAWQA Designated Urban Areas

1999 Studies	2001 Studies
Anchorage, Alaska	Atlanta, Georgia
Birmingham, Alabama	Reno-Sparks, Nevada
Boston, Massachusetts	Dallas-Fort Worth, Texas
Chicago, Illinois	Denver, Colorado
Cincinnati-Dayton, Ohio	Raleigh-Durham-Chapel Hill, North Carolina
Los Angeles, California	Milwaukee, Wisconsin
Philadelphia, Pennsylvania	Portland-Salem-Eugene, Oregon
Trenton, New Jersey	
Salt Lake City, Utah	

Figure 16.5 Fifteen metropolitan areas studied in the NAWQA program (Wilber and Couch, 2002).

- What is the threshold of urbanization (impervious area) where stream ecosystems begin to degrade?
- How are physical characteristics of the watershed—such as basin slope, geology, and soils—related to hydrologic, chemical, and biological responses to urbanization?
- How does water quality degradation from urbanization affect the susceptabilities of certain species?
- What are the best water quality monitoring measures for urbanizing watersheds?

USGS is sampling these watersheds to be able to find the answers to the questions above, as shown in Figure 16.6. Preliminary findings showed that rapid degradation of stream ecosystems (invertebrates) occurs early in the process of watershed urbanization. For example, invertebrate communities that are sensitive to pollution and habitat modifications declined when approximately 5 percent of land cover in the watershed was converted to impervious area (roads, parking lots, and houses) in the Anchorage, Alaska, study. Other studies (Schueler, 1994) showed that streams begin to degrade after conversion to as low as 10 percent impervious area. Early results also showed that rapid water quality and ecosystem degradation is associated with processes, such as deforestation, that alter hydrology, stream temperature, and habitat, not just impervious area. In some areas, these physical factors se-

Figure 16.6   USGS biologists collecting fish for tissue analysis and community status assessment (USGS, 2003).

verely degrade biological communities before nutrients and other contaminants from nonpoint sources reach concentrations that may further degrade the communities.

Another preliminary finding was that the effect of urbanization has a greater impact on forested and rangeland watersheds than on agricultural areas planted in row crops, most likely because of disturbances from agriculture already inherent in the streams.

USGS provides its data on current stream flow conditions, water quality of streams and groundwater, and water levels in wells through the National Water Information System Web (NWISWeb) "Data for the Nation" site at http://waterdata.usgs.gov/nwis.

16.4.2   Environmental Protection Agency STORET Program

The U.S. EPA maintains two data management systems containing water quality information for the nation's waters: the Legacy Data Center (LDC), and STORET. STORET is the acronym for STOrage and RETrieval, a repository for water quality, biological, and physical data. Both the LDC and STORET contain raw surface- and groundwater, biological, chemical, and physical data for the United States, Canada, and Mexico, collected by federal, state, and local agencies, Native American tribes, academic institutions, and volunteers. The difference is that LDC is the older system, with data collected through 1998. Data collected since 1999 are stored in STORET, along with some LDC data that have been migrated and properly documented (U.S. EPA, 2003).

Sampling locations are identified by latitude and longitude, and therefore can be easily brought into the GIS system. Other descriptors include sponsoring organization; USGS Hydrologic Unit Code (HUC); state, county; a short description of the site; date and time of sampling; type of medium sampled (fish tissue, water, sediment); why the data were gathered; sampling and analytical methods used; the laboratory used to analyze the samples; the quality control checks used when sampling and handling the samples, and analyzing the data; and the personnel responsible for the data. The data contained in the LDC and STORET systems are available on the Web at www.epa.gov/storet/about.html.

16.4.3 The EPA's Watershed Assessment, Tracking and Environmental Results (WATERS) Program

For the EPA to administer the Clean Water Act, several programs are required to determine water quality, including:

- States adopt goals and standards for their water bodies under the Water Quality Standards Database (WQSDB), which contains information on the uses that have been designated for water bodies.
- Scientists sample, test, and monitor the water quality data and place it in the STORET system.
- Streams are classified as "good" (the water body fully supports its intended uses) or "impaired" (the water body does not support one or more of its intended uses), and the results are stored in the National Assessment Database (NAD). Assessed waters are classified as either "fully supporting," "threatened," or "not supporting" their designated uses.
- The impaired waters are targeted by pollution control programs to reduce the discharge of pollutants into those waters. The results are stored in the Total Maximum Daily Load (TMDL) Tracking System, which contains information on waters that are not supporting their designated uses.

The WATERS program integrates the databases of all these separate information systems and helps the EPA Office of Water (OW) to monitor the quality of the nation's surface waters. With WATERS, the databases are connected to the National Hydrography Dataset (NHD), allowing one program database to reach another with information to be shared across programs. WATERS empowers the EPA Office of Water to meet its goal of achieving "clean and safe water" under the Clean Water Act.

16.4.4 Programs Administering Grants

There are many programs (too numerous to list here) available to agencies, organizations, and private landowners to aid in sustainable water resource and watershed management. The National Sustainable Agriculture Information Service (2003) has an excellent Web site that describes numerous programs

that watershed organizations could pursue (http://attra.ncat.org/guide/.htm).
A number of the programs on that list are given here, followed by a more in-depth description of four of them.

Conservation Reserve Enhancement Program (CREP)
Commission for Environmental Cooporation (CEC)
Environmental Quality Incentives Program (EQIP)
Farmland Protection Program
Forestry Incentives Program (FIP)
Forest Legacy Program
Partners for Fish and Wildlife
Pesticides (Section 406: RAMP, CAR, Pesticide Management Centers)
Water Quality (Section 406)
Wetlands Reserve Program (WRP)
Wildlife Habitat Incentives Program (WHIP)
Emergency Watershed Protection Program (EWP)
Environmental Finance Program
Drinking Water State Revolving Fund (DWSRF)

16.4.4.1 *Conservation Reserve Enhancement Program* (CREP). The
Conservation Reserve Enhancement Program is a federal voluntary program
for agricultural landowners who receive annual payments for keeping riparian
buffers in conservation plantings. State participation is required for federal
funds to be released. The CREP goals are to protect water quality by reducing
nutrients and sediment from entering adjoining waterways and to improve
wildlife habitat, particularly for quail, pheasant, and songbirds. Contracts run
from 10 to 15 years and can provide landowners with more income from that
portion of their property than what they could earn from farming the land.
Maryland, the first state in the country to establish the CREP, also has a
program to purchase permanent easements so that the land can never be developed. Maryland has also developed a 300-foot buffer called the CP-22
Trees and Shrubs Concept, shown in Figure 16.7. The first 100 feet of the
buffer, measured from the stream bank, is planted in trees and shrubs, with
weeds growing in between. The middle 100 feet is planted in trees and shrubs
with warm-season grasses, including species such as partridge pea planted
shortly after the trees are planted. Partridge pea provides a food source for
quail and other wildlife. The outside of the buffer is planted in warm-season
grasses only. This combination has been found to provide tremendous pollution filtering capabilities and provides wildlife with a food source, water,
cover, and breeding ground.

16.4.4.2 *Commission for Environmental Cooperation* (CEC). The North
American Free Trade Agreement (NAFTA), signed by Canada, Mexico, and

1. - trees and shrubs with weeds growing in between.
2. - trees and shrubs with warm-season grasses
3. - warm season grasses only.

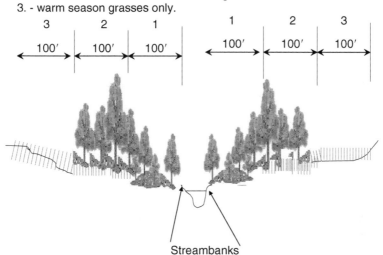

Figure 16.7 CREP recommended buffer.

the United States, created the North American Commission for Environmental Cooperation (CEC). The CEC is an international organization that facilitates cooperation and public participation to foster conservation, protection, and enhancement of the North American environment for present and future generations. It is a council of cabinet-level environment officials from the three countries, the Joint Public Advisory Committee (a group of five citizens from each country), and a Secretariat, staffed with environmental experts.

The CEC sponsors programs such as the North American Pollutant Release and Transfer Register (PRTR) project, which was initiated in 1995. The primary aim of the PRTR project is to promote public access to information on pollutant releases and transfers in North America, in order to:

- Enhance understanding of sources and handling of toxic substances.
- Promote the use of PRTRs by citizens, communities, industry, government, and other interested parties.
- Provide an informed basis for stakeholder dialogue and priority-setting.
- Foster pollution reduction efforts.

The publication of the annual "Taking Stock" report on North American pollutant releases and transfers is the cornerstone of the project. Other objectives of the North American PRTR project are to:

- Facilitate access/use of PRTR data through development of a North American PRTR Web site.

- Gain further insight into pollution-related issues of particular interest in North America by undertaking special analyses of PRTR data and other relevant information.
- Enhance comparability among the North American PRTR systems.
- Provide support for the further development of the PRTR program in Mexico.
- Facilitate coordination of North American PRTR-related activities with similar international activities.

The CEC has also developed the North American Fund for Environmental Cooperation (NAFEC), which provides grants for programs such as environmental cooperation and environmental monitoring and assessment related to human health.

16.4.4.3   *Forest Land Enhancement Program (FLEP).* The U.S. House Agriculture Committee approved a farm bill, the Agricultural Act of 2001 (HR 2646), on July 27, 2001. The bill proposes a Forest Land Enhancement Program (FLEP), which would replace the Forestry Incentives Program (FIP) and the Stewardship Incentives Program (SIP) (USDA Forest Service, 2002; Virginia Department of Forestry, 2002). Through FLEP, state forestry agencies can provide a wide array of services to ensure that our nation's nonindustrial private forests (NIPF) and related resources continue to provide forest products and safeguard the health of our water, air, and wildlife. The program is being managed by the Forest Service through the state forestry agencies, to:

- Establish, manage, maintain, enhance, and restore NIPF lands.
- Enhance the productivity of timber, fish and wildlife habitat, soil, water, air quality, wetlands, and riparian buffers of these lands.
- Assist owners and managers to more actively manage NIPF lands and related resources.
- Reduce the risk and help restore, recover, and mitigate the damage to forests caused by fire, insects, invasive species, disease, and damaging weather.
- Encourage the use of state, federal, and private-sector resource management expertise, financial assistance, and educational programs.

The FLEP program provided funding in the amount of $20 million for each fiscal year 2003 through 2007, which can be used for financial, educational, or technical assistance to NIPF owners to achieve a broad array of objectives, including:

- Forest stewardship plan development
- Afforestation and reforestation

- Forest stand improvement
- Agroforestry implementation
- Water quality improvement and watershed protection
- Fish and wildlife habitat protection
- Forest health and protection
- Invasive species control
- Wildfire and catastrophic risk reduction
- Wildfire and catastrophic event rehabilitation
- Special practices

16.4.4.4 Forest Legacy Program (FLP). Congress created the Forest Legacy Program (FLP) to help landowners, state and local governments, and private land trusts identify and protect environmentally important forest lands threatened by present and future conversion to nonforest uses. FLP's success depends on private landowners who want to conserve the special values of their forested land for future generations. FLP buys certain land use rights from the landowner at fair market value, such as the right to develop land, with the dual aims of promoting effective forest land management and protecting the land from conversion to nonforest uses. Priority is given to forest lands that possess important scenic, cultural, and recreation resources, fish and wildlife habitats, water resources, and other ecological values.

Each state drafts a forest conservation plan identifying "environmentally important" forest lands that include:

- Forest buffers that stabilize soil and provide a natural filter for potential pollutants that might get into streams and rivers
- Recreational resources such as places to swim, hike, and camp
- Fish and wildlife habitat such as clear, cold brooks for trout and dense forest cover to help deer survive cold, snowy winters
- Sites of cultural value such as unique geology, historic settlements, and artifacts
- Areas with beautiful scenery to enjoy from trails or roads
- Areas where timber and wood fiber grow rapidly
- Areas having other ecological value, such as clean air and water

The FLP is a volunteer program and provides federal funds for:

- Purchase of conservation easements
- Fee acquisition
- Grants to states for the preceding activities

- Surveys, title work, and other activities to facilitate donations of easements for FLP purposes
- State FLP planning and administration

Willing owners who are accepted into the program can sell all or part of their ownership rights, such as the right to develop the land, to the federal government. The government will pay for these rights at full fair market value.

16.5 GREENWAYS AND RIVERS CONSERVATION PLANS

Greenways are corridors of land, usually following a river course or hiking trail established to permanently conserve and preserve the corridor's heritage and recreational opportunities. Establishment of a greenway and/or conservation plan includes GIS mapping of the same physical features as needed in a watershed assessment, however, additional layers such as cultural resources (historic buildings or structures, etc.), access points, linkage to other greenways or conservation areas, amenities, etc. may exist. There are several "national" or large-scale greenways such as the Missouri–Mississippi Rivers Confluence Greenway, the East Coast Greenway, and the Appalachian Trail, as well as regional greenways such as the Hudson River Valley Greenway and the Susquehanna River Partnership. Localized watershed plans could evaluate the possibility of establishing a more local greenway and connectivity to one of these larger systems. Establishment of greenway systems includes public participation and input.

16.6 LAND TRUSTS

Large parcels worthy of preservation should be identified in a watershed assessment and protection plan. The source of this data is the county tax office. Many *tax parcel* data is not currently in digital format, however if it is, it can easily be brought into the GIS system and overlaid with other features such as wetlands, floodplains, endangered species, etc. to identify parcels worthy of protection. These parcels can then be targeted for preservation thorough a variety of measures such as land trust organizations, private lands conservation, land acquisition, conservation easements, or conservation buyer projects. Agencies such as the Nature Conservancy can be contacted to aid in such measures and/or to advise on acquisition methods. The nature conservancy utilizes an approach called "Conservation by Design." This is a strategic, science-based planning process that helps identify the highest-priority places—landscapes and seascapes that, if conserved, promise to ensure biodiversity over the long term. The process is similar to the one described in this text and evaluates the same physical parameters.

16.7 SUMMARY

The Clean Water Act is the primary regulation developed to improve the quality of our nation's waters. State laws should look at a more comprehensive approach to water resource management in the future. Local ordinances that protect the public health, safety, and welfare are an excellent means of protecting land and water resources. Many innovative land use and management programs are being developed to aid in preservation of land that would impact water quality if developed.

REFERENCES

American Farmland Trust. *Soil Conservation in America: What Do We Have to Lose?,* (Washington, DC: American Farmland Trust), 1984.

City of Monterey, California, City of Santa California Coastal Commission, Monterey Bay National Marine Sanctuary, Association of Monterey Bay Area Governments, Woodward-Clyde, Central Coast Regional Water Quality Control Board. "Model Urban Runoff Program: A How-To Guide for Developing Urban Runoff Programs for Small Municipalities" (City of Monterey, CA: Public Works Department), July 1998.

Cook, G. Dennis, and Robert H. Kennedy. "Water Quality Management for Reservoirs and Tailwaters," Report 1, in "Reservoir Water Quality Management Techniques," Technical Report E-89-1 (Vicksburg, MS: U.S. Army Corps of Engineers Waterways Experiment Station), January 1989.

Environmental Media Services, Clean Water Act, http://www.ems.org/cleanwater/information.html, Web site accessed January 28, 2004.

Federal Register. "Rules and Regulations," vol. 64, no. 235, pp. 68721-68770, December 8, 1999, from the Federal Register Online via GPO Access, www.gpoaccess.gov/fr/index.html.

Marsh, D. "When a Watershed Is Gripped by Nonpoint Source Pollution," *All Riled Up: Nonpoint Source Pollution: The Point of No Return* (Madison, WI: Wisconsin Department of Natural Resources), undated.

Massachusetts Adopt-a-Stream Program Storm Drain Mapping Project. Field Manual, draft, Riverways Program, Department of Fisheries, Wildlife and Environmental Law Enforcement, January 2002.

National Sustainable Agriculture Information Service, Building Better Rural Places Web page, http://attra.ncat.org/, http://attra.ncat.org/guide/.htm, accessed December 22, 2003.

New Jersey Department of Environmental Protection. *Stormwater BMP Manual,* (Trenton, NJ: New Jersey Department of Environmental Protection), 2003

Pennsylvania Department of Environmental Protection, Pennsylvania Code, Title 25 Environmental Protection, Chapter 93, "Water Quality Standards," December 15, 2001.

Pennsylvania Department of Environmental Protection, Protocol (Harrisburg, PA: Pennsylvania Department of Environmental Protection), 2002.

Pennsylvania Department of Environmental Protection, Comprehensive Stormwater Management Policy, Document ID # 392-0300-002 (Harrisburg, PA: Pennsylvania Department of Environmental Protection), September 28, 2002.

Public Law 104-182 104th Congress. An act to reauthorize and amend title XIV of the Public Health Service Act, commonly known as the "Safe Drinking Water Act," and for other purposes, August 6, 1996.

Schueler, Tom. "The Importance of Imperviousness," *Watershed Protection Techniques, Quarterly Bulletin,* vol. 1, no. 3, Fall 1994.

Tchobanoglous, George, and Edward D. Schroeder. *Water Quality* (Reading, MA: Addison-Wesley), 1987.

USDA Soil Conservation Service. "Water Quality Indicators Guide: Surface Water," SCS-TP-161, USDA-SCS, Washington, DC, September 1989.

USDA Forest Service, Forest Land Enhancement Program Briefing Paper, www.fs.fed.us/spf/coop/FLEPbriefing.pdf, November 21, 2002.

U.S. Environmental Protection Agency (U.S.EPA). Watershed Protection: TMDL Note 2, "Bioassessment and TMDLs," EPA 841-K-94-005a (Washington, DC: U.S. Environmental Protection Agency), 1994a.

———. "National Water Quality Inventory," 1992 report to Congress, EPA 841-R-94-001 (Washington, DC: U.S. Environmental Protection Agency, Office of Water), 1994b.

———. "Guidelines for the Preparation of the Comprehensive State Water Quality Assessments 305[b] Reports, EPA 841-B-97-002A (Washington, DC: U.S. Environmental Protection Agency, Office of Water), 1997.

———. *Rapid Bioassessment Protocols for Use in Wadeable Streams and Rivers, Periphyton, Macroinvertebrates, and Fish,* 2nd ed., 841-B-99-002 (Washington, DC: U.S. Environmental Protection Agency, Office of Water), July 1999.

———. "Lake and Reservoir Restoration Guidance Manual," EPA 440/5-88-002 (Washington, DC: U.S. Environmental Protection Agency), February 1988.

———. Watershed Assessment, Tracking and Environmental Results (WATERS), www.epa.gov/waters/, last updated on January 3rd, 2003, accessed June 12, 2003.

———. STORET Database Access, http://www.epa.gov/storet/dbtop.html, last updated on Monday, December 15th, 2003, accessed December 23, 2003.

U.S. Forest Service. Cooperative Forestry, Forest Legacy Program National Program Office, Washington, DC, 2002.

U.S. Geological Survey. "Effects of Urbanization on Stream Ecosystems," Fact Sheet 042-02, May 2002.

U.S. Geological Survey, National Water Quality Assessment Program (NAWQA), web site http://water.usgs.gov/nawqa/, accessed July 25, 2003.

Virginia Department of Forestry. "The Forest Land Enhancement Program (FLEP): A New Forest Service Program," www.dof.state.va.us/mgt/cip-fact-flep.shtml, August 2, 2002.

Wilber, W.G., and C.A. Couch. "Assessing Five National Priorities in Water Resources," *Water Resources Impact,* vol. 4, no. 4, pp. 17–21, 2002.

CHAPTER 17

SYSTEMATIC APPROACH TO WATERSHED ASSESSMENT: THE DIGITAL WATERSHED, WATERSHED ANALYSIS, GIS, AND MODELING

17.0 INTRODUCTION

As described in Chapter 10, a wealth of GIS data is available, and more is being made accessible every day. Many watershed planners obtain this data, compile it into maps for the watershed, and redisplay it in the watershed plan or report. Unfortunately, this process is a waste of time, effort, and funds. For example, the typical procedure followed by the inexperienced person is to obtain existing geologic data and produce a geology map for the watershed. But these maps are already published, and could be copied for publication into the report. Thus, digital data should not be viewed as a means to re-create a map, but as a valuable tool to analyze trends and patterns, and as a means for analysis in the watershed assessment. Layers should be used in conjunction with each other to analyze these patterns, and the graphic results utilized as a tool to portray the trends, as described in the modeling and overlaying sections in this chapter. All that said, the true root of the GIS power is the attributes of the features, which allow computational analyses and selection.

17.1 GIS APPLICATIONS

Recent advances in GIS have enabled planners, watershed managers, and hydrologic engineers to expand their capabilities for watershed management. Recent increases in computer speed and storage, coupled with reduced costs of the machines themselves, have opened up a whole new "watershed" of opportunity for watershed management professionals to store, manipulate, and analyze watershed GIS data. Although mainframes and workstations had the computational power to perform data-intensive analyses, they were typically too expensive for the small engineering firm or watershed district. Consider that, in 1990, a typical PC consisted of 16 megabytes of RAM, a 386 processor running at 80 megahertz, and a 100-meg hard drive, whereas in 2003, machines (selling at a lower price than the 386 at its time) have allowed the common practitioner to develop new and innovative watershed management tools. These hardware advances have also enabled the general practitioner to load large volumes of cartographic data required for watershed management.

Great strides have also been made in recent years to tailor GIS technology and to develop data and specific GIS applications for various facets of watershed management. Advances in hardware, software, data development, and specific applications have concomitantly contributed to progress in watershed management capabilities. This chapter focuses initially on advances in GIS data and then concentrates on the software and applications that utilize these data.

With the advent of faster and more powerful computers, particularly with the capability to store large volumes of cartographic data, development of new and innovative GIS data, software packages, and applications has exploded. This translates to the ability of watershed managers to better and more efficiently analyze watershed physical features in a manner that allows them to manage watersheds to meet their particular concern, whether it is stormwater management, floodplain management, water quality control, or conservation planning.

A word of caution is, however, in order here: Existing GIS data or models must be used with care, to ensure their accuracy for the intended purpose. To that end, this chapter provides a systematic approach to watershed assessment utilizing digital data.

17.2 COMPILING THE DIGITAL WATERSHED

One of the first steps in a watershed assessment process is to collect and compile as much available digital spatial (GIS) data on the watershed to begin the assessment process. This, of course, will depend on the funds available, goals and objectives, availability of required data, and time frame of the study.

It is highly recommended that any GIS work be developed by one central party that is deemed responsible for the GIS; this party would serve as the central GIS clearinghouse and would also be responsible for quality control. It is also highly recommended that the central GIS clearinghouse be responsible for coordinating the GIS effort with other ongoing studies. Enormous amounts of federal and state funds are wasted because various parties recreate the same GIS data for similar projects. For instance, one watershed had concurrent projects by three different organizations and three different consultants: a stormwater management plan, a TMDL project, and a rivers conservation plan. All three were developing the same base map and land use, soils, and geology datasets. More than $100,000 was wasted in duplicative efforts. Another problem with this watershed was that each project required land use data in a slightly different format, which could have been compiled to meet all three of the project's needs. Other problems with various parties developing datasets are issues of conflicting projections, quality control procedures, and accuracy.

Once the comprehensive GIS data collection effort is complete, but before overlaying and analysis is performed, trends and patterns can be developed through attribute manipulation and display, as described in subsequent sections here. The goal of the GIS effort is *not* to reproduce maps, but to determine trends and patterns from the physical features of the watershed.

The general procedure to establish the digital watershed is described next. Note, however, that the order of the steps may have to be rearranged based on available data, scope of the project, and funding.

17.3 DATA ANALYSIS

17.3.1 Step 1: Planimetric Base Map

Once the project coordinate system has been established, the planimetric base map data, which includes transportation, hydrography (see step 2), and political boundaries, should be brought into the GIS. The base map should be developed from the most accurate available local data at the scale required for the objective at hand. It is best to utilize *local data* (municipal or county) if available, as long as it meets the quality control check, as it is developed on a large scale, and hence should have a better accuracy than a small-scale dataset. Keep in mind that small-scale data may not accurately overlay on more accurate large-scale data, and thus their potential use should be evaluated before implementing them.

Next, available national data should be obtained to assess trends. The data should then be refined with as much regional and local data as economically possible, depending on the goals of the assessment. Compiling national data is fairly straightforward utilizing the data sources described in Chapter 10. The biggest obstacle to utilizing data from past studies is that each dataset was developed under different projections and/or coordinate systems, requiring a considerable investment of time to *reproject* it. Most government data has well-developed metadata, which makes the process easier, because the original projection is known. Fortunately, this procedure is becoming much easier thanks to advances in software. Features such as ESRI's ArcGIS "projection on the fly" reproject data to the project coordinate system as they are brought into the project. ArcToolbox, another module of ArcGIS, allows menu-driven reprojections, which previously had to be performed through cumbersome programming language.

17.3.2   Step 2: Hydrography

17.3.2.1   Streams. Of primary importance in a watershed assessment is the hydrography data layer, which should be established next in the GIS. Although locally developed stream data may be more accurate, it may not have the connectivity required for modeling purposes. Therefore, it is recommended that this data be from the National Hydrography Dataset (NHD), which has stream connectivity for hydrologic and hydraulic modeling if required. This dataset also includes the Strahler stream order, which aids in FGM and management criteria, as shown in Figure 17.1.

17.3.2.2   Lakes, Reservoirs, Wetlands. Lakes, reservoirs, and wetlands are an important layer to aid in the watershed assessment due to their impact on water quality, storage, and possible attenuation capability of flood flows. Lake and reservoir data layers can be obtained from a variety of sources. Typically, state transportation departments create a lake layer when develop-

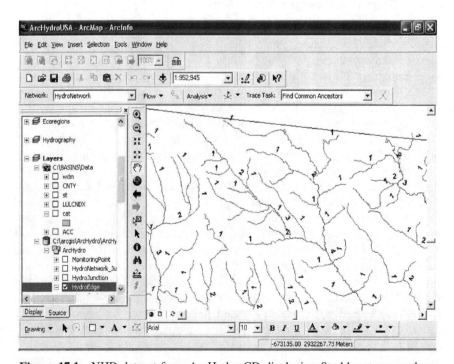

Figure 17.1   NHD dataset from ArcHydro CD displaying Strahler stream orders.

ing road centerline data. The USGS DRGs also show lakes. Lakes are labeled as water in the USDA soils database attributes. Another source of potential lake locations is the state agency responsible for dam safety, which typically has the dams located within a GIS—although this data may not be available to the public since the terrorist attacks of September 11, 2001.

17.3.3 Step 3: DOQs and DRGs

DOQs and DRGs should be brought into the GIS next, to aid in the analysis. These datasets provide a valuable base map set, which can be used in tandem with the planimetric data. The DRGs and DOQs will provide valuable data on land use, and, when coupled with the water quality sampling data from the streams, will aid in determining trends and potential pollution sources.

17.3.4 Step 4: Physiographic Provinces, Ecosystems, and Climate

It is recommended that the physiographic provinces, EPA, and Forest Service ecosystem provinces and climatic zones be obtained and displayed. This will provide valuable data on trends and what to expect based on the descriptions of the region in which the watershed is located.

17.3.5 Step 5: DEMs and TINs

As described in Chapter 10, DEMs provide a valuable resource in analyzing a watershed's relief. The DEMs make it possible to visualize general elevation differences in the watershed with elevation classes, which can be displayed in two dimensions (Figure 17.2), three-dimensional (3-D) relief as viewed from overhead (Figure 17.3), and three dimensions. The 3-D DEMs are analyzed either in grayscale, also referred to as a *hillshade* (Figure 17.4), or with elevation classes (Figures 17.5a and b). DEMs also allow contours to be developed (Figure 17.6).

The USGS DRGs can be overlaid onto the DEM to develop a highly detailed map of relief and surface features and to provide the user with a very valuable analysis tool, as shown in Figure 17.7.

Triangulated irregular networks (*TINs*) can be developed from DEMs, and vice versa. TINs are useful in hydrologic modeling, for visual analysis of the terrain. Other physical features, discussed later, can be overlaid onto the TIN to aid in watershed analysis and visualization. A typical TIN with elevation classes is shown in Figure 17.8.

17.3.6 Step 6: Soils

The valuable part of the digital soils database is that most of the data in the standard USDA Soil Survey tables and the attribute data are attached to the soil polygons. This allows for a variety of analysis and queries. For instance,

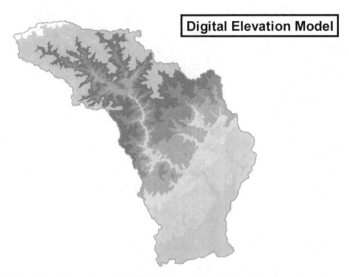

Figure 17.2 Two-dimensional DEM of a watershed with elevation classes. (DEM developed by Paul DeBarry, courtesy of Borton-Lawson Engineering, 2003.)

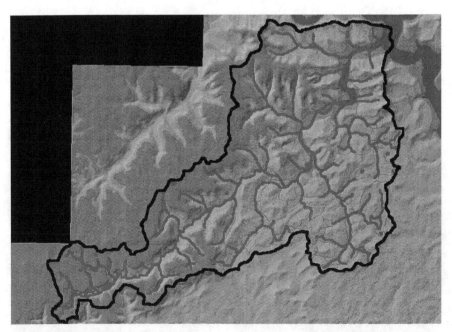

Figure 17.3 Three-dimensional DEM of a watershed with elevation classes; view from directly overhead. (Developed by Paul DeBarry, courtesy of Borton-Lawson Engineering, 1999e.)

Figure 17.4 Three-dimensional grayscale hillshade DEM of a watershed. (Developed by Paul DeBarry, courtesy of Borton-Lawson Engineering, 2001.)

Figure 17.5a Three-dimensional DEM of a watershed with elevation classes. (Developed by Paul DeBarry, courtesy of Borton-Lawson Engineering, 2003.)

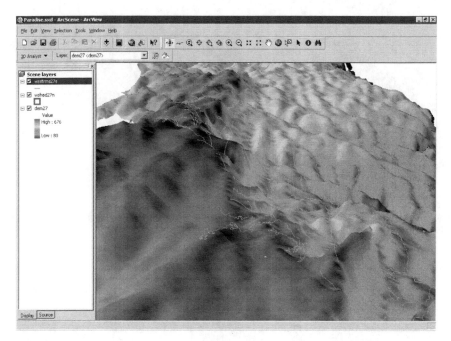

Figure 17.5b   Three-dimensional DEM of a watershed with elevation classes.

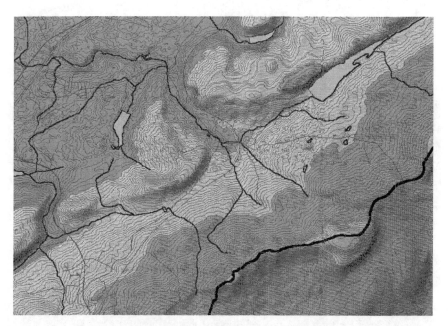

Figure 17.6   Contours developed from a DEM showing the watershed boundary. (Developed by Paul DeBarry, courtesy of Borton-Lawson Engineering, 2002.)

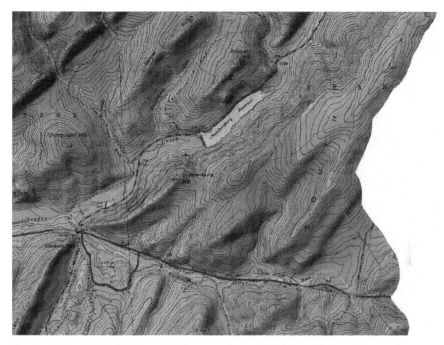

Figure 17.7 USGS DRG overlaid onto a three-dimensional DEM. The overlaying of the two layers provides a detailed display of the topography and its influence on man-made features in the watershed. (Developed by Paul DeBarry, courtesy of Borton-Lawson Engineering, 2002.)

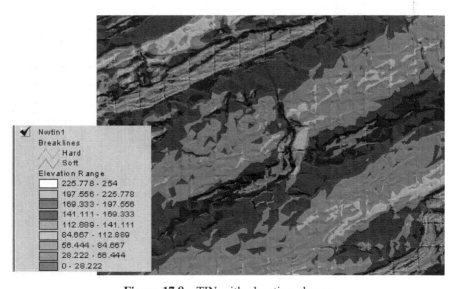

Figure 17.8 TIN with elevation classes.

if erosion is a problem in a particular watershed, a soils erodibility map can be generated by querying the erodibility index of the soils. Overlaying this on a DEM provides a very early indication of where erosion problems would occur in a watershed, as shown in Figure 17.9. This can then lay the groundwork for targeting field exploration of these potential erosion problems.

Another useful application of the soils data through attribute analysis would be to display the pH of the soil or its buffering capacity, if there is an acid rain problem or acidic streams. The buffering capability of the soils could be plotted to visualize the potential effectiveness of lime treatment (see Figure 17.10).

One of the most useful soil map applications, particularly for stormwater management, is to develop a hydrologic soil group map from the soil series. The hydrologic soil groups can then be incorporated into the GIS watershed models for hydrologic modeling. A typical hydrologic soil group map is shown in Figure 17.11. Related to the hydrologic soils groups is the permeability of the soils, which can aid in developing a water budget to be utilized as a management tool. A typical soil permeability map can be found in Figure 17.12.

17.3.7 Step 7: Watershed Slopes

Slopes are an important factor in watershed management, not only for assessing erodibility and the surface hydrology of the watershed, but also for

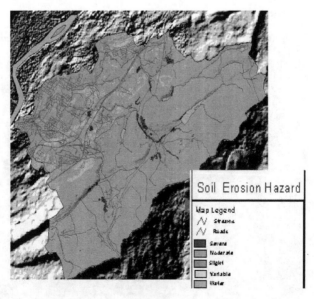

Figure 17.9 Erodible soils graphically displayed in a hillshade of a watershed. Notice the erodible soils are found in the stream valleys. (Developed by Paul DeBarry, courtesy of Borton-Lawson Engineering, 1999.)

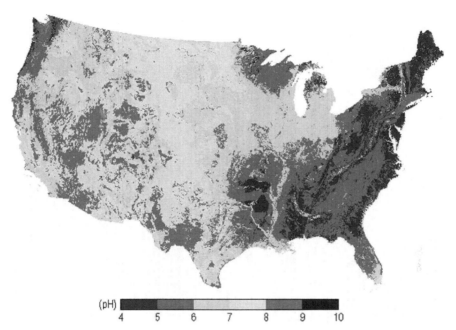

Figure 17.10   Soil pH. (*Source:* "Soil Information for Environmental Modeling and Ecosystem Management," Penn State's Earth System Science Center, 2003, after D.A. Miller and R.A. White, "A Conterminous United States Multi-Layer Soil Characteristics Data Set for Regional Climate and Hydrology Modeling," *Earth Interactions,* vol. 2, paper 2, 1998.)

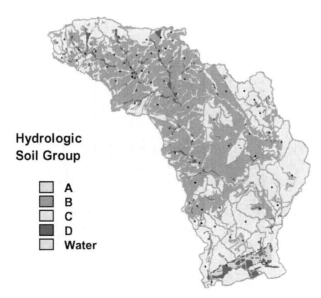

Figure 17.11   Hydrologic soil groups. (Developed by Paul DeBarry, courtesy of Borton-Lawson Engineering, 2004.)

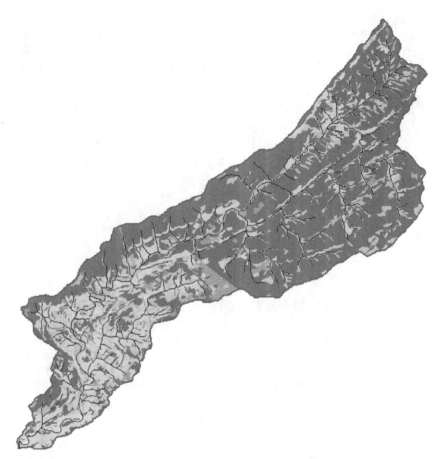

Figure 17.12   Variations in soil permeability in a watershed. The actual soil permeability determination is complex due to the differences in permeabilities of the soil horizons. (Developed by Paul DeBarry, courtesy of Borton-Lawson Engineering, 2003.)

the planning and management of the watershed. Typically, no "slopes" coverage or layer is readily available for use or download, but slopes can be determined through a variety of sources. The classification of the soils in USDA soil surveys includes slope classifications in the third parameter of the series name; that is, A, B, C, D, E, and F. A slope map can be developed from the slope classification. In many instances, however, the slope definitions overlap and would have to be resolved in the development of this slope map. A typical soils-derived slope map can be found in Figure 17.13.

A slope analysis map can also be developed utilizing the DEM, as shown in Figure 17.14. Slope classifications could be based upon one of the objectives; for instance, the suitability for septic system disposal sites typically ranges from 0 to 8 percent, 8 to 12 percent, 12 to 15 percent, 15 to 25 percent

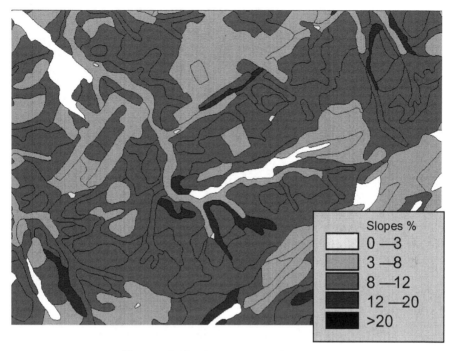

Figure 17.13 Slopes derived from soils.

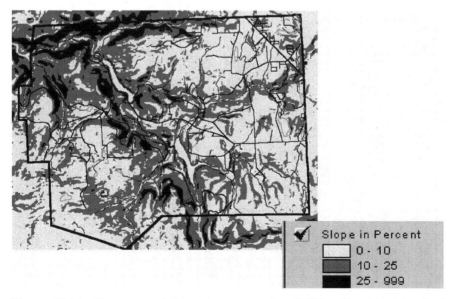

Figure 17.14 Slope map derived from processing DEMs. (Developed by Paul DeBarry, courtesy of Borton-Lawson Engineering, 1998.)

and greater than 25 percent. Minimal slopes can also be calculated in the DEM when developing channels for routing in hydrologic models.

Slopes can also be overlaid on TINs, to aid in the development of the opportunities and constraints analysis (see Figure 17.15).

17.3.8 Step 8: Land Use/Land Cover

Almost any watershed assessment will require a land use/land cover (LULC) layer. As described in Chapter 10, there are many available sources of LULC data, and each one has varying *attributes, format, scale,* and *accuracy.* If the watershed assessment does not require detailed hydrologic modeling, the smaller-scale land use data can be utilized. If modeling will be required, the model should be consulted prior to developing a land use layer, to ensure the attributes and data format are consistent with the model input requirements. If raster LULC is not detailed enough for the model, land cover can be digitized from digital aerial photography on the screen to develop a land cover layer, as shown in Figure 17.16. Both existing and future land use layers should be developed for the watershed. Comparing land uses with soil erodibility, water quality sampling results, nonpoint source pollution loads, stream bank erosion problems, and flooding problems may provide some answers to the potential cause of those problems.

Figure 17.15 Slopes overlaid on TIN to aid in opportunities and constraints analysis and visualization. (Developed by Paul DeBarry, courtesy of Borton-Lawson Engineering, 2002.)

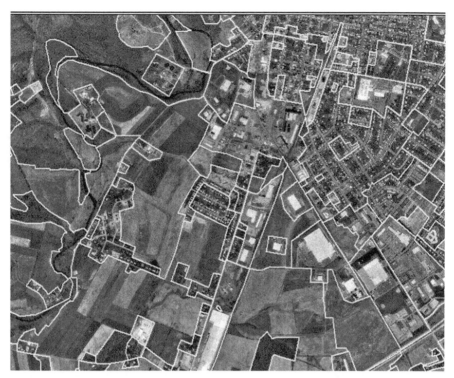

Figure 17.16 Land cover being digitized from digital aerial photography. (Courtesy of Borton-Lawson Engineering.)

17.3.9 Step 9: Other Physical Features of the Watershed

At this juncture, other physical features of the watershed not yet described should be brought into the project file. Some of these features include:

Surface and bedrock geology (including limestone areas)
Floodplains
Satellite imagery
Aerial photography
Habitat and endangered species
Other applicable layers

17.3.10 Step 10: BASINS Data

Next, the BASINS data, specified in Chapter 10, should be perused. The specific data layers necessary to aid in evaluating the processes that affect the goals and objectives of the watershed plan should be entered into the watershed's GIS database.

17.3.11   Step 11: Watershed-Specific Collected Data

Once the readily available GIS data have been entered into the GIS, the data collected from the watershed-specific assessment can be entered into the GIS. This could include the following list of items, as summarized from Chapter 9:

Problem areas
Areas of disturbance
FGM channel reaches and
   designations
Obstructions to flow (bridges,
   culverts, etc.)
Stormwater collection systems
Stormwater control facilities
Flood control projects
Flood flows from FEMA FISs
Illicit discharge points
Industrial discharge points

Sewage treatment plant discharge
   points
Sewer service areas
Potential contamination sites
Stream water quality sampling data
Biological sampling data
Water system service areas
Public and private well data
Wellhead protection zones (3)
Stream flow and rainfall gages
Other watershed-specific data
   required to achieve the goals of
   the watershed assessment

17.4   THE GIS OVERLAYING PROCEDURE

Although analyzing the individual GIS layers described will tell the user a lot about a watershed, perhaps more important is the capability of the GIS to overlay the various data layers to develop a combined set of data. One example would be the development of an *opportunities and constraints analysis* of the watershed. By choosing the constraints for development—such as steep slopes, wetlands, floodplains, erodible soils, wellhead protection areas, and others—and then overlaying all of these layers, one can develop an opportunities and constraints map. The opportunities would be areas where development could proceed with appropriate authorizations, whereas the constraints would be the areas that should be preserved (through conservation easements, for example). A typical opportunities and constraints map is shown in Figure 17.17. This can be created on a large scale, for example on a large watershed basis or on the basis of a particular site development. The accuracy of the opportunities and constraints would of course depend on the accuracy of the data utilized.

Another example of a composite map that can be developed by overlaying a number of GIS layers is the pollution vulnerability map, described in Chapter 22.

17.5   GIS IN HYDROLOGIC AND HYDRAULIC ANALYSES

Several advances have become apparent since the advent of the PC powerful enough to run a GIS. Several procedures have been developed to incorporate

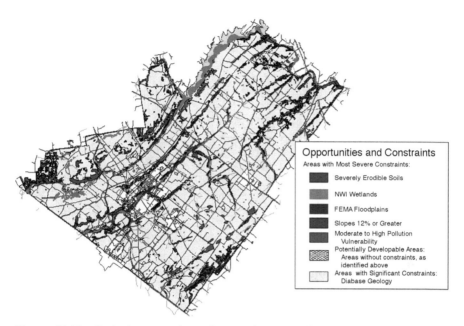

Figure 17.17 Typical opportunity and constraints map. (Developed by Paul DeBarry, courtesy of Borton-Lawson Engineering, 2002.)

GIS into watershed applications (DeBarry, 1990, 1991, 1996, 1999). These GIS applications improve efficiency and accuracy and cut costs in the hydrologic parameter calculation methodology required by hydrologic models. Many subroutines have been developed to analyze the terrain and hydrologic processes from the grid cells of DEMs. Some of the hydrologic subroutines include, in order of popularity: *flow direction, subwatershed or drainage basin boundary determination, flow accumulation,* and *stream channel determination.* The GIS hydrologic operations are based on the premise that water flows downhill in the direction of steepest descent, and the elevations of the grid cells dictate this direction (Nelson et al., 1997; Maidment, 2002). The processes described here are inherent in the ESRI ArcGIS and ArcView products or extensions.

Flow direction is a computation wherein each individual cell is evaluated in relation to the surrounding eight cells and the lowest flow direction vector is defined in the direction of the steepest slope to the centroid of the adjoining cells. This is known as the *eight-direction pour point model,* and is shown in Figure 17.18 (Puecker and Douglas, 1975). As seen in this figure, the cells adjoining the cell of interest are encoded with a value of 1 for east, 2 for southeast, 4 for south, and so on, with each previous value multiplied by 2. The length of flow, which is greater for the diagonal cells than for the adjacent cells, is taken into account

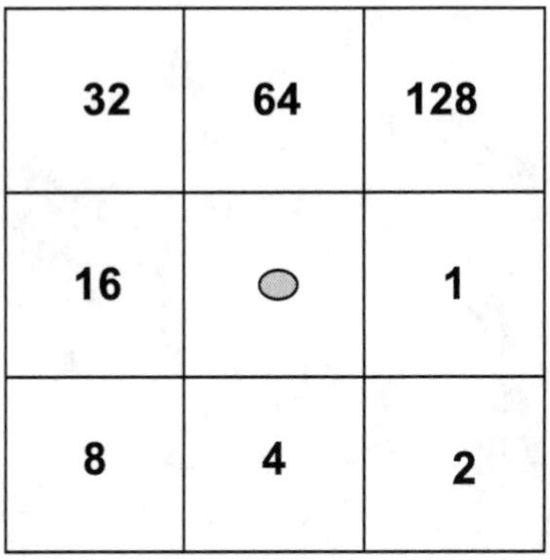

32	64	128
16	⬭	1
8	4	2

Figure 17.18 Eight-direction pour point model (modified from Puecker and Douglas, 1975).

in the computations. In flat areas where all adjacent and adjacent diagonal cells have the same elevation as the cell being processed, the analysis will extend to the next outer subset of cells. The direction of flow will then be determined for that grid cell, as shown in Figure 17.19. This continues for each cell in the watershed until the flow direction grid is determined, as shown in Figure 17.20 (DeBarry, 1996). Thus, every cell in the watershed will "know" which direction its water will flow. An actual flow direction grid for a watershed is shown in Figure 17.21.

Flow accumulation utilizes the precalculated flow direction grid and looks at all the cells above a particular cell that would flow into it, and accumulates and records the total number of cells, as shown in Figure 17.22. Essentially, the number of cells multiplied by the cell size provides the drainage area to the cell of interest. Flow accumulation then connects these cells to form a flow accumulation stream network, as shown in Figure 17.23.

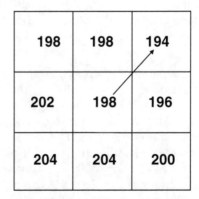

Figure 17.19 Flow direction grid cell analysis (modified from Maidment, 2002).

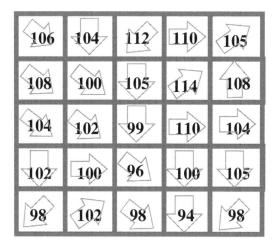

Figure 17.20   Typical flow direction grid (DeBarry, 1996).

Watershed delineation functions work essentially the same way, but in the reverse direction. The highest cell adjacent to the cell being analyzed is "flagged" as the ridge and the process continues until the watershed or subwatershed is delineated, as shown in Figure 17.24, with a three-dimensional portrayal of the process shown in Figure 17.25. Subwatershed boundaries can be determined at any point of interest through many of the hydrologic packages, such as the ESRI hydrologic functions in ArcGIS and ArcView (ESRI, 2003), WMS (Nelson, 2001), the GIS/HEC-1 Interface Module (DeBarry, 1996), and PrePro (Olivera, 1999). A watershed with many delineated subwatersheds can be seen in Figure 17.26.

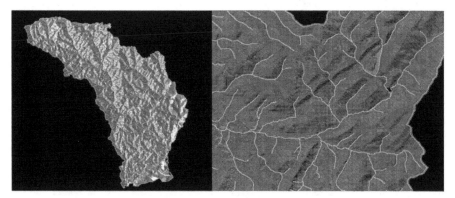

Figure 17.21   Watershed flow direction grid. (Developed by Paul DeBarry, courtesy of Borton-Lawson Engineering, 2003.)

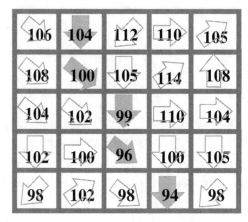

Figure 17.22 Flow accumulation grid (DeBarry, 1996).

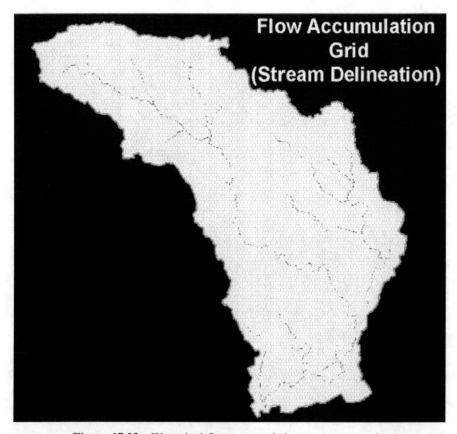

Figure 17.23 Watershed flow accumulation stream network.

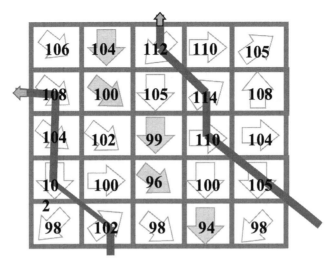

Figure 17.24 Watershed boundary delineation process.

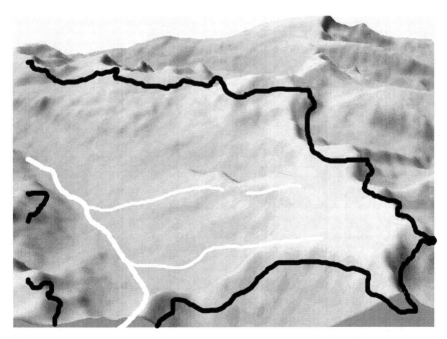

Figure 17.25 Finalized DEM-delineated watershed boundary and stream network (DeBarry, 1996).

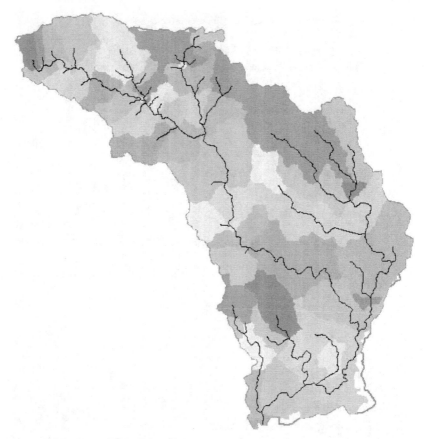

Figure 17.26 Watershed with DEM-delineated subwatersheds. (Developed by Paul DeBarry, courtesy of Borton-Lawson Engineering, 2003.)

It is important to note that raw DEM data often contain *sinks,* where the cell in question has the lowest value of any of its neighbors. These cells are important to locate so that potential recharge areas can be identified and preserved. However, cells are often identified as sinks due to the resolution of the DEM. In these cases, the sinks need to be filled, or raised, to at least the elevation of the lowest adjacent grid cell so that a *pour point* is formed. A pour point allows the accumulated water to flow, or pour, in a particular direction. Preprocessing algorithms to fill sinks have been developed by J. Garbrecht and L.W. Martz (1995) in the TOpographic PArameteriZation (TOPAZ) routines.

These hydrologic geoprocessing advances have been applied to develop required input requirements such as watershed *time-of-concentration, lag,* or *routing parameters* for models such as the HEC-1, HEC-HMS, PSRM, SWMM, and other hydrologic and water quality computer models, and for distributed modeling (DeBarry, 1990; DeBarry, 1996; DeBarry et al., 1999). A typical procedure is as follows:

1. Develop land use, hydrologic soil groups, and subwatershed boundaries in the GIS.
2. Determine subwatershed drainage areas.
3. Overlay the preceding layers to obtain composite NRCS curve numbers and/or percent impervious areas for each subwatershed.
4. Determine flow direction, flow accumulation, and channel segmentation for each reach in each subwatershed.
5. Develop times-of-concentration and resultant lag time for each subwatershed.
6. Load data into respective computer model.

This process has also been applied on an international level by the U.S. Geological Survey's (USGS) Earth Resources Observation Systems (EROS) Data Center to develop a national dataset, the HYDRO1k. HYDRO1k was developed from the USGS's recently released 30 arc-second (0.008333333333333 degrees or a resolution of 1 km) DEM of the world (GTOPO30). This geographic database provides comprehensive and consistent global coverage of topographically derived datasets. The available data, described previously, shown in Figure 17.27, can be downloaded from the USGS Web site, http://edcdaac.usgs.gov/gtopo30/hydro/namerica.html (USGS, 2003). These data are useful for large-scale hydrologic basin modeling or water balance computations (Maidment, 2003). However, for smaller watersheds, hydrologic digital processing on 10-meter DEMs, or more accurate data, is suggested, if available.

The American Society of Civil Engineers (ASCE) Task Committee on GIS Modules and Distributed Models of the Watershed completed a survey of the advances in this field. The committee was composed of members known in the field for development of GIS models and processes for hydrologic modeling. Many models—such as the GIS/HEC-1 interface model, WMS, CASC2D, TOPAZ, and others—have significantly advanced the field of hydrologic modeling for watershed management. A summary report of the Task Committee's findings was published in 1999 (DeBarry et al., 1999).

17.6 GIS AND FLOODPLAIN MANAGEMENT

GIS is now being used to map floodplains for floodplain management purposes. From its origins of digitizing floodplains from Flood Insurance Studies (FISs) on an as-needed basis to putting the floodplains into countywide coverage called Q3 format, GIS has recently taken on an even more important role in floodplain management: The U.S. Army Corps of Engineers Hydro-

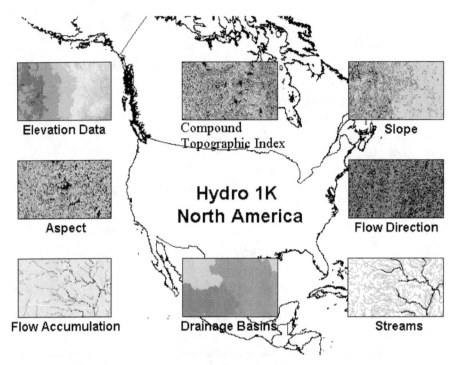

Figure 17.27   Available HYDRO1k hydrologic data (USGS, 2003).

logic Engineering Center has incorporated options in HEC-RAS to import GIS-generated stream networks and cross sections, analyze the data in 3-D perspectives, and export the cross sections with computed water surface elevations and floodplain boundaries as a set of polygons (HEC, 1997).

In another application of GIS for floodplain management, remote sensing data with sensors in the visible/near-infrared and Synthetic Aperture Radar (SAR) can be imported into the GIS and used for flood delineation and mapping based on vegetation or soils changes, monitoring, and flood damage assessment.

17.7 GIS AND GROUNDWATER MANAGEMENT

GIS can be utilized in a number of applications, from development of pollution vulnerability mapping based on physical feature overlaying (Reese and DeBarry, 1993) to actual GIS/modeling interfaces with such programs as MODFLOW. In the development of pollution vulnerability maps, features that would affect the vulnerability of groundwater to potential contamination can be overlaid. These features would include geology (fractures, limestone, etc.), soils (permeability), land slope, and streams (inflow or outflow). A pollution vulnerability map can then be utilized for siting new wells, developing emergency spill response plans, and planning transportation (by overlaying major commercial highways).

Another issue that should be considered in the development of watershed management plans in critical water supply areas is individual well recharge areas or zones of influence. These land areas can be mapped into the GIS from MODFLOW or related groundwater programs, and specific recharge evaluation criteria can be developed for these areas.

17.8 GIS AND WATER QUALITY

With an increased emphasis on managing the quality of stormwater runoff and controlling nonpoint source pollution through the use of best management practices (BMPs), GIS can serve as a valuable tool for watershed managers to analyze pollutant loads and target priority management areas. As early as 1991, GIS was used to compute nonpoint source pollution loads via the unit aerial loading approach based on the National Urban Runoff Program (NURP) and local area land use/loading data on a subwatershed basis (DeBarry, 1991). Those subwatersheds contributing the most pollution for existing and proposed conditions were then displayed graphically in the GIS to develop priority management areas, or those areas requiring BMPs. The GIS has recently been used in a similar approach whereby event mean concentrations (EMCs) were multiplied by the expected runoff to determine pollutant loads by grid (Quenzer and Maidment, 1998). Water quality models have also been programmed into ArcView (Quenzer and Maidment, 1998). The effort to develop total maximum daily loads (TMDLs) through the use of the EPA's Better Assessment Science Integration Point and Non-Point Sources (BASINS) computer program is also moving forward.

17.9 MANAGEMENT DISTRICT DETERMINATION

In order to properly manage the land and water resources in a watershed, the watershed should be divided into *management districts* based on *subwatershed* areas. Subwatershed areas, also known as *subareas,* are smaller watersheds of the watershed being studied. Several factors that should be considered when delineating subwatersheds include: potential subareas for stormwater management, ecoregions, physiographic provinces, geology, soil associations (or the newly established STATSGO data), locations of stream gages, problem areas, obstructions, dams, and lakes—hence, the development of many of the GIS data layers mentioned previously.

For stormwater management purposes, subareas are delineated based on *points of interest* determined during the study process. Points of interest are areas where a known or modeled flow is required, and are determined based on problem areas, obstructions, tributary confluences, stream gage locations, and/or dams. Subareas for stormwater management purposes typically average about 1 to 2 square miles, with variations depending on land use (urbanization) and the frequency of the factors mentioned previously. Subwatersheds for urban areas are typically smaller in size to better analyze the situation.

This may be too small for other management purposes, but it is recommended that the subareas be delineated for stormwater management and flow prediction first. Then the common subwatershed boundaries can be implemented to develop the larger management units. Watershed management districts are discussed in more detail in Chapter 18. To begin the process of determining management districts, the following data layers should be imported:

- Stormwater management subwatersheds (discussed in Chapter 18)
- Ordered streams and their respective watersheds
- Ecoregions
- Physiographic provinces
- Surface geology
- Soils associations
- Existing and future landcover/land use

Each layer should be projected to the proper coordinate system and overlaid into the GIS. One should begin developing the management areas by locating downstream points of subwatershed areas that coincide with physiographic province boundaries and ecosystem region boundaries, keeping in mind the accuracy of the data (see Chapter 10 on GIS data). Other factors that will then aid in determining management areas to achieve the goals and objectives of the plan are:

- Stream gage locations
- Stormwater and flooding problem areas
- Areas of disturbance
- Obstructions from the field survey
- Streambank erosion sites from the FGM assessment
- Water quality problems from the water quality assessment
- Stressed habitat locations from the GIS and field survey
- Groundwater divides/sources
- Tributary confluences
- Dams, lakes, wetlands, estuaries, bays, etc.
- Point and nonpoint sources of pollution
- Political influences

17.10 DISPLAYING THE RESULTS

Essential to the success of the watershed assessment and management plan is how well the information is displayed to the public. Again, a regurgitation of mapped data is not the solution. Trends, patterns, and overlaying results provide a basis from which to make decisions and develop planning documents.

17.11 SUMMARY

Applying GIS data layers in a watershed assessment should be a methodical, stepped process. The true advantage to using GIS is its analytical capabilities; displaying maps of the data is just an added benefit. Illustrating the interaction of the different data layers in relation to each other, such as in an opportunities and constraints map, is also a powerful feature of the GIS. The GIS allows detailed and efficient modeling capabilities, provides for long-term database management, and serves as a tool to portray ideas to the stakeholders of the watershed.

REFERENCES

Anderson, J.R., E.E. Hardy, J.T. Roach, and R.E. Witmer. "A Land Use and Land Cover Classification System for Use with Remote Sensor Data," U.S. Geological Survey, Professional Paper 964, Reston, VA, p. 28, 1976.

DeBarry, Paul A., and J. Carrington. "Computer Watersheds" in *Civil Engineering* (Washington, DC: American Society of Civil Engineers [A.S.C.E.]), June 1990.

DeBarry, Paul A. "GIS Applications in Storm Water Management and Nonpoint Source Pollution," Proceedings from the ASCE National Conference on Hydraulic Engineering, Nashville, TN, 1991.

———. "Results of a GIS/HEC-1 Interface Module," Water & Environment Congress, Anaheim, CA, June 1996.

DeBarry, Paul A., R.G. Quimpo, J. Garbrecht, T. Evans, L. Garcis, L. Johnson, J. Jorgeson, V. Krysanova, G. Leavesley, D. Maidment, E. James Nelson, F. Ogden, F. Olivera, T. Seybert, W. Sloan, D. Burrows, E. Engman, R. Binger, B. Evans, F. Theurer. "GIS Modules and Distributed Models of the Watershed," Report from the Task Committee on GIS Modules and Distributed Models of the Watershed, ASCE, Reston, VA, 1999.

DOI USGS. "Land Use and Land Cover and Associated Maps Fact Sheet" (Reston, VA: U.S. Geological Survey), 1991.

———. "Land Use Land Cover Digital Data from 1:250,000 and 1:100,000-Scale Maps," *Data User Guide 4* (Reston, VA: U.S. Geological Survey), 1986.

ESRI, Environmental Systems Research Institute, ArcGIS Software version 8.3, Environmental Systems Research Institute, Redlands, CA, 2004.

Garbrecht, J., and L.W. Martz. "TOPAZ: An Automated Digital Landscape Analysis Tool for Topographic Evaluation, Drainage Identification, Watershed Segmentation, and Subcatchment Parameterization: Overview," U.S. Department of Agriculture, Agricultural Research Service, ARS Publication NAWQL 95-1, October 1995.

Hydrologic Engineering Center (HEC). "HEC-RAS River Analysis System User's Manual" (CPD-68) (Davis, CA: Hydrologic Engineering Center), 1997.

Loelkes, G.L., Jr., G.E. Howard, Jr., E.L. Schwertz, Jr., P.D. Lampert, and S.W. Miller. "Land Use/Land Cover and Environmental Photointerpretation Keys" (Reston, VA: U.S. Geological Survey), 1983.

Maidment, David R., Ed. *Arc Hydro—GIS for Water Resources* (Redlands, CA: ESRI Press), 2002.

Maidment, David R., Francisco Olivera, Seann M. Reed, Zichuan Ye, Sandra Akmansoy, and Daene C. McKinney. "Water Balance of the Niger River Basin in West Africa," Presented at the 17th Annual ESRI User Conference, San Diego, CA, July 1997.

Mitchell, W.B., S.C. Guptill, K.E. Anderson, R.G. Fegeas, and C.A. Hallam. "GIRAS: A Geographic Information and Analysis System for Handling Land Use and Land Cover Data," U.S. Geological Survey, Professional Paper 1059, Reston, VA, p 16, 1977.

Nelson, James E., Norman L. Jones, and Christopher Smemoe. "From a Grid or Coverage to a Hydrograph: Unlocking Your GIS Data for Hydrologic Applications," ESRI Users' Conference, San Diego, CA, 1997.

Nelson, James E. "Hydrologic Analysis and Design with the Watershed Modeling System," National Highway Institute, training manual for course 35080, January 2001.

Olivera, Francisco, and David Maidment, "GIS Tools for HMS Modeling Support," Proceedings of the 1999 ESRI User Conference, San Diego, CA, July 1999.

Puecker, T.K., and D.H. Douglas. "Detection of Surface-Specific Points by Local Parallel Processing of Discrete Terrain Elevation Data," *Computer Graphics and Image Processing,* vol. 4, pp. 375–387, 1975.

Quenzer, Ann M., and David R. Maidment. "Constituent Loads and Water Quality in the Corpus Christi Bay System," Water Resources Engineering 1998, ASCE, Nashville, TN, 1998.

Reese, Geoffery, and Paul A. DeBarry. "The GIS Landscape for Wellhead Protection," Proceedings of the ASCE National Hydraulics Conference, San Francisco, CA, 1993.

U.S. Geological Survey (USGS), Land Processes Distributed Active Archive Center, HYDRO1k, http://edcdaac.usgs.gov/gtopo30/hydro/namerica.html, accessed March 16, 2003.

CHAPTER 18

STORMWATER MANAGEMENT ON A WATERSHED BASIS: REGIONAL STORMWATER MANAGEMENT

18.0 INTRODUCTION

As watersheds continue to become more urbanized, the amount of rainfall that runs off the paved areas, as opposed to naturally infiltrating into the ground, causes a number of stormwater-related problems, including: an increase in frequency, magnitude, and duration of flooding; accelerated erosion and resultant sedimentation of our streams; decreased water quality; and groundwater and stream base flow augmentation reduction. As evidenced by many of the older urban areas with limited hydraulic capacities of storm sewers, bridges, and channelized streams, unmanaged development upstream has caused severe environmental degradation and economic losses from flooding, as shown in Figure 18.1. In the past, flooding problems were controlled by constructing expensive dikes, levees, or channels. Research has shown, however, that these structures have limited design capacity, and as watersheds become developed, they too can't handle the runoff volume, and eventually overflow.

In order to properly manage stormwater runoff from new development, the comprehensive picture, or holistic approach—that is, analyzing the hydrology of an entire watershed—must be followed. An example of legislative foresight to manage stormwater runoff on a watershedwide basis can be found in Pennsylvania's Storm Water Management Act of 1978 (Act 167). After severe flooding on the Susquehanna River and its tributaries in 1972, caused by Hurricane Agnes, the Pennsylvania General Assembly wisely passed Act 167, with the purpose of developing comprehensive watershedwide management plans to prevent the pitfalls of uncontrolled stormwater runoff. Such comprehensive stormwater management plans can encourage groundwater recharge,

Figure 18.1 Two effects of unmanaged urbanization: stream bank erosion in Crum Creek, Pennsylvania, and flooding in Darby Creek, Pennsylvania.

prevent stream bank erosion and flooding, and resolve stormwater-related problems through the use of best management practices (BMPs), or nonstructural and structural land use measures designed to keep stormwater as close to its source as possible.

A comprehensive watershed stormwater management plan provides reasonable regulation of development activities to control accelerated runoff quantity, quality, and recharge promotion to protect the health, safety, and welfare of the public. The plan includes recognition of the various rules, regulations, and laws at the federal, state, county, and municipal levels. Once implemented, the plan will aid in reducing these adverse effects and costly flood damages by reducing the source and cause of local uncontrolled runoff. The plan will make municipalities and developers more aware of comprehensive planning in stormwater control and will help maintain the quality of the stream and its tributaries and sustain their designated use.

Although specific goals and objectives of watershed plans will vary by watershed, depending on location, urban conditions, existing problems, reasons for developing the plan, and so on, the general goal of all watershed plans should be to maintain the *hydrologic regime,* which will also aid in preventing flooding and stormwater-related problems and in maintaining high water quality. The natural hydrologic regime is altered by land development, among other factors, which increases impervious areas, which in turn increase runoff, reduce recharge, and decrease water quality. Successful management of stormwater runoff on a watershedwide basis depends on the ability to:

- Predict the effects of increased urban development on stormwater runoff.
- Define the response of the drainage system to a variety of storm events.
- Select the most cost-effective and optimum stormwater control system for a particular watershed.

This must be accomplished through a detailed hydrologic model. To that end, this chapter outlines various options that can be incorporated into a comprehensive watershed plan, thereby enabling the reader to choose the alternative that is best suited to the watershed in question.

18.1 URBANIZATION/LAND DEVELOPMENT

The factor that has the greatest impact on certain watersheds is land development, in particular the increase in impervious areas. Various aspects of the land development process need to be examined to fully appreciate the changes wrought by this land alteration.

Perhaps as detrimental to our nation's waterways as sediment is the amount of impervious area. Urbanization has detrimental effects on both water quantity and quality. The main problem of urbanization is the increase in impervious area. Increase in an impervious area increases runoff and decreases infiltration, which in turn decreases groundwater storage and stream base flow.

18.2 STORMWATER MANAGEMENT AND LAND DEVELOPMENT

The key to properly managing a watershed is proper stormwater management. Stormwater management can be applied to new development, or stormwater facilities in existing development can be retrofitted to obtain multiple objectives. The land development process typically starts with the planning phase, which is undertaken by a professional planning/engineering firm. The land development process enumerated here is typical, but will vary by local governing body and states:

1. *Sketch phase.* The roads, lots, parking areas, stormwater appurtenances, and so on are laid out on a plan.
2. *Preliminary plan phase.* Design of the land development takes place, incorporating road profiles; cross sections; lot design with bearing and distances; stormwater, water, and sewer appurtenances; and erosion and sedimentation measures.
3. *Final plan phase.* Plans are prepared for recording in the county courthouse for taxation purposes. Construction documents are prepared.

In between each phase, submissions are given to the governing body for review and approval.

Designs are made in conformance with local ordinances, which often dictate how roads, lots, and stormwater appurtenances are laid out. Many ordinances are out of date, however, originating in the 1960s or 1970s, with only minor revisions to comply with new research. Many ordinances, for example, still have requirements to remove the runoff from the property as quickly as possible via storm sewers or swales. Doing so has the net effect of decreasing times-of-concentrations, increasing peak flows, and providing direct conduits of nonpoint source pollution to receiving waters.

In addition, even those ordinances that do contain stormwater runoff requirements may also be outdated, and include the typical "reduce post-development peak flows to predevelopment peak flow rates" for a specified storm or storms. What is actually meant by predevelopment in these ordinances is existing conditions. In other words, the intent is not to reduce flows to the predevelopment or undeveloped condition, but to the exiting conditions of the site, which may include some developed portions. In order to maintain the natural hydrologic regime, however, predevelopment conditions would actually be preferable. This distinction should be spelled out in the ordinances.

Some ordinances, in an attempt to be conservative, include requirements to "reduce the peak 100-year postdevelopment flow to the peak 100-year predevelopment flow," ignoring storms smaller than the 100-year flow. This requirement would dictate detention basins that had outlet structures that would barely attenuate the storms of lesser frequency than the 100-year flow. For example, it may allow the postdevelopment 25-year storm to flow through the basin undetained, as the outlet structure was designed to limit outflow to the predevelopment 100-year flow.

The most common practice when applying stormwater management measures in a watershed is to require developers to install detention facilities on their particular site to control the postdevelopment runoff to predevelopment rates for a particular design storm or storms, typically the 5-, 10-, or 25-year storm frequency with a 24-hour duration. This is commonly referred to as *on-site detention*. However, as explained earlier in this book, placing this type of control in the wrong location in the watershed can actually *increase* peaks further downstream. Moreover, controlling only for, say, the 10-year storm, allows the 5-year storm to discharge an increase over predevelopment conditions, which may cause, aggravate, or accelerate problems further downstream. Uncoordinated on-site detention will reduce peak flows to predevelopment flows from the site, but doing so has many disadvantages, hence is not recommended for all the reasons listed previously and later in this chapter.

The design process, which, as stated, should begin in the land development plan sketch phase, often is where the original mistakes are made for improper stormwater management. A designer's clients are typically the developers, who want to maximize their economic rate of return on their property. (A typical land development process, and the recommended land development process that incorporates conservation design, can be found in Chapter 19.)

To better understand why other alternatives for nonstructural (conservation design) stormwater management should be considered, this chapter takes a closer look at how a detention basin works. The basin is designed to have an outlet structure and volume to reduce the postdevelopment flow. The result is an outflow hydrograph that is attenuated in time, as shown in Figure 18.2. This attenuation may actually increase flows in the main tributary or stem if

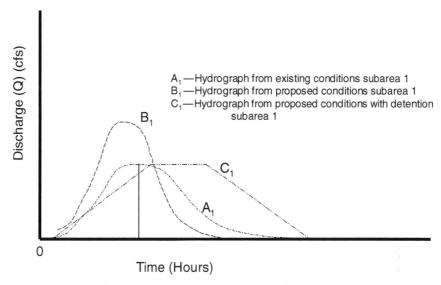

Figure 18.2   Attenuated outflow hydrograph from a detention basin.

the attenuated outflow from the basin corresponds to the time-to-peak of the watershed, as shown in Figure 18.3. Though the site hydrographs are exaggerated in the figures—one would not locate a regional basin where this would occur—most detention facilities are located on private development sites, and the site hydrograph could also represent an accumulated hydrograph from many sites in the vicinity.

This process can be analogized to a bathtub, with the brim of the bathtub as the watershed boundary. If one were to take two cups of water and simultaneously empty one near the drain (representing the watershed outlet or drain) and the other at the upper portion of the tub (representing the headwaters), the water from the cup emptied over the drain would be down the drain before the headwaters reached the drain. But, if one were to plug the drain halfway, representing the attenuation in a detention facility, and did the same thing, the "detention" of the lower water would hold back the water. The headwaters would then merge with the detained water, raising the elevation of the water level in the bathtub, or the elevation of streams in the watershed.

This situation, as with the bathtub, is typical of the lower portions of the watershed. To prevent flooding, the goal is to manage on-site flooding by keeping separate the hydrographs from the various portions of the watershed. Timing of the tributaries is of primary importance in achieving this. The timing of the general areas of a watershed can be found in Figure 18.4. As is apparent, the central portion of the watershed typically contributes the most to the peak of the overall watershed. The lower portion drains quickly out of the watershed "system" while, due to floodplain and channel attenuation, the

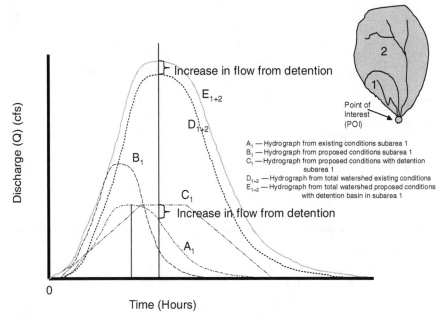

A₁ — Hydrograph from existing conditions subarea 1
B₁ — Hydrograph from proposed conditions subarea 1
C₁ — Hydrograph from proposed conditions with detention subarea 1
D₁₊₂ — Hydrograph from total watershed existing conditions
E₁₊₂ — Hydrograph from total watershed proposed conditions with detention basin in subarea 1

Figure 18.3 Effects on watershed from detention facility placed in the wrong location.

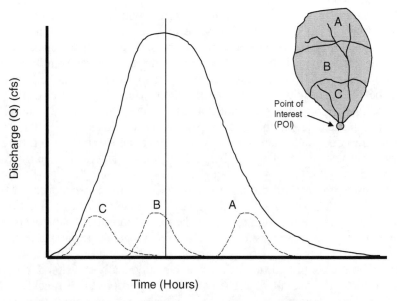

Figure 18.4 General timing of various portions of the watershed.

headwaters typically take a while to reach the outlet point. A variety of watershed factors will have an impact on this timing, including watershed shape, average stream slope, headwater versus outlet slope, floodplain storage, and others.

Several methods of stormwater management could be implemented to remedy this situation, each with its advantages and disadvantages. These methods are described in the following sections.

18.3 WATERSHED-LEVEL STORMWATER RUNOFF CONTROL MANAGEMENT PHILOSOPHY

Stormwater management requirements should be tailored to the particular area of the watershed. Standards set forth in a watershed plan should vary depending on the location of the watershed, which is influenced by problem areas, existing flood control facility capacities, and watershed tributary timing and existing flood control facilities such as flood control dams. This text divides these variable standards into three categories: the *release rate concept,* the *variable stormwater management district concept,* and the *distributed storage concept,* each of which is discussed a little later in the chapter.

18.4 CONVENTIONAL ON-SITE RUNOFF CONTROL VERSUS WATERSHED LEVEL RUNOFF CONTROL

An increase in development and, in turn, in impervious surfaces not only increases runoff peaks but also runoff volume. Figure 18.5 shows hydrographs for subwatershed 1 alone (hydrograph A_1) and for the total watershed or subwatersheds 1 and 2 (hydrograph D_{1+2}) under existing conditions. As the figure illustrates, with post-development conditions for hypothetical subwatershed 1, the runoff peak increases from 300 to 500 cubic feet per second (cfs) over existing conditions. Also, the total volume of runoff, or the area under the hydrographs, hypothetically would increase from 345 to 506 acre-feet. Under standard detention basin systems, this difference in runoff volume is passed through the basin, although lagged or detained, and is lost infiltration. This increase in volume of flow is what causes problems downstream.

The primary difference between the on-site runoff control philosophy and the watershed-level philosophy is the manner in which runoff volume is managed. The conventional on-site control philosophy has as its goal control of the runoff peak from the site. Although numerous volume controls can be implemented on-site—such as infiltration basins, porous pavement, and so on—typically it is done to control the runoff peak. Any volume control provided by these measures would be an added benefit. Only under very unusual circumstances (e.g., a very small development) could the total volume of runoff be kept at the level of existing conditions. In contrast, the proposed

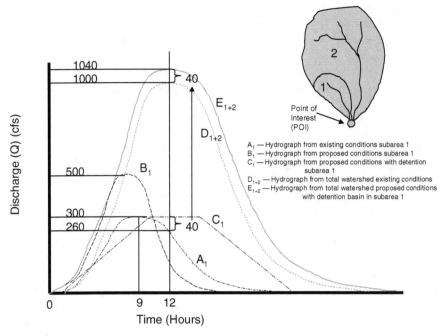

Figure 18.5 Increase in watershed hydrograph from detention facility placed in the wrong location.

watershed-level runoff control philosophy seeks to manage the increase in runoff volumes such that the peak rates of runoff throughout the watershed are not increased. The basic goal is, therefore, the same for both on-site and watershed-level philosophies; again, it is the means by which this is achieved that are different. The results can be very different, as illustrated in Figure 18.5.

By reducing the postdevelopment peak for watershed 1 (hydrograph B_1) to the existing condition peak (hydrograph A_1) using conventional on-site detention, the detained peak flow (hydrograph C_1) is extended for a period of time due to the increase in runoff volume. This lagged peak could then contribute to the peak of the entire watershed, subwatersheds 1 and 2 in Figure 18.5.

The total volume of runoff for the area under the hydrographs remains the same for hydrographs B and C, or the post-development flow with and without detention. As can be seen from Figure 18.5, the peak flow for the predevelopment hydrograph is 300 cfs. The peak flow for the postdevelopment conditions without detention and with detention at hour 12 is 500 and 300 cfs, respectively. It can also be seen that the amount of flow from the predevelopment hydrograph contributing to the total watershed peak at hour 12 is 260 cfs. By providing detention in the wrong location, and detaining the outflow of the basin, the outflow of the basin coincides with the peak of the total watershed. The basin is actually allowing 40 more cfs out of the basin at the time-to-peak, or hour 12. This would have the effect of increasing the peak of the overall watershed by that same 40 cfs, as shown in hydrograh E_{1+2}. Thus, supplying detention to reduce postdevelopment runoff rates to predevelopment rates actually could increase the peak in other areas of the watershed.

Through this example, it is clear that the watershed-level stormwater management approach, which manages water volumes, would be a more practical management strategy than typical on-site detention.

18.5 RELEASE RATE PERCENTAGE CONCEPT

One strategy developed to regulate the rate of release from a detention facility while considering the timing and effect of this release on the flows in the watershed of which the detention facility is located and on the hydrologic response of the entire watershed is the *release rate percentage concept* (Cherry et al., 1982). The allowable release or outflow from a subwatershed and, in turn, a developer's detention facility, would be the ratio of runoff contributing to a point of interest (POI) at that point's time-of-peak to the maximum runoff rate for predevelopment conditions. For instance, as Figure 18.5 showed, the runoff from subwatershed 1 (existing conditions) contributing to the point of interest at hour 12 is 260 cfs. The maximum rate of runoff for existing conditions is 300 cfs at hour 9. Therefore, the allowable release rate for subwatershed 1 and any detention basin installed within subwatershed 1 would be $260 \div 300 = 87$ percent. A developer required to install a detention basin would, therefore, have to limit the outflow from the detention basin to 87 percent of the peak value obtained for predevelopment conditions to be consistent with the plan. Conventional on-site detention philosophy would detain postdevelopment runoff flows to 100 percent of predevelopment flows. The release rate is described in equation form by:

Release Rate

$$= \frac{\text{Subwatershed predevelopment contribution to point-of-interest peak}}{\text{Subwatershed predevelopment peak flow}}$$

(18.1)

A hypothetical example of how release rates can be developed for a watershed is as follows:

1. A 12-square-mile watershed is divided into 12 subwatersheds averaging 1 square mile in size, as shown in Figure 18.6. Subwatersheds with more intense land use are typically smaller in size than more undeveloped watersheds. Subwatersheds are selected based on stream conflu-

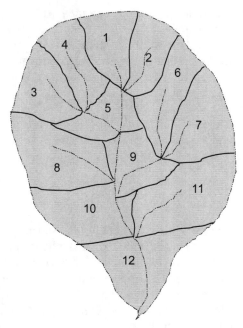

Figure 18.6 Hypothetical watershed divided into 12 subwatersheds.

ences, problem areas, dams and obstructions identified in the watershed assessment, and physical characteristics.

2. The watershed is then modeled using an appropriate hydrologic computer model. The Penn State Runoff Model (PSRM) (Aron and Lakatos, 1986) has a release rate analysis routine that supplies peak flow presentation tables and release rate tables, which will be described later. The initial concept for stormwater management is to ensure that future peak flows do not increase at key points of interest. By selecting particular points of interest, some areas in the watershed are left with no stormwater control. Therefore, the best way to provide comprehensive stormwater management would be to treat the outlet of every subwatershed as a point of interest. Using this technique, peak flows should not increase at any point in the watershed in the future.

3. Overall watershed runoff values and the interaction of these flows are analyzed for several duration and frequency storms utilizing peak flow presentation tables (supplied in the PSRM output).

The model can be run for the 1.5-, 2-, 2.33-, 5-, 10-, 25-, 50-, and 100-year storms. Hydrograph timing and peak flow contributions can then be plotted and developed for all subwatersheds delineated, as shown in Figure 18.7. This is a time-consuming task, but one that can be automated from

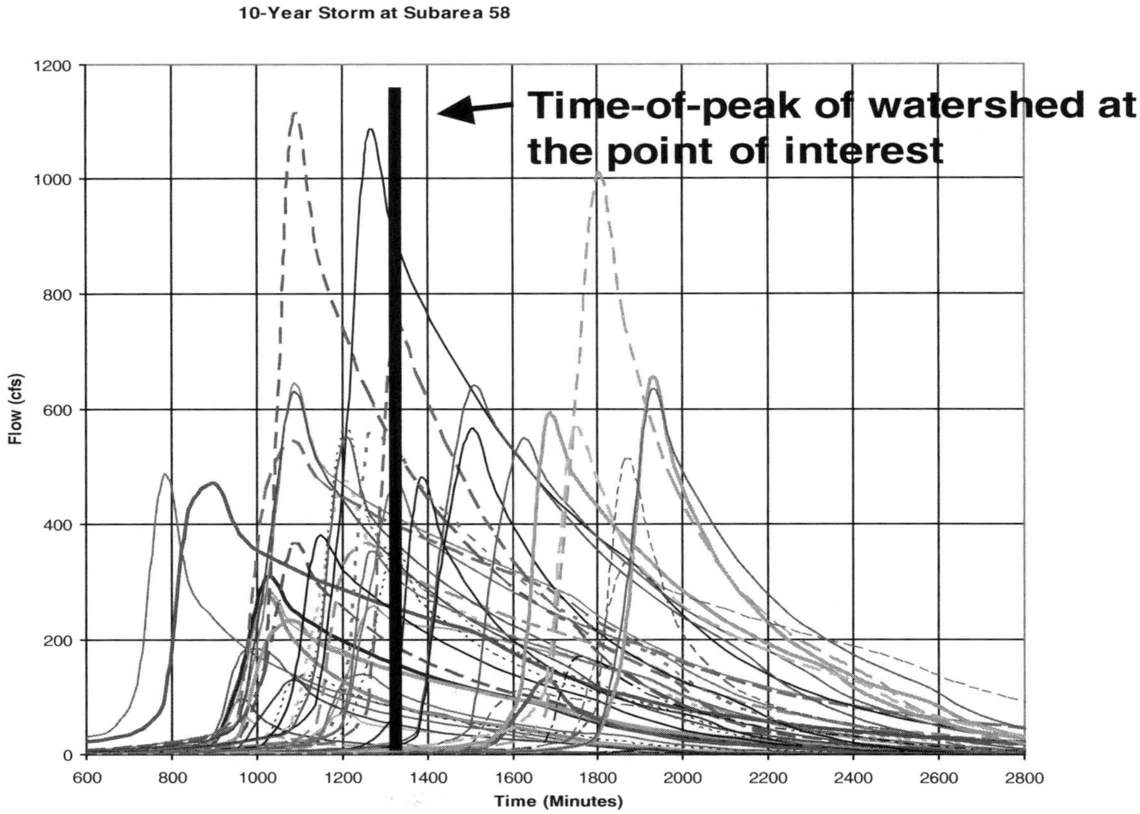

10-Year Storm at Subarea 58

Figure 18.7 Routed hydrograph's contributions to a particular point of interest.

509

output from most hydrologic computer models. The respective release rates can be automatically tabulated by PSRM for all subwatersheds and their relationship to all other points in the watershed. The most restrictive storm's release rate is then established as a release rate adapted for that particular subwatershed. The most restrictive release rate would be that which is calculated at the most critical point in a subwatershed flow path.

The release rate computation can be explained through the next series of tables. Table 18.1 summarizes the peak flows by subwatershed (subarea) and total contributing watershed areas. For instance, subarea 10 peaks at time 1355 with 129 cfs.

Table 18.2 presents the peak flow for subarea 12. Each upstream subarea would have its own peak flow presentation table to each downstream subarea it contributed to. This displays a portion of each subwatershed's hydrographs in relation to the peak of the entire watershed at the bottom of subarea 12, or POI 12. (Note that the peaks in Table 18.2 for each subarea may not exactly correspond to the peak in Table 18.1 due to the computer program's ability to display only at certain time increments.) The program keeps track of the actual peaks. The release rate is computed by the ratio of the flow of the individual subarea contribution (117 cfs) at the peak of the overall hydrograph (time 1380) to the total peak of the individual subarea (129) from Table 18.1. Subarea 10 would have a release rate of 117 ÷ 129, or 91 percent.

Release rates are computed for each subarea in the same manner at every point of interest or subarea for which it flows into. The results can be displayed in Table 18.3.

As can be seen from Table 18.3, subarea 10 has a more restricted release rate at subarea 11 than at subarea 12. The minimum reasonable release rate is, therefore, displayed in the last row of Table 18.3. A release rate map of

TABLE 18.1 Peak Outflow Summary

Subarea Number	Subarea Peak Runoff: cfs	Subarea Time of Peak: min.	Runoff Volume: cfs-hrs	Combined Peak Outflow: cfs	Watershed Time of Peak: min.
1	128	1410	445	128	1410
2	239	1350	500	239	1350
3	2	1350	2	332	1350
4	113	1355	337	380	1380
5	266	1350	716	266	1350
6	33	1350	44	581	1380
7	72	1350	188	624	1380
8	303	1350	715	749	1380
9	380	1355	1182	380	1355
10	129	1355	373	129	1355
11	2	1350	2	1158	1380
12	285	1350	604	1256	1380

TABLE 18.2 Peak Flow Presentation for Subarea 12

Sub Area	Travel Time (min.)	Combined Peak Flow Subarea 12 (cfs)											Release Rate
		50	57	68	87	146	797	Peak 1260	1256	1181	1103	1025	
							Time (min.)						
		1200	1230	1260	1290	1320	1350	1380	1410	1440	1470	1500	
12	0	11	12	15	21	41	285	138	110	98	89	83	48
11	18.5	0	0	0	0	0	0	2	0	0	0	0	62
10	20.7	3	3	3	6	11	50	117	92	86	81	78	91
9	20.1	8	9	11	14	24	146	348	306	294	282	269	92
8	18.5	8	9	12	15	27	132	257	162	143	132	122	85
7	27.3	2	2	2	2	3	14	69	50	47	42	39	95
6	39	0	0	0	0	0	0	23	14	5	3	2	100
5	42.2	5	6	6	8	12	24	170	221	177	161	147	100
4	40	2	2	2	3	5	9	78	98	86	81	75	100
3	55	0	0	0	0	0	0	0	2	0	0	0	100
2	57.1	6	6	8	9	12	18	53	227	111	90	81	100
1	56	0	0	0	0	2	2	20	111	126	128	122	100

TABLE 18.3   Release Rate (RR)

		Source											
		1	2	3	4	5	6	7	8	9	10	11	12
	1	100	0	0	0	0	0	0	0	0	0	0	0
	2	0	100	0	0	0	0	0	0	0	0	0	0
	3	100	100	100	0	0	0	0	0	0	0	0	0
	4	100	77	50	80	0	0	0	0	0	0	0	0
	5	0	0	0	0	100	0	0	0	0	0	0	0
Destination	6	100	79	53	81	74	18	0	0	0	0	0	0
	7	100	100	92	88	85	48	69	0	0	0	0	0
	8	100	100	100	93	93	69	76	56	0	0	0	0
	9	0	0	0	0	0	0	0	0	100	0	0	0
	10	0	0	0	0	0	0	0	0	0	100	0	0
	11	100	100	100	93	94	71	77	58	83	75	0	0
	12	100	100	100	100	100	100	95	85	92	91	62	48
	Min RR	100	77	50	80	74	18	69	56	83	75	62	48

the watershed could then be displayed as shown in Figure 18.8. It is important to note that the watershed in this figure is hypothetical, hence the release rates shown in Table 18.3 don't exactly correlate for this example.

The release rate concept provides an effective and flexible tool for comprehensive watershed stormwater management. In order for it to work, how-

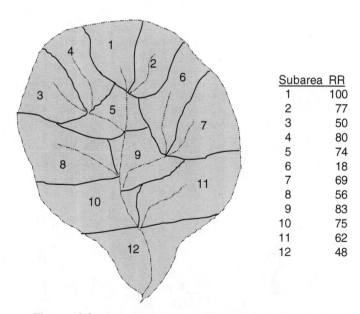

Subarea	RR
1	100
2	77
3	50
4	80
5	74
6	18
7	69
8	56
9	83
10	75
11	62
12	48

Figure 18.8   Release rate map of hypothetical watershed.

ever, it must be used for *all* development within a subwatershed that would result in an increase in the postdevelopment runoff rate. Using this approach, a developer can evaluate which stormwater control measures are most appropriate for each particular site. The methodology would be as follows:

1. Locate the subwatershed for which the site is located and determine the applicable release rate.
2. Calculate the predevelopment and postdevelopment runoff for the site using one of the methods and the parameters specified in this plan for the design storms specified. If the postdevelopment runoff peak is greater than the predevelopment runoff peak, stormwater management techniques will be required. Stormwater management techniques increase infiltration and reduce runoff.
3. Recalculate postdevelopment runoff with stormwater management techniques incorporated using the same method as in step 1. If the postdevelopment runoff is still greater than the predevelopment runoff, stormwater detention will be required.
4. Using the allowable release rate percentage indicated for the subwatershed in which the site is located, calculate the allowable peak outflow or release rate by multiplying the peak obtained for existing conditions by the release rate percentage.

To apply this to a development scenario, if a developer were to propose a development in subarea 10, he or she would be required to reduce the postdevelopment flow to 75 percent of the predevelopment flow for the required design storm. If the design storm were the 10-year storm, and the predevelopment 10-year peak flow from the site were 100 cfs, the developer would have to reduce the postdevelopment conditions flow from the site to 75 cfs.

It should be noted that a release rate for a subarea to its own point of interest would be 100 percent. Also, those subwatersheds typically located in the headwaters, whose routed peak rate of flow occurs after the peak of the watershed, might not require a release rate, because the peak rate of flow is already timed after the watershed's peak.

18.6 MINIMUM REASONABLE RELEASE RATE DETERMINATION

Calculations of the release rates yielded some fairly low release rates, typically for subareas in the lower portions of the watershed. These subwatersheds exhibit peak flows that occur substantially before the peak flow of the total watershed and contribute only a small amount from the falling limb of the hydrograph to the total watershed peak. Supplying detention to meet a required release rate of 10 or 20 would be economically infeasible and not justifiable. It might, in fact, be more advantageous to provide no detention and allow the runoff to "escape" the watershed before the watershed peaks—

as long as the resulting quantity, velocity, and direction of stormwater runoff otherwise adequately protects health and property from possible injury. Also, it might be advantageous to establish a minimum release rate that would balance costs and still meet the comprehensive stormwater management intent. As release rates decline below 50 percent, the cost-effectiveness of applying a specific release rate diminishes. Conversely, as release rates increase from 50 percent, allowing less control for certain subwatersheds increases the chance of creating problems downstream.

The exact cutoff point to provide the proper balance between economics and proper control cannot be identified, although previous studies (Lehigh Valley Planning Commission, 1987) have concluded that the minimum release rate to use as a breakpoint would be this 50 percent point. This is due to the fact that it is the median value, and hence represents a direct compromise between runoff control cost and absolute control effectiveness; it does not preclude the no-increase goal as documented by the watershed modeling analysis. Establishment of this minimum release rate indicates that subwatersheds with calculated release rates of less than 50 could possibly contribute to the peak rate of runoff at some point in the watershed; however, taking the following factors into consideration, the contribution can be termed insignificant according to the plan:

- The release rate established for each subwatershed is the minimum generated for the return periods evaluated. In many subwatersheds, the calculated release rate was greater than 50 for certain design storms.
- Those subwatersheds that peaked significantly before the entire watershed was evaluated and were deemed able to use "no detention."
- These options exist for the subbasins with release rates below 50: no detention, a 50 percent release rate, or the calculated release rate. Those subwatersheds where no detention was required were indicated, which leaves the 50 percent or calculated release rate options remaining. The plan could not economically justify requiring a control more stringent the 50 percent. In many instances, the calculated release rate was not significantly less than 50.

18.7 VARIABLE STORMWATER MANAGEMENT DISTRICTS

The release rate or percentage of predevelopment flow for a particular subarea can be equated to a design storm. For instance, if the 10-year predevelopment flow for a subwatershed were 100 cfs, and the release rate 75 percent, the allowable flow would be 75 cfs. This may equate to, for example, the 5-year predevelopment flow. Therefore, in developing performance standards, to require the 10-year postdevelopment flow to be reduced, to the 5-year predevelopment (or existing conditions, however defined) flow would accomplish the same goal as the release rate.

This can be illustrated in Figures 18.9 and 18.10. The amount of flow that the subarea analyzed as contributing to the point of interest is equal to the five-year predevelopment flow (Figure 18.9). Therefore, the 10-year postdevelopment flow would be required to be reduced to the 5-year predevelopment flow, as shown in Figure 18.10.

The variable stormwater management district concept classifies the watershed into three management districts:

Standard detention district (district A)

Alternate storm district (district B)

Water quality and recharge only district (district C)

Each of these districts is described in turn in this section.

The three districts are developed using the same timing concepts as the release rate process, based on tributary timing, and apply to location in the watershed. As shown earlier in Figure 18.4, the runoff generated from the subwatersheds in the upper portion of the watershed, or the headwaters, takes a while to be routed through the floodplains and reach the mouth, whereas the lower portion of the watershed drains quickly and exits the system before the watershed peaks. These three portions of the watershed can be categorized

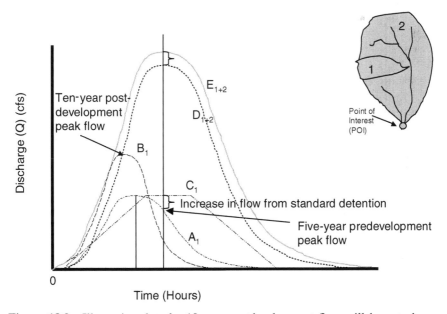

Figure 18.9 Illustration that the 10-year postdevelopment flow will have to be reduced to the 5-year postdevelopment flow to avoid increasing flows at the point of interest.

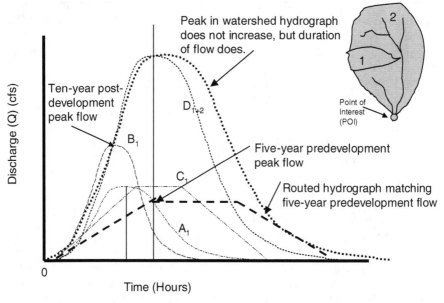

Figure 18.10   Reducing the 10-year flow to the 5-year flow will keep the peak flow of the point of interest at existing conditions, but will increase the duration of flow.

into the three management districts for water quantity control, based on where the subwatershed peak flow occurs, as shown in Figure 18.11. These determinations can be accomplished only through detailed hydrologic modeling and knowledge of all the variables that might affect the results.

The *standard detention district* is basically the old concept of limiting the postdevelopment runoff to the predevelopment rate for a particular design storm frequency. Standard detention may be applicable to certain portions of the watershed, which, again, can be determined only through detailed hydrologic modeling of the watershed. In general, however, it can be said that these standards generally would apply to the upper portion of the watershed, or headwaters, and might be applicable to certain other areas. They should be applied only after evaluating the timing of the watershed. The design storm must be determined by evaluating downstream effects of such criteria and by taking into account frequency of problems and capacities of existing downstream culverts, bridges, or obstructions of existing flood control projects. In all cases it is recommended that some type of water quality criteria be applied, while assuring they do not affect the water quantity goals. The post to pre criteria may be applied to the entire range of storm frequencies—that is, 1-, 2-, 5-, 10-, 25-, 50-, and 100-year storms—or may be applied only to several of these, again depending on the factors previously mentioned. In theory, the developer should be required to limit his or her postdevelopment runoff to the predevelopment rate for all the design storms. However, if there are no

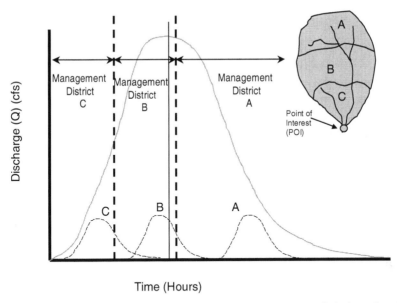

Figure 18.11   Stormwater management districts based on general timing of various portions of the watershed.

constraints downstream, this standard could be relaxed to include only certain storms to be controlled.

When evaluating the hydrology on a watershedwide basis, taking into account channel storage, timing, and various impoundments, the *alternate storm district concept* may be justified. The alternate storm district concept as applied here refers to limiting the postdevelopment runoff from a storm frequency to a predevelopment runoff of a lesser storm frequency; for example, limiting the 25-year postdevelopment flow to a 10-year predevelopment flow, as described previously. This requires establishing criteria for the various subwatersheds within a watershed, where these should be applied. Another common application of the alternate storm concept, particularly for water quality, is limiting the two-year postdevelopment flow to the one-year predevelopment discharge. Comparing the timing of various discharges in relation to the entire watershed, as described in the discussion of the release rate concept, the alternate storm concept accomplishes the same results as the release rate concept by separating peaks from various subareas, that is, preventing them from coinciding to maintain existing flow conditions throughout the watershed.

Applying the alternate storm concept to a particular site does not cause a significant increase in storage volume required for a developer, yet accomplishes comprehensive watershed planning for flood control and stormwater management. Applying the standard detention concept does not accomplish these goals; it reduces peaks yet does not increase volumes required tremendously. Therefore, the variable rate concept is justifiable in areas of the watershed where the concept accomplishes the goals of maintaining existing flows and not increasing flows in stream segments as the watershed becomes developed.

As we have seen, the lower portion of the watershed, or District C, or near the drain of the bathtub, typically drains before the watershed peaks. It is, therefore, advisable not to hold back this water through the use of detention facilities, but to allow it to drain freely out of the watershed system. However, this would not preclude the need for water quality and groundwater recharge management requirements and benefits. Therefore, stormwater management measures in those areas would manage the smaller storms and allow the larger storms to flow undetained to the receiving stream.

Hydrographs showing the timing of the individual subareas and the contribution to the peak at a particular point of interest (subarea 110), and how they are classified into the management districts, are shown in Figure 18.12. The analysis is performed for many key points of interest in the watershed to ensure, theoretically, no increase in stream flows anywhere in the watershed. Figure 18.13 shows how the watershed is divided into the three management districts based on the subarea flow contributions. Figure 18.14 shows the criteria that were developed, or the requirements that the developer would have to follow when proposing a development within the watershed. Every watershed responds differently, and the criteria will vary based on the modeling results. Often, more than the three districts area required.

It should be noted that although most developers utilize detention basins to accomplish these criteria, alternate and innovative design methods, such as conservation design, low-impact development, infiltration, and so on, may be utilized to accomplish these goals.

As explained, locating individual site detention basins in the wrong location may actually increase peak flow rates in the stream, if the timing of the release of the basin corresponds to the time of peak of the watershed. Therefore, detention basins may not be appropriately located in these "water quality and recharge only" districts, but should be placed in the most hydraulically advantageous location in the watershed. Simply defined, the *distributed storage concept* is the process of utilizing the hydraulically most suitable areas or sites for regional or subregional detention facilities. Distributed storage may "distribute" the storage between on-site detention facilities and regional facilities.

This concept requires coordination among developers and local and county officials, and is best implemented when a "watershed authority" or similar governing agency is well established. Often, a detention basin is placed on each individual subdivision, as shown in Figure 18.15. The disadvantage to this method is that the cumulative effect of the basins is difficult to determine. Instead of each developer placing a basin on each site, under an organized

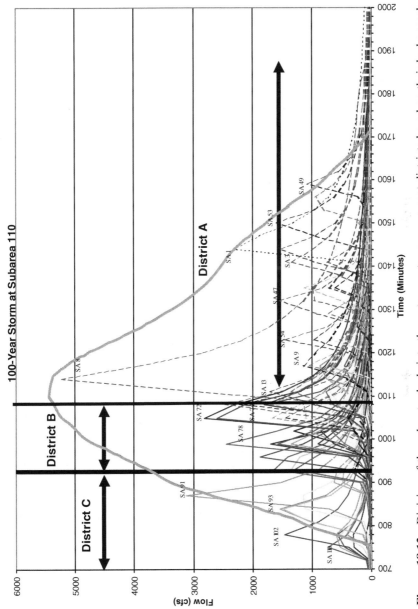

Figure 18.12 Division of the subwatersheds into three stormwater management districts based on their hydrograph timing.

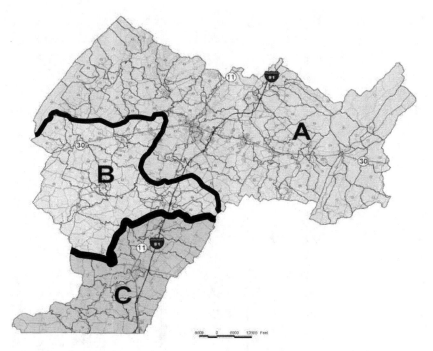

Figure 18.13 Watershed with three management districts. The headwaters are in district A, in the northeast portion of the watershed. The mouth is at the south-central portion of District C. (*Source:* Franklin County Planning Commission, 2003.)

stormwater management program developers could pool their capital resources and build a subregional basin in the location that will most benefit or have the greatest hydrologic impact, as shown in Figure 18.16.

The third option would be to construct a regional basin in the most hydrologically advantageous location, as shown in Figure 18.17, where upstream and possibly downstream developers would not be required to install stormwater quantity controls. They would, however, still have to implement water quality and infiltration BMPs. Regional facilities are those that would detain a large drainage area or a large enough portion of the watershed to have an impact on the resultant hydrograph.

Locations for regional basins should be evaluated in any watershed assessment and/or stormwater management plan through hydrologic modeling. They should be placed in the most hydrologically advantageous position, that is, where they will have the greatest impact on "dampening," or separating, the hydrograph from the peak on the stream and the other tributaries contributing downstream of the site. Several parameters should be considered when evaluating regional basin locations:

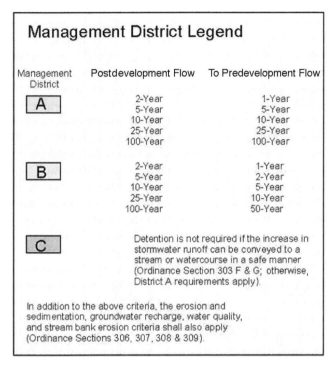

Figure 18.14   Stormwater management criteria required to maintain peak flows in the watershed to existing conditions levels. (*Source:* Franklin County Planning Commission, 2003).

- Site location's influence on the total watershed hydrology
- Available undeveloped land
- Ownership of the land
- Topography
- Avoidance of environmentally sensitive locations (wetlands)
- Total drainage area and percent of total contributing area to the basin location
- Absence of environmental degradation

The final selection of which storage area to construct should be made by assessing the advantages and disadvantages of the identified storage locations. Existing dams could also act as regional basins if hydrograph attenuation is required in this area of the watershed. Dam heights could be extended to provide additional storage volume, or spillways could be modified to reduce outflow, to achieve the desired results.

Regional basins are not popular among environmentalists because, if improperly constructed, they can be detrimental to the environment and down-

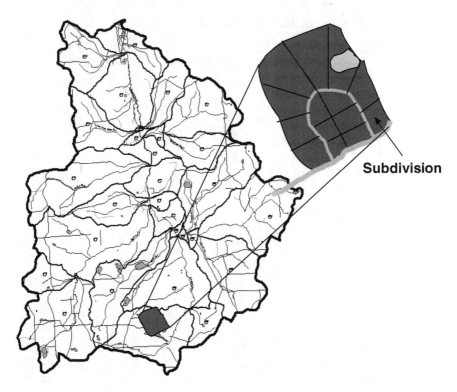

Figure 18.15   On-site, detention facility locations.

stream reaches. However, when properly constructed, they can provide benefits to humans, streams, ecosystems, water supplies, and so on. For example, if a regional facility were designed not to channel or detain any storms less than a 1½-year storm, or what a study indicated was the bankfull capacity, then the downstream environmental detriments usually associated with dams could be minimized, if not eliminated. The facility could be designed to detain storms larger than the 1½-year storm and prevent downstream flooding and stormwater problems.

Wetlands are often associated with the best locations to build basins. Because "on-line" construction of dams, or those that traverse the stream, do cause riparian loss and wetland disturbance, if present, another option to minimize these impacts would be to construct bypass basins, whereby only the flows associated with the larger storms that would cause downstream damage—that is, the 50- or 100-year storms—would be bypassed into an adjacent basin to store the excess flows and attenuate downstream damage-causing flows, as shown in Figure 18.18. The actual storms to divert would depend on the frequency of flooding downstream. This is one area where it might be

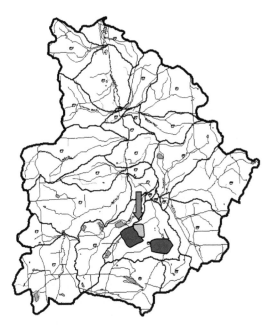

Figure 18.16   Subregional, combined site, or shared detention facility location.

possible to achieve the best of both worlds, hence should be an area to target research.

The combination of on-site detention and regional basins significantly improves the capability of developers and communities to manage stormwater on a watershed basis. The goal would be the development and use of cost-effective and environmentally sensitive stormwater runoff controls. Factors such as capital costs, operation and maintenance (O&M) costs, environmental protection and enhancement, public convenience, and other community development objectives make a strong case for regional facilities. For instance, recreational areas can be utilized for temporary storage of stormwater, or a regional "wet" storage facility can be used for recreation. Construction and operation and maintenance are often financed through stormwater utilities, which can collect fees, typically based on the amount of impervious area of a site.

18.8   SUMMARY

All water resources in a watershed are related to the hydrologic cycle. Natural watersheds are in equilibrium. As watersheds become developed or urbanized,

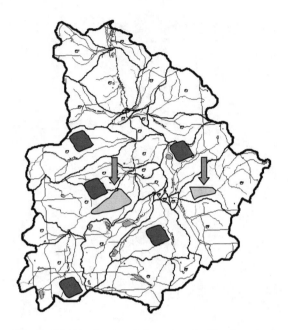

Figure 18.17 Regional detention facility location.

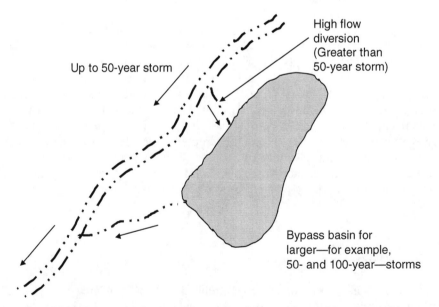

Figure 18.18 Bypass basin to mitigate flooding problems downstream in larger storms.

or are influenced by other human changes, this equilibrium becomes unbalanced. The result can be increased stormwater and flooding problems; diminished stream base flow and water supplies; accelerated stream bank erosion; and resultant deposition, aquatic stress, and loss of critical habitat or other areas that should be preserved. The increase in impervious area increases stormwater runoff. Stormwater runoff and streams do not follow political boundaries, but physical features of the watershed; therefore stormwater must be managed on a watershed level and must be an integral—if not the most important—part of any watershed assessment and plan.

Regulation of stormwater runoff can be achieved through stormwater management ordinances. Many ordinances, when referring to predevelopment conditions as the goal to match flows, are actually referring to existing conditions. In the explanations in this chapter, the goals of reducing proposed conditions flows to existing conditions flows was used. In actuality, to best achieve the natural hydrologic regime of a watershed, matching the predevelopment or natural conditions runoff rates and volumes should be the goal at proposed development sites, not just to match existing conditions (which may be partially developed). Ordinances should make this distinction. Predevelopment conditions will be utilized as the goal in subsequent chapters of this book.

REFERENCES

Aron, Gert, and David F. Lakatos. "Penn State Runoff Model (PSRM) User's Manual" (University Park, PA: Department of Civil Engineering, Pennsylvania State University), 1986.

Cherry, Paul, Julie Kascal, William A. Sprague, David Lakatos, and Alan Zeigler. "Release Rate Percentage Concept for the Control of Stormwater Runoff," 1982 International Symposium on Urban Hydrology, Hydraulics, and Sediment Control, University of Kentucky, Lexington, KY, July 27–29, 1982.

DeBarry, Paul A. "Regional Stormwater Management," American Society of Civil Engineers (ASCE) Short Course, New York, NY, and Houston, TX, 1985.

Franklin County Planning Commission and Borton-Lawson Engineering. "Conococheague Creek Watershed Storm Water Management Plan," report (Chowbergburg, PA: FCPC), 2003.

Lehigh Valley Planning Commission. Little Lehigh Creek Watershed Stormwater Management Plan, Lehigh Valley Planning Commission, Allentown, PA, 1987.

CHAPTER 19

ON-SITE STORMWATER MANAGEMENT AND CONSERVATION DESIGN

19.0 INTRODUCTION

The ultimate goal of watershed assessment and management should be to maintain or bring the hydrologic balance back into equilibrium. As with streams, discussed in Chapter 5, watersheds are in an equilibrium with precipitation, surface runoff, evapotranspiration, groundwater recharge, interflow, and so on, otherwise known as the *water* or *hydrologic balance*. The impact of urbanization drastically alters the values in the hydrologic budget equation.

Stormwater management entails bringing surface runoff caused by precipitation events under control. In past years, stormwater control was viewed only on a site-specific basis. Recently, national state, regional, and local perspectives and policies have changed, with the realization that proper stormwater management can only be accomplished by evaluating the comprehensive watershed approach (i.e., by analyzing the adverse impacts that a development located in a watershed's headwaters may have on flooding downstream). Consequently, on-site stormwater management criteria are often components of a regional watershed management plan. Land development and improper stormwater management increase stormwater flows, velocities, and volumes. This not only increases the risk of flooding downstream but also causes erosion and sedimentation problems, reduces stream quality, increases soil and stream bank erosion and sedimentation, raises the temperature of the streams, impairs the aquatic food chain, and reduces the base flow of streams (which is imperative for aquatic life during the drier summer months). Therefore, stormwater management entails evaluating the entire hydrologic regime of a watershed, not just controlling runoff peak rates from development sites to predevelopment rates using detention basins, as is commonly performed today.

Stormwater management requires cooperation among the federal, state, county, regional, and local jurisdictions, and among developers, engineers, and citizens; it involves proper planning, engineering, construction, operation, and maintenance. This entails educating the public and local officials and requires program development, financing, policy revision, development of workable criteria, and adoption of new ordinances and revisions to outdated ones. A comprehensive watershed stormwater management plan will enable continued development to occur within the watershed, utilizing both structural and nonstructural measures to properly manage stormwater runoff in the watershed. The plan should be a diversified multiple-purpose plan, one that addresses the full range of hydrologic consequences resulting from development, rather than simply focusing on controlling site-specific peak flow without consideration of tributary timing of flow volume reduction, base flow augmentation, water quality control, and ecological protection.

Managing stormwater runoff on a site-specific basis without consideration of the site's location within the watershed does not meet the objectives of watershed-based planning. The timing of flood peaks for each subwatershed (or subbasin) within a watershed contributes greatly to the flooding potential of a particular storm. Each stormwater control site within a subbasin should be managed by evaluating the comprehensive picture.

19.1 TRADITIONAL PROCEDURE FOR SITE DESIGN STORMWATER MANAGEMENT

As discussed in previous chapters, impervious areas are the main culprits in affecting or altering the hydrologic cycle. Human development of the land alters the drainage patterns and imperviousness of the land surface and applies reverse logic in solving stormwater problems. For instance, it is amazing how someone will buy a wooded lot to build a house and cut down all the trees to provide sunshine on the yard, but down the road his neighbor will buy a lot that was once a farm field and plant trees to provide shade.

Stormwater runoff, in the past, was managed by controlling and conveying the stormwater in concrete channels or storm sewers to the bottom of a development site, where it was detained in a detention basin. It was geared toward the visible, increased peak rates of runoff and was often accomplished through ordinances that required post-development peak rates of runoff be reduced to existing condition (or predevelopment, depending on ordinance language) peak rates of runoff for some specified design storm, for example, the 25-year storm. This usually was accomplished using the standard detention basin with an outlet structure designed to reduce the inflow to the predevelopment flow. Detention basins typically attenuate flood peak flows from the larger magnitude storms, such as the 10-, 25-, 50-, and 100-year storms.

Managing stormwater runoff peaks, however, does little to improve the quality of stormwater runoff discharged from the basin or to recharge the groundwater, and, in fact, may actually increase flows and stream bank erosion downstream.

As a consequence, in many design firms the typical land development design follows this procedure:

1. Determine maximum number of lots/impervious areas that the site can handle.
2. Design site layout to maximize the number of lots. (This step typically leads to maximum road lengths and impervious areas.)
3. Examine drainage pattern to see where runoff concentrates. A detention basin is then located at the lowermost portion of the property to manage stormwater runoff.
4. Design a detention basin to control postdevelopment flow to predevelopment rates (typically losing a lot or a corner of the parking lot for this purpose).

This procedure may be typical, but it is not highly recommended because of the disadvantages associated with it, which include:

- *Concentration of flow to one discharge point.* Although the design may reduce postdevelopment flows to predevelopment rates from the property, the original predevelopment flows were, in many cases, not concentrated to discharge at the point of outflow from the detention basin, but were diffused across the property line. In such cases, flows being released from the basin are greater at that point than they were prior to development.
- *Increased volume of runoff.* Standard stormwater detention basins do not reduce the increased volume of postdevelopment runoff, and this increased volume of runoff could overtax stream bank capacity, and increase flows in the stream downstream. (This will be discussed in greater depth later.)
- *Erosion at outlet.* The point discharge from the basin outlet pipe may cause erosion downstream of this point, if the predevelopment flow was not concentrated at this point. In addition, there may not always be a receiving stream immediately downstream of the outlet, hence the discharge may flow over an adjacent property owner's land or alongside a road swale, which may now be overtaxed.
- *Attenuation of outflow hydrograph.* This attenuation could increase downstream flows, as was discussed earlier.
- *Unwanted cost for developer.* These basins are typically an unwanted cost for the developer, who not only is losing a saleable lot but also has to provide the funds to construct the basin.
- *Unsightly control facility.* These basins, which are usually "crammed" into the lowermost corner lot, are often unsightly, with their steep side slopes, and do not conform to the topography of the area. Thus, they

detract from the aesthetics of the land development and depreciate the surrounding properties.

- *Maintenance and maintenance responsibilities.* Detention basins require maintenance. Who will maintain these basins after the developer is gone is a common question and source of disagreement.
- *Water quality.* Standard detention basins do little to manage the water quality of the discharge. Detention times are usually less than 24 hours, meaning that most solids do not settle, but flow freely out the discharge pipe, eventually ending up in a receiving water body.
- *Groundwater recharge.* Basins, unless specifically designed, do little to recharge the groundwater, which is essential for maintaining stream base flow. Often, basins are lined, or the bottoms are compacted during construction.

In the case of a detailed watershed plan, the management criteria are obtained from the subwatershed where the site is located, based upon the hydrologic modeling and the timing of the various tributaries. This runoff control can be obtained in a number of different ways. Various measures can be applied to reduce or delay stormwater runoff, and there are advantages and disadvantages to each type of runoff control measure. It is typically left to the developer or the developer's engineer to select the technique that is the most appropriate to the type of project and the physical characteristics of the site.

Much more appropriate designs can be applied to stormwater management facilities. Furthermore, utilizing innovative nonstructural stormwater management measures such as open space, low-impact development (LID), or conservation design (discussed later in this chapter), could eliminate entirely the need to construct a detention basin, as well as provide the added benefit of water quality improvement and groundwater recharge. Conforming the stormwater management schema into the land development design to make it an amenity, as opposed to a distraction, has been shown to increase property values and improve appreciation of the aesthetically and environmentally designed land development.

More recently, therefore, there has been greater emphasis on stormwater management best management practices (BMPs), which treat the water and help recharge it.

19.2 BEST MANAGEMENT PRACTICES

In developing the stormwater management component of the watershed management plan, it is necessary to consider a water quality control management scheme for the stormwater runoff to minimize the effects of nonpoint source pollution—in other words, to treat the quality of the stormwater runoff. This is typically achieved through the application of best management practices (BMPs). Stormwater-related BMPs can be classified as *nonstructural* and *structural,* although there is some crossover. Nonstructural BMPs are alternate site layout designs, either open space preservation or something as simple as housekeeping measures or development of educational material. Nonstructural stormwater management or best management practices, through land use planning tools, offer means other than building structures for managing stormwater. Structural BMP includes constructing a measure to control water quality. Some of the most popular concepts will be discussed below.

In determining the measures or combination of measures to install, the following parameters should be considered:

- Soil characteristics (hydrologic soil group: permeability, erodibility, etc.)
- Subsurface conditions (depth to seasonal high water table, bedrock, etc.)
- Topography (steepness of slope, earthcut)
- Existing drainage patterns (nearby streams, swales, and flooding potential)
- Economics
- Advantages and disadvantages of each technique

BMPs can be incorporated into the land development design to achieve the natural hydrologic regime of the site, which can be achieved through the five-phase approach to stormwater management.

19.3 FIVE-PHASE APPROACH TO STORMWATER MANAGEMENT

This book has repeatedly stressed that, in order to achieve sustainable water resources, a comprehensive approach to stormwater management is necessary. Sustainable water resources would include storm peak flow management, replenished base flow, fluvial geomorphologic stability, and nondegradation of water quality. That said, actually managing a watershed, with a watershed authority, is currently politically and physically very difficult—although it may become easier in the future. Land development and urbanization will continue to occur, often regulated on a political boundary (local municipality or county basis). Therefore, managing the land development process on the local level has been, and continues to be, a necessary means of managing stormwater runoff.

One solution, practiced in watershed plans in the past (DeBarry, 1989; DeBarry, 1993) but formalized in the Maryland Storm Water Manual (2000), looks at a five-phase approach to stormwater management, called the "Unified Stormwater Sizing Criteria." This approach requires that five factors be considered and met in the land development process and design of stormwater management facilities. It does not necessarily pertain to "structural" solutions, rather it promotes nonstructural solutions to achieve these goals, with structural solutions being implemented only if the goals cannot be achieved through nonstructural means. The five factors are:

- Manage a groundwater recharge volume.
- Manage a water quality volume.
- Manage a channel protection storage volume (typically around the 1.5-year storm).
- Manage an overbank flood protection peak and volume (5-, 10-, and 25-year storms).
- Manage an extreme flood peak (50- and 100-year floods).

These factors are illustrated in Figure 19.1.

Although the latter two are easily managed, using the traditional postdevelopment peak flow to predevelopment peak flow approach, the first three address a stormwater runoff volume issue. Therefore, each criterion must be treated separately and will be described in greater detail in this section. The accomplishment of all five should be possible through innovative design, nonstructural stormwater management (conservation design), and best management practices—at least, this should be the goal. Of course, there may be extenuating circumstances making this goal impossible to achieve.

1—Groundwater recharge volume
2—Water quality volume
3—Channel protection storage volume (typically around the 1.5-year storm)
4—Overbank flood protection peak and volume (5-, 10-, and 25-year storms)
5—Extreme flood peak (50- and 100-year floods)

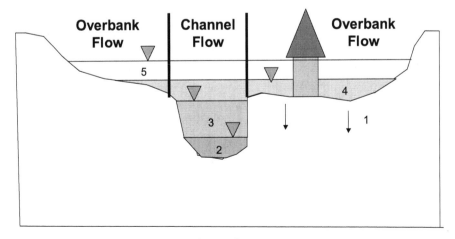

Figure 19.1 Five phases of stormwater management.

19.3.1 Infiltration: Groundwater Recharge

Recharging rainfall into the ground replenishes the groundwater that, in turn, provides base flow to streams, a process that keeps streams flowing during the dryer summer months, and maintains groundwater for drinking purposes. As development occurs, increasing the impervious area, and less rainfall reaches the groundwater systems, lower base flows and reduced groundwater supplies may result. Thus, in highly developed watersheds, it is not uncommon to see dry streams during periods of drought, along with severally depleted groundwater drinking supplies. Stormwater management measures, such as porous pavement with underground infiltration beds and infiltration/ recharge structures or other BMPs, can be designed to promote groundwater recharge. These measures are encouraged, particularly in hydrologic soil groups A and B, and should be utilized wherever feasible. Innovative approaches are encouraged to aid in meeting the applicable recharge requirements.

Managing the recharge volume is often a controversial subject. The difficulty in managing stormwater runoff is that, in the water budget equation, the infiltration volume can be recharged and the stormwater runoff peaks controlled, but the portion of the water budget that cannot be easily replicated on a storm event basis is the evapotranspiration (ET)—which, in many cases, may be up to 50 percent or more of the annual water budget. The difficulty lies in how to manage this ET volume. If it is recharged, the soils may not have the capability to accept this volume of water; sinkholes may develop in karst topography; stormwater recharge facility costs may be high in order to provide the required surface area or temporary storage volume for recharge; and finally, groundwater pollution potential is always a concern. If the ET volume is passed downstream, it may exacerbate stream bank erosion problems. Moreover, a developer often cannot visualize how the natural recharge from its site can supply a stream's base flow five miles down the watershed, and thus aid in preventing the stream from drying up. After explaining the cause/effect relationship, typically everyone agrees that it must be accomplished; thereafter, however, the physical methodology and exact volume to recharge is often a cause of controversy.

Infiltration is becoming more and more important in stormwater management as areas urbanize and people realize their stream base flows are being diminished. Many studies have been performed on infiltration stormwater management measures, and there is a long list of infiltration BMPs. It is evident that in areas of urbanization—such as eastern Pennsylvania, New Jersey, and Maryland, which have experienced deplenished base flow conditions during the late summer months—recognize the need to bring the water budget back into balance.

Several states have or are in the process of developing infiltration requirements for stormwater management (Pennsylvania Association of Conservation Districts, 1998; PADEP 2004; New Jersey Storm Water Management Regulations, 2004; MDE, 2000), most of which are based on maintaining the natural hydrologic budget. Several of these methods are discussed throughout the chapter.

19.3.1.1 *Method 1: Infiltrating Volume Difference between Storms.*
This method requires infiltrating the difference in volume between the 1½- or 2-year postdevelopment storm and the 1½- or 2-year predevelopment storm, respectively. The basin storage volume can be determined in two ways, either by determining the difference in runoff volume between the two hydrographs (Figure 19.2) or by routing a hypothetical detention basin to reduce the postdevelopment flow to predevelopment rates, as shown in Figure 19.3. These volumes have been proposed not only to recharge the difference in infiltration between postdevelopment and predevelopment conditions, but also the difference in evapotranspiration. In addition to replenishing base flow, the volume is significant enough to aid in preventing the increased flows that would occur from eroding stream banks. This, however, creates a large volume and may not be applicable to certain areas. The volume determined by the method in Figure 19.2 generally is greater than in the Figure 19.3 method.

19.3.1.2 *Method 2: Maryland Procedure.*
In the "Maryland Stormwater Design Manual" (2000), equations have been developed based on drainage area, percent impervious area, and hydrologic soil groups to compute a required recharge volume. These equations have been tailored to Maryland's physiographic and hydrologic conditions. The required size of the recharge facility is based on the following equation:

$$Re_v = [(S)(R_v)(A)] \div 12 \qquad (19.1)$$

Where:

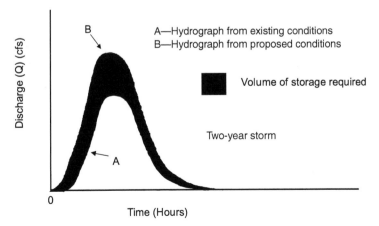

Figure 19.2 Difference in hydrograph volumes.

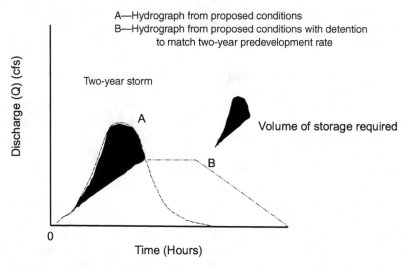

Figure 19.3 Difference in hydrograph volumes through the attenuated outflow hydrograph from a detention basin method.

Re_v = Recharge Volume (acre-feet)
S = Soil-specific recharge factor (inches)
R_v = Volumetric runoff coefficient
A = Site area contributing to the recharge facility (acres)

And:

$$R_v = 0.05 + 0.009 \times (I) \qquad (19.2)$$

Where:

I = percent impervious area

S is based on hydrologic soil group in Table 19.1 (from Section 2.2, of the "Maryland Manual").

TABLE 19.1 Soils-Specific Recharge Factor Based on the "Maryland Manual"

Hydrologic Soil Group	Soil-Specific Recharge Factor (S)
A	0.38 inches
B	0.25 inches
C	0.13 inches
D	0.06 inches

TABLE 19.2 Estimation of Recharge Rates Based on the "Maryland Manual"

Hydrologic Soil Group	Average Annual Recharge Volume (in./year)
A	18
B	12
C	6
D	3

The soil-specific recharge factor (which is local to Maryland) is obtained from dividing the values in Table 19.2 of the "Maryland Manual" by the annual rainfall; or, in the case of Maryland, 42 inches. (Note: To date, no source has been found for, nor has it been possible to reproduce, the values in Table 19.2.)

19.3.1.3 Method 3: PA BMP Manual Method. The State of Pennsylvania, in its "Best Management Practices for Developing Areas Manual" (Pennsylvania Association of Conservation Districts, 1998), devised the following methodology for infiltration. (To date, it is not mandatory, only suggested.) The difficulty with this methodology lies in determining the exact definition of *percent capture storm* for legal purposes. The runoff capture storage is based on equations 19.3 and 19.4:

$$R_s = (V - 0.2*S)^2 \div ((V - 0.2 \times S) + S) \tag{19.3}$$

$$S = (1000 \div CN) - 10 \tag{19.4}$$

Where:

R_s = Runoff capture storage (inches, over the drainage area)
V = Runoff capture volume (inches, over the drainage area)
S = Potential maximum retention after runoff begins (inches, over the drainage area)
CN = NRCS runoff curve number

The runoff capture volume can be determined from Table 19.3 (from the "Pennsylvania Best Management Practices Manual for Developing Areas," Appendix F, 1998).

19.3.1.4 Method 4: SCS (NRCS) Curve Number Method. The Soil Conservation Service (SCS, now the NRCS) curve number method can be utilized to estimate the amount of infiltration required to maintain the hydrologic regime. It is important to point out, however, that the SCS curve number equations were developed to predict runoff from single-peaking, large storms, not for small storms or for infiltration requirements. Nevertheless, it can be

TABLE 19.3 24-hour Storm Volumes Representing 60, 75, and 90 Percent of Annual Rainfall

Rainfall Region	60 Percent Storm (in.)	75 Percent Storm (in.)	90 Percent Storm (in.)
1	0.46	0.69	1.13
2	0.59	0.88	1.48
3	0.63	0.97	1.60
4	0.79	1.19	1.95
5	0.83	1.24	2.04
Average	0.66	0.99	1.64

Rainfall Source: Aron et al., 1986.

used as a guide, and further research may make it, or a modification to it, applicable for small storm recharge prediction.

Setting Q to 0 in the preceding equation would represent zero runoff from the site. Solving the equation for P, while holding Q equal to 0, would result in an equation that represents the rainfall volume that would represent the initial abstractions (Ia) and retention from a site with a particular (or composite) CN under existing conditions. Initial abstractions and retention include canopy and depression storage, evapotranspiration (ET), and infiltration. To accurately predict recharge, the canopy and depression storage and evaporation would have to be extracted from the Ia value. However, during a short duration storm event for which stormwater BMPs are designed, ET is negligible and depression storage will eventually infiltrate over time. Therefore, the retention parameters can be lumped to determine a conservative required infiltration or recharge value. This equation can take the form:

$$\text{For zero runoff: } P = I = (200 \div CN) - 2 \qquad (19.5)$$

Where:

$P = I$ equals the portion of rainfall with the surface runoff subtracted (in inches)

$CN = $ SCS Curve Number

This equation can be displayed graphically, as shown in Figure 19.4.

All that must be done to determine the infiltration requirement is to determine the CN, go up until the line is intersected, then go to the left to determine the inches of recharge required. The goal, then, is to maintain the natural predevelopment conditions hydrologic regime of a site by recharging that portion of rainwater that recharged under natural conditions. Therefore, mul-

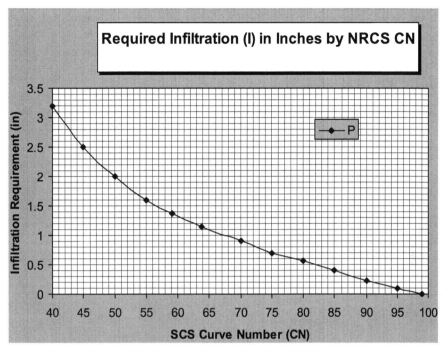

Figure 19.4 Infiltration requirement based on NRCS curve number.

tiplying the infiltration (in) by the site area would be the required site's recharge, or runoff capture volume, Re_v.

In most cases, existing sites may have varying land uses and hydrologic soil groups (HSG). For these, obtaining a composite CN for the existing site conditions, and then obtaining the infiltration requirement, is not the proper way of determining the required recharge. Instead, it is necessary to determine the infiltration requirement for each unique land use/hydrologic soil group combination and then sum the infiltration values to obtain the recharge volume, as shown in the sample in Figure 19.5.

From Equation 19.5, or Figure 19.4, the required infiltration (I) can be obtained:

Portion of Site 1: $CN = 58$, $I = 1.44$ in. $\times$ 2 Ac impervious $\times$ 1 ft/12 in. $= 0.24$ Ac-ft

Portion of Site 2: $CN = 71$, $I = 0.82$ in. $\times$ 4 Ac impervious $\times$ 1 ft/12 in. $= 0.27$ Ac-ft

Portion of Site 3: $CN = 60$, $I = 1.33$ in. $\times$ 1 Ac Impervious $\times$ 1 ft/12 in. $= 0.11$ Ac-ft

Portion of Site 4: $CN = 73$, $I = 0.74$ in. $\times$ 2 Ac Impervious $\times$ 1 ft/12 in. $= 0.12$ Ac-ft

$Re_v = $ Total required recharge $= 0.74$ Ac-ft

This process also indicates that the more permeable soils (HSG A and B), where the most rainfall would recharge the groundwater, would be maintained in their natural conditions. It is advantageous to develop on the less permeable soils (C and D), and keep the more permeable soils (A and B) as the recharge areas. If one were to run through the calculations, one would find that developing on or disturbing the permeable soils would require a lot more recharge volume than the other way around. This process, therefore, promotes the preservation of recharge areas.

19.3.1.5 Method 5: Water Budget Method. If stream flow data is available, base flow, which equates to annual recharge, can be separated from surface flow to determine the average watershed recharge rate. Land development should then strive to recharge this annual recharge amount from all impervious surfaces to aid in maintaining the hydrologic regime of the watershed. The USGS (2003) has developed the HYSEP computer program (Sloto, 1988) to provide an automated and consistent method of separating the groundwater or base flow component from the remaining hydrograph. The program uses the fixed-interval, sliding-interval, or local minimum hydrograph-separation techniques of Pettyjohn and Henning (1979).

The water budget method for determining recharge requirements methodology can be shown through the following example.

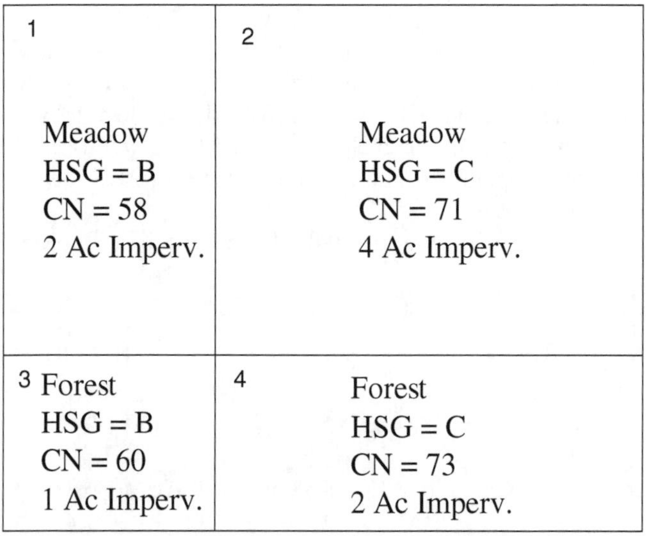

Figure 19.5 Hypothetical undeveloped site.

Starting with a version of the water budget equation detailed in Chapter 3:

$$P = R_{sro} + R_e + WL \qquad (19.6)$$

Where:

P = Precipitation (rain and snow)
R_{sro} = Overland runoff
R_e = Groundwater recharge/discharge (mean annual baseflow)
WL = Water loss (primarily evapotranspiration)

For the Pocono Creek, a tributary of the Brodhead Creek Watershed located in the Poconos, Pennsylvania, the water budget (average) at two nearby available stream gages is shown in Table 19.4 (DRBC, 2003).

Rounding the recharge to 17.5 inches for this watershed, approximately 63 percent of the total runoff is groundwater recharge. Therefore, about 37 percent would be overland runoff. When evaluating this annual recharge value to actual rainstorms, rainfall for the period of record of the closest reliable gage should be obtained; thereafter, a plot of the average of all the storms can be developed. A quicker way to perform this analysis is to obtain an annual series of rainfall representing the average rainfall year and plotting these values, as shown in Figure 19.6; the results are not significantly different. As can be seen in Figure 19.6, once plotted, the storm volumes that would need to be recharged to meet the 17.5 inches per year of the water budget can be determined—in this case, the 0.6-inch storm.

Assuming, then, that a site is developed so that undisturbed areas are not compacted, the designer would be required to recharge 0.6 inches of rainfall from all the impervious sites. As depicted in Figure 19.7, this "captures" approximately 80 percent of the storms, the same storms that produce the first flush. Caution is required so as not to pollute the groundwater from

TABLE 19.4 Water Budget for Two Gages in the Brodhead Creek Watershed as Developed by the USGS

Gage	Brodhead Creek Near Analomink	Brodhead Creek Near Minisink Hills
Drainage Area (Square Miles)	65.9	259
Mean Daily Stream Flow (cfs)	136	562
Mean Annual Stream Flow (in.)	28.01	29.45
Mean Daily Base Flow (cfs)	85	336
Mean Annual Base Flow (in.)	17.51	17.61
Surface Runoff (in.)	10.51	11.84
Groundwater contributed (percent)	63	60

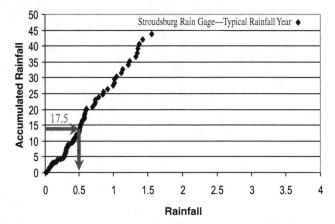

Figure 19.6 Translation of the annual water budget to rainfall average required to be recharged to maintain the watershed's hydrologic regime.

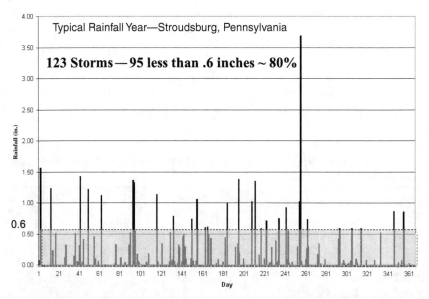

Figure 19.7 Annual rainfall series for a typical rainfall year showing that 80 percent of the storms produce 0.6 inches of rain or less.

capturing the first flush. To that end, a "treatment train," or series, of water quality BMPs (wet pond, artificial wetland), recharge BMPs (infiltration basins, trenches, porous pavement), and quantity BMPs (detention basins) might be required to ensure proper functioning of these facilities. Various BMPs will be described later in this chapter.

19.3.1.6 Method Volume Comparison. Borton-Lawson Engineering (2003) compared the various volume requirements produced from each of the aforementioned methodologies and determined the following volume requirements for a hypothetical commercial land development site, along with estimated costs. These are given in Table 19.5.

The cost of each method is based on a 24-inch perforated pipe with 6 inches of number-3 rock on the sides and 20 inches of number-3 rock below. This rock has a void ratio of 0.33, and the infiltrator system has the capability to store 6.23 ft^3 of volume per foot of pipe. The cost is estimated at $60 per foot of pipe and includes the excavation for the pipe and the number-3 rock.

19.3.2 Water Quality

Pollutants accumulate on impervious surfaces between rainfall events or during dry weather. Pollutant concentrations in runoff from developed land, therefore, tend to be greatest at the beginning of the storm event, or the first half to 1 inch of runoff, a phenomenon commonly known as the *first flush*. It has also been found that 30 to 90 percent of rainfall events are 1 inch of rainfall or less, storms that essentially simulate this first flush. As can be seen from Figure 19.7, the majority of the storms are 1 inch of rainfall or less. The majority of the nonpoint source pollutants, therefore, are being washed into streams during the smaller storms. Capturing these first flushes and/or smaller storms makes it possible to detain the stormwater, thereby allowing pollutants to settle, thus resulting in a "cleaner" outflow.

TABLE 19.5 Summary Comparison of the Five Methods

Volume Required (ft^3)							
Two-Year Post to Predetention Volume		Two-Year Post to Predetention Total Volume		MD-BMP Standards	PA-BMP Standards		
Rational	TR-55	Rational	TR-55	—	60% Capture	75% Capture	90% Capture
8,200	17,200	14,200	25,000	2,600	42	934	8,650
Cost Estimation (Dollars)							
$78,972	$165,650	$136,757	$240,770	$25,040	$404	$8,995	$83,306

Traditional detention basins typically have relatively short detention times. Many older ordinances require that detention basins drain completely within 24 hours from the end of the storm. This ideology promotes very little settling time, as well as "flushing" of the pollutants directly though a detention basin and into a receiving watercourse. A certain volume of water should be held for a sufficient period of time to allow solids to settle. Therefore, innovative best management practices, such as constructed wetlands, wet basins, and extended detention basins, allow for this extended period of storage so that the solids can settle. Such practices, coupled with the ability of aquatic vegetation to utilize nutrients, enable treatment of the stormwater runoff.

Various methods have been developed to compute this required water quality volume, two of which will be discussed here.

19.3.2.1 Method 1: The "Maryland Stormwater Design Manual." To determine a water quality volume that must be detained, the "Maryland Stormwater Design Manual" utilizes these equations:

$$WQv = [1.0 \; Rv \; A] \div 12 \text{—Eastern Zone} \qquad (19.7)$$

$$WQv = [0.9 \; Rv \; A] \div 12 \text{—Western Zone} \qquad (19.8)$$

Where:

WQv = Water quality volume in acre-feet
Rv = 0.05 + 0.009 (I), and I is percent impervious cover
A = Site area in acres

19.3.2.2 Method 2: Peak Flow Attenuation. Another methodology, one based on the concept that postdevelopment peak flow should be reduced to predevelopment peak flow, provides an extended volume of storage time. This standard involves reducing the postdevelopment two-year peak rate of stormwater runoff to the predevelopment peak rate of runoff from the one-year storm, as shown in Figure 19.8. Additionally, provisions are usually made to extend the detention time, such as adding a small orifice at the bottom of the outlet structure so that the postdevelopment one-year storm takes a minimum of 24 hours to drain from the facility from a point at which the maximum volume of water from the one-year storm is captured (i.e., the maximum water surface elevation is achieved in the facility). In some instances, where a water quality basin is located in the lower portions of the watershed and may detain the stormwater, it may be applicable to detain only the two-year storm, and not attenuate the larger storms, as was discussed in Chapter 18.

Release of water can begin at the start of the storm (i.e., the invert of the water quality orifice is at the invert of the facility). The designer of the facility should consider, and minimize, the clogging and sedimentation potential. Orifices smaller than 3 inches in diameter are not recommended. However, if

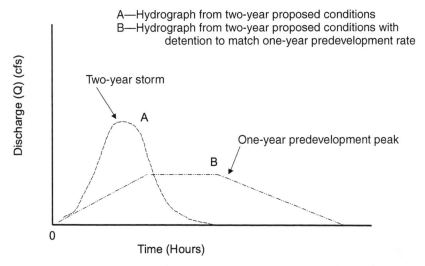

Figure 19.8   Reducing the postdevelopment two-year storm to the predevelopment one-year storm provides a volume and attenuation, to allow pollutants to settle.

the designer can provide proof that the smaller orifices are protected from clogging by use of trash racks and so on, smaller orifices may be permitted.

19.3.2.3   *Design Considerations.* Often, if a channel flows directly into the water quality treatment facility, such as a wet pond, the larger storms will tend to churn the sediments that had accumulated in the basin from the previous, or the current, storm event. These *suspended sediments* may be flushed out of the facility and, thereby, defeat the purpose of the basin. This can be averted by constructing a bypass or channel to divert only the "water-quality storm" (0.6 inches from impervious area, two-year flood, etc.) and lower storms into the basin, and divert storm flows greater than the water quality storm into a separate stormwater quantity control basin (standard detention basin), as shown in Figure 19.9.

19.3.3   Stream Bank Erosion Protection (Stormwater Management to Control Stream Bank Erosion)

Although detention basins can reduce the postdevelopment peak rate of flow to the predevelopment rate, the increased volume of runoff still will be passed downstream unless special provisions are designed into the basin to recharge this increase. Soils typically have a *critical velocity,* at which point they will begin to erode. The critical velocity is a function of the amount of load suspended in the water and the soil particle size distribution of the bed. In turn, the *bed sediment volume* eroded is directly related to the time duration

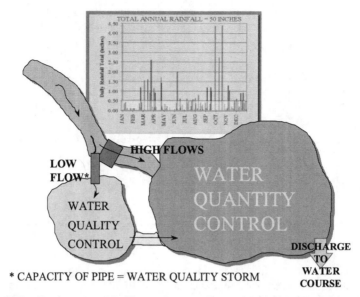

Figure 19.9 Dual-purpose facility: water‑quality storms being diverted into water quality treatment. This will treat 80 percent of the storms, or those that produce the first flush. Larger storms overflow the diversion and flow directly into the water quantity facility.

over which the steam velocity is greater than the critical velocity (McCuen et al., 1987) As storm flows increase, the velocities in streams also increase, thus exacerbating stream bank erosion problems. As noted, it has also been found that stream bank capacities are equivalent to approximately the 1½-year storm (Dunne and Leopold, 1978), and stream banks begin to erode when flows approximate this depth, that is, the critical velocity, as shown in Figure 19.10 (in this case, the one-year storm).

Once flows exceed this bankfull capacity, the flows extend out into the floodplain; therefore, the velocities within the channel bank may not increase significantly, even through the peak discharge does. For stream bank erosion, it is therefore important to manage for these higher-frequency smaller storms, such as the one- to two-year storms. Therefore, stream flows kept to near the one-year storm flow would minimize stream bank erosion. Detaining the smaller, more frequent events, where feasible, would, thus, minimize the number of storms causing stream bank erosion.

Applying the water quality criteria described previously will also help address the stream bank erosion problem. Detaining the two-year postdevelopment storm to the one-year predevelopment storm, and detaining the one-year postdevelopment storm a minimum of 24 hours, would, therefore, minimize the number of storms causing stream bank erosion. This is the same management criterion that has been recognized to also improve the water quality from stormwater runoff.

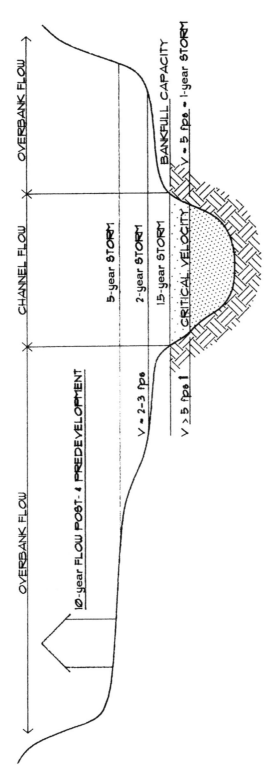

Figure 19.10 Cross-section showing relation of bankfull to critical velocity of one stream. This relationship will vary by stream, region, and soils. Theoretically, if the flow was never greater than 5 feet per second, the stream would remain in stable condition (DeBarry and Stolinas, 1994).

Stormwater management approaches that use conventional detention facilities to control postdevelopment peak discharge to predevelopment rates attenuate flows; therefore, the duration of near-peak flows and, in turn, velocities, are increased. Thus, the use of detention basins in the wrong location in a watershed may increase stream bank erosion due to attenuated flows above the critical velocity. A stormwater management plan being developed to prevent future stream bank erosion follows, as a case study.

19.3.3.1 Case Study. A watershed in Bradford County, in the northern tier of north-central Pennsylvania, was experiencing severe stream bank erosion. The landscape of the watershed includes deep-seated streams and rolling hills. A major portion of the landscape was formed during the Pleistocene Epoch in the Wisconsin Stage of glaciation. The Laurentide Ice Sheet, which covered most of the area, retreated some 10,000 to 15,000 years ago (Strahler and Strahler, 1978). As the glacier retreated, it deposited large amounts of glacial till, outwash, and other geologic features such as kames and moraines.

Most of the soils were a Volisia/Mardin/Lordstown series, which are deep and poorly drained with a seasonal high water table. Underneath the soil lies stony glacial till that was deposited hundreds of thousands of years ago. Over time, local streams began to cut into the till, and stream banks were severely eroding, with erosion rates exacerbated by anthropogenic activities, primarily in the form of increased runoff from agricultural and developed areas. As a result of the stream banks' erosion, large amounts of suspended alluvium and gravel, which aid in scouring stream banks and channels, were being generated. These materials progress downstream until the water level stabilizes or velocities decrease, leaving gravel bars along many portions of Bradford County streams. When trees and buildings began falling into the creek bed, citizens became motivated to correct this obvious problem.

Peak discharge control methods using conventional detention facilities are most effective for addressing floodplain flooding by regulating the peak rate of discharge. The majority of the stormwater problem areas indicated by the municipalities in the watershed were related to stream bank erosion and resultant deposition, as opposed to flooding. A hydrologic model was developed, and existing and proposed hydrographs were developed. Computer sequences done for future conditions assumed a 10-year buildout projection; they indicated that projected peak flows did not increase significantly over existing conditions. This was attributed to the very rural nature of the watershed, the absence of any driving force for development—for example, a commercial strip with public sewer and water—the relatively poor soils to support subsurface disposal, and the lack of density to support central utilities. Because flooding was not a primary problem in the watershed, and because the projected flows did not increase significantly over the existing conditions, the main emphasis was to evaluate stormwater management alternatives for stream bank erosion and sediment deposition problems.

If the attenuated peak flows from a detention basin increase the duration of the velocity above the critical velocity, this can increase the stream bank erosion over postdevelopment conditions without detention. However, the absence of detention may increase flooding downstream; therefore, actually reducing peak rates of runoff to a certain percentage of predevelopment flows, or another alternative, may be the best way to maintain noncritical velocities. This is illustrated in Figure 19.11.

The existing condition two-year storm hydrograph for a subarea is plotted, along with a future hydrograph and future conditions with detention. The two-year storm was utilized because it is just over the bankfull capacity; and, as discussed earlier, overbank flows typically do not affect the stream banks. A future hydrograph was utilized to project uncontrolled development.

Channel velocities corresponding to respective flows were also plotted for all three scenarios. Based on the alluvial silts, fine gravel, and graded loam to cobbles that comprise the stream banks, a critical velocity of 5 feet per second (fps) was determined (Fortier and Scobey, 1926). This corresponded to a discharge of 450 cubic feet per second (cfs) as determined through data collected for the computer modeling and Manning's formula. Note that the channel capacity was 592 cfs, whereas the peak two-year flow is 727 cfs. The time period that each scenario is above the critical velocity is summarized here:

Scenario	Time Greater Than Critical Velocity
Existing conditions	2.9 hours
Proposed conditions without detention	3.5 hours
Proposed conditions with detention	4.1 hours

The ideal situation would be to control runoff velocities to below the critical velocity level; that said, the majority of the stream reaches currently have velocities greater than the critical velocity during the two-year storm. A management scheme was, therefore, devised to try to maintain or reduce existing stream velocities. Any proposed development or site alteration activity governed by the watershed stormwater management ordinance was required to employ one of the infiltration methods described previously and detain runoff according to the following criteria:

- One-year postdevelopment flow to predevelopment conditions
- Two-year postdevelopment flow to the one-year predevelopment conditions flow
- Five-year postdevelopment flow to the two-year predevelopment conditions flow
- Ten-year postdevelopment flow to the 10-year predevelopment conditions

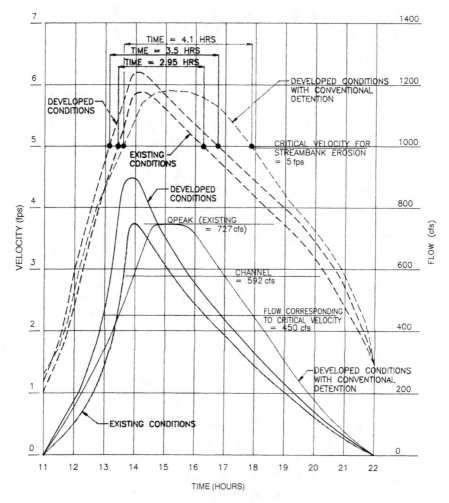

FIGURE COMPARISON OF FLOWS AND VELOCITIES WITH CRITICAL VELOCITY
FOR SUB AREA NO. 92 FOR THE 2-YEAR STORM

KEY:
– – – – VELOCITY (F.P.S.)
———— FLOW (C.F.S.)

Figure 19.11 Relationship of critical discharge, velocity, and duration with and without detention (DeBarry and Stolinas, 1994).

The design of any detention basin intended to meet the requirements can be accomplished through use of a multistage outlet structure (multiple orifices), as shown in Figure 19.12. This should be verified by routing the design storm hydrograph through the proposed basin using an acceptable routing methodology.

CRITERIA

DESIGN STORM	DETAIN TO PREDEVELOPMENT FLOW	STORM	Q PRE	Q POST	Q OUT OF BASIN
1 year	1 year	1	5	7	5
2 year	1 year	2	8	12	5
5 year	2 year	5	14	23	8
10 year	10 year	10	20	30	20
100 year	N/A*	100	50	70	70

* DO NOT HAVE TO DETAIN FOR 100-YEAR STORM, BUT BASINS MUST PASS THE FLOW.

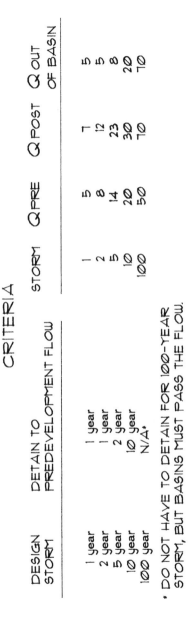
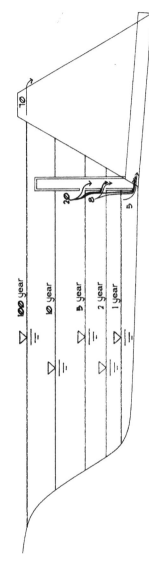

Figure 19.12 Multistage outlet structure to manage multiple design storms (DeBarry and Stolinas, 1994).

19.3.4 Overbank Flood Protection

Flooding and stormwater problems are caused by excess stormwater quantity. Storm events that result in water exceeding the natural bank of a stream are termed *overbank events* and are typically defined with an expected frequency of occurrence of 5 to 25 years. Much study has been done to ascertain the frequency at which most streams exceed the natural bank of the stream, called the *bankfull flow*. From this research, it has been determined that most natural stream channels have just enough capacity to handle approximately the 1½- to 2-year flood before spilling into the floodplain. Therefore, events greater than the two-year storm result in overbank flooding. Although some overbank and extreme flooding events are inevitable, the goal is to control their frequency, such that the level of overbank flooding is the same over time, so that damage to existing infrastructures is not exacerbated by upstream development. This can be accomplished through standard detention, innovative BMP design, and infiltration.

19.3.5 Extreme Flood Management

Extreme flooding events are defined by the severity of damage they cause. Typically, events such as the 50- and 100-year events or greater are labeled as extreme events. Standard detention and routing procedures can reduce the postdevelopment peak rate of flow to the predevelopment peak rate of flow for these storms. BMPs, infiltration structures, and other innovative designs can also be developed to reduce the postdevelopment flows.

As we have seen, it is advantageous to manage this stormwater closest to its point of origination, or where the raindrop hits the ground. It is also far better to prevent or minimize the amount of that raindrop that turns into runoff; or, in other words, minimize the amount of impervious area on a site. This can be accomplished through nonstructural stormwater management measures, also called *conservation design* or *low impact development (LID)*. Site design should be focused on minimizing runoff; then, if still required, structural BMPs should be designed to accomplish the five phases of stormwater runoff.

19.4 NONSTRUCTURAL BMPs AND STORMWATER MANAGEMENT

Nonstructural stormwater management concepts not only minimize the quantity of runoff but have been shown to reduce the amount of nonpoint source pollution, as shown in Table 19.6.

Table 19.7 lists nonstructural nonpoint source pollution source control measures, several of which will be discussed in more detail in this section.

TABLE 19.6 Runoff Effects of Sprawl and Town

	Undeveloped	Town	Sprawl
Runoff (inches)	0.04	0.23	0.33
Sediment (tons)	1.29	4.72	17.36
Total Sediment Nitrogen (lbs/ac)	0.02	0.04	0.13
Total Sediment Phosphorus (lbs/ac)	0.01	0.02	0.06
Chemical Oxygen Demand (COD) (lbs/ac)	0.5	1.68	2.86

Source: South Carolina Coastal Conservation League (SCCCL), 1995

19.4.1 Preservation of Environmentally Sensitive Areas

Preservation of environmentally sensitive areas aids in achieving goals of maintaining the hydrologic balance of a watershed. The following list offers a number of environmentally sensitive areas to use as starting point for open space planning or site-specific conservation design (though the features and their priority will vary by region or site):

- *Prominent forest areas.* Preserving large contiguous tracts of woodland associated with or adjacent to the other open space is important to prevent damage to groundwater, interference with wildlife migration, and degradation of overall esthetic value. When possible, these large tracts should be retained in their natural state; when that is impossible, impacts should be minimized by blending development into these natural areas.

TABLE 19.7 Nonstructural Stormwater Management Measures

Preservation of environmentally sensitive areas	Fertilizer application control
Regional planning measures	Pesticide use control
Site design and planning measures (conservation design, low impact development)	Proper household hazardous material disposal
Zoning	Lawn debris management
Street sweeping	Pet waste disposal
Solid waste collection and disposal	Illicit discharge detection and elimination
Highway deicing compound control	General community outreach
Nonpoint source pollution control on construction sites	Education (brochures, workshops, Web pages)
Automotive product disposal	Floodplain management (not building in floodplains, or compensating for storage volume loss)
Commercial and retail space good housekeeping	Steep slope restrictions
Industrial good housekeeping	Impervious area reduction
Storm drain inlet stenciling	

- *Areas of steep slopes (greater than 20 percent).* Slope-density provisions decrease allowable development densities as slope increases. The rationale behind such provisions is that, as slope increases, so does the potential for environmental degradation. Limiting development according to slope shifts development into areas with the least potential for suffering environmental damage. Furthermore, aesthetic values can be maintained, with each municipality utilizing the same basic concept, adjusting the provisions to meet their own specific concerns and needs.

  Additionally, the erodibility of soils on steep slopes should be taken into consideration in the site-planning phase. Depending on soil types, certain soils have a high erodibility, even on flatter slopes. Disturbing these soils can create situations where high soil washoff is likely, and the reestablishment of vegetation to control this washoff is ineffective.

- *Wetlands.* Wetlands provide unique habitat for both plants and wildlife. Their hydrology also provides for significant areas of flood storage, groundwater recharge, and pollutant-filtering capabilities. For these and other reasons, including protection of natural wetlands areas is critical in the early site planning stages. In case of high quality (HQ) or exceptional values (EV) wetlands, avoidance of these areas is recommended. In areas of smaller, lower-quality wetlands, mitigation of the wetland may be appropriate. In certain instances, wetland purchasing and leasing by nonprofit organizations will protect critical or sensitive areas or permanently preserve them.

- *Karst geology areas.* These areas are composed of highly erodible rock structures, such as limestone, dolomite, or marl carbonate rocks, thus they are highly susceptible to development of voids and underground channels, where groundwater has dissolved the rock structure. Karst areas, therefore, should be avoided when locating stormwater conveyance or control facilities. Where karst areas cannot be avoided, stormwater control facilities such as detention basins should be lined to prevent infiltration to the groundwater, which can potentially lead to sinkhole formation. In any case, concentration of runoff should be minimized through such techniques as filter strips. Recharge can be promoted as long as the volume to be recharged is spread out so as not to overtax the natural recharge rate.

- *Recharge areas.* Depressions with permeable soils are often overlooked in the development process. These areas collect and recharge significant amounts of water back into the groundwater, hence should be preserved or incorporated into the design. Often, these areas are simply ignored in the land development process and graded or filled in. In addition, although under existing conditions there would be little or no runoff from these depressions, engineering calculations ignore this when determining the existing condition or predevelopment runoff rate. The result: detention basin outlet structures that allow more than existing conditions or predevelopment runoff rates to discharge.

- *High flooding areas.* Floodplain regulation is a zoning measure whereby areas adjacent to water bodies and subject to frequent flooding are zoned to restrict their use. Normally, public and private recreational development such as parks, day camps, picnic grounds, and the like are permitted in floodplain districts, provided that they do not require substantial structures, fill, or storage of materials and equipment. Water-related uses and activities such as docks, boat rentals and launching, and swimming areas are usually permitted by special exceptions only.
- *Trail preservation.* Trail preservation provisions could designate existing regional trail corridors on the zoning map and establish provisions to restrict development within a certain distance of the trail (setbacks).
- *Stream corridors.* Stream corridors include both the actual waterways and adjacent riparian lands. These areas are proven habitats for fish, aquatic plants, and benthic organisms. Disturbance of such areas can destroy these natural habitats and result in stream bank instability problems, which can convey excess sediment to the stream channel downstream. Due to their importance, these areas should be identified early in the planning process and be protected through the use of setbacks and buffers as discussed in Chapter 20.
- Other features worth evaluating and identifying on the site would be:
  - Natural stormwater regime
  - Watershed divides
  - Water features
  - Farmland
  - Open space/greenways/riparian buffers
  - Historic, archaeological, and cultural features
  - Significant wildlife habitat
  - Soils (hydric, erodible, HSG A and B)
  - Views into and from the site

Other items of local significance may, of course, be added to this list. As stated, the GIS is a valuable tool to aid in determining and portraying these areas. A schematic of how this would work is shown in Figure 19.13.

The surest way to minimize disturbances to sensitive areas and natural features is to avoid them. However, absolute avoidance is not always practical. Further, avoidance alone may not be sufficient for protecting beneficial functions, and buffers surrounding them may be required. Site planners and designers, in cooperation with local zoning officials and plan reviewers, can implement planning concepts that both protect the resource and add to the value of the development and the community because they fully understand the critical functions of sensitive areas. Some of the concepts most useful for protecting sensitive areas are included in Table 19.8.

Planning for protection of environmentally sensitive areas should take place in a community comprehensive plan and in the early stages of a development

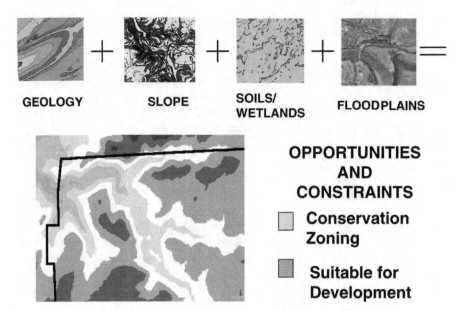

Figure 19.13 Developing site opportunities and constraints.

design. Identification of these areas is important early in a project so that planners can incorporate the effort into their overall development plan. In the majority of cases, these areas are inappropriate for infrastructure investment, due to environmental and economic reasons, so development should be strongly discouraged in these areas. Failing to take this step may result in environmental impacts, such as soil instability, erosion and sedimentation, and associated restrictive environmental capacities. Where development is permitted, it should be strictly regulated to ensure that proper precautions have been taken to guard against the aforementioned hazards. In short, conservation of these areas should be maximized, as will be discussed below.

19.4.2 Regional Planning Measures

Preservation of environmentally sensitive areas that aid in maintaining the natural hydrologic regime of a watershed begins with the leadership of the

TABLE 19.8 Useful Concepts for Protecting Sensitive Areas

Open space requirements	Conservation easements
Density limitations	Impervious surface reduction
Relax road requirements	Disconnecting impervious areas
PRDs	Conservation design/low-impact development
Cluster development	Retentive grading/grading reduction
TRDs	Historic area protection

community or regional planning commission. *Open space* planning is developing a preservation plan, usually on a community, municipality, or county level, to preserve desired open space. The open space planning concept has been around for many years. Ian McHarg described it in his book, *Design with Nature,* first published in 1967. The concept has since been expanded by Randall Arendt of The Natural Lands Trust (1999). But just because the concept has been known for many years does not mean it has been, or is, always applied. In fact, the benefits of applying it as an aid in maintaining the hydrologic regime are just now being realized (DeBarry, 1996–2004; Scheuler, 1987).

The synopsis given here is intended to explain and clarify what open space planning is, to give an overview of the planning process and the benefits of such planning, and to offer suggestions for municipalities that seek to encourage more widespread use of open space planning within their boundaries.

In general, many local jurisdictions have adopted ordinances that place little, if any, significance on preservation of natural features or on creation of usable permanent open space other than floodplains, wetlands, and steep slopes. The result of such conventional development is the conversion of every bit of "buildable" land into an extension of suburbia. Nature is subverted and subdivided for house lots, shopping centers, and parking lots; none of the original landscape remains.

Open space planning, in contrast, is a practice that preserves a significant portion of the existing landscape while meeting the development needs of the community. A very simplistic explanation of open space development is: increased development density on roughly half of the buildable land in the community. The remaining buildable and unbuildable land is designated as permanent open space. Unbuildable land is the environmentally sensitive areas. Buildable land, therefore, is any area that does not fall within the environmentally sensitive area boundaries. The Natural Lands Trust refers to this as "developing a community resource inventory," and lists nine pertinent elements, reproduced in Table 19.9.

Erodible soils, hydrologic soil groups A and B, or other locally significant features may be added.

TABLE 19.9 Nine Elements for a Community Resource Inventory

Wetlands and their buffers
Floodways and floodplains
Moderate and steep slopes
Groundwater resources and their recharge areas
Woodlands
Productive farmland (capability class I soils)
Significant wildlife habitat
Historic, archeological, and cultural features
Scenic viewsheds from public roads

This mapping can be transformed into land use zoning overlays. This process can be tailored specifically for water resources protection. By mapping the areas such as floodplains, recharge areas, hydrologic soil groups A and B, and wetlands, that, if developed, would impact the quality or quantity of water, overlay zoning could be developed and land development, water withdrawal, or other restrictions applied against these areas to preserve the hydrologic features.

19.4.3   Site Design Measures

These planning concepts developed on a regional level can be applied to the land development design process. Various aspects of how this can be incorporated into the development process are described in the following subsections.

19.4.3.1   *Conservation Design/Low-Impact Development* (LID). A practical approach to managing stormwater is to incorporate conservation design in the land development process. This important tool has many worthwhile benefits. *Conservation design* "rearranges the development on each parcel as it is being planned so that half (or more) of the buildable land is set aside as open space" (Arendt, 1997). Although this definition preserves 50 percent of "buildable" land, conservation design practices can be utilized to preserve environmentally sensitive areas. *Low-impact development* is a comprehensive land planning and engineering land development design approach geared specifically toward maintaining and enhancing the predevelopment hydrologic regime of the watershed for which the site is located.

These procedures include developing a constraints map of the the environmentally sensitive areas on a site slated for land development, preserving those areas and developing on the *areas of opportunity,* or those areas most suited for development. If an opportunity and constraints map has been developed with the GIS as part of a watershed assessment or regional comprehensive plan, the data will aid in educating the developer. Specifically, much of the data, such as hydrologic soils groups A and B, can be utilized by the developer to begin the conservation design process. Items more specific to the site, such as steep slopes, may need to be redeveloped from more accurate, locally derived site-specific data. Applying the five-phase approach to stormwater management in conjunction with conservation design practices discussed in this chapter will meet the goals of low impact development. A discussion of other concepts that can be used in conjunction with conservation design to minimize the impact of stormwater runoff follows.

19.4.3.2   *Density Restrictions.* These are restrictions placed on the density level to which land can be developed, based on the number of lots per acre,

the number of units per acre, and the minimum percentage impervious area per acre.

19.4.3.3 *Planned Residential Development (PRD).* PRD is a mechanism for flexibility in land use controls, to provide greater opportunities for better housing and recreation. By allowing flexibility and innovation in residential development, the PRD provisions provide for a greater percentage of a site to be maintained as common open space and for recreation. This measure is similar to *cluster zoning* (see next item), only on a larger scale, and nonresidential uses may be permitted. The developer is given more freedom in arranging buildings on the site, in exchange for a greater amount of land being dedicated to open space and recreation uses. PRD is a valuable way of meeting open space and recreation needs for communities. It ensures that the developer, who is creating the demand, provides the recreation service, rather than burdening the community with the responsibility.

19.4.3.4 *Cluster Development.* Cluster development, a form of PRD, in conjunction with substantial open space requirements, can promote imaginative, well-designed subdivisions that preserve open space and respect the physical and environmental qualities of the land. Clustering allows greater flexibility in the location of lots on the tract, which results in the ability to concentrate and group buildings on the least sensitive portions of the site. This, in turn, allows for the preservation of the most critical natural features (e.g., steep slopes, the ridgeline, scenic vistas, prime timber stands) of the tract, as shown in Figure 19.14. The open space provisions associated with cluster regulations, which require a certain percentage of the total tract be permanently preserved, should be mandatory and can range from 15–50% of the gross area of the tract. This common open space should be permanently set aside for the purposes of recreation (active or passive) and/or the conservation of natural features.

19.4.3.5 *Transferable Development Rights (TDRs).* Each parcel of land within a jurisdiction would be assigned a certain number of development rights, probably in proportion to its current market value. The land would then be regulated, with some owners allowed to develop and others restricted (see Figure 19.15). Under TDRs, those who were granted development permission would be required to buy a certain number of rights from those in the restricted class. TDRs are particularly effective in preserving agricultural or environmentally sensitive areas.

19.4.3.6 *Conservation Easements.* A *conservation easement* is a legal agreement between a qualified conservation organization or government agency and a land owner that permanently limits certain specified uses on all or a portion of a property for conservation purposes while leaving the property

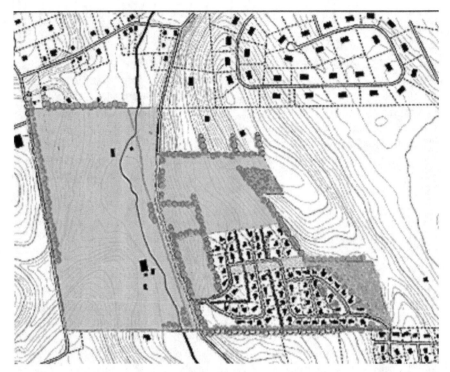

Figure 19.14   Cluster housing allows flexible site design to preserve recharge areas, steep slopes, or other areas that might affect the hydrology of the site (© Penn State Department of Landscape Architecture).

in the landowner's ownership. Conservation easements can be tailored to the conservation requirements of a particular property and to the desires of the landowner and the conservation organization alike.

19.4.3.7   Setbacks. A *setback* is a protected area between a development site (roads, building, parking lots, etc.) that separates these areas from environmentally sensitive areas such as wetlands or stream channels. Typically, these areas are undeveloped, except for recreational or low impact access activities such as walkways, foot bridges, bike paths, lawns, or parks. Reducing setbacks, allowing building closer to the road, minimizes driveway and sidewalk lengths and reduces impervious area.

19.4.3.8   Riparian Buffers. A nonstructural BMP would be, for example, forested or typically vegetated buffers that tend to filter pollutants as they flow through the buffers. Riparian buffers are discussed in more detail in Chapter 20.

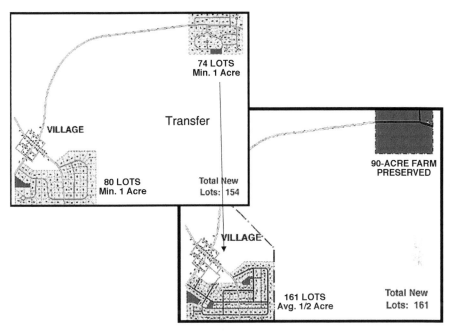

Figure 19.15 Transferring development rights allows preservation of hydrologically sensitive areas—floodplains and recharge areas (© Penn State Department of Landscape Architecture).

19.4.3.9 *Impervious Surface Reduction.* Asphalt and concrete, the most common types of driving surfaces, are very impervious. For most commercial development, and some residential development, paved parking or road surfaces account for a large percentage of impervious area increases. Alternative surfaces are generally more pervious than asphalt or concrete. The less runoff, the fewer pipes and storage systems are needed to prevent flooding. These reduced-impervious areas can be achieved through many means, including: paving blocks (pervious overload parking; see Figure 19.16), porous pavement, permeable parking areas (see Figure 19.17), reduced parking spaces, narrower streets, draining from impervious to pervious, sloping sidewalks away from streets, coordinating parking with multistoried parking, cluster development, taller structures, eliminating curbs and sidewalks, minimizing driveways, and implementing buffers.

Relaxing road width requirements is an effective way of reducing impervious surfaces within a development site, as shown in Figure 19.18. Often, ordinances require a certain number of parking spaces based on the proposed land use type or building roof areas. Many times, these ordinances are outdated, requiring much more impervious area than is necessary, as shown in Figure 19.19.

Relaxing parking requirements can minimize the impervious area and, in turn, reduce the number of required stormwater management facilities. Such

Figure 19.16 Paving blocks allow infiltration. (*Source:* Maryland Department of the Environment.)

Figure 19.17 Innovative permeable paving areas reduce runoff.

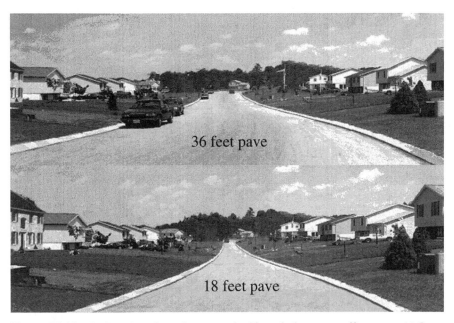

Figure 19.18   Reduced road requirements significantly lower runoff amounts (© Penn State Department of Landscape Architecture).

Figure 19.19   Many ordinances require too many parking spaces. (Photo courtesy of Borton-Lawson Engineering, 2000.)

practices must, however, be coordinated closely with officials, as most municipalities require certain minimums to be set. Limiting sidewalks is another effective method of reducing impervious surfaces in residential developments. In low-traffic residential neighborhoods, sidewalks can be limited to one side of a roadway, or possibly be eliminated altogether, to be replaced by biking or walking trails, constructed with pervious material, which could follow the back of lot lines. Sloping the sidewalks toward the grassed lawns, as opposed to toward the impervious road, allows stormwater to infiltrate.

19.4.3.10 *Disconnecting Impervious Areas.* In many circumstances, impervious area reductions cannot be applied, but an effective alternative is to manage impervious areas by redirecting runoff from conveyance systems and toward open pervious areas. In this way, rainfall that had not been allowed to naturally infiltrate due to the introduction of impervious surfaces would be at least partially infiltrated when dispersed over open pervious areas. Minimizing directly connected impervious surface areas involves a variety of practices designed to limit the amount of stormwater runoff that is directly connected to the storm drainage system. Runoff is instead directed to landscaped areas, grass buffer strips, and grassed swales, to reduce the velocity of runoff, reduce runoff volumes, attenuate peak flows, and encourage filtration and infiltration of runoff (UDFCD, 1992; DeBarry, 1996–2004). By incorporating these principles into site designs, the size and number of conventional BMPs such as ponds and constructed wetland systems can be significantly reduced. Minimizing directly connected impervious surfaces incorporates both nonstructural and structural control measures.

19.4.3.11 *Disconnecting Rooftop Runoff.* Rooftop runoff should be disconnected from the road or storm sewers, as shown in Figure 19.20, and be directed over a pervious area where it may either infiltrate the soil or filter over it. Grading the site to promote overland flow or providing bioretention on single-family residential lots is also recommended.

19.4.3.12 *Grading/Area of Disturbance Reduction.* In many development scenarios, rather than conforming a plan design to the existing topography, sites are extensively graded in order to fit the needs of a proposed plan. This approach results in massive grading; is expensive; requires stripping, stockpiling, and replacing the topsoil; and results in compaction of the soil, destruction of natural drainageways, and loss of site diversity. By varying lot sizes and building styles, and by using at least limited clustering, the need for mass grading can be reduced substantially.

19.4.3.13 *Retentive Grading.* As mentioned, in the past stormwater has been translocated along street shoulders or swales or storm sewers to the lowest location of a subdivision or land development, where a detention basin or a best management practice is typically installed to manage the stormwater

Figure 19.20 Connecting downspouts directly to impervious areas or storm sewers does not make it possible for the rainwater to infiltrate into a pervious surface. More-over, doing so also speeds the flow of water to the systems, decreasing the time-of-concentration, which increases flow peaks.

runoff. If, instead, intermediate depressions, referred to here as *retentive grad-ing,* were placed throughout the site, this might eliminate or at least reduce the size of any required detention basins. These depressions may be parabolic, with a depth of only 6 inches or so and a top width of 30 feet, so that they blend into the landscape. They also promote groundwater recharge and im-prove the quality of the runoff, as shown in Figure 19.21.

Figure 19.21 Retentive grading manages stormwater close to its source, promotes infiltration, and minimizes the size of stormwater management facilities (DeBarry, 1996–2004).

19.5 THE ROLE OF SOILS IN LOW IMPACT DEVELOPMENT

The spatial layout of soils, and their permeability, are typically overlooked in the layout design of and land development and location of stormwater management structures. Moreover, if the site were composed of both hydrologic soil group A and hydrologic soil group C, the runoff from development of the site would be dramatically different than if it were located on one soil or the other. For instance, utilizing the SCS TR55 runoff equations of development on a C soil versus on an A soil would have the results shown in Table 19.10. Note, in particular, that the difference in impervious or percolation infiltration between the A soils and pavement is a lot greater than the difference in runoff under existing conditions between a C soil and pavement. Therefore, ideally, it would be advantageous to develop on the C soils and route the runoff toward the A soils where the ability to infiltrate is greatest.

Utilizing the SCS curve number equation (equation 19.9) to determine runoff, the following comparison shows the difference in runoff (inches) and storage volume for various land uses on A soils versus on C soils, assuming a 50-acre site and 5 inches of rain (approximately a 10-year storm).

$$Q = \frac{(P - 0.2S)^2}{(P + 0.8S)} \tag{19.9}$$

What these numbers show is, given the same amount of rainfall, how much more rainfall infiltrates in the A soil than in the C soil. (Note that the SCS equation includes evapotranspiration and depression storage in the initial ab-

TABLE 19.10 Designing Based on Hydrologic Soil Groups

	Hydrologic Soil Group A Volume of Runoff		
	Curve No.	Q inches	Ac-Ft
Meadow	30	.005	.02
1/2 Ac Resid.	54	0.9	3.8
1/4 Ac Resid.	61	1.4	5.7
	Hydrologic Soil Group C Volume of Runoff		
	Curve No.	Q inches	Ac-Ft
Meadow	71	2.1	8.9
1/2 Ac Resid	80	2.9	12.0
1/4 Ac Resid.	83	3.2	13.4

straction; however, for all intents and purposes, for this demonstration, these would be fairly constant for the A and C soils.) Developing on A soils turns an area with a large recharge potential into an impervious area. Developing on C soils does not have as much of an impact, as the soils are not that permeable. Also, compare the required detention of developing a 50-acre tract currently in meadow conditions into half-acre lots on strictly A or strictly C soils: 3.8 acre-feet of storage would be required to reduce the postdevelopment flow to the predevelopment flow on A soils, and only 3.1 acre-feet if the C soils were developed. That means 20 percent less storage would be required by developing on C soils versus A soils. This can be seen in the difference of these curve numbers:

POST -PRE				POST -PRE			
A SOIL	61	-30	Difference = 31	C SOIL	83	-71	Difference = 12

19.6 RECOMMENDED PROCEDURE FOR SITE DESIGN STORMWATER MANAGEMENT: THE LOW IMPACT DEVELOPMENT DESIGN PLANNING PROCESS

Randall Arendt's aforementioned enhancements to Ian McHarg's concept of conservation design, published in 1997, were geared primarily to open space preservation. More recently, these ideas have been applied to water resource sustainability. The major goal of this procedure as it applies to site design stormwater management is to maintain the existing conditions hydrograph or hydrologic regime, including recharge areas, floodplains, and flow paths, not just manage the peak flow. Thus, it incorporates preserving the environmentally sensitive areas described earlier. The process, which will aid in achieving these goals, is delineated step by step in the following subsections (DeBarry, 1996–2004).

19.6.1 Site Analysis and Ranking

The first step in the land development process is to evaluate the layout of the land on a hydrologic basis using available maps or digital products (DeBarry, 1996–2004). The list of environmentally sensitive areas already described should be included in the mapping and an environmentally sensitive area map (ESAM) developed (MCDC, 2004).

Each of the features is mapped on the site, preferably using the GIS, as shown in Figure 19.13, Section 19.4.1. Following the evaluation, the next step is to perform a field verification and analysis. Next, all of the items or classifications are given a ranking based on their environmental and/or social significance to the site.

Using this ranking system in conjunction with the map overlays, unbuildable (constraints) and buildable (opportunities) areas are defined, as displayed in Figure 19.22. The constraints are the features worth preserving. Constraints can be ranked as *primary,* those where no building should take place, and *secondary,* those areas where building could occur. This, obviously, may vary by region or site. For example, groundwater recharge areas may be of primary importance on Long Island, New York, where saltwater intrusion of groundwater supplies is a concern, whereas the recharge area may not be a concern in the Mississippi Valley.

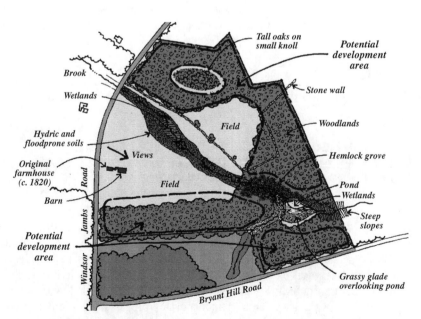

Figure 19.22   Opportunities and constraints site plan (Randall Arendt, 1997).

Specific goals and objectives for the plan should be developed. These goals might be to maintain open space, prevent flooding, control erosion and sedimentation, or to focus on water quality, base flow augmentation, groundwater recharge, or stream bank erosion. To aid in identifying goals and objectives, consider these factors:

- Public health, safety, and welfare
- Environmental benefits
- Cost/benefit analysis
- Long-term impacts
- Aesthetic appeal
- Regulations (state and federal)
- Regulations (local)

The resulting buildable area can then be reduced by approximately 50 percent; although this value will vary, 50 percent is recommended in order to provide undisturbed areas of sufficient size to preserve natural function, as shown in the following example:

Total site acres	= 250 acres
Including: floodplain	−70 acres
steep slopes	−20 acres
wetland (entirely within the floodplain)	−30* acres
Total buildable area	= 160 acres
Allowed buildable area = 160 acres × 0.5	= 80 acres

*Not included in the calculation

Constraints within certain other constraints, such as wetlands in floodplains, are not counted twice, as shown in this example.

The total buildable area is then evaluated for secondary constraints; for example, the rankings may be:

Highest	Stormwater regime
	Woodlands, old growth in good health
	Aquifer recharge
	Significant wildlife habitat
Moderate	Historic features
	Farmland
Lowest	Soils
	Views

The remaining area, or the area on which to build, is obtained by removing buildable areas with the highest significance rankings. Using the overlay mapping in conjunction with the significance rankings, the 50-acre allowed building area is defined. Any areas overlapped by high significance classifications are removed first; areas where high significance overlaps moderate significance are removed second; and so on, until a 50-acre allowed buildable area is defined.

19.6.2 Yield Plan

"The second step of the planning process is to prepare a yield plan, utilizing the conventional development patterns as controlled by the zoning, subdivision, and land development ordinances" (Arendt, 1997; MCPC, 2004). Factors that affect the hydrologic regime, including floodplain, steep slopes, and wetlands, are removed from the buildable area. A conventional plan is prepared on the entire buildable area. The number of lots possible on the plan dictates the yield that will be allowed on the conservation plan. The yield can be either a straight one-to-one or, if the municipality wishes to encourage conservation plans, the ratio can be changed to allow conservation plans more units (higher yield) than conventional developments.*

19.6.3 Building Sites

The third step of the process is to select the building sites. Based on the proposed use, locations are identified for the proposed buildings. For residential purposes, views are important, whereas for commercial uses, access and visibility are more common concerns. Instead of channelizing water to the lowermost portion of the site, retentive grading or other "at the location of origination" stormwater management measures should be employed.*

19.6.4 Road and Trail Layout

Roads and trails are located on the site. Locations should be selected to provide adequate access to each use while respecting and avoiding unbuildable areas and buildable areas of high significance. The length of the road system should be kept to a minimum, and water features should be avoided.*

19.6.5 Lot Layout

The final step in the design process is the lot layout. Depending on the type of use—residential, commercial, or industrial—lot layouts will vary significantly. The important thing to focus on is the proposed use and its physical requirements, not ordained requirements. The lot shapes should follow the hydrologic function rather than a strict adherence to conventional lot dimension requirements. These lots must be smaller and narrower, in most instances, than conventional zoning or subdivision/land development ordinances permit.

---

*This test is taken from MPC, 2004.

The advantages and benefits of preserving large areas of undisturbed and uninterrupted open space far outweigh the regulatory conveniences lost with the replacement of conventional lots/development standards.*

All the steps in the recommended process incorporating the concepts set forth in this chapter to achieve the LID goals are as follows:

- Develop the ESAM (constraints).
- Develop the yield plan on remaining land.
- Determine building sites and incorporate nonstructural stormwater management measures to minimize runoff.
- If not met by nonstructural means, meet the recharge, water quality, and streambank erosion objectives by installing BMPs as close to the source of the rainwater as possible.
- Meet the overbank and extreme event stormwater requirements through the use of BMPs as close to the source of the rainwater as possible.
- Construct detention basins as a last resort—even detention basins constructed for water quantity control can be designed as a BMP to have water quality and recharge benefits.

19.7 ADVANTAGES AND BENEFITS OF LID DESIGN

Because the advantages and benefits of conservation design techniques over conventional methods are numerous, for ease of display, they are grouped as either *ecological* or *economic* in Table 19.11.

19.8 FUTURE IMPLEMENTATION

To realize these benefits, low impact development must occur. Unfortunately, until recently, conventional development was the norm; and in many municipalities this is still the case, often because of existing ordinances, not choice. To developers who would prefer to take a conservation design approach, these regulations, in effect, erect additional hurdles in the form of additional hearings and approvals processes, resulting in higher costs and time delays. To encourage the use of open space planning and conservation design, municipalities, at a minimum, must make the approval/review process the same as or easier than for conventional development. At a maximum, lot or unit bonuses could be provided as incentives to encourage open space design plans.

The implementation of the runoff control strategy for new development should be through municipal adoption of the appropriate ordinance provisions.

---

*This text is taken from MPC, 2004.

TABLE 19.11 **Advantages and Benefits of Conservation/LID Design**

Ecological	Economic
Protects upland buffers along wetlands, water bodies, and watercourses; preserves habitat for flora and fauna.	Construction costs are reduced because the improvements are concentrated rather than spread thinly all over the site.
Preserved tree canopies shade water bodies. Cool water temperatures are critical for many types of aquatic life during the heat of the summer.	Utility runs are shortened.
	Road construction is reduced.
Runoff is less than with conventional development because there is less disturbance.	Less earth disturbance is required, resulting in fewer erosion control measures.
Native/natural vegetation in buffers helps to filter stormwater. Filtering removes/reduces pollutants, suspended sediments, and excess nutrients.	Runoff is reduced, resulting in fewer/smaller stormwater control structures.
Vegetation and debris reduces runoff velocities.	Maintenance costs are reduced because of reduced infrastructure. Savings could be used by the municipality to upgrade services, improve amenities, or reduce taxes.
Buffers/open spaces provide aquifer recharge benefits.	Land values, especially residential, appreciate faster in conservation design than in conventional development.
Preserves visual and aesthetic resources.	
Provides opportunities to utilize environmentally sensitive sewage treatment and disposal methods, which are not possible with conventional development. These systems aid aquifer recharge rather than dumping treated water into other surface water bodies.	

As part of the preparation of a watershed management plan, a model municipal stormwater management ordinance should be prepared that would implement the plan provisions. This could be adopted essentially as-is by the municipalities. Provisions would also be required in the municipal Subdivision and Land Development Ordinance (SALDO) and the municipal building code to ensure that activities regulated by the ordinance were appropriately referenced.

19.9 STRUCTURAL BMPs

If the goals of maintaining the natural hydrologic regime cannot be achieved though nonstructural measures, then structural stormwater runoff quality and quantity controls must be implemented. Several BMPs have been evaluated for their pollutant removal efficiency, with data continually being collected (ASCE, 2004). The most common are described in this section; but before selecting which are appropriate to the situation at hand, designers should factor in the determinants listed in Table 19.12.

Numerous best management practices manuals are available, typically developed on a statewide or regional governing agency basis. Therefore, the objective of this section is not to attempt to reiterate this extensive research and data, but to summarize the various alternatives for easy reference. Many of these are available for download from the Web; and for convenience, the U.S. EPA site contains a link to online-accessible BMP manuals at http://www.epa.gov/owow/nps/bestnpsdocs.html and http://yosemite.epa.gov/R10/water.nsf/0/17090627a929f2a488256bdc007d8dee?OpenDocument. Designers should, of course, study carefully the manual(s) appropriate to the watershed under development to determine the type of measures applicable in each case.

Table 19.13 summarizes the various structural management measures available for use in establishing standards for a particular watershed plan. Note that many of the BMPs listed here could fall under one or more categories, depending on their design.

Several of the BMPs just listed are summarized in the following subsections (MDE, 2000). Their advantages and disadvantages are described in detail in their respective manuals.

19.9.1 Surface Storage (Detention)

There are many design configurations and variations of the BMPs described here, including the addition of treatment forebays to the facility, and many unique designs can be drawn by combining these facilities. When several of these are designed in a series, the result is referred to as a *treatment train,* to provide several layers of treatment.

TABLE 19.12   Determining Factors for Designing BMPs

Total contributing area	Stream bank erosion
Permeability and infiltration rate of the site soils	Efficiency of the BMPs to mitigate potential water quality problems
Slope, and depth to bedrock	Volume of runoff that will be effectively treated
Seasonal high-water table	
Proximity to building foundations and well heads	Nature of the pollutant being removed
Erodibility of soils	Maintenance requirements
Land availability and configuration of the topography	Creation/protection of aquatic and wildlife habitat
Peak discharge and required volume control	Recreational value
	Enhancement of aesthetic and property values

TABLE 19.13 Popular Structural BMPs

Building-related	Surface Storage	Infiltration Facilities	Grassed Facilities	Parking	Collection and Treatment	Underground
Rooftop gardens	Dry detention basin	Porous pavement	Grassed waterways and channels	Ponding and detention on pavement	New sewer system control	Cisterns and covered ponds
Ponding on roof; constricted downspouts	Extended detention basin	Infiltration basins	Vegetated filter strips	Concrete grid and modular pavement	Storm sewer system storage	
Increased roof roughness	Wet detention basin	Infiltration trenches	Seepage areas	Porous asphalt pavement	Flow regulators	
Disconnection of downspouts from storm sewers	Constructed wetlands	Dry wells	High-delay grass and routing flow over lawns		Treatment	
	Bioretention/ bioinfiltration	Surface sand			Water quality inlets (oil and grease separators)	
		Underground vault sand filter				

Dry detention basin. A typical detention basin (see Figure 19.23) remains dry between periods of rain events. Its primary purpose is to reduce the peak rate of runoff to that which occurred prior to development. The short ponding time during a storm event does not allow a portion of the pollutants to settle out. Detention basins, because they do not treat water quality, often are not considered a BMP.

Extended detention basin. Extended detention basins are designed to allow an extended ponding time, thus allowing a larger volume of pollutants to settle out. These basins are typically designed to reduce peak rates of runoff for a much greater frequency storm, for example, the one-year storm. Figure 19.24 shows this type of basin.

Wet detention basin (wet ponds). A wet detention basin or wet pond is essentially a pond that has a permanent pool of water (see Figure 19.25). The pool allows an extended detention time, to allow pollutants to settle. Aquatic plants and organisms utilize the nutrients in the water, preventing escape of those pollutants. Providing a shallow depth bench around the perimeter or at one end of the basin allows vegetation to grow, for additional biological treatment. Some concerns have been raised about wet ponds in regard to their providing breeding grounds for mosquitoes. However, some preliminary findings at the University of Villanova, Villanova, Pennsylvania, indicate that the biologic activity

Figure 19.23 "Standard" detention basin for water quantity control. The concrete low-flow channel and the large-diameter orifice will allow pollutants to exit the basin untreated.

Figure 19.24 Extended detention basins typically have large storage volumes.

Figure 19.25 In this wet pond, the ducks have inhabited the facility (even though they are plastic decoys placed there by the DOT employees).

associated with wet basins keeps the mosquito population in check (Traver, 2003).

Rain barrels. One way to get citizens involved and keep their awareness of water resource conservation active would be to develop a program to supply each individual in the watershed with rain barrels, a practice that farmers utilized years ago when water supply wasn't as readily accessible as it is today. Rain barrels capture rainfall from downspouts and store it until a time when it is needed. The citizens can utilize the water to water lawns, flush toilets, or for other uses in times of drought, thereby reducing the demand on the aquifer or public reservoir. They can also be tailored to act as localized "detention basins," storing a volume of water and releasing it after the highest intensity of the storm has passed.

Constructed wetlands. Constructed stormwater wetlands systems are used to remove a wide range of pollutants through settlement and particulate matter, in addition to the biological uptake of pollutants—primarily nutrients—by the wetland plants. They are designed to maximize pollutant removal from stormwater runoff through plant uptake settling and retention; to that end, they usually have some sort of water level regulator, such as the weir shown in Figure 19.26. In order to maintain a permanent pool of water, constructed wetlands should have a minimum drainage area specified.

Figure 19.26 Constructed wetlands designed for water quality control. Some studies have shown that during vegetation die-off, wetlands can actually be a nutrient source.

A variety of constructed stormwater systems can be designed, including shallow marshes, extended detention wetlands, and pond wetlands. They are typically excavated basins into which wetland vegetation is planted to enhance pollutant removal. Although similar to wet detention basins, they are typically much shallower, to allow rooted vegetation to grow. Much larger surface areas are therefore required to store the required volume of stormwater runoff. Design criteria for wetland systems will vary by state, and design considerations may include setbacks from septic systems, property lines, and wells; soils' suitability to retain water; and pollution potential to groundwater. Other design considerations include establishment of localized adapted perennial species of vegetation, mulch-enriched soil in the upper surface of the wetland, and various outlet configurations, operation, and maintenance. Some studies have shown that during vegetation die-off, wetlands can actually be a nutrient source.

19.9.2   Infiltration/Recharge Facilities

When water is allowed to infiltrate into the surrounding soil, infiltration systems can be an excellent source of groundwater recharge

Bioretention/Bioinfiltration. Bioretention is achieved via a shallow surface storage area basin or swale that is planted with native vegetation above an excavation back filled with an undergrained soil sand mixture, as shown in Figure 19.27. Its primary purpose is to remove suspended solids, nutrients, metals, hydrocarbons, and bacteria from stormwater runoff. A bioretention system is designed to mimic the functions of a natural forest ecosystem for treating stormwater runoff. It is a variation of a surface sand filter, with the sand filtration media replaced by a planted soil bed. Stormwater flows into the bioretention area, ponds on the surface, and gradually infiltrates into the soil bed. Pollutants are removed by a number of processes, including adsorption, filtration, volatilization, ion exchange, and decomposition (Prince George's County, MD, 1993). Treated water is allowed to infiltrate into the surrounding soil, or is collected by an under-drain system and discharged to the storm sewer system or directly to receiving waters. The components of a bioretention system include: grass buffer strips, ponding area, organic mulch layer, planting soil bed, sand bed, and plants.

Bioretention systems also provide quantity control and groundwater recharge. However, they are most effective for receiving runoff as close as possible to the source. They are intended to filter the stormwater runoff from impervious areas from residential and commercial areas, and can be installed in median strips, parking islands, or unused areas. Prince Georges County, Maryland (1993), has developed an excellent manual on bioretention.

Figure 19.27 Bioretention area during a rainstorm at Villanova University, Villanova, PA. (Photo courtesy of Dr. Robert Traver, 2003.)

Infiltration basins. An infiltration basin is an excavated impoundment with a relatively permeable bottom soil, as seen in Figure 19.28. Their purpose is to temporarily store the surface runoff for a selected design storm and then allow the stored water to infiltrate into the groundwater. This method prevents surface water pollution; however, care is necessary to prevent groundwater pollution.

Infiltration trenches. Trenches excavated in porous soils and filled with aggregate allow runoff from small drainage areas to infiltrate into the ground. They can often be placed between parking aisles, space that is not utilized by vehicles. Figure 19.29 shows a typical infiltration trench.

Dry wells. Dry wells, pits excavated in porous soils and filled with aggregate, are typically used to control roof runoff.

Surface sand filter. The surface sand filter was developed for sites that could not infiltrate runoff or were too small for effective use of detention systems. The surface sand filter system usually incorporates two basins. Runoff first enters a *sedimentation basin,* where coarse particles are removed by gravity settling. This basin can be either wet or dry. Water then flows over a weir or through a riser into the *filter basin.* The filter bed consists of sand with gravel and a perforated pipe under-drain system to capture the treated water.

Underground vault sand filter. The underground vault sand design incorporates three chambers. The first chamber and the throat of the sec-

Figure 19.28   Infiltration basin.

Figure 19.29   Infiltration trench.

ond chamber contain a permanent pool of water, to function as a sedimentation chamber and an oil and grease and floatables trap, as well as to provide for temporary runoff storage. A submerged opening or inverted elbow near the bottom of the dividing wall connects the two chambers. This submerged opening provides a water seal that prevents the transfer of oil and floatables to the second chamber, which contains the filter bed.

19.9.3 Grassed Facilities

Grassed waterways and seepage areas. Grassed waterways and seepage areas reduce runoff velocities, enhance infiltration, and filter runoff pollutants, thus improving runoff quality.

Grass filter strips. Grass filter strips are densely vegetated, uniformly graded areas that intercept sheet runoff from impervious surfaces such as parking lots, highways, and rooftops. These strips can accept sheet flow directly from impervious surfaces, or concentrated flow can be distributed along the width of the strip using a gravel trench or other level spreader. Grass filter strips are designed to trap sediments, to partially infiltrate this runoff, and to reduce the velocity of the runoff. They are frequently used as a "pretreatment" system prior to stormwater being treated by BMPs such as filters or bioretention systems.

Vegetated swales. Vegetated swales are broad, shallow channels with a dense stand of vegetation covering the side slopes and channel bottom. They are designed to slowly convey stormwater runoff, and in the process trap pollutants, promote infiltration, and reduce flow velocities. *Dry swales* are used where standing water is not desired, such as in residential areas. *Wet swales* can be used where standing water does not create a nuisance problem and where the groundwater level is close enough to the surface to maintain the permanent pool in interevent periods. Wet swales can also include a range of wetland vegetation to aid in pollutant removal, an added benefit.

19.9.4 Parking

Concrete grid and modular pavement. Concrete grid and modular pavement systems promote infiltration and retard runoff, thereby improving runoff quality. These are typically promoted in overflow parking areas, as discussed in the nonstructural section.

Porous asphalt or concrete pavement. Special asphaltic or concrete paving material allows stormwater to infiltrate the pavement through an aggregate base and into the soil, thus reducing runoff and, in turn, pollutant washoff to streams. Porous pavement systems rely on storage

Figure 19.30 Porous pavement. Clogging of the surface can become a problem if not maintained.

volumes, typically stone with large voids, beneath the pavement surface to temporarily store and infiltrate the runoff volume. Overflows should also be supplied. Runoff temperatures are also reduced from conventional pavement because the initial rainfall (which typically generates the warmest runoff) infiltrates, instead of running off into receiving waters.

Porous asphalt pavement systems should be proposed only with appropriate filters and where strict construction and material specifications are in place. Often, inlets with perforated pipes underneath standard pavement serve the same purpose and eliminate the concern that the asphaltic materials will leach into the groundwater system. A typical porous pavement system is shown in Figure 19.30.

19.9.5 Miscellaneous and Vendor-Supplied Systems

A wide variety of miscellaneous and proprietary devices are available for urban management use. Many are "drop-in" systems that incorporate some combination of filtration stormwater media, hydrodynamic sediment removal, oil and grease removal, or screening to remove pollutants from stormwater.

The treatment train. Treating the runoff as close to the source as possible slows velocities and allows for more efficient treatment due to manageable volumes of stormwater. Applying the retentive grading concept

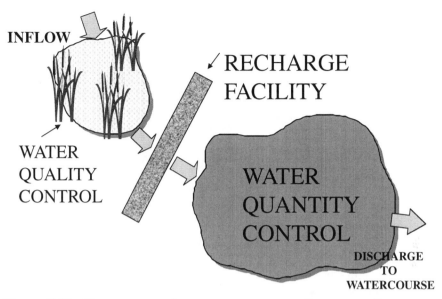

Figure 19.31 The treatment train to manage stormwater for water quality, promote groundwater recharge, and prevent flooding.

Figure 19.32 Grate collection system, drop box (below grade), and sediment forebay leading to the wet pond.

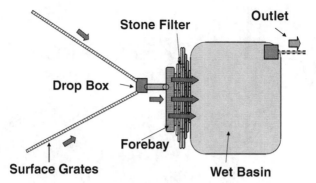

Figure 19.33   Treatment train concept before a wet basin: grate collection system, drop box (below grade), and sediment forebay leading to the wet pond.

keeps pollutants near the source and may be supplemented downslope with other BMPs. Although one BMP may do an adequate job of improving the water quality of the runoff, a number of BMPs in a series may do a much better job. This is the same principle as a tertiary sewage treatment plant. As shown in the example in Figure 19.31, the train is designed to first remove pollutants, next infiltrate to replenish base flow, then manage flooding.

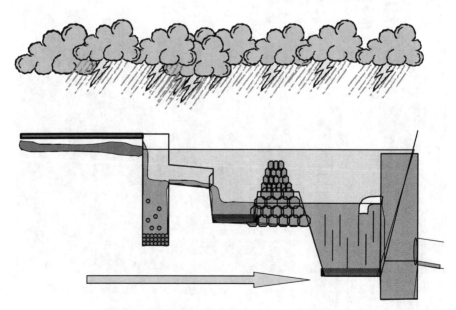

Figure 19.34   Profile of treatment train system.

Figure 19.35 Aerial view of water quality BMP incorporated into the land development design.

Referring back to the wet basin in Figure 19.25, it can be seen that it, too, incorporates a treatment train concept, primarily for water quality control. The wet basin, which improved water quality and provided detention, is actually the third stage of this treatment system, with a drop box and stone filter serving as the primary and secondary treatment devices, trapping and settling sediment and associated pollutants. In this particular wet pond, treatment was provided in a series similar to that in a sewage treatment plant. First, the stormwater was collected by grates in the pavement surface and routed to a drop box, which allows larger particles to settle and prevents the secondary treatment system from filling too soon. The secondary system is composed of a forebay with a rock filter. Water has to filter through this before entering the final wet basin, as shown in Figures 19.32, 19.33 and 19.34. Calculating pollutant removal from a treatment train is not straightforward, and future research should concentrate on this area to provide methodologies to determine pollutant removal rates.

19.10 SUMMARY

Implementation of BMPs need not be a financial burden to developers. By incorporating BMPs into the conservation design process, and making them

Figure 19.36 Water quality BMP incorporated into the land development design as seen by the residents.

aesthetic amenities, as opposed to viewing them as distractions, developers can increase their revenues. One sample is shown in Figures 19.35 and 19.36, where the developer, by making a wet pond a feature in the development, obtained $7,500 more for the 16 pond-front units than from those without a pond view, netting an additional $60,000.

REFERENCES

American Society of Civil Engineers (ASCE), International Stormwater Best-Management Practices (BMP) Database, ASCE, http://www.bmpdatabase.org/, Washington, DC, 2004.

Arendt, Randall. *Growing Greener: Putting Conservation into Local Codes.* (Media, PA: Natural Lands Trust), November 1997.

Aron, Gert, David J. Wall, Elizabeth L. White, Christopher N. Dunn, and David M. Kotz. *Field Manual of Pennsylvania Department of Transportation Storm Intensity-Duration-Frequency Charts* (University Park, PA: Pennsylvania Department of Transportation and Federal Highway Administration, Pennsylvania State University), May 1986.

Borton-Lawson Engineering. "Estimating Runoff Capture," Internal Report, Wilkes-Barre, PA, October 2003.

Bradford County Planning Commission. "The Wysox Creek Watershed Storm Water Management Plan," Bradford County Planning Commission, Towanda, PA, 1992.

DeBarry, Paul A. "Regional Stormwater Management," American Society of Civil Engineers (ASCE) Short Course, New York, NY, and Houston, TX, 1985.

———. "Brodhead Creek Watershed Stormwater Management Plan," Monroe County Planning Commission, Monroe County Conservation District, Stroudsburg, PA, 1989.

———. "Alternatives in Stormwater Management," Proceedings of the U.S.EPA, Stormwater/Wetlands Technical Workshop, Annapolis, MD, November 8–9, 1993.

———. "Innovative Stormwater Management" in *Computational Methods in Stormwater Management,* Penn State Short Course, University Park, PA, 1996–2004.

DeBarry, Paul A., Durla N. Lathia, Craig Todd. "Comprehensive Quantity, Quality & Ecosystem Management Plan for the Tobyhanna Creek Watershed" in *Proceedings of Stormwater Management: Planning the Future! An Ecosystem Based Approach,* Maritime Institute, Linthicum Heights, Maryland, November 9, 1995.

DeBarry, Paul A., and Raymond Stolinas, "The Geography of a Rural Watershed: A Citizen/Consultant Approach to Solving Accelerated Stream Bank Erosion through Act 167," in Chin Y. Kuo (ed.), *Stormwater Runoff and Quality Management* (University Park, PA: Penn State University), 1994.

Delaware River Basin Commission (DRBC). "Pocono Creek Pilot Project Tailoring Watershed Protection to Meet Community Needs—Final Report," Delaware River Basin Commission, West Trenton, PA, 2003.

Dunne, Thomas, and Luna B. Leopold. "Water in Environment Planning" (San Francisco, CA: W.H. Freeman and Co.), 1978.

Fortier, Samuel, and Fred C. Scobey. "Permissible Canal Velocities," Transactions, American Society of Civil Engineers (ASCE), volume 89, paper no. 1588, pp. 940–984, 1926.

Maryland Department of Environment (MDE). "Stormwater Design Manual," Baltimore, MD, 2000.

McCuen, Richard H., Glenn E. Noglen, Elizabeth W. Kistler, and Penny C. Simpson. "Policy Guidelines for Controlling Stream Channel Erosion with Detention Basins" (College Park, MD: Department of Civil Engineering, University of Maryland), 1987.

McHarg, Ian L. *Design with Nature, 2nd ed.* (New York: John Wiley & Sons, Inc.), 1992.

Monroe County Planning Commission (MCPC), Monroe County Conservation District (MCCD), and Borton-Lawson Engineering. "Brodhead/McMichaels Creek Watersheds Stormwater Management Plan," report (Stroudsburg, PA: MCPC), 2004.

Natural Lands Trust. *Growing Greener: A Conservation Planning Workbook for Municipal Officials in Pennsylvania* (Media, PA: Natural Lands Trust), January 1999.

New Jersey Department of Environmental Protection. *Revised Manual for New Jersey: Best Management Practices for Control of Nonpoint Source Pollution from Stormwater,* Trenton, NJ, fifth draft, May 3, 2000.

New Jersey Storm Water Management Regulations, New Jersey Register 35 N.J.R.169(a), January 5, 2004.

Pennsylvania Blueprints: "Best Land Use Principles & Results," Department of Landscape Architecture, Penn State University, University Park, PA, June 1997.

Pennsylvania Association of Conservation Districts (PACD). *Pennsylvania Handbook of Best Management Practices for Developing Areas* (Harrisburg, PA: PACD), 1998.

Pennsylvania Department of Environmental Protection (PADEP). *Stormwater Best Management Practices Manual PADEP* (Harrisburg: PADEP), 2004.

Pettyjohn, W.A., and Roger Henning. "Preliminary Estimate of Ground-Water Recharge Rates, Related Streamflow and Water Quality in Ohio: Ohio State University Water Resources Center Project Completion," Report Number 552, 1979.

Prince Georges County, Maryland, Department of Environmental Resources. "Design Manual for Use of Bioretention in Stormwater Management," June 8, 1993.

Schueler, T.R. *Controlling Urban Runoff* (Washington, DC: Metropolitan Washington Council of Governments [MWCoG]), July 1987.

Sloto, R.A. "A Computer Method for Estimating Ground-Water Contribution to Streamflow Using Hydrograph-Separation Techniques," report given at National Computer Technology Meeting, Phoenix, AZ, 1988.

South Carolina Coastal Conservation League (SCCCL). "Getting a Rein on Runoff: How Sprawl and the Traditional Town Compare," no. 7, Fall 1995.

Strahler, Alan H., and Arthur N. Strahler. *Modern Physical Geography* (New York: John Wiley & Sons, Inc.), 1978.

Traver, Dr. Robert. personal conversation, October 2003.

Urban Drainage and Flood Control District. *Urban Storm Drainage Manual.* (Denver, CO: Denver Regional Council of Governments [DRCOG]), 1992.

USDA Soil Conservation Service (SCS). "Soil Survey for Bradford and Sullivan Counties, Pennsylvania," (National Cooperative Soil Survey), 1986.

U.S. EPA. State Storm Water Best Management Practice Manuals, Region 10, http://yosemite.epa.gov/R10/WATER.NSF/0/17090627a929f2a488256bdc007d8dee? OpenDocument, accessed June 27, 2003.

USGS. "Water Resources Applications Software: Summary of HYSEP, http://water.usgs.gov/cgi-bin/man_wrdapp?hysep, 2003.

CHAPTER 20

FLOODPLAIN MANAGEMENT AND RIPARIAN BUFFERS

20.0 INTRODUCTION

Floodplain management has been evolving, not only in the United States, but throughout the world. The process began in 1804 when Thomas Jefferson commissioned Captain Meriwether Lewis to provide education in scientific observation and collection at the American Philosophical Society's Philadelphia headquarters before setting off with Captain William Clark to map out the Missouri River. In 1816 a peacetime army of engineers and the Topographic Bureau was established. (Schubert, undated). Floods of 1849 and 1850 in the Mississippi River Valley caused widespread damage and spurred national interest in controlling the Mississippi River.

Between 1851 and 1874, various authorities developed various plans for improving the Mississippi River, and semi-accurate maps were compiled. A comprehensive report on a "management plan" for the Mississippi River was filed on August 5, 1861, with the Bureau of Topographical Engineers of the U.S. Army (Humphreys and Abbot, 1861). However, there was a general lack of a unified policy and central direction for mapping the West. During the Civil War, the U.S. Corps of Engineers took on a major role in understanding the nation's topography (Schubert, undated).

On June 28, 1879, the Mississippi River Commission (MRC) was formed by an Act of Congress to direct and complete surveys, examinations, and investigations (topographical, hydrographical, and hydrometrical) of the river and its tributaries; to correct, permanently locate, and deepen the channel and protect the banks; and to report in full upon the practicability, feasibility, and probable cost of the various plans known as the jetty system, the levee system, and the outlet system, as well as on other plans deemed necessary (USACE, 2004).

Human nature is to "tame the beast," so in the past the theory has been to construct dikes, levees, and flood walls to contain the flood of a certain magnitude within its banks. This confines the stream and/or river and prevents natural floodplain inundation.

Floodplains, in urban areas in particular, have succumbed to human changes for economic reasons. But as has been realized (in the past and more recently since the Midwest floods of 1993 and 1995), floodplain containment has a specific lifespan, due to increase in development of the up-gradient watersheds, thereby producing floods of higher frequency and occurrence. In addition, containment also reduces the natural storage capacity of floodplains, thus compounding downstream problems. Filling the floodplains actually has similar effects to channelization, increasing velocities and "pushing" the water downstream.

Floods can be influenced by a number of factors, including rainfall intensity and duration, soil conditions, frozen ground or snow pack, topography or slope of the watershed, and land cover. There are two main types of floods: the *standard slow-rising,* long-duration flood (e.g., 24-hour storms), and the *flash flood,* a quick-response flood caused by a dam, levee, or ice jam break or a high-intensity, short-duration storm (a few minutes to a few hours). *Convectional precipitation storms* can cause flash floods, whereas *cyclonic precipitation* can cause the long-duration flood. Flash floods are the number-one cause of weather-related fatalities in the United States (National Oceanic and Atmospheric Administration (NOAA), 1992). For instance, the National Weather Service (NWS) reported that the June 9, 1972, flash flood in Black Hills Rapid City, South Dakota, was caused by 15 inches of rain in 5 hours and resulted in 238 fatalities and $164 million in damages.

Floodplain management begins with management within the watershed. The history of flooding and the type and frequency of flooding should be researched and catalogued; and management measures, such as land use controls or proper stormwater management, should be implemented to minimize or prevent future flooding. Responsible floodplain management must be included in all watershed plans, for two reasons: to preserve the storage capacity of existing floodplains, and thereby reduce downstream damages, and because it is less expensive than constructing large flood control projects. Floodplain management also preserves the natural integrity of floodplains and riparian buffers, thus improving or maintaining water quality. Although exclusion of floodplain development may be considered by some as a taking issue, or the taking of land by a governmental entity without just monetary compensation. Each situation should be analyzed individually and appropriate floodplain management measures incorporated into the watershed plan, as appropriate.

Factors to consider when devising a floodplain management plan include the taking issue, existing use, potential use, and compounded impact of improper floodplain development. Floodplain management, in the past, has concentrated on flood damage prevention; but as this chapter will explain, flood-plains are much more than areas that occasionally get flooded. They provide riparian wildlife habitat, groundwater recharge, flow attenuation, and stream bank erosion protection. That is the reason both floodplain and riparian buffers are discussed in the same chapter.

20.1 FEDERAL EMERGENCY MANAGEMENT AGENCY'S FLOOD INSURANCE PROGRAM

Floodplain management has been implemented in the United States through the *National Flood Insurance Program* (*NFIP*) (44 CFR Parts 59-78), which was established in 1968 following passage of the *National Flood Insurance Act*. The act is administered by the Federal Insurance Administration (FIA), a division of the Federal Emergency Management Agency (FEMA). It was further defined by the Flood Disaster Protection Act of 1973, which made available flood insurance to residents who lived in communities that admin-istered a floodplain management program to mitigate future flood damage. The second goal of the program is to reduce future flood damage by requiring local regulation, typically through local ordinances, of new development in the floodplain. The goals of the act were driven by economics, specifically to minimize losses due to floods, thereby balancing regulation and private prop-erty taking.

A major achievement of the act was the establishment of floodplain areas within the United States through the *Flood Insurance Study* (*FIS*) program. Through this process, floodplains were mapped, which gave communities the tools required to manage new development in the floodplains. Development is defined by FEMA as "any man-made change to improved or unimproved real estate, including but not limited to buildings or other structures, mining, dredging, filling, grading, paving, excavation, or drilling operations." *Flood-prone areas* are referred to as *special flood hazard areas,* those areas inun-dated by a 100-year flood. The maps were incorporated into the *Flood Insurance Rate Map* (*FIRM*), which is distributed to federal, state, and insur-ance agencies and to the municipalities, to aid in establishing insurance premiums.

Initially, three mapped products were developed: the FIRM, a Flood Hazard Boundary Map (FHBM), and a Flood Boundary and Floodway Map (FBFM). The FHBM is not based on a detailed hydrologic and hydraulic study, so the floodplain boundaries are approximate. The FBFM is developed from detailed hydrologic and hydraulic analysis and shows more detail than the FIRM, including cross-section location, floodway, elevations, and more. FEMA's re-vised FIS guidelines, published in January 1995, combined the FBFM with the FIRM (Federal Emergency Management Agency, 1995). Included in the FIS is a description of the community, flood elevation, and discharge tables;

flood profiles; and descriptive figures. The program also established the 100-year and 500-year floodplains (and, sometimes, the 10- and 50-year floodplains), 100-year floodway, and flood hazard and/or flood risk zones within each area.

The 100-year floodplain would be defined as the outer boundary from a flood with a 1 percent chance of occurrence in any one year. The elevation of the 100-year flood is known as the *base flood elevation* (*BFE*) and has been adopted as the regulatory flood. Areas within the 100-year flood boundary are known as *special flood hazard areas* (*SFHA*), whereas areas between the 100-year and 500-year floodplain boundaries are termed *areas of moderate flood hazard* (*AMoFH*). These flood depths are typically within a few feet. Areas outside the floodplain are termed *areas of minimal flood hazard* (*AMiFH*). The floodplain can be further classified into the *floodway* and the *flood fringe*.

The floodway is the area near the center of the stream that has the greatest velocities and greatest discharge. To be more specific, the *regulatory floodway* is defined by FEMA as "the channel of a river and the adjacent land areas that must be reserved in order to discharge the base flood without cumulatively increasing the water surface elevation more than a designated height." This "designated height" is typically 1 foot, although some states are more restrictive. For example, New Jersey allows only a 0.2-foot rise in BFE (N.J.A.C. 7:13). The BFE is determined by allowing an encroachment in the flood fringe, starting from the floodplain boundary and working toward the stream, up to a point at which the constriction in the floodplain raises the elevation of the base flood a maximum of 1 foot. The horizontal point at which this 1-foot rise occurs, or the remaining open waterway, becomes the floodway, as shown in Figure 20.1.

The determination of the floodway is easily accomplished using computer programs such as HEC-2 or its newer version, HEC-RAS. It is important to

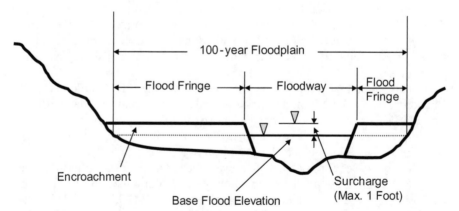

Figure 20.1 A 100-year floodplain, flood fringe, floodway, and base flood elevation.

keep this area free of debris and obstructions so that floodwaters can flow downstream unimpeded. Regulations must be in place to restrict encroachments, including fill, new construction, substantial improvements, and other developments within floodways that would cause any rise in base flood elevation. The floodway concept is a trade-off, allowing some regulated floodplain development while reserving an open area to allow floodways to pass unobstructed. Obstructing the floodway would cause a *backwater,* or *surcharge,* and additional upstream flooding. Development in the floodway is discouraged, and many community ordinances prohibit floodway development, while others allow exceptions such as additions to structures in the existing floodway that will not increase the BFE. (Note: The details of floodway regulations are beyond the scope of this text.)

The *flood fringe* is the flooded area between the floodway and the outer boundary of the floodplain. It is sometimes also referred to as the *floodway fringe,* but this tends to be confusing, and so is not used here. These relationships can be found in Figure 20.1.

Under the NFIP, floodplains are mapped to various degrees, according to the following criteria:

Approximate. Approximate study methods are those that produce a 100-year floodplain, but do not produce the 100-year flood elevation or base flood elevation. These floodplains are approximated from land features, soils, and/or general topography.

Detailed. Detailed methods include, at a minimum, surveying, hydrologic, and hydraulic analyses to determine an accurate floodplain boundary based on obtaining the 100-year flood elevations and depths, and are displayed on the FIRM. Detailed studies usually terminate where the floodplain becomes less than 200 feet wide or where the drainage area becomes less than 1 square mile.

Determining whether a new study will be approximate or detailed is based on development pressure. Developing areas are defined by FEMA as "areas where industrial, commercial, or residential growth is beginning, and/or where subdivision is underway and where these trends are likely to continue, and include areas likely to be developed within five years following the study."

In the FIS, each floodplain or special flood hazard area is divided into *flood insurance rate zones* to aid insurance agents with their actuarial flood insurance rates. The zones are defined by FEMA as (FEMA, 1995):

Zone A. This is the flood insurance rate zone that corresponds to the 100-year floodplains as determined by FEMA and reported in the FIS by approximate methods. Because detailed hydraulic analyses are not performed for such areas, no BFEs or depths are shown within this zone.

Zone AE. This is the flood insurance rate zone that corresponds to the 100-year floodplains determined by FEMA and reported in the FIS detailed methods. In most instances, BFEs derived from the detailed hydraulic analyses are shown at selected intervals within this zone.

Zone AH. This flood insurance rate zone corresponds to the areas of 100-year shallow flooding with a constant water-surface elevation (usually, an area of ponding) where average depths are between 1 and 3 feet. The BFEs derived from the detailed hydraulic analyses are shown at selected intervals within this zone.

Zone AO. This flood insurance rate zone corresponds to the areas of 100-year shallow flooding (usually, sheet flow on sloping terrain) where average depths are between 1 and 3 feet. The depth should be averaged along the cross section and then along the direction of flow to determine the extent of the zone. Average depths derived from the detailed hydraulic analyses are shown within this zone. In addition, alluvial fan flood hazards are shown as Zone AO on the FIRM.

Zone A99. This zone corresponds to areas of the 100-year floodplain that will be protected by a federal flood protection system where construction has reached specified statutory milestones. No BFEs or depths are shown within this zone.

Zone AR. This zone corresponds to areas of special flood hazard that result from the decertification of a previously accredited flood protection system that is determined to be in the process of being restored to provide a 100-year or greater level of flood protection.

Zone V. This zone corresponds to the 100-year coastal floodplains that have additional hazards associated with storm waves. Because approximate hydraulic analyses are performed for such areas, no BFEs are shown within this zone.

Zone VE. Zone VE corresponds to the 100-year coastal floodplains that have additional hazards associated with storm waves. BFEs derived from the detailed hydraulic analyses are shown at selected intervals within this zone.

Zone X. Zone X corresponds to areas outside the 100-year floodplain and areas of 100-year sheet flow flooding where average depths are less than 1 foot, areas of 100-year stream flooding where the contributing drainage area is less than 1 square mile, or areas protected from the 100-year flood levees. No BFEs or depths are shown within this zone.

Zone D. Zone D corresponds to unstudied areas where flood hazards are undetermined, but possible. This designation may not be used in Flood Insurance Studies unless otherwise approved by the FEMA's Regional Project Officer (PO).

FEMA has also developed the *Digital Flood Insurance Rate Map* (*DFIRM*), which includes digital base map information, graphics, text, shading, and other geographic and graphic data required to create the final hard-copy FIRM product, based on the "FEMA Guidelines and Specifications for Flood Hazard Mapping Partners." The DFIRM provides the database from which the Digital Line Graph (DLG) thematic product of the flood risks is extracted to create the DFIRM-DLG. When DFIRMs are available, usually on a countywide format and at a 1:24,000 scale, they are reviewed by the community and serve as the official flood map until it is updated through FEMA's *Map Modernization Program.*

20.1.1 FEMA's Map Modernization Program

FEMA's Map Modernization revised its vision of the DFIRM products to produce a product that will combine flood hazard data in vector format with a vector or raster base map. FEMA's goal is to have DFIRMs for all communities participating in the National Flood Insurance Program. (Federal Emergency Management Agency, 2001).

The goal of FEMA's Map Modernization Plan, according to its Web site, www.fema.gov/mit/tsd/mm_ahh1.htm (2003), is to upgrade its 100,000 panel flood map inventory by:

- Developing up-to-date flood hazard data for all flood-prone areas nationwide to support sound floodplain management and prudent flood insurance decisions.
- Providing the maps and data in digital format to improve the efficiency and precision with which mapping program customers can use this information.
- Fully integrating FEMA's community and state partners into the mapping process to build on local knowledge and efforts.
- Improving processes to make it faster to create and update the maps.
- Improving customer services to speed processing of flood map orders and to raise public awareness of flood hazards.

FEMA's Map Modernization Program automated *hydrologic and hydraulic* (*H&H*) techniques take advantage of capabilities within Geographic Information Systems (GIS) or a computer-aided design and drafting (CADD) environment, other software applications, and database structures to support the production of DFIRMs. The automated procedure provides tools that have graphical user interfaces (GUIs), to facilitate H&H model development, perform data processing and storage tasks, and improve graphics and visualization of modeling results. Some of the tools are pre- and postprocessors for industry-standard H&H models, while others are fully integrated systems such as distributed modeling that are used to compile and edit digital data and perform the analyses from start to end. The automated H&H tool list, shown in Table 20.1 (FEMA, 2003), provides information on the data requirements and automated techniques used to produce and revise FIRMs, DFIRMs, and other NFIP products. As H&H modeling analysis tools, they can also be utilized in whole or in part for other watershed assessment applications such as stormwater management and FGM analyses.

A schematic of relationships between automated H&H and the DFIRM spatial database can be seen in Figure 20.2. The main thrust of the Map Modernization Program is utilization of georeferenced remote sensing and GIS data, integrated with the H&H models and connected to the DFIRM database that is the central engineering and mapping data storage and visualization tool. This will be key to future watershed assessment and management programs since, once completed, the new FISs will have established a hydraulically connected digital watershed.

20.1.2   FEMA's Map Store

FEMA's FIS flood maps are now available online though the Map Service Center's (MSC) Map Store. The Map Store runs through an ArcIMS application where users can zoom to the area where flood maps are required. The opening page of the Map Store is shown in Figure 20.3.

20.2   LAND DEVELOPMENT PROCESS IN THE FLOODPLAIN

When a new development is proposed that may conflict with the floodplain, a developer must determine the location of the proposed development in relation to the flood fringe and floodway based on existing maps. In the absence of the maps or data, the base flood elevation, floodplain, and floodway must be determined through detailed surveying and hydrologic and hydraulic studies. If the area has been mapped and the existing maps show the proposed development to be within the flood fringe, but detailed survey shows that it is not, then the developer must submit a *Letter of Map Amendment* (*LOMA*) to FEMA through the community to be exempted from flood insurance purchase requirements. To request that the maps affecting the site be officially revised to more accurately portray the floodplain, the owner would submit a *Letter of Map Revision* (*LOMR*) to FEMA, again through the community. This may occur if there is an error in the original mapping, more detailed information is supplied by the developer, or some factor changes the floodplain.

TABLE 20.1 Preliminary Automated H & H Tool List

Title	Tool Function	Model Basis	Platform	Proprietary	FEMA Category
Watershed Modeling System (WMS)	Hydrology, Floodplain Mapping	Terrain modeling, hydrologic input parameterization, and interfaces with HEC-HMS, HEC-1, TR-20, TR-55, Rational Method, and National Flood Frequency Method	Stand-alone program that operates on Windows platform.	Yes	1
HEC-GeoHMS	Hydrology	Terrain modeling, hydrologic input parameterization, and interface with HEC-HMS	Third-party add-on to ArcView 3.x; also requires Spatial Analyst.	No	1
Hydrologic Modeling Extension	Hydrology	Terrain modeling and hydrologic input parameterization. No interfaces to third-party H&H models.	Provided with Spatial Analyst for ArcView 3.x GIS.	No	1
GIS Weasel	Hydrology	Terrain modeling, hydrologic input parameterization, interfaces with the Modular Modeling System (MMS) from USGS	Workstation Arc/Info 8.x for Windows NT or UNIX; also requires ArcGrid.	No	3
BASINS	Hydrology	Terrain modeling, hydrologic input parameterization; interfaces with HSPF.	Third-party add-on to ArcView 3.x; some functionality requires Spatial Analyst.	No	1
GISHydro	Hydrology	USGS Regression Equations for Maryland; interfaces with TR-20.	Third-party add-on to ArcView 3.x GIS; also requires Spatial Analyst.	No	1

TABLE 20.1 (*Continued*)

Title	Tool Function	Model Basis	Platform	Proprietary	FEMA Category
HydroCAD	Hydrology	Hydrologic input parameterization; interfaces with TR-20, TR-55, and SBUH	Stand-alone program operating on Windows platform.	Yes	1
RiverTools	Hydrology	Terrain modeling and watershed characterization. No interfaces to third-party H&H models.	Stand-alone program for Apple, Windows, or Unix platforms.	Yes	1
HEC-GeoRAS	Hydraulics, Floodplain Mapping	Terrain modeling, hydraulic input parameterization; interfaces with HEC-RAS.	Third-party add-on to ArcView 3.x or Workstation ArcInfo 7.x GIS; also requires 3D Analyst for AV, or ArcTIN for AI.	No	1
PC-SWMM	Hydrology, Hydraulics	Hydraulic input parameterization; interfaces with USEPA SWMM.	Internal GIS with importing features from other GIS formats for Windows.	Yes	1
RMS (River Modeling System)	Hydraulics, Floodplain Mapping	Terrain modeling, hydraulic input parameterization; interfaces with HEC-2 and HEC-RAS.	Third-party add-on to AutoCAD.	Yes	1
RiverCAD	Hydraulics, Floodplain Mapping	Terrain modeling, hydraulic input parameterization; interfaces with HEC-2 and HEC-RAS.	Stand-alone program operating on Windows platform.	Yes	1
HEC-RAS/12D	Hydraulics, Floodplain Mapping	Terrain modeling, hydraulic input parameterization; interfaces with HEC-RAS.	Stand-alone program operating on Windows platform.	Yes	1

Tool	Model Basis		Platform	Proprietary	FEMA Category
SMS (Surface Water Modeling System)	Hydraulics, Floodplain Mapping	Terrain modeling, hydraulic input parameterization; interfaces with WSPRO, FESWMS-2DH, RMA2, and RMA4.	Stand-alone program operating on Windows platform.	Yes	1
Mike 11 GIS	Hydraulics, Floodplain Mapping	Terrain modeling, hydraulic input parameterization; interfaces with Mike 11.	Third-party add-on to ArcView 3.x GIS; also requires Spatial Analyst.	Yes	1
WISE (Watershed Information SystEm)	Hydrology, Hydraulics, Floodplain Mapping	Terrain modeling, H&H input parameterization, USGS Regression Equations for North Carolina; interfaces with HEC-1, HEC-2, and HEC-RAS.	Stand-alone program operating on Windows platform.	Yes	1

Preliminary Automated H&H Tool List Legend:

- *Model Basis.* Describes the general types of automation and third-party H&H models supported.
- *Platform.* Describes the specific GIS software or computer platform for the tool.
- *Proprietary.* Describes whether the tool is a proprietary product of an entity.
- *FEMA Category.* Describes the level of FEMA review required when using the tool. Category 1: Tool uses existing preapproved FEMA model and has detailed documentation. Most analyses will not require additional review. Category 2: Tool uses same methods as existing preapproved FEMA model, but uses custom software to perform methods. Most analyses will require additional model runs with existing software for comparison. Category 3: Tool uses nonapproved model and will require extensive review.

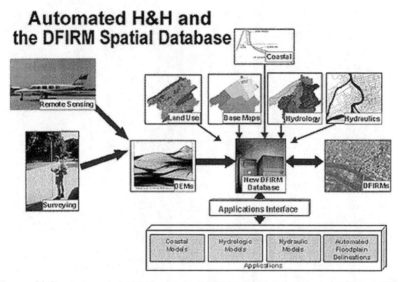

Figure 20.2 Automated H&H and the DFIRM spatial database (FEMA, 2003).

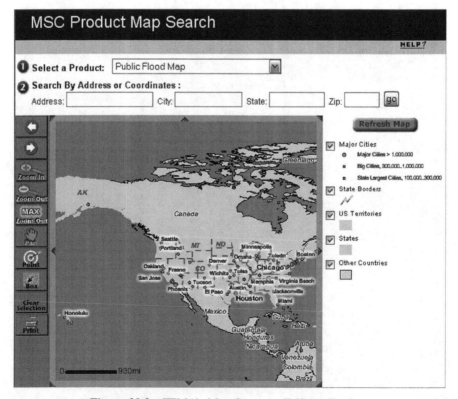

Figure 20.3 FEMA's Map Store ArcIMS application.

The general procedure when proposing land development in the floodplain is as follows (U.S. Army Corps of Engineers, 1988):

1. Investigate community floodplain maps.
2. Locate proposed development in relation to floodplain and floodway.
3. Perform floodplain calculation for existing and proposed conditions. If in the floodway, no construction is allowed except under limited variance from FEMA. If in the flood fringe, evaluate or flood-proof structures and/or fill to a minimum of the 100-year flood elevation (BFE). Most state regulations require that first floors be at an elevation of 1 to 1½ feet above the BFE.
4. Submit to community for review against FEMA NFIP regulations and approval. FEMA must review encroachments in the floodway. It is up to the community to decide whether a FEMA review of the flood fringe encroachment calculations is required.

20.3 ROOM FOR IMPROVEMENT

Although the Flood Insurance Program is doing an admirable job of minimizing future flood risks and providing insurance to potential flood victims, it still allows development and fill in the floodplain (flood fringe) right up to the floodway boundary. In fact, to elevate structures above the base flood elevation, the program actually promotes the filling of floodplains to an elevation at or above the 100-year base flood elevation. Filling of floodplains, when viewed on a site-by-site basis, does not appear to have a major impact on river or flood conditions as long as the regulations are followed. However, in the long term, as development pressure along waterways increases, continually filling the floodplain over long reaches of rivers or streams can have adverse cumulative impacts, as will be discussed shortly. Filling the floodplains, as well as channelization, dikes, levees, floodwalls, and so on, opposes the natural river/stream processes of flow, erosion, bedload transportation, sedimentation, and stream bank migration, and is prone to failure (Kusler, 1982).

As demonstrated repeatedly in the past, as watersheds become increasingly developed, with the percentage of impervious area also increasing, the flows and volumes of stormwater runoff will continue to increase from the same rainfall event. In addition, the frequency of a runoff event will continue to occur more frequently. For example, the elevation of structures placed 1½ feet above the 100-year base flood elevation, based on flows at the time of the study—say, 1985—may not actually be at an elevation above the floodplain from the 100-year flood in 2050. In other words, the 100-year flood may be 20,000 cubic feet per second (cfs) at a site in 1985 when the watershed is 4 percent impervious; however, in 2050, when the watershed is 25 percent impervious, the flows may be 25,000 cfs, with corresponding increased flood elevations.

Although the FIP offers support through its goals of providing flood insurance and minimizing future damage, a comprehensive approach to floodplain management to achieve longstanding riparian buffers, preventing nonpoint source pollution washing into waterways during floods, and preserving floodplain ecological systems is really not addressed by the act. Permanently preserving the floodplains as open space has many hydrologic, hydraulic, physical, ecological, and economical long-term benefits (Kusler, 1982), each of which is discussed in the following subsections.

20.3.1 Hydrology

Floodplains act as natural buffers, filtering pollutants before they reach the waterways. Preserving the floodplain in open space keeps impervious and urban areas away from the waterway. This has the advantage of keeping the alluvial soils, which have a texture and structure (or lack thereof) that are typically very good at recharging the groundwater and, in turn, stream base flow, in their natural conditions. In addition, it prevents the pollutant accumulations on impervious areas from being washed directly into streams and pushes the runoff to flow over riparian land first, allowing some pollutant filtering. Nutrients such as nitrogen and phosphorus also tend to have a better chance of being utilized by floodplain vegetation before being washed into the stream. Pathogens and other contaminants also may be filtered.

Another major factor in maintaining floodplains is the capability to lower the temperature of runoff from impervious areas as it flows across the natural landscape. Floodplains provide natural storage of water during flood events, acting as natural detention basins, temporarily storing floodwaters and detaining water to reduce downstream flows. For example, 1 acre of floodplain with floodwater 2 feet deep would store 650,000 gallons of water. A filled floodplain would force this water downstream at a time when this additional water might be detrimental. Floodplains also typically have an abundant groundwater source, due to the permeability of the alluvial material, and are prime, regional groundwater areas that should be preserved.

20.3.2 Hydraulics

Filling the floodplain reduces the hydraulic storage capacity. Under natural conditions, velocities slow in the floodplain fringes due to the increased wetted perimeter and "drag" on velocities. Floodplain plants slow the velocity of water, allowing sediment to settle that would otherwise be washed downstream. Filling the floodplain is essentially developing a "chute," increasing velocities through the encroached section, reducing travel time through the river reach, eliminating floodplain meander ratio, and translocating the higher velocities downstream.

Although the required calculations must show that a backwater is not created to cause damage upstream, each proposed land development project is reviewed individually. The cumulative impacts, and uncoordinated review of projects, over many years could increase elevations upstream.

20.3.3 Physical

As noted, rivers and streams have an equilibrium bedload. A portion of this bedload is deposited in the floodplains where velocities slow in each and every flood. These soils are typically very rich in nutrients and have a texture suitable for agriculture. Filling the flood fringe forces the same flow into a smaller area, increasing velocities. These higher velocities keep much of the material flowing through the constricted area and passes it downstream where it will eventually settle in an area, possibly where it shouldn't, such as under bridges or in navigable waters. Moreover, the increased velocities tend to erode stream banks below the constriction of channelization or bridge projects. Archeological sites such as Native American encampments and historic sites are also often associated with floodplains, and can be preserved by proper floodplain management.

20.3.4 Ecological

Floodplains, due to their unique hydrologic and soils features, typically have unique and highly productive ecosystems. Wetlands, salt marsh hay, wild rice, cranberries, blueberries, and rare and endangered species all are often associated with floodplains. Although wetlands are protected under Section 404 of the Clean Water Act, encroachment on floodplains frequently stresses the unique species that inhabit these ecosystems, due to the absence of any buffer setback criteria. Many migratory flyways and waterfowl and salamander breeding habitats are associated with floodplains, which often form *vernal ponds* during spring thaws.

Floodplains not only provide valuable habitat, but serve as a major food source for many species of wildlife. A number of notable floodplains and associated wetlands that house many of the threatened and endangered (T&E) species in the United States, such as the Midwest potholes, Mississippi floodplains and bayous, and Florida Everglades and mangroves, are under increasing pressure from development.

20.3.5 Economic

On the surface, preventing development in the floodplain would not seem to have an economic benefit, particularly for the landowner of the property. But the economic *disadvantages* to floodplain development include the risk of increasing flows and sediment downstream, eventual and possible long-term flooding of the allowed development, and lost economics of the destroyed ecology. In contrast, a natural greenbelt increases surrounding property values and raises the value of living in the adjoining urban area. Federal buyout of open-space floodplain land is much cheaper than a federal buyout of homes and businesses in developed floodplains after a severe flood. For example, one small community containing 74 moderate homes, after repeated floods, was purchased for $6.2 million and the land area made permanent open space. Furthermore, floodplains typically are prime agricultural areas, providing valuable cash crops, thanks to the prime soils, abundance of water supply, flat terrain, and replenishment of nutrient-rich soils during occasional floods. Floodplains also offer exceptional recreational opportunities for such activities as boating/canoeing, hiking, camping, bird-watching, and hunting.

In the long term, if left unfilled, floodplains could become a source of the commercial products mentioned in the ecological section, providing nutrient uptake and biomass for alternate fuel sources (Kusler, 1982). In addition, they may be a source of organic, sand, and gravel material, if properly removed. Instead of being permanently lost, if filled and built upon, floodplains could become a resource offering unique hydrologic and physical properties, resulting in future economic benefits.

In summary, federal floodplain regulations attempt to achieve a balance between economics and regulation; thus they should be regarded as the *minimum* required to prevent future flood losses and provide flood insurance to those in the floodplain. Local jurisdictions need to go further, by coordinating their floodplain management ordinances with riparian buffer, scenic river, greenway, open space, agricultural preservation, stormwater management, coastal zone management, zoning, conservation easement, and acquisition programs and initiates. In actuality, Section 404 of the Clean Streams Law (Clean Water Act) could be interpreted to prevent filling of floodplains if they serve pollution control functions.

20.4 FLOODPLAIN AND RIPARIAN BUFFER MANAGEMENT

The United States Department of Agriculture (USDA, 1998) defines a *riparian area* as "the aquatic ecosystem and the portions of the adjacent terrestrial ecosystem that directly affect or are affected by the aquatic environment. This includes streams, rivers, lakes, and bays and their adjacent side channels, floodplains, and wetlands. In specific cases, the riparian area may also include a portion of the hillslope that directly serves as streamside habitats for wildlife." The EPA *Coastal Waters Guidance Manual* (1993) defines riparian areas as "vegetated ecosystems along a water body through which energy, materials, and water pass. Riparian areas characteristically have a high water table and are subject to periodic flooding and influence from the adjacent water body. These systems encompass wetlands, uplands, or some combination of these landforms. They will not in all cases have all the characteristics necessary for them to be classified as wetlands." A *buffer* is an area managed to reduce the impact of an adjacent land use. The USDA Natural Resources

Conservation Service (NRCS) defines a riparian forest buffer as "an area of trees and/or shrubs located adjacent to and up-gradient from water bodies" (USDA NRCS, 1996). As the definitions indicate, riparian buffers are influenced by, and influence, the aquatic and hydrologic regimes of the water body. Development of the riparian areas, then, has an impact on both.

A buffer acts as the "right-of-way" for a stream, and functions as an integral part of the stream ecosystem. Stream buffers are undisturbed vegetated networks surrounding streams and lakes, which add to the quality of the stream and the community in many diverse ways, as summarized in Table 20.2 (modified from Schueler, 1995). Much of the pollutant removal observed in rural and agricultural buffers appears to be due to relatively slow transport of pollutants across the buffer in or under sheet flow in the shallow groundwater system. In both cases, this relatively slow movement promotes greater pollutant removal by soils, roots, and microbes.

Floodplains and riparian buffer areas have long been established to store floodwaters, which reduces flooding downstream, allows sediment to settle, provides groundwater recharge, keeps stream temperatures cooler, removes nutrients, provides important wildlife habitat, and preserves stream banks. In contrast, filling riparian buffers and floodplains, in accordance with current floodplain ordinance criteria, not only is a detriment to the natural functions listed, but also acts to channel the flood flow volumes and sediment downstream, causing increased flooding and sedimentation problems. In addition, structures and fill in floodplains are prone to damage. Proper watershed management, therefore, includes riparian and floodplain preservation.

Forest riparian buffers provide shade to aid in keeping stream temperatures low; filter, absorb, and adsorb pollutants; provide an area for sediment deposition; promote microbial decomposition of organic matter and nutrients;

TABLE 20.2 Twenty Benefits of Urban Stream Buffers

1. Reduces watershed imperviousness by 5 percent.	11. Mitigates stream warming.
2. Distances areas of impervious cover from the stream.	12. Protects associated wetlands.
3. Reduces small drainage problems and complaints.	13. Prevents disturbance to steep slopes.
4. Stream "right-of-way" allows for lateral movement.	14. Preserves important terrestrial habitat.
5. Provides effective flood control.	15. Opens corridors for conservation.
6. Offers protection from stream bank erosion.	16. Provides essential habitat for amphibians.
7. Increases property values.	17. Reduces barriers to fish migration.
8. Increases pollutant removal.	18. Discourages excessive storm drain enclosures/channel hardening.
9. Becomes foundation for present or future greenways.	19. Provides space for stormwater ponds.
10. Provides food and habitat for wildlife.	20. Allows for future restoration.

minimize or prevent stream bank erosion; provide terrestrial, stream bank, and aquatic habitat and species biodiversity; open wildlife corridors; provide infiltration, which replenishes groundwater and cool stream base flow; and provide flood flow attenuation. Fallen trees and logs create *pools,* which provide microhabitat for macroinvertebrates and other aquatic life. The *tree canopy* provides shade, thus keeping stream temperatures low. Maximum summer stream temperatures in a deforested stream can be 10 to 20 degrees Fahrenheit higher than a forested stream (USDA, 1998). Temperatures of 4 to 10 degrees Fahrenheit have been shown to alter the macroinvertebrate composition of a stream (USDA, 1998). Cooler streams contain more oxygen, providing better support for aquatic life. Unshaded streams, partly because of the increase in sunlight and increased stream temperature, promote undesirable *filamentous algae,* whereas shaded streams support the advantageous *diatomatious algae.* The food source of many macroinvertebrates is the diatoms that attach to the stream bottom and cannot consume filamentous algae. Riparian forest soils, because they are typically saturated, contain anaerobic microbes that *denitrify* nitrate to nitrogen gas and adsorb phosphorus. Even without being saturated, usually enough decomposition of organic matter occurs to create conditions devoid of oxygen that allow anaerobic microbes to denitrify nitrogen.

The "Chesapeake Bay Riparian Handbook" (USDA, 1998) classifies forest buffers into forest, agriculture, suburban/developing, and urban landscapes, with each described based on the land use influence on the water body, and refers to them as *Riparian Forest Buffer Systems* (*RFBS*). The handbook describes the value of RFBSs in terms of temperature and light, habitat diversity and channel morphology, and food webs and species diversity, all topics covered elsewhere in this book. The handbook further describes how forest buffers facilitate removal of nonpoint source pollution, including nitrate removal, plant uptake by nutrients, microbial processes, and removal of surface-borne pollutants.

The capability of a riparian buffer to remove pollutants will, naturally, vary with vegetation type, slope, soils, hydraulic conditions, and so on. For instance, a riparian buffer with highly permeable sandy soils will not provide as much denitrification as less permeable soils and will typically allow the nitrates to be transported into the stream through groundwater flow. Buffers are most effective where the tree canopy completely covers the stream (USDA, 1998).

20.5   RIPARIAN BUFFERS AND STREAM ORDERS

As the stream order increases, its relative size or dimensions also increase. The "Chesapeake Bay Riparian Handbook" points out that "As streams increase in size, the integrated effects of adjacent riparian ecosystems should decrease relative to the overall water quality of the stream." In other words, higher-order streams are more influenced by land use within the watershed than the riparian buffer conditions. First-order streams, or smaller intermittent streams, have little up-gradient contributing drainage area, and short contributing flow paths; therefore, the condition of the riparian buffer may have a significant impact on the water quality of the stream. Also, as demonstrated, as stream order increases, the chance that the entire tree canopy will cover the stream diminishes. The generalized effect of stream order on various buffer functions is shown in Figure 20.4 (USDA, 1998).

One way to make a preliminary prediction of the relative effectiveness of riparian buffers is to determine the watershed *drainage density,* or the length of channel per unit drainage area of the watershed. The higher the drainage density, the lesser the influence of the watershed landmass and land use, the greater the importance of riparian buffers.

20.6 RIPARIAN BUFFER ASSESSMENT

A watershed assessment should include an inventory of existing forested buffers to assess their adequacy. This can be accomplished through use of the GIS. Naturally, the assessment will only be as accurate at the data utilized in the inventory. Using the "buffer commands" in the GIS, a buffer width, specified by the user, can be displayed, buffering each stream segment. Overlaying this onto an existing land use map, either in vector (polygons) or raster (grid) format, the total amount and percentage of forested buffer can be determined. Locationally, this allows the user to target areas where forested buffers should

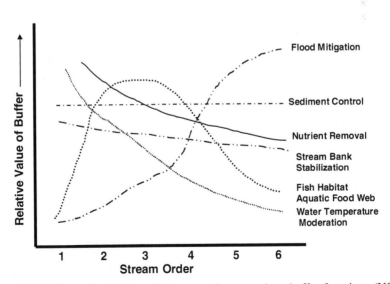

Figure 20.4 Generalized effect of stream order on various buffer functions (USDA, 1998).

be maintained or restored; it may also indicate a source of sediment pollution (those areas without forested buffers), provide a baseline for future inventory, and will aid in policy and management decisions.

Utilizing a similar methodology, a study conducted by the Land Analysis Laboratory at Penn State University determined that 59.8 percent of streams, rivers, and shorelines in the Chesapeake Bay Watershed had at least 100 feet of forest buffer on one side (Day and Brooks, 1997). The results were based on USGS 8- and 11-digit hydrologic units (HUC-8 and HUC-11) and had an error of 5 to 8 percent. When trying to determine the amount of buffer between 100 feet and 300 feet, the error went up to 18 percent.

20.7   RIPARIAN BUFFER MANAGEMENT

As described in Chapter 9 and shown here in Figure 20.5, unfenced stream corridors in pasture areas exacerbate removal of bank-stabilizing vegetation and allow trampling of the stream banks, which causes bank erosion. They also free direct nutrient-laden runoff to enter the stream; and the absence of shade causes the warming of water. Many public funds are made available for restoration of stream banks that have been damaged by unrestrained livestock grazing. These funds, such as those from the Conservation Reserve Enhancement Program (CREP), often compensate the landowner who fences

Figure 20.5   A riparian management program should be an integral part of any watershed management plan, to reduce the damaging effects of pasture conditions.

the riparian buffer area and allows the vegetation process to develop a natural riparian buffer, thereby eventually providing stabilizing roots, shade, and pollutant-removal capabilities, as shown in Figure 20.6. Access to water could be provided in a controlled manner to prevent the damaging effects described. Urban stream corridors need to be evaluated in an entirely different light, taking into account the increased runoff from urban areas. In any case, upstream stormwater runoff should be managed through a watershedwide stormwater management plan.

There are, of course, those situations where actual structural restoration techniques would be warranted, and the riffle and pool effect required for fish habitat. The goals and objectives of the study should be analyzed, and the cost/benefits of structural versus nonstructural restoration techniques determined, before undertaking a restoration project. A riparian buffer management program should be an integral part of any watershed management plan.

The "Chesapeake Bay Riparian Handbook" helps to guide the forest buffer planning process by dividing the buffer into three zones (see Figure 20.7). Zone 1 should contain undisturbed forest, stretch from the edge of the stream upland, and provide stream bank erosion protection and aquatic habitat. Roots in this zone hold soil in place and offer erosion protection by resisting the forces of flowing water. The purpose of Zone 1 is to maintain a stable ecosystem, provide shade to the stream, maintain a root system on the banks to prevent stream bank erosion, enable some nutrient removal, and provide the necessary *detritus* (dead organic matter) and large woody debris to the stream

Figure 20.6 The root mass from a forested buffer helps prevent stream bank erosion.

Figure 20.7   Three-zone buffer. (Illustration courtesy of the Chesapeake Bay Program, 1998.)

ecosystem. The U.S. Forest Service (USDA, 1998a) and the USDA Natural Resources Conservation Service (NRCS) (USDA NRCS, 1996) recommends this width to be 15 feet. It also recommends keeping livestock out of Zone 1.

Zone 2 should comprise managed forest and shrubs and be located upslope of Zone 1. Vegetation in this zone will utilize nutrients, particularly nitrogen and phosphorus from surface flow and shallow groundwater flow being absorbed through the vegetation's roots. Overland flow through the vegetation will generally remove sediment and other pollutants through filtration. Studies have shown that Zone 2 can remove up to 80 percent of the sediment in runoff (USDA, 1998). Zone 2 soils and microbes also remove nutrients, as described for Zone 1. The U.S. Forest Service recommends this width to be a minimum of 60 feet (USDA, 1998a), whereas the NRCS (1996) recommends Zone 2 be a minimum of 20 feet. Both measurements are proposed to be perpendicular to the water body. However, the NRCS recommends a combined width of 100 feet for Zones 1 and 2, or 30 percent of the geomorphic floodplain, whichever is less, but not less than 35 feet. It also recommends keeping livestock out of Zone 2.

Zone 3 should contain porous grass-covered land and be located upslope of Zone 2. Zone 3 grasses allow for maximum infiltration and spread the flow into sheet flow entering Zone 2, allowing the forest buffer to maximize its potential. Zone 3 can include man-made facilities to enhance filterability and spreading of flow to sheet flow entering Zone 2. Grass buffers, also known as grass filter strips, have been found to remove up to 50 percent of the sediment in runoff (USDA, 1998). Grass buffers may be supplemented with other best management practices (BMPs) such as infiltration trenches or basins, particularly in urban areas. The U.S. Forest Service recommends this minimum width to be 20 feet (USDA, 1998a). General environmental advantages of the three zones are shown in Figure 20.8.

Factors that affect the effectiveness of pollutant removal include slope, flow length, depth to high water table, permeability and amount of organic matter of the soils, entrance velocity of the overland flow, and vegetative cover. Buffer effectiveness can be enhanced by incorporating level spreaders or biofiltration swales upgradient of the buffer and through routine maintenance.

The wider the buffer width, the greater the opportunity for the buffer to perform its aforementioned functions. "The Chesapeake Bay Riparian Handbook" has identified the following recommended minimum buffer widths to aid in the following functions:

Stream bank stabilization: 10 to 40 feet

Water temperature moderation: 10 to 60 feet

Nitrogen removal: 25 to 140 feet

Sediment removal: 40 to 160 feet

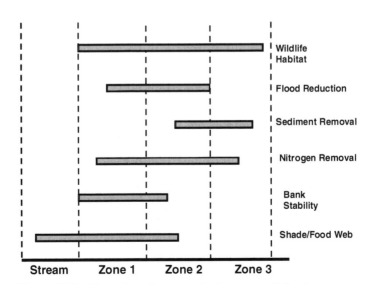

Figure 20.8 General environmental advantages of the three zones.

Water quality: 50 to 100 feet

Flood mitigation: 70 to 200 feet

Wildlife habitat: 40 to 300 feet

The USDA (1998a) also recommends that all hydrologic soil groups (HSG) D and C subject to frequent flooding be included in the buffers, even if beyond the distance recommended. Furthermore, the USDA recommends buffer widths be added to the 60-foot measure given for Zone 2 based on the following soil capability classes:

Capability Class	Buffer Width (feet)
III e/s, IV e/s	25
VI e/s, VII e/s	75

The "e" stands for risk of erosion, while the "s" stands for soil that is shallow, droughty, or stony.

In establishing riparian buffers, they need to be tailored to the *structure* and *dynamics* of the particular ecosystem (USDA, 1998). In addition, the *horizontal* and *vertical diversity* should be considered, the basis for the three-zone system. It has been found that various species inhabit different layers of the tree stand, or the vertical structure, depending on distribution of varying tree heights, tree species, and crown characteristics. Likewise, microhabitats will form based on horizontal differences in habitat species.

20.8 SUMMARY

Floodplains are floodplains for a reason. Undisturbed floodplains and associated riparian buffers attenuate flood flow, recharge groundwater, provide valuable habitat, prevent stream bank erosion, and help improve water quality. Although developmental pressures and economics have dictated floodplain regulations that allow reasonable development in floodplains, year after year, many millions of dollars continue to be spent on flood recovery efforts, often in the same community. These dollars would be better spent in buying out the affected areas and replenishing the floodplains.

New regulations should be put in place that take into consideration more restrictive floodplain requirements for new development. Many communities have already recognized the importance of this effort and have instituted floodplain and riparian buffer ordinances to restrict development not only in the floodway, but in the floodplain as well.

REFERENCES

Day, Rick L., and R. Brooks. "Chesapeake Bay Riparian Forest Buffer Inventory Final Report" (University Park, PA: Penn State University), 1997.

Federal Emergency Management Agency (FEMA). "Flood Insurance Study Guidelines and Specifications for Study Contractors," January 1995.

———. "Modernizing FEMA's Flood Hazard Mapping Program: A Progress Report," www.fema.gov/mit/tsd/dl_mpmod.htm, May 2001, accessed July 14, 2003.

———. "Modernizing Map Modernization Objective: Automated Hydrology and Hydraulics (H&H)," www.fema.gov/mit/tsd/mm_ahh9.htm, April 2003, accessed July 15, 2003.

Humphreys, A.A., and H.L. Abbot. *Report Upon the Physics and Hydraulics of the Mississippi River; Protection of the Alluvial Region Against Overflow; and Upon the Deepening of the Mouths: Based upon Surveys and Investigations* (Washington, DC: Government Printing Office), 1861.

Kusler, John A. *Regulation of Flood Hazard Areas to Reduce Flood Losses,* Vol. 3, prepared for the U.S. Water Resources Council, March 1982.

National Oceanic and Atmospheric Administration (NOAA) National Weather Service (NWS). "Flash Floods and Floods...the Awesome Power! A Preparedness Guide," U.S. Department of Commerce, Washington, DC, July 1992.

New Jersey, Division of Water Resources. Flood Hazard Area Regulations, N.J.A.C. 7:13, May 7, 1984.

Schubert, Frank N. *Vanguard of Expansion—Army Engineers in the Trans-Mississippi West 1819–1879* (Washington, DC: U.S. Government Printing Office), undated.

Schueler, Thomas R. "Site Planning for Urban Stream Protection," Environmental Land Planning Document Series, Metropolitan Washington Council of Governments and the Center for Watershed Protection, December 1995.

USDA Forest Service. "Chesapeake Bay Riparian Handbook: A Guide for Establishing and Maintaining Riparian Forest Buffers," Chesapeake Bay Program, June 1998.

———. "Chesapeake Bay Riparian Handbook: A Guide for Establishing and Maintaining Riparian Forest Buffers," Appendix 1, USDA Forest Service Specification Riparian Forest Buffer, Chesapeake Bay Program, June 1998a.

USDA Natural Resource Conservation Service (NRCS). "NHCP, Conservation Practice Standard," Riparian Forest Buffer Code 391, May 1996.

U.S. Army Corps of Engineers. "Mississippi River Commission History," on-line report, http://www.mvd.usace.army.mil/MRC/history.php, accessed February 7, 2004.

U.S. Army Corps of Engineers, Hydrologic Engineering Center (HEC). *Flood Plain Hydraulics: HEC-2 Users Manual* (Davis, CA: Hydrologic Engineering Center), September 1982.

———. *Floodplain Management Information Series: A Special Report—Procedures for Compliance with Floodway Regulations* (Davis, CA: Hydrologic Engineering Center), September 1988.

————. *River Analysis System: HEC-RAS Users Manual, Version 3.1.1* (Davis, CA: Hydrologic Engineering Center), May 2003.

U.S. Environmental Protection Agency. *Guidance Specifying Management Measures for Sources of Nonpoint Pollution in Coastal Waters, 840-B-92-002* (Washington, DC: Office of Water), January 1993.

CHAPTER 21

STREAM AND LAKE MANAGEMENT

21.0 INTRODUCTION

As noted previously, the management plan should include application of the fluvial geomorphological (FGM) assessment techniques described in Chapter 13. To help with that aspect of plan development, this chapter provides a brief summary of restoration techniques. Lakes are an integral component of the watershed, and how they are managed affects not only the lake's quality but also the quality of the stream down from the lake. A lake management plan should, therefore, be a component of the watershed assessment plan.

21.1 STREAMBANK MANAGEMENT USING FGM TECHNIQUES

The goal of an FGM assessment of a stream is twofold: to determine its natural geomorphologic characteristics and to develop management and/or restoration practices to bring the stream back into "natural" conditions. This effort is often made in the absence of an assessment of the watershed up-gradient to a specific site in question and the effects that development and resultant increased flows in the watershed will have on the stream. Therefore, in many situations, a plan to manage stormwater from new development should be established before restoration techniques are employed.

The stream channels, surrounding area, and up-gradient watershed should be classified as *urban, wooded,* or *agricultural* (pasture), and the appropriate management technique applied. Urban stream channels or watersheds may require structural restoration techniques, whereas for channels running through agricultural land, it may be better to provide a nonstructural approach, fencing the riparian area, as discussed in Chapter 20, and allowing the natural vegetation to reestablish itself. After performing the FGM assessment, a cost/ benefit analysis should be performed, to compare the nonstructural and structural alternatives and ultimate goals: that is, reestablishment of bank and channel habitat for fish, or reestablishment of the riparian area to prevent streambank erosion, remove pollutants, and reduce warming of the water. Often, all of these may together comprise the ultimate goal, as all are intertwined.

Structural restoration techniques involve forcing the water flow path, often toward the center of the channel, to produce pools for fish, and in a manner that prevents it from hitting stream banks, thereby preventing streambank erosion. Structural techniques can focus on preventing stream bank erosion or improving fish habitat. Several restoration/habitat improvement structures are listed in Table 21.1 (modified from Rosgen, 1996, and Stormwater Center, 2003), and a typical improvement structure is shown in Figure 21.1.

Details of these structures are outlined in the references. In the past, riprap and gabions were used for stream bank erosion protection. These structures do a fine job for that purpose; however, unless otherwise incorporated into a natural channel design feature, they do little to create avenues for fish habitat.

21.2 LAKE MANAGEMENT

The best shoreline for lakes is a natural one. Humans continually try to improve nature, notably the Mississippi River, the Florida Everglades, and many other examples; often, the intervention just does not work. Natural shoreline protection, such as native species woody growth and scrub/shrub species,

TABLE 21.1 Sample List of Structural Channel Restoration Techniques

Channel Stability/Habitat Improvement		
Vortex rock weir	W-weir	V-weir
J-hook vein	Rock cross vane	Rock vortex weirs
Step pools	Log drops	V-log drops
Native material revetment (root wads, vegetative transplants, rock spur, log spur)		

Habitat Enhancement		
Submerged shelters	Migration barrier	Defectors (single and double wing)
Boulder placement	Channel constrictor	Check dam (low or medium stage)
Bank cover	Floating log cover	Migration barrier

Spawning Habitat		
V-shaped gravel traps	Log sill gravel traps	Gravel placement

Figure 21.1 Typical channel restoration/fish habitat improvement structure, which could be classified as a rock cross vane or modified V-weir. Notice how the structure "channels" the flow away from the stream banks and creates a pool just below the structure.

provides habitat for terrestrial and aquatic species. It also provides a "buffer," to filter pollutants as they get washed over the vegetation during rainfall runoff events. But probably the most compelling reason for maintaining natural vegetative shorelines is their capability to withstand shoreline erosion. The natural root system holds the soils in place during wave action. After woody or scrub/shrub vegetation is removed, the root systems eventually decay, leaving the soil unstructured and susceptible to erosion. Erosion from shorelines carries sediment into the lake, adding "filling" to the lake bottom in the form of sediment. Sediment carries phosphorus, which, when in the water, becomes soluble. Phosphorus is the limiting nutrient in natural waters, and when it occurs in overabundance, it causes algal blooms and, eventually, lake eutrophication.

Perhaps the worst lake shoreline is the lawn. Geese tend to feed in short-grassed areas, where they can spot insects readily while observing their surroundings for predators. Grass does not have the extensive root system that woody vegetation does, hence makes the shoreline a prime target for erosion. Any good lake management program should make the maintenance of a natural, woody lake shoreline a priority.

21.2.1 Lake Water Quality

Fecal coliforms are bacteria found in the feces of mammals. They constitute one of three bacteria commonly used to measure possible contamination from human or animal waste. The extent to which fecal coliforms are present in the water can indicate the general quality of that water and the likelihood that the water is fecally contaminated.

The appearance of coliforms in water also may be a sign of the presence of potentially harmful contaminants that may cause disease. Waters that are polluted may contain several different disease-causing organisms, commonly called *pathogens*. Enteric pathogens (i.e., those that live in the human intestine) can carry or cause a number of different infectious diseases. Swimmers in sewage-polluted water, for example, could contract any illness that is spread by ingestion of fecal-contaminated water.

In most locations where fecal coliforms are present, the water is also receiving nutrients, which may overly enrich the lake, causing a condition called *eutrophication*. Elevated levels of nutrients in water can occur naturally or may be the result of nutrient inputs from human activities. The major nutrients are phosphorus and nitrogen. They often promote excessive growth of algae and other aquatic organisms. Algal blooms and proliferation of water plants reduce light and may release toxic compounds into the water, often killing aquatic organisms. They may also affect fish, through overutilization of oxygen, and impact boating, fishing, and recreational activities.

Water that flows on the earth's surface is exposed to many microorganisms that can cause diseases. Water flowing down streams and into lakes gathers biological impurities, which include plant particles and animal matter in various states of decay. Runoff from roads and yards can carry animal wastes to streams and lakes. Leaking and/or malfunctioning septic tanks and cesspools are the major source of fecal coliform contamination and nutrient loadings to lakes.

21.2.2   Sampling

Water samples can be collected from a boat. Sample locations just alongside the shore are logical sites for sampling, as these are areas where high concentrations of bacteria will appear if there is a problem. Areas suspect for fecal contamination (such as just below a pipe or dark-green grassed area) should be targeted, along with general locations further from the shoreline to get a comparison.

21.2.3   Evaluating Results

An example of the results of the sampling can be shown graphically with the aid of the GIS, as shown in Figure 21.2. This figure was generated by placing the sample locations obtained from a GPS into the GIS. The gaps in the range of coliform counts in the key indicate that there were no samples within this range. Fecal coliforms are generally reported as the number of bacteria colonies grown in a petri dish per 100 milliliters (mL) of sample. Coliform counts of less than 200/100 mL are typically suitable for swimming.

Figure 21.2   Portraying sampling results in the GIS provides a quick visual analysis of where problem areas (pollutant sources) may be occurring.

Contamination could be coming in from the inlet or from another source along the shore in that general area. This source could be septic systems or natural runoff.

A nice Web-enabled interactive watershed and waterbody assessment tool is Watershedss (**WATER, S**oil, and **H**ydro- **E**nvironmental **D**ecision **S**upport **S**ystem) developed by the North Carolina State University Water Quality Group as part of a grant from the U.S. Environmental Protection Agency, Office of Research and Development, National Exposure Research Laboratory, Ecosystems Research Division, Athens, GA (EPA Project #CR822270/ Grant Cooperative Agreement 818397011), *Understanding the Role of Agricultural Landscape Feature Function and Position in Achieving Environmental Endpoints.* It has an on-line decision support tool where one can, for instance, compare the ammonia levels in a lake for various temperatures and pH levels to acute or chronic levels for bass, as shown in Table 21.2.

21.2.4 Recommendations

Correcting fecal contamination problems is often more difficult because of political, rather than physical, reasons. Many shorelines are private property, and those with leaking septic systems need to be corrected. Often, the number of residents surrounding the lake does justify the cost of installing a central

TABLE 21.2 Chronic Criteria for Total Ammonia (mg/l): Warmwater Fish

Temp C	pH												
	6.6	6.8	7.0	7.2	7.4	7.6	7.8	8.0	8.2	8.4	8.6	8.8	9.0
4	2.5	2.5	2.5	2.5	2.5	2.5	2.1	1.5	0.9	0.6	0.4	0.3	0.2
6	2.4	2.4	2.4	2.4	2.4	2.4	2.1	1.5	0.9	0.6	0.4	0.2	0.2
8	2.3	2.3	2.3	2.3	2.3	2.4	2.0	1.4	0.9	0.6	0.4	0.2	0.2
10	2.3	2.3	2.3	2.3	2.3	2.3	2.0	1.4	0.9	0.6	0.4	0.2	0.2
12	2.3	2.3	2.3	2.3	2.3	2.3	2.0	1.4	0.9	0.6	0.4	0.2	0.2
14	2.2	2.2	2.2	2.2	2.2	2.2	2.0	1.4	0.9	0.6	0.4	0.2	0.2
16	2.2	2.2	2.2	2.2	2.2	2.2	1.9	1.4	0.9	0.6	0.4	0.2	0.2
18	2.2	2.2	2.2	2.2	2.2	2.2	1.9	1.3	0.9	0.6	0.4	0.3	0.2
20	2.1	2.2	2.2	2.2	2.2	2.2	1.9	1.3	0.9	0.6	0.4	0.3	0.2
22	1.9	1.9	1.9	1.9	1.9	1.9	1.6	1.2	0.8	0.5	0.3	0.2	0.2
24	1.6	1.6	1.6	1.6	1.6	1.6	1.4	1.0	0.7	0.4	0.3	0.2	0.1
26	1.4	1.4	1.4	1.4	1.4	1.4	1.2	0.9	0.6	0.4	0.3	0.2	0.1
28	1.2	1.2	1.2	1.2	1.2	1.2	1.1	0.8	0.5	0.3	0.2	0.2	0.1
30	1.1	1.1	1.1	1.1	1.1	1.1	0.9	0.7	0.5	0.3	0.2	0.2	0.1

Adapted from Missouri State Regulations, 1994, http://www.water.ncsu.edu/watershedss/index3.html.

sewer system and sewage treatment plant (STP). The malfunction should be corrected as soon as conceivably possible. This will most likely require a system repair. The other source of contamination should also be explored further.

It is also recommended that other sources of direct sewage discharge into the lake be investigated, which will require gaining permission from property owners to explore these possible sources of contamination. Although many property owners may initially resist having their surroundings explored in this fashion, they will find it preferable to contributing raw sewage to the lake.

21.2.5 Lake Management

The goal of any good watershed management plan is to maintain an undisturbed "buffer" around the lake, that is, the exact area that comprises private property, as shown in Figure 21.3. One lake stewardship program recommended harvesting the trees surrounding the shoreline to prevent the organic matter from fallen leaves from entering the lake. The remainder of the watershed, beyond the community's control—not the lake shoreline—would be the area in which to implement the tree-harvesting recommendations. In addition, selective harvesting typically removes the trees with burrows, which provide nesting opportunities for squirrels, owls, and many other wildlife species. Leaving trees hanging over the lake, or eventually falling into the lake, provides for good wildlife and fish habitat.

Sediment-laden runoff carries with it phosphorus, which is the main culprit behind lake eutrophication. The sediment itself fills the lake and increases the

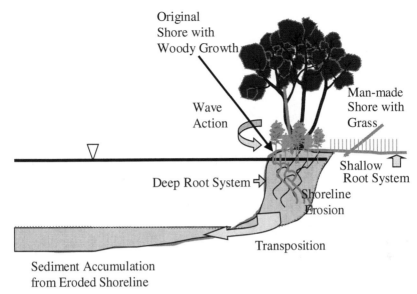

Original Shore with Woody Growth

Wave Action

Man-made Shore with Grass

Deep Root System

Shallow Root System

Shoreline Erosion

Transposition

Sediment Accumulation from Eroded Shoreline

Figure 21.3 Buffers also protect shorelines from erosion.

water temperature, both of which are detrimental to lake water quality. When making required road improvements, ways to minimize erosion and sedimentation entering the lake should be employed.

21.2.6 Beaver Ponds

In general, a beaver pond tends to shift a stream from a running-water ecosystem to more of a shallow lake environment, as shown in Figure 21.4. Locally, the beaver ponds trap sediments and organic matter and increase algal productivity. Beaver ponds also help retain and store small floods; however, the dams can wash out during extreme floods and thereby increase downstream flood damage. The dams often raise the local water table, too, and create a greater connection with the floodplain. Beaver activity breaks the forest canopy, but the ponding water often kills other trees whose roots cannot tolerate inundation. These conditions, in turn, favor the growth of riparian tree species such as alders and willows, which are a preferred food source for the beaver. The patches, edges, and dead-standing trees can result in a threefold increase in songbird species (Medin and Clary, 1990) and can dramatically enhance amphibian and mammal habitat (Olson and Hubert, 1994).

Beaver dams function very much like stormwater ponds, and exert a similar influence on downstream water quality. For example, Maret and colleagues (1987) found that beaver pond complexes in a Wyoming stream sharply reduced total suspended solid concentrations and reduced phosphorus and nitrogen by 20 to 50 percent. Beaver ponds are usually an effective buffer, and tend to increase the pH of water.

Figure 21.4   Beaver dams change the ecosystem from a stream to a "lake" environment, but recharge groundwater in the process.

At the same time, beaver ponds increase downstream water temperature, which can adversely affect trout populations at lower elevations and latitudes. In addition, decomposition and microbial action occurring within the beaver pond typically lower the dissolved oxygen content downstream. The aquatic insect community often becomes less diverse, both within and below beaver ponds, with running-water species being replaced by pond taxa (Smith, et al., 1991).

The effects of beaver dams are not temporary. Even though their construction looks a little shoddy in comparison to a stormwater pond, a typical dam and "lodge complex" is maintained for about 10 years before it is abandoned (Hammerson, 1994). The dams slow the flow of water, minimizing soil erosion and scouring. In some cases, they help restore drought areas by raising the water table and creating lush meadows (Stuebner, 1994). Table 21.3 (adapted from Hammerson, 1994) lists the effects of local or downstream changes caused by beaver dams. Some are beneficial; others are damaging to the watershed and its resources.

21.3 SUMMARY

Stream restoration requirements should be included in the watershed management plan, but only after a stormwater management plan has been devel-

TABLE 21.3 Local or Downstream Changes Caused by Beaver Dams

Number	Local or Downstream Impact
1	Storage of precipitation; gradual release during dry weather
2	Reduced current velocity
3	Increase in wetted surface area of channel by several orders of magnitude
4	Increased water depth
5	Higher elevation of the local water table
6	Decrease in amount of forest canopy
7	Loss of habitat for species that depend on live deciduous trees
8	Enhanced or degraded fish habitat and fisheries
9	Creation of habitat for species that prefer ponds, edges, and dead trees
10	Shift of aquatic insect taxa within pond to collectors and predators, and away from shredders and scrapers
11	Increase in aquatic insect emergence, per unit length of "stream"
12	Increase in algal productivity
13	Increase in trapping of sediment, and decreased turbidity
14	Favorable conditions for willow and alder
15	Increased movement of carbon, nitrogen, and other nutrients into stream
16	Reduced stream acidity (i.e., higher pH)
17	Lower oxygen levels in the spring and early summer due to decomposition
18	Increased resistance to ecosystem perturbation

oped. Lakes affect water quality and temperature. Lakes with surface discharge release water that has been heated by the sun. Lakes with a lower release allow cooler water to be released. Both affect the biotic community downstream. How the lake is managed, its shoreline, use, treatment, and so on, all have an effect on the water quality of the lake and downstream watercourses. Stream and lake management should be included in a watershed management plan.

REFERENCES

Dunne, T., and Luna L. Leopold. *Water in Environmental Planning* (New York: W.H. Freeman), 1978.

Hammerson, Geoffrey A. "Beaver (*Castor canadensis*): Ecosystem Alterations, Management, and Monitoring," *Natural Areas Journal* 14 (1994), pp. 44–57.

Maret, T.J., M. Parker, and T.E. Fannin. "The Effect of Beaver Ponds on the Nonpoint Source Water Quality of a Stream in Southwestern Wyoming," *Water Research*, 21(3) (1987), pp. 263–268.

McCandless, Tamara L., and Richard A. Everett. "Maryland Stream Survey: Bankfull Discharge and Channel Characteristics of Streams in the Piedmont Hydrologic Region," U.S. Fish and Wildlife Service, CBFO-SO2-01, Annapolis, MD, 2002.

Medin, D.E., and W.P. Clary. "Bird Populations in and Adjacent to a Beaver Pond Ecosystem and Adjacent Riparian Habitat in Idaho," Intermountain Forest and Range Experimental Station, USDA Forest Service, Ogden, UT, 1990.

———. "Small Mammals of a Beaver Pond Ecosystem and Adjacent Riparian Habitat in Idaho," Intermountain For. and Range Exp. Sta., USDA Forest Service, Ogden, UT, 1991.

Olson, R., and W.A. Hubert. "Beaver: Water Resources and Riparian Habitat Manager" (Laramie, WY: University of Wyoming), 1994.

Rosgen, David. *Applied River Morphology* (Pagosa Springs, CO: Wildland Hydrology), 1996.

Smith, M.E., C.T. Driscoll, B.J. Wyskowski, C.M. Brooks, and C.C. Costentini. "Modification of Stream Ecosystem Structure and Function by Beaver," *Canadian Journal of Zoology* 69 1 (1991), pp. 55–61.

Stormwater Center, Stream Restoration: Grade Control Practices, On-Line Fact Sheet www.stormwatercenter.net / Assorted%20Fact%20Sheets / Restoration / grade_control. htm, accessed September 12, 2003.

Stuebner, S. "Bullish on Beavers," *National Wildlife,* 32 (1994), pp. 24–27.

White, Kirk E. "Regional Curve Development and Selection of a Reference Reach in the Non-Urban, Lowland Sections of the Piedment Physiographic Province, Pennsylvania and Maryland," Water-Resources Investigations Report 01-4146, U.S. Geologic Survey and Pennsylvania Department of Environmental Protection, New Cumberland, PA, 2001.

CHAPTER 22

GROUNDWATER AND INTEGRATED WATER RESOURCES MANAGEMENT

22.0 INTRODUCTION

Managing the nation's groundwater quantity and quality is as important, if not more so, than managing surface water, because groundwater supplies are more difficult to replenish, and take longer and are more expensive to clean up once contaminated. Residential, commercial, and industrial growth has tremendous implications and puts tremendous pressures on groundwater resources. Groundwater withdrawals increase with each new well drilled, possibly affecting the water table. Subsurface wastewater disposal systems, leaking underground storage tanks, improperly stored road salt, and pesticide applications, among others, all could pose a threat to drinking water supplies. As with surface water, groundwater is best managed through preventative measures, which include water supply and wellhead protection programs to preserve the integrity of existing water supply sources.

This chapter addresses the main points that should be evaluated during the development of a water supply and/or wellhead protection plan. Section 1428 of the Federal Safe Drinking Water Act Amendments, passed in 1986, requires states to establish programs to safeguard public water supplies from contamination. These programs may be in the form of state incentive grants to ensure the public's health, safety, and welfare by maintaining water supply quality.

22.1 WATER SUPPLY PLAN

Water supply may come from groundwater wells and surface water reservoirs or river/stream intake. Various classifications of water supply systems include small *community systems* (sometimes defined as those serving 25 persons on a regular basis, or with more than 15 service connections), *nontransient community systems* (serving individuals for at least six months of the year), and *noncommunity systems* (supplying transient populations). Community systems would include municipal housing developments, single-family homes, mobile home parks, and nursing care facilities. A water supply plan involves assessing the water source quantity and quality of existing water suppliers and developing an action plan for future water supplies. It may also involve conducting a regionalization study, an assessment of the advantages and disadvantages of combining various water supply and/or distribution systems.

A water supply plan should, therefore, address the following elements:

- Source-specific quality and quantity information
- Groundwater versus surface water sources
- Single source or multiple sources
- Quality and quantity projections
- Balance of aquifer recharge and withdrawal
- Nonpoint source pollution assessment
- Drought analysis and planning
- Storage capability (surface or tanks)
- Infrastructure analysis—capital improvements program
- Deficiency identification
- Water conservation and reuse measures
- Multisystem, municipal, and agency coordination
- Operation and financial analysis
- Monitoring program
- Wellhead protection program
- Hazardous materials management plan
- Proposals for meeting identified drinking water needs
- Contingency, emergency, and security planning
- Conservation land development practices to maintain the hydrologic balance
- Integration of surface water, groundwater, wastewater, and land use management controls and policies
- Formal policy of expansion review to ensure hydrologic balance
- Possible incentives programs
- Encouragement of industries that utilize water-conserving measures
- Possible water-based zoning to preserve water supplies
- Public education

All water systems should be mapped, along with their sources. Areas contributing groundwater and surface water to the sources should also be mapped in the GIS and incorporated into the watershed assessment for possible future protection and preservation.

22.2 WELLHEAD PROTECTION STUDY

The wellhead protection study identifies various zones of contribution to the well and water supply based on hydrogeologic modeling. The premise of the wellhead protection plan is to identify these *wellhead protection areas* (*WHPAs*), then preserve the recharge characteristics of these areas through land use or management controls. The wellhead protection plan often includes a pollution vulnerability assessment, a contingency management plan, a model wellhead protection ordinance, and various pollution-prevention recommendations.

The steps in developing a wellhead protection plan are detailed in the Environmental Protection Agency's report "Wellhead Protection: A Guide for Small Communities," and would include the steps delineated in the following subsections (modified from U.S. EPA, 1993).

22.2.1 Step 1: Form a Community Planning Team

This team could be part of the watershed assessment committee or a subcommittee of that committee, and may include a technical advisory committee.

22.2.2 Step 2: Delineate the Wellhead Protection Area

Groundwater flow systems can be identified by their underground travel time, or the time it takes to get from the recharge area to the discharge area. Local groundwater systems are those whose recharge and discharge areas are relatively close and from which it takes only a few days to a few years to travel the distance. In contrast, intermediate flow takes a few decades, and regional flow systems may take millennia (Toth, 1963). A generalized groundwater flow map can be constructed for unconfined conditions by obtaining static water levels and generalizing the altitude of the groundwater surface to mimic the land surface with varying degrees of depth, based on recharge potential.

Methods to delineate WHPAs vary with available data, which will dictate the complexity of the method chosen. *The Arbitrary Fixed Radius (AFR)* method entails drawing a circle around a wellhead using a fixed radius, determined from geologic and sociopolitical factors. The *Calculated Fixed Radius (CFR)* method utilizes an accepted formula to calculate the radius, based on hydrogeologic conditions, in contrast to the most intense method, which utilizes an accepted hydrogeologic model such as MODFLOW (McDonald and Harbaugh, 1988). Deciding which method to choose is typically based on geologic and aquifer conditions, amount of available detailed data, budget, and intended future management tools and use. Wellhead protection area delineation and conceptual groundwater flow models should be based on geology and aquifer type, and consider water-level data from existing water supply or observation wells, aquifer geometry, and structural fabrics of the bedrock.

Wellhead protection areas are typically classified into three zones, as follows (U.S. EPA, 1993).

22.2.2.1   Zone I. Zone I is the protective area immediately surrounding the well, well field, or spring, and is typically an arbitrary calculated fixed radius, ranging from 100 to 400 feet. The radius would vary depending on site-specific hydrogeologic and aquifer conditions. This zone is often the radius of the *cone of depression* if this data is available. The cone of depression is the depression of hydraulic heads caused by the pumping or withdrawal of water.

Contamination events occurring within this zone could travel quickly to the well, and thus Zone I is a contingency action zone where response time in the event of a contamination event will be short. Storage or use of contaminants should be restricted in this zone, and any spills reported immediately to the water supplier.

22.2.2.2   Zone II. Zone II encompasses the portion of the aquifer through which water is diverted to the well, well field, or spring, and travel time to the wellhead is important. This time is typically determined through detailed hydrogeologic modeling using finite difference; for example, MODFLOW (McDonald and Harbaugh, 1988) coupled with Visual MODFLOW Pro v3.1 (Waterloo Hydrogeologic Software, 2003); finite element, for example FEFLOW v5.0 (Waterloo Hydrogeologic Software, 2003); or analytic, for example WHPA (U.S. EPA, 1991) methods.

Typical data required for modeling wellhead protection zones includes:

- Aquifer type(s): confined, semiconfined, unconfined
- Hydraulic conductivity
- Storativity of the aquifer
- Storativity of the confining units
- Hydraulic gradient
- Aquifer porosity
- Aquifer-saturated thickness
- Confining unit thickness
- Recharge rate
- Aquifer geometry and boundaries
- Directional and aerial variation in hydrogeologic parameters
- Well-pumping test data and rates

In the absence of data to accurately determine Zone II, an arbitrary calculated fixed radius averaging about a half mile is often used. A time-of-travel for a contaminant to go from the upper reaches to the well is determined; the zone is determined from this basis. A typical time-of-travel would be 10 to 25 years (U.S. EPA, 1993). Contaminant management and remedial action activities are given special attention in this zone.

22.2.2.3 *Zone III.* Zone III is the area that contributes water supply to Zones I and II and is typically delineated as the watershed boundary to the point of interest (well), including any downslope area that contributes ground-water to the well, possibly due to the angle and direction of a cleavage plane. If the groundwater divide is known to extend beyond the watershed boundary, the groundwater divide would be utilized as the upper boundary.

A typical wellhead delineation is given in Figure 22.1.

22.2.3 Step 3: Conduct a Pollution Vulnerability Assessment

Groundwater resources must be protected. One means to aid in this effort is to determine the vulnerability of groundwater to potential contamination should a pollution event, such as a chemical spill, occur. Following such an

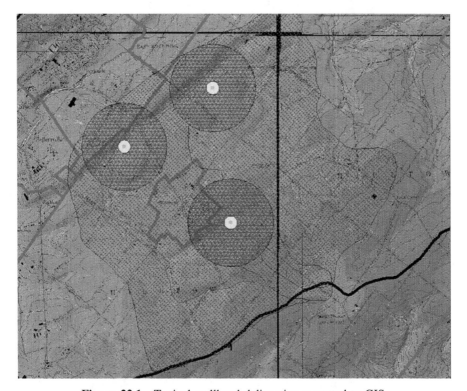

Figure 22.1 Typical wellhead delineation exported to GIS.

event, the various soil and geologic material, land slopes, and distance to streams or recharge areas will factor into how quickly, and to what extent, the contaminant will affect the groundwater.

The pollution vulnerability assessment should include much of the same information required for watershed surface water assessment, that is: physiography, geology, soils, land use, and so on. (Reese and DeBarry, 1993; Vbra and Zaporozec, 1994). The development of this data is conducive to the GIS, as discussed in previous chapters. The geology, rock type, folding, faulting, and any other physical features that may affect groundwater flow direction, rate, and quality also should be analyzed. Hydrogeologic units, or aquifers based on groundwater flow characteristics—as well as whether these are consolidated or unconsolidated materials—should be identified. Stratification from drilling logs can be identified; and it may be possible to determine well yield from pumping tests. Groundwater flow systems, *local, intermediate,* or *regional,* can then be identified by analyzing static well conditions in conjunction with topographic conditions and geology, and used to assess the lengths of groundwater flow paths.

Although actual pollution vulnerability can be determined only through in-depth site-specific analysis, a pollution vulnerability assessment can be performed within the GIS on a regional basis by overlaying many of the layers described in Chapters 10 and 12. By applying attribute information and overlaying the various factors that affect or influence the capability of a pollutant to reach the groundwater system, a pollution vulnerability map can be generated within the GIS (Reese and DeBarry, 1993).

Rating the parameters just given, and overlaying the layers in the GIS, Reese and DeBarry (1993) delineated a procedure to develop a pollution vulnerability map for Lehigh and Northampton Counties in east-central Pennsylvania, a process they named an *Environmental Landscape Classification System (ELCS).* The ELCS was based on the premise that hydrogeologic elements (i.e., geology, soils, surface water, and groundwater) are interconnected and exhibit predictable and regional patterns. Although predictable, however, the patterns will vary within a region, and the ELCS can aid in pinpointing these differences based on the interrelationships of the landscape features. Three parameters were used to assess vulnerability of bedrock aquifers, namely *glacial deposits and bedrock type, soils,* and *topography* (slopes). A brief description of how each influences pollution vulnerability follows.

Geology. In areas of glacial till, the depth and texture of the till needs to be analyzed. Till may provide an additional layer of filtration between the surface and the bedrock aquifers. Alluvial deposits, including outwash and kame deposits of sand and gravel, often serve as shallow aquifer water supplies. These deposits are highly vulnerable, regardless of the soil cover. A localized or perched water table within the till should also be considered in the vulnerability assessment. Rock type and *transmissivity* can be assigned to each geologic unit. Transmissivity is the capacity of an aquifer to transmit water that is equal to the *hydraulic conductivity* of the aquifer times its thickness (square feet per day). Hydraulic conductivity is the capacity of a material to transmit water, expressed in terms of the volume of water that will move per unit time under a unit *hydraulic gradient* through a unit area measured at right angles to the direction of flow (gpd/sf). Hydraulic gradient is the slope of the water table—or more generally, *potentiometric surface*—and can be determined by measuring the static water elevation in a number of wells located in the same aquifer (ft/ft). Wells and quarries may provide a direct conduit for surface contamination to enter the groundwater system, and should also be provided with a buffer in the GIS and given a "high" vulnerability rating.

Soils. The capability of soils to filter or adsorb pollutants depends primarily on texture, structure, composition, and depth. Permeability affects how quickly the pollutants will travel through the soils. Permeability can be assigned within the GIS to each soil series polygon. Another option would be to assign the hydrologic soil groups to give a relative indication of permeability. Natural protection of bedrock aquifers is provided by soil and sediment cover. Highly permeable soils (hydrologic soil groups A and B) provide little protection, whereas less permeable soils provide progressively greater levels of protection. Soil depth can vary from very thin, discontinuous, to very thick. This data can also be obtained from the MUIR tables.

Slopes. Utilizing the DEM, slope classifications can be developed. Steep terrain typically has shallower soils that would filter or adsorb pollution due to erosion and downslope sediment transport. Steep terrain also affects the rate and flow of surface runoff, therefore a contaminant spill on a steep slope would be of concern in regard to surface water. Flatter slopes generally will have more infiltration than steeper slopes (depending on soils, of course) and, therefore, raise the potential for pollutants to enter the groundwater system; however, water tables may be near or at the surface, providing a conduit for direct pollution migration to the groundwater system. This would be representative of recharge areas. That said, where the water bodies are classified as "gaining," contaminated groundwater would contribute to the surface water. After evaluating hydrogeologic conditions, Reese and DeBarry established a "buffer" of 75 meters within the GIS to indicate high vulnerability. They classed slopes of greater than 15 percent and within 75 meters of a stream as high vulnerability, and all other areas as low-pollution vulnerability. Steeper slopes (greater than 15 percent), also have less development potential.

Once this GIS data has been developed and fully attributed, a pollution vulnerability map can be drawn. One study (Monroe County Planning Commission, 1997) assigned the following classifications through the overlaying process. (Note: The stream buffer, which provides an additional layer of assessment, can be visually displayed in the GIS.)

Very High Vulnerability	Valley alluvium and glacial till aquifers
Very High Vulnerability	Hydrogeologic Units 1, 6, 7—HSG A, B
High Vulnerability	Hydrogeologic Units 1, 6, 7—HSG C
High to Moderate Vulnerability	Hydrogeologic Units 1, 6, 7—HSG D
Moderate to High Vulnerability	Hydrogeologic Unit 2—HSG A, B
Moderate Vulnerability	Hydrogeologic Unit 2—HSG C
Moderate to Low Vulnerability	Hydrogeologic Unit 3, 4, 5, 8—HSG A, B
Low to Moderate Vulnerability	Hydrogeologic Unit 3, 4, 5, 8—HSG C
Low Vulnerability	Hydrogeologic Unit 3, 4, 5, 8—HSG D

A numbered listing of the hydrogeologic units is given here, along with a brief description, to enable the reader to visualize the classification scheme.

1. Unconsolidated sediments: Sediments not combined into a single or solid mass deposited by gravity or water.
2. Catskill Formation: Conglomerate, sandstone, siltstone, and shale with a well-developed joint system; includes seven varying members, 0 to 1,825 feet thick.
3. Trimmers Rock Formation: Fissile siltstones and minor shales; possesses a well-developed joint system, 600 to 17,00 feet thick.
4. Mahatango Formation: Siltstone and shale; bedding largely masked by cleavage.
5. Marcellus Formation: Carbonaceous fissile shale; joints have no pronounced orientation.
6. Undifferentiated Devonian and Silurian Rocks: Complex sequence of carbonates, shales, siltstones, and minor calcareous sandstones and conglomerate; includes 14 varying members, 0 to 270 feet thick.
7. Bloomsburg Formation: Sandstone and conglomerate, with some siltstone and shale; 1,500 feet thick
8. Shawangunk Formation: Quartzose sandstone and conglomerate, some siltstone and sale; 1,400 feet thick.

Hydrogeologic units 2 though 8 were ranked based on transmissivity. Generally, wells located in bedrock with low transmissivity will have smaller time-of-travel capture zones per unit drawdown, and thus a lower pollution vulnerability. A Pollution Vulnerability map is shown in Figure 22.2.

The final pollution vulnerability map illustrates the relative vulnerability of water supplies to pollution from surface or near-surface release of contaminants. This can be created and managed in the GIS, and can provide a spatial relationship between geology, glacial deposits, soils, topography, land use

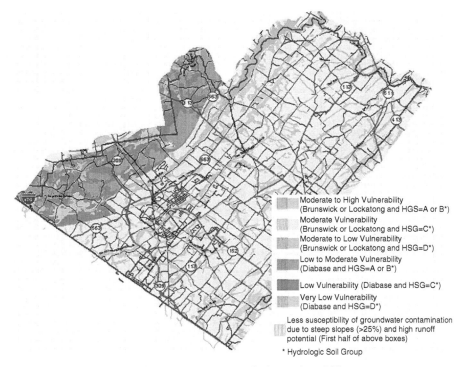

Moderate to High Vulnerability
(Brunswick or Lockatong and HGS=A or B*)
Moderate Vulnerability
(Brunswick or Lockatong and HSG=C*)
Moderate to Low Vulnerability
(Brunswick or Lockatong and HSG=D*)
Low to Moderate Vulnerability
(Diabase and HGS=A or B*)

Low Vulnerability (Diabase and HSG=C*)
Very Low Vulnerability
(Diabase and HSG=D*)

Less susceptibility of groundwater contamination
due to steep slopes (>25%) and high runoff
potential (First half of above boxes)

* Hydrologic Soil Group

Figure 22.2   GIS-generated pollution vulnerability map.

activities, contamination sites, wells, and groundwater systems. It then becomes a tool for defining critical areas surrounding wells and water supplies to be protected, aids in land use and zoning decisions, allows water suppliers to assess relative vulnerability of their well-water supplies, and aids in new well siting. The map also aids in contingency planning. Furthermore, combining the vulnerability map with land use activities and patterns, future growth patterns, and contaminant locations and types enables identification of possible impacts on groundwater quality and supplies from these various activities. This information aids in long-term water supply protection, preservation, and development efforts.

This concept can be expanded to include such applications as potential contamination from septic systems, if they were mapped and categorized. Threats to groundwater can then be mapped to determine potential future problems. Other threats might include transportation routes (spills), manufacturing facilities, dry cleaners, gas stations, salt storage areas, railroads, quarries, stormwater recharge facilities, and others. Buffers can be placed to surround the point or line features in the GIS; and, again, overlaying with the wells' zone of contributions will allow for a visual assessment of future potential contamination problems.

22.2.4 Step 4: Identify and Locate All Potential Sources of Contamination

These potential sources of contamination would include those listed in Table 22.1. All specific sites identified should be located in the GIS and attributed based on use.

22.2.5 Identify and Apply Management Tools to the Delineated WHPAs

This task involves analyzing all the regulatory and nonregulatory means of managing the wellhead protection area for pollution prevention measures. Regulatory controls may include revising the zoning ordinance, citing violations, and so on. Nonregulatory controls may include appointing clean-up days, disturbing educational brochures, conducting workshops, and the like. A model wellhead protection ordinance should be included in the wellhead protection plan. Wellhead protection tools are listed and described in Table 22.2 (Horsley and Witten, 1989).

22.2.6 Develop a Contingency Management Plan

Contingency management planners should examine such factors as alternate sources of water, terrorism safeguards and security issues, water quality preservation, and so on.

22.3 INTEGRATED WATER RESOURCES MANAGEMENT

As established throughout this text, the processes used for data collection and analysis, to set goals and objectives, and to establish management procedures are all very similar and integrated, whether they are used for watershed assessment and restoration, stormwater or floodplain management, or water supply and groundwater management. Likewise, the Clean Water Act and Safe Drinking Water Act, though each has specific goals, integrate surface water and groundwater management. But much of the regulatory framework, which categorizes specific objectives, promotes the separate evaluation of the various disciplines, often resulting in duplication of tasks, efforts, and spending. It would make better sense to link all phases of the hydrologic cycle into a truly integrated water resources assessment and management plan. Doing so is the best way to obtain sustainability of our water resources.

One such plan was developed for the Pennridge Area Coordinating Committee, a collection of seven municipalities located in Bucks County, Pennsylvania. This truly progressive group, in acknowledgment of the development pressures that were being imposed on their land and water resources, obtained grants, with the aid of the County Planning Commission, to develop the Penn-

TABLE 22.1   Potential Sources of Contamination (*Source:* U.S. EPA, 1997)

Agriculture
  Crop-related sources
  Irrigated crop production
  Nonirrigated crop production
  Specialty crop production (e.g., horticulture, citrus, nuts, fruits)
  *Grazing-related sources**
  Pasture grazing—riparian and/or upland*
  Pasture grazing—riparian*
  Pasture grazing—upland*
  Range grazing—riparian and/or upland*
  Range grazing—riparian*
  Range grazing—upland*
  *Intensive animal feeding operations**
  Concentrated animal feeding operations (CAFOs; permitted; PS)*
  Confined animal feeding operations (NPS)*
  Aquaculture*
Atmospheric deposition
Collection system failure
Combined sewer overflow*
Construction
  Highway/road/bridge construction
  Land development
Contaminated sediments
Debris and bottom deposits
Domestic wastewater lagoon*
Erosion from derelict land
Groundwater loadings
Groundwater withdrawal
Habitat modification (other than hydromodification)
  Removal of riparian vegetation
  Bank or shoreline modification/destabilization
  Drainage/filling of wetlands
Highway maintenance and runoff
Hydromodification
  Channelization
  Dredging
  Dam construction
  Upstream impoundment
  Flow regulations/modification
Industrial Point Sources
  Major industrial point sources
  Minor industrial point sources
Internal nutrient cycling (primarily lakes)

TABLE 22.1 *(Continued)*

Land disposal
   Sludge*
   Wastewater*
   Landfills*
   Inappropriate waste disposal/wildcat dumping*
   Industrial land treatment
   On-site wastewater systems (septic tanks)*
   Hazardous waste
   Septage disposal*
Leaking underground storage tanks
Marinas and recreational boating
   In-water releases*
   On-land releases*
Municipal point sources*
   Major municipal point sources—dry and/or wet weather discharges*
   Major municipal point sources—dry weather discharges*
   Major municipal point sources—wet weather discharges*
   Minor municipal point sources—dry and/or wet weather discharges*
   Minor municipal point sources—dry weather discharges*
   Minor municipal point sources—wet weather discharges*
   Package plants (small flows)*
Natural sources (e.g., arsenic, radon, wildlife)*
Other
Recreation and tourism activities (other than boating)
   Golf courses
Resource extraction
   Surface mining
   Subsurface mining
   Placer mining
   Dredge mining
   Petroleum activities
   Mill tailings
   Mine tailings
   Acid mine drainage
   Abandoned mining
   Inactive mining
Salt storage sites
Sediment resuspension
Sewer lines (leaking)*
Silviculture
   Harvesting, restoration, residue management
   Forest management (e.g., pumped drainage, fertilization, pesticide application)
   Logging road construction/maintenance
   Silvicultural point sources
Sources outside state jurisdiction or borders*

TABLE 22.1 *(Continued)*

Spills (accidental)*
Unknown source*
Urban runoff/storm sewers*
    Nonindustrial permitted*
    Industrial permitted*
    Other urban runoff*
    Illicit connections/illegal hookups/dry weather flows*
    Highway/road/bridge runoff
    Erosion and sedimentation
Waste storage/storage tank leaks (above ground)*

* Indicates sources of microbial contaminants.

Note: EPA's 305(b) guidance also asks states to identify major sources of contamination of waters designated for drinking water supply use. States are urged to coordinate their source water and 305(b) information and programs.

ridge Water Resources Plan. This plan evaluated the area's stormwater, flood water, surface water, and groundwater resources, as well as the anthropogenic factors, such as groundwater withdrawal, nonpoint source pollution, and stream bank erosion, to develop a long-term integrated management plan on a subwatershed basis (DeBarry and Livrone, 2001). The goal of the plan is to "address concerns about future growth and development in a proactive manner by developing a water resources and growth management plan, addressing both quantity and quality issues." The committee strived to develop a plan that would maintain the hydrologic balance and hydrologic regime of the watershed in the Pennridge area through "good science and smart planning," as Eric Schefhauser, one of its members, put it.

A detailed GIS analysis was performed to evaluate natural and man-made factors affecting the watersheds; and a management, monitoring, and growth management plan was devised to ensure water resource sustainability. Gaging station locations were proposed to monitor water quantity and quality. Stormwater management criteria were defined in conjunction with a concurrent watershed stormwater management plan, and incorporated into ordinances, to preserve water quality and groundwater recharge and to prevent stream bank erosion, as well as to manage stormwater quantity. Incentives for water conservation by industry, and a water-based zoning initiative, were also explored. The committee applied for additional grants to delineate wellheads, develop a surface water protection program, and conduct a nonpoint source pollution assessment; a strong educational component and Web site (http:// hilltown.org/pacc/) were included as part of the plan.

One of the goals of the plan is to collect data on water quality and static water levels and yields in wells, then develop an interactive live three-dimensional GIS model of the area showing existing groundwater levels in relation to the hydrologic balance, as shown in Figure 22.3. Proposed with-

TABLE 22.2 Summary of Local Wellhead Protection Tools

Tools	Applicability to Wellhead Protection	Land Use Practice	Legal Considerations	Administrative Considerations
		Regulatory: Zoning		
Overlay Groundwater Protection Districts	Used to map wellhead protection areas (WHPAs). Provides for identification of sensitive areas for protection. Used in conjunction with other tools that follow.	Community identifies WHPAs in practical base/zoning map.	Well-accepted method of identifying sensitive areas. May face legal challenges if WHPA boundaries are based solely on arbitrary delineation.	Requires staff to develop overlay map. Inherent nature of zoning provides "grandfather" protection to preexisting uses and structures.
Prohibition of Various Land Uses	Used within mapped WHPAs to prohibit groundwater contaminants and uses that generate contaminants.	Community adopts prohibited uses list within their zoning ordinance.	Well-organized function of zoning. Appropriate techniques to protect natural resources from contamination.	Requires amendment to zoning ordinance. Requires enforcement by both visual inspection and on-site investigations.
Special Permitting	Used to restrict uses within WHPAs that may cause groundwater contamination if left unregulated.	Community adopts special permit "thresholds" for various uses and structures within WHPAs. Community grants special permits for "threshold" uses only if groundwater quality will not be compromised.	Well-organized method of segregating land uses within critical resource areas such as WHPAs. Requires case-by-case analysis to ensure equal treatment of applicants.	Requires detailed understanding of WHPA sensitivity by local permit-granting authority. Requires enforcement of special permit requirements and on-site investigations.

Technique	Purpose	Description	Comments	Administration
Large-Lot Zoning	Used to reduce impacts of residential development by limiting numbers of units within WHPAs.	Community "down zones" to increase minimum acreage needed for residential development.	Well-recognized prerogative of local government. Requires rational connection between minimum lot size selected and resource protection goals. Arbitrary large-lot zones have been struck down without logical connection to Master Plan or WHPA program.	Requires amendment to zoning ordinance.
Cluster/PUD Design	Used to guide residential development of WHPAs. Allows for "point source" discharges that are more easily monitored.	Community offers cluster/PUD as development option within zoning ordinance. Community identifies areas where cluster/PUD is allowed (i.e., within WHPAs).	Well-accepted option for residential land development.	Slightly more complicated to administer than traditional "grid" subdivision. Enforcement/Inspection requirements are similar to "grid" subdivision.
Growth Controls/Timing	Used to time the occurrence of development within WHPAs. Gives communities the opportunity to plan for wellhead delineation and protection.	Community imposes growth controls in the form of building caps, subdivision phasing, or other limitations tied to planning concerns.	Well-accepted option for communities facing development pressures within sensitive resource areas. Growth controls may be challenged if they are imposed without a rational connection to the resource being protected.	Generally complicated administrative process. Requires administrative staff to issue permits and enforce growth control ordinances.

TABLE 22.2 *(Continued)*

Tools	Applicability to Wellhead Protection	Land Use Practice	Legal Considerations	Administrative Considerations
Performance Standards	Used to regulate development within WHPAs by enforcing predetermined standards for water quality. Allows for aggressive protection of WHPAs by limiting development within WHPAs to an accepted level.	Community identifies WHPAs and established "thresholds" for water quality.	Adoption of specific WHPA performance standards requires sound technical support. Performance standards must be enforced on a case-by-case basis.	Complex administrative requirements to evaluate impacts of land development within WHPAs.
		Regulatory: Subdivision Control		
Drainage Requirements	Used to ensure that subdivision road drainage is directed outside of WHPAs. Used to employ advanced engineering designs of subdivision roads within WHPAs.	Community adopts stringent subdivision rules and regulations to regulate road drainage/runoff in subdivisions within WHPAs.	Well-accepted purpose of subdivision control.	Requires moderate level of inspection and enforcement by administrative staff.
		Regulatory: Health Regulation		
Underground Fuel Storage Systems	Used to prohibit underground fuel storage systems (USTs) within WHPAs and to regulate USTs within WHPAs.	Community adopts health/zoning ordinance prohibiting USTs within WHPAs. Community adopts special permit or performance standards for use of USTs within WHPAs.	Well-accepted regulatory option for local government.	Prohibition of USTs require little administrative support. Regulating USTs requires moderate amounts of administrative support for inspection follow-up and enforcement.

Privately Owned Wastewater Treatment Plants (Small Sewage Treatment Plants)	Used to prohibit small sewage treatment plants (SSTP) within WHPAs.	Community adopts health/zoning ordinance within WHPAs. Community adopts special permit or performance standards for use of SSTPs within WHPAs.	Well-accepted regulatory option for local government.	Prohibition of SSTPs requires little administrative support. Regulating SSTPs requires moderate amount of administrative support of inspection follow-up and enforcement.
Septic Cleaner Ban	Used to prohibit the application of certain solvents, septic cleaners, known groundwater containments, within WHPAs.	Community adopts health/zoning ordinance prohibiting the use of septic cleaners containing 1,1,1-trichloroethane or other solvent compounds within WHPAs.	Well-accepted method of protecting groundwater quality.	Difficult to enforce even with sufficient administrative support.
Septic System Upgrades	Used to require periodic inspection and upgrading of septic systems.	Community adopts health/zoning ordinance requiring inspection and, if necessary, upgrading of septic systems on a time basis (e.g., every two years) or upon title/property transfer.	Well-accepted as within purview of government to ensure protection of groundwater quality.	Significant administrative resources required for this option.
Toxic and Hazardous Materials Handling Regulations	Used to ensure proper handling and disposal of toxic materials/waste.	Community adopts health/zoning ordinance requiring registration and inspection of all businesses within WHPA using toxic/hazardous materials above certain quantities.	Well accepted as within purview of government to ensure protection of groundwater.	Requires administrative support and on-site inspections.

TABLE 22.2 (Continued)

	Applicability to Wellhead Protection	Land Use Practice	Legal Considerations	Administrative Considerations
Tools				
Private Well Protection	Used to protect private on-site water supply wells.	Community adopts health/zoning ordinance to require permits for new private wells and to ensure appropriate well-to-septic-system setbacks. Also requires pump and water quality testing.	Well accepted as within purview of government to ensure protection of groundwater.	Requires administrative support and review of applications.
Nonregulatory: Land Transfer and Voluntary Restriction				
Sale/Donation	Land acquired by a community within WHPAs, either by purchase or donation. Provides broad protection to the groundwater supply.	As nonregulatory technique, communities generally work in partnership with nonprofit land conservation organizations.	There are many legal consequences of accepting land for donation or sale from the private sector, mostly involving liability.	There are few administrative requirements involved in accepting donations or sales of land from the private sector. Administrative requirements for maintenance of land accepted or purchased may be substantial, particularly if the community does not have a program for open space management.
Conservation Easements	Can be used to limit development within WHPAs.	Similar to sales/donations, conservation easements are generally obtained with the assistance of nonprofit land conservation organization.	Same as above.	Same as above.

Technique	Description		
Limited Development	As the title implies, this technique limits development to portions of a land parcel outside of WHPAs.	Land developers work with community as part of a cluster/PUD to develop limited portions of a site and restrict other portions, particularly those with WHPAs.	Similar to those noted in cluster/PUD under zoning.	Similar to those noted in cluster/PUD under zoning.
		Nonregulatory: Other		
Monitoring	Used to monitor groundwater quality within WHPAs.	Communities establish groundwater monitoring program within WHPAs. Communities require developers within WHPAs to monitor groundwater quality down-gradient from their development.	Accepted method of ensuring groundwater quality.	Requires moderate administrative staffing to ensure routing sampling and response if sampling indicates contamination.
Contingency Plans	Used to ensure appropriate response in cases of contaminant release or other emergencies within WHPA.	Community prepares a contingency plan involving wide range of municipal/county officials.	None	Requires significant up-front planning to anticipate and be prepared for emergencies.
Hazardous Waste Collection	Used to reduce accumulation of hazardous materials within WHPAs and the community at large.	Communities, in cooperation with the state, regional planning commission, or other entity, sponsor a "hazardous waste collection day" several times per year.	There are several legal issues raised by the collection, transport, and disposal of hazardous waste.	Hazardous waste collection programs are generally sponsored by government agencies, but administered by a private contractor.

TABLE 22.2 (*Continued*)

Tools	Applicability to Wellhead Protection	Land Use Practice	Legal Considerations	Administrative Considerations
Public Education	Used to inform community residents of the connection between land use within WHPAs and drinking water quality.	Communities can employ a variety of public education techniques ranging from brochures detailing their WHPA program to seminars to involvement in events such as hazardous waste collection days.	No outstanding legal considerations.	Requires some degree of administrative support for programs such as brochure mailing to more intensive support for seminars and hazardous waste collection days.
Legislative				
Regional WHPA Districts	Used to protect regional aquifer systems by establishing new legislative districts that often transcend existing corporate boundaries.	Requires state legislative action to create a new legislative authority.	Well-accepted method of protecting regional groundwater resources.	Administrative requirements will vary depending on the goal of the regional district. Mapping of the regional WHPAs requires moderate administrative support, while creating land use controls within the WHPA will require significant administrative personnel and support.
Land Banking	Used to acquire and protect land within WHPAs.	Land banks are usually accomplished with a transfer tax established by state government empowering local government to impose a tax on the transfer of land from one party to another.	Land banks can be subject to legal challenge as an unjust tax, but have been accepted as a legitimate method of raising revenue for resource protection.	Land banks require significant administrative support if they are to function effectively.

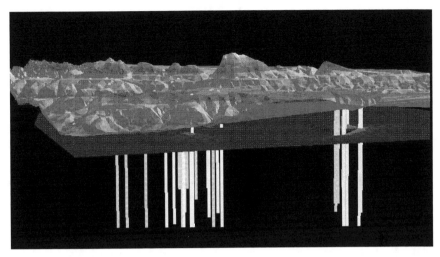

Figure 22.3 Interactive GIS water table model of the Pennridge area (DeBarry and Livrone, 2001).

drawals can then be plugged into the system to determine its impact on water levels and supply before the situation becomes stressed.

Another example of an integrated water resources management plan is called Watersheds, designed for Chester County, Pennsylvania, and its watersheds (Chester County Water Resources Authority, 2001). This plan also strives to sustain land and water resources, and proposes methods and ordinance provisions to maintain the hydrologic regime. This plan can be found on the Web at www.chesco.org/water/ExecSummary/toc.pdf.

The landmass between Philadelphia, Trenton, New Jersey, and New York City, most of which is in the Delaware River Basin, is one of the fastest-growing communities in the nation. Realizing this, the Delaware River Basin Commission (DRBC) developed the Southeastern Pennsylvania Ground Water Protected Area (GWPA) and associated regulations (GWPAR). The regulations established numerical groundwater withdrawal limits on a subwatershed basis that fall partially or within this GWPA; they are more restrictive than those applying to the remainder of the Delaware River Basin.

Because the Delaware River supplies drinking water to New York City, Trenton, and Philadelphia, maintaining adequate base flow in the river is imperative, to prevent saltwater migration upstream and to maintain water supply to Philadelphia and Trenton in times of drought. The typical monitoring scenario is shown in Figure 22.4. The goal is to prevent the depletion of groundwater and at the same time, protect the interests and rights of lawful users of the same water source. In addition to concerns about the river, recent droughts have raised concerns about localized streams and groundwater supplies.

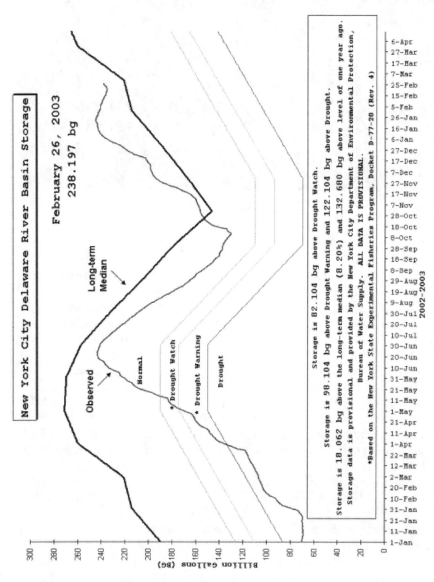

Figure 22.4 Delaware River storage.

DRBC has also devised a protocol and guidelines for developing Integrated Resource Plans (IRP) (Delaware River Basin Commission, 2001). The IRP guidelines purposes are to:

- Evaluate and develop management objectives and strategies on a subbasin basis to ensure that groundwater and surface water withdrawals are managed in a manner that protects both instream and withdrawal uses in the subbasin.
- Evaluate the adequacy of existing groundwater and surface water resources to meet all existing and future needs in the subbasin, and assess options for meeting those needs.
- Engage stakeholders as active participants in developing effective, long-term water resource management objectives and strategies.
- Consider the interrelationship of water quality and water availability for current and future water uses in a subbasin.
- Assist planners to better integrate water resources protection in land use planning. Almost all land use decisions affect water resources. Recognizing that growth will occur in most subbasins, Integrated Resource Plans can assist in better managing *how* that growth occurs. The availability of groundwater or surface water, individually, may not be a limiting factor for growth, as a combination of both, or sources of water outside of the subbasin, may exist. By evaluating all water resources options, existing and future needs may be met while simultaneously protecting the resources and supporting other uses, including instream flow needs.

The IRP guidelines require an integrated approach according to the following outline, which can be found at www.state.nj.us/drbc/Res2002-7.htm.

I. Incorporate public participation.
II. Assess water resources and existing uses of water.
  A. Existing water resources assessment: The water resources assessment should include an evaluation of the following:
    - Subbasin water resources
    - Subbasin water resources for water supply purposes
  B. Existing water use assessment
    - Water use for water supply purposes
    - Water use to maintain instream needs
III. Estimate future water demands and resource requirements.
  A. Water use for water supply purposes
    - Planning period
    - Future population and water users in the subbasin
    - Location of future water use

- Water conservation
- Uncertainties
- Results

B. Water use to maintain instream needs

IV. Assess the capacity of the subbasin to meet present and future demands for withdrawal and nonwithdrawal uses, such as instream flows.

V. Evaluate supply-side and demand-side alternatives to meet withdrawal needs.

VI. Assess options for wastewater discharge to subsurface formations, streams, and other surface waters.

VII. Consider stormwater and floodplain management.

VIII. Identify potential conflicts and problems, and outline plans and programs to resolve conflicts and meet needs.

- Select alternatives and consider multiple objectives.
- Develop implementation plans.

Although the DRBC's primary concern is water supply, its members realize that all facets of the hydrologic cycle are integrated, and that integrated, holistic water and land use assessment and management for sustainable resources must not be just a dream for the future. Areas experiencing rapid growth and urbanization are today showing the effects of human-induced changes on natural resources, so now, not sometime in the future, is the time to act to maintain sustainability.

22.4 SUMMARY

Groundwater is an intricate and complicated science, hence it is often very expensive and time-consuming to properly analyze its movement, supply, and quality. But surface water and groundwater are one resource, so groundwater cannot be ignored in a watershed assessment and management plan. Measures can be put into place, on a watershedwide basis, to prevent future contamination of aquifers and other water supplies. A pollution vulnerability map is a valuable tool in this effort, both for preserving and exploring future water supplies. Wellhead protection zones also help in protecting the quality of sustainable water resources.

REFERENCES

Chester County Water Resources Authority. *WATERSHEDS: An Integrated Water Resources Management Plan for Chester County, Pennsylvania and Its Watersheds*

Report (West Chester, PA: Chester County Water Resources Authority), February 4, 2003.

DeBarry, Paul A., and Dennis P. Livrone. "Integrated Water Resources for Eight Municipalities in Bucks County Using GIS," Preceedings of the 2001 Pennsylvania Storm Water Management Symposium, Villanova, PA, 2001.

Delaware River Basin Commission (DRBC). "Guidelines for Developing an Integrated Resources Plan Under the Delaware River Basin Commission Southeastern Pennsylvania Ground Water Protected Area Regulations," Draft, June 20, 2001.

Horsley and Witten, Inc. "Tools and Options for Action at the Local Government Level," Aquifer Protection Seminar, Barnstable Village, MA, 1989.

McDonald M.G., and A.W. Harbaugh. "A Modular Three-dimensional Finite Difference Ground Water Flow Model," in *U.S. Geological Survey Techniques of Water Resources Investigations,* Book 6, Chapter A1, U.S. Government Printing Office, Washington, D.C., 1988.

Monroe County Planning Commission. "Water Supply and Model Wellhead Protection Study," June 1997.

Reese, Geoffery A., Paul A. DeBarry, and Chin Y. Kuo (ed.). "GIS Landscape: A Solution for Wellhead Protection," Proceedings of the Symposium of Engineering Hydrology, San Francisco, CA, July 25–30, 1993.

Toth, J. "A Theoretical Analysis of Groundwater Flow in Small Drainage Basins," *Journal of Geophysical Research* 68 pp. 4795–4812 (1963).

U.S. EPA. "Wellhead Protection: A Guide for Small Communities," EPA/625/R-93/002 (Washington, DC: Office of Research and Development, Office of Water), February 1993.

———. "State Source Water Assessment and Protection Programs Guidance—Final Guidance," EPA 816-R-97-009. (Washington, D.C.: U.S. EPA Office of Water), August 1997.

———. WHPA: "Modular Semi-Analytical Model for the Delineation of Wellhead Protection Areas," Version 2.0 (Washington, DC: Office of Ground Water Protection), 1991.

Vbra, J., and A. Zaporozec (eds.). "Guidelines on Mapping Groundwater Vulnerability," *International Contributions to Hydrology,* vol. 16 (Hannover: Verlag Heinz Heise), 1994.

Waterloo Hydrogeologic Software. *Visual MODFLOW Pro Version 3.1, Users Manual* (Waterloo, Ontario, Canada: Waterloo Hydrogeologic Software), 2003.

———. *FEFLOW Version 5.0, Users Manual* (Waterloo, Ontario, Canada: Waterloo Hydrogeologic Software), 2003.

CHAPTER 23

SUMMARY

Over the long term, natural watershed processes are in equilibrium, but in the short term they are dynamic. Human influence on the land and water alters this equilibrium and natural hydrologic regime. Rainwater, surface water, and groundwater are one and the same, connected through the hydrologic cycle. Maintaining the natural hydrologic regime and water resource sustainability can only be achieved through careful planning and management of our watersheds.

Before a watershed can be managed for water resource sustainability, one must have a thorough understanding of the physical processes that occur in a natural and anthropogenic watershed, and how affecting one process affects the other. Watershed processes involve many disciplines, including geology, hydrogeology, biology, chemistry, hydrology, hydraulics, and fluvial geomorphology, and affect and are affected by planning, socioeconomic, and political factors. Understanding how these various disciplines interrelate is critical to the effective assessment and management of water and land resources for a sustainable future.

Once the processes are understood, a project team should be established and goals and objectives for watershed management/restoration devised. The goals will vary by watershed depending on all the factors discussed in this book. The assessment should then be conducted to determine the health and issues of the watershed. The assessment should include, depending on the watershed, an analysis of the water chemistry, biology, habitat, hydrology, hydrogeology, geology, soils, fluvial geomorphology, lakes, dams, land use, problem areas, obstructions, and how existing ordinances are or are not working to preserve the natural hydrologic regime. GIS and modeling are extremely valuable tools in evaluating watershed patterns and trends. These tasks should be performed by highly qualified individuals.

The watershed management plan should be a holistic one, managing not just one aspect or problem in the watershed, but looking at the comprehensive picture and realizing that groundwater and surface water are one resource. For instance, at first glance, it is difficult to visualize how floods may affect groundwater supply, but as this book points out, the watershed is one system, and improper management of one component of the system affects both water quality and quantity. There are many tools available to aid in managing a watershed, including planning, smarter development design, regulations, and ordinances. Humans tend to be reactive, creating a structure to fix a problem or a regulation to prevent the problem from happening in the future. A watershed plan should be developed and should be proactive, knowing the intricacies of the watershed processes. It should maintain and/or improve water supply, base flow, water quality, and habitat, as well as prevent stormwater problems, streambank erosion, and flooding. Every watershed should have a long-term plan to achieve these goals, taking into account all the physical, funding, and regulatory tools available. A comprehensive water resource ordinance, possibly including zoning based on water resource protection, is one means to achieve these goals and will gain popularity in the future.

To their detriment, existing regulations and funding programs often promote segregation of these disciplines, with the goals of one program sometimes counterproductive to the goals of another. Similarly, programs often duplicate efforts—such as data collection, GIS mapping, and management plan action items—causing the waste of public funds. It's important to recognize that all landmass is in one watershed or another, and that water flow does not follow municipal boundaries, meaning that resource management is best performed on a watershed basis, not according to political boundaries. Fortunately, watersheds are natural management units, so after assessing natural and anthropogenic influences of a watershed, or more refined subwatersheds, a management plan for long-term sustainability can be developed. But this requires: stakeholder involvement and qualified professional guidance; in-field data collection, state-of-the-art GIS, and modeling tools; and well-defined goals and objectives tailored to the watershed.

Water is powerful. Water is awesome. Like an undisciplined child, water will attempt to go where it wants, seeking the path of least resistance. This may be by creating meanders to naturally increase its length, thereby reducing its slope and, in turn, the erosive velocities. It may be by spreading out into a floodplain, storing its waters for natural attenuation, groundwater recharge, and soil replenishment. To tame the awesome power of flowing water, the human tendency is to build—for example, levees and dikes—to enable development in the floodplain. But in sharp contrast to water, which has been flowing eternally, structures have a limited lifespan, particularly as the watershed becomes developed, flows increase, and the frequency of obtaining flows greater than design flows increases. Public funding is then necessary to pay for insurance claims, clean up, and rebuild towns and levees when the latter are overtopped or break, and the natural floodplain again takes up its natural path. One must learn to work with the power of water, not against it.

Are we making the same mistake with channel restoration? To answer that question raises others: Are funds best spent using nonstructural measures such as fencing riparian areas and allowing natural vegetation to succeed and stabilize the disturbed areas in a watershed? This option may include compensating farmers for the loss of productive land. Or are funds better spent to implement structural, natural channel design measures? Structural solutions have four problems: They are very expensive on a linear foot basis; they benefit only a short segment of stream; they have a limited lifespan; and, eventually, they may also need to be replaced. The past has taught that dikes and levees don't work, but that hasn't stopped humans from building them. Common sense would indicate, instead, that the most economical and environmentally beneficial way to manage the stormwater runoff in the watershed is in a proactive manner, by providing nonstructural, natural progression measures for stream corridor restoration.

SELECTED LIST OF GIS DATA AND RELATED SITES

ESRI's Downloadable Data—Access to many popular GIS data sets.
http://www.esri.com/data/download/index.html

ESRI's Geography Network—Network of global geographic information users and providers.
http://www.geographynetwork.com/

Federal Geographic Data Committee—Policies, standards, and procedures for organizations to cooperatively produce and share geographic data.
http://www.fgdc.gov/

GIS Data Depot—Digital GIS data.
http://data.geocomm.com/

National Spatial Data Infrastructure, Geospatial One-Stop—Geospatial One-Stop helps improve access to geospatial information.
http://www.geo-one-stop.gov/

NSDI Clearinghouse Nodes—References from registered NSDI Clearinghouse Nodes to provide additional information and context for geospatial data.
http://clearinghouse4.fgdc.gov/registry/clearinghouse_sites.html

USGS EROS Data Center—National map site featuring sites and products.
http://edc.usgs.gov/

USGS Water Resources of the United States—Water Resources Maps and GIS Information.
http://water.usgs.gov/maps.html

APPENDIX B

ACRONYMS

1Q20 — Lowest one-day flow recorded over a 20-year period, 72

30Q2 — Lowest mean flow recorded over 30 consecutive days over a two-year recurrence interval, 72

AFO — Animal Feeding Operations, 202, 449

AFR — Arbitrary Fixed Radius, 625

AFWS — Automated Flood Warning System, 286

ALERT — Automatic Local Evaluation in Real Time, 286

AMC — Antecedent Moisture Condition, 316

AMD — Acid Mine Drainage, 4, 71, 146, 202

AMiFH — Areas of Minimal Flood Hazard, 590

AMoFH — Areas of Moderate Flood Hazard, 590

ANSI — American National Standards Institute, 239

APSRS — Aerial Photography Summary Record System, 246

APWA — American Public Works Association, 170

ARC — Antecedent Runoff Condition, 316

ASCE — American Society of Civil Engineers, 51, 56

ASCII — American Standard Code of Information Interchange, 229, 239

ASCS — Agricultural Stabilization and Conservation Service, 197

AVHRR — Advanced Very-High Resolution Radiometer, 240, 242

BASINs — Better Assessment Science Integrating Point and Nonpoint Sources, 173, 236, 255, 485, 495

BAT — Best Available Treatment, 437

BBI — Beck's Biotic Index, 143

BCI — Bird Community Index, 147

BFE — Base Flood Elevation, 590

BMP — Best Management Practices, 8, 65, 173, 187, 208, 209, 413, 442, 526–583

BOD$_5$ — Biochemical Oxygen Demand (5-day), 167, 414

BTPT — Best Technology in Process and Treatment, 437

CAFO — Concentrated Animal Feeding Operations, 202, 449

CAR — Crops at Risk, 464

CBF — Cartographic Boundary Files, 229, 230

CBP — Chesapeake Bay Program, 455

CCMP — Comprehensive Conservation and Management Plan, 456

CD ROM — Compact Disk Read-Only Memory, 228, 240

CEC — Cation Exchange Capacity, 42

CEC — Commission for Environmental Cooperation, 464

CERCLA — Comprehensive Environmental Response, Compensation, and Liability Act, 455

CFR — Calculated Fixed Radius, 625

CN — Curve Number, 329, 537

COD — Chemical Oxygen Demand, 167

CPD — Coastal Program Division, 456

CREP — Conservation Reserve Enhancement Program, 464

GLOSSARY

Accuracy   In mapping, the reduction of positional and attribute errors based on information sources and data input instruments. Data precision must reflect, not exceed, its accuracy.[3]

Acre-feet (AF)   An expression of water quantity. One acre-foot of water will cover 1 acre of ground 1 foot deep or other equivalent.[6]

Acre-foot   *See* Acre-feet.

Algae   Chlorophyll-bearing nonvascular, primarily aquatic species that have no true roots, stems, or leaves. Most algae are microscopic, although some species can be as large as vascular plants.[7]

Alluvial aquifer   A water-bearing deposit of unconsolidated material (sand and gravel) left behind by a river or other flowing water.

Alluvium   Deposits of clay, silt, sand, gravel, or other unconsolidated material deposited in a river or streambed, on a floodplain or delta.[7]

American Standard Code for Information Exchange (ASCII)   A popular standard for the exchange of alphanumeric data.[3]

Ammonia   A compound of nitrogen and hydrogen (NH_3) that is a common by-product of animal waste. Ammonia readily converts to nitrate in soils and streams.[7]

Annual flood   The maximum discharge peak during a given water year (e.g., October 1 to September 30).[5]

Anomalies   As related to fish, externally visible skin or subcutaneous disorders, including deformities, eroded fins, lesions, and tumors.[7]

Anthropogenic   Occurring because of, or influenced by, human activity.[7]

Appropriation Doctrine   Water laws developed in the arid western states, where water supplies are limited and often inadequate, that award a water right to a person actually using the water. The Appropriation Doctrine has two fundamental principles: One, "first in time of use is first in right"; that is, the earliest appropriator on a stream has the first right to use the water. Two, application of the water to a beneficial use is the basis and measure of the right.[6]

Aquatic guidelines Specific levels of water quality that, if reached, may adversely affect aquatic life. These are nonenforceable guidelines issued by a governmental agency or other institution.[7]

Aquatic-life criteria Water-quality guidelines for protection of aquatic life. They often refer to U.S. Environmental Protection Agency water-quality criteria for protection of aquatic organisms. *See also* water quality guidelines, water quality criteria, and freshwater chronic criteria.[7]

Aquifer A water-bearing layer of soil, sand, gravel, or rock that will yield significant quantities of water to wells or springs.[7]

ARC/INFO A software package that provides a menu and key-in operator interface with commands for generating, editing, and analyzing graphics and data. ARC/INFO is vector georelational software from the Environmental Systems Research Institute, Redlands, California.[3]

ARC Macro Language (AML) A fourth-generation programming language with facilities to use named variables, perform logical branching and loops, perform arithmetic and trigonometric operations, make calls and pass variables to other programs, and perform selected Geographic Information Systems operations. AML also provides facilities to design and create custom menus, and is used to refer to a program written using the ARC Macro Language.[3]

ArcView GIS and mapping software that provides data visualization, query, analysis, and integration, along with the capability to create and edit geographic data.[1]

Argillite A compact sedimentary rock composed largely of clay materials.[1]

Artesian aquifer An aquifer that is completely filled with water under pressure and that is overlain by material that restricts the movement of water, the pressure of which allows discharge to the land surface.

Artificial recharge Augmentation of natural replenishment of groundwater storage by some method of construction, by spreading of water, or by pumping water directly into an aquifer.[7]

Atmospheric deposition The transfer of substances from the air to the surface of the earth, either in wet form (rain, fog, snow, dew, frost, hail) or in dry form (gases, aerosols, particles).[7]

Attribute Descriptive characteristic or quality describing a map feature. Attributes may be represented by locational or nonlocational descriptive information about a feature. The attributes of a hydrologic unit include its code, size, or name. Commonly used in GIS applications.[3]

Background concentration A concentration of a substance in a particular environment that is indicative of minimal influence by human (anthropogenic) sources.[7]

Bank   The sloping ground that borders a stream and confines the water in the natural channel when the water level, or flow, is normal.[7]

Bankfull stage   An established river stage at a certain point along a river that is intended to represent the maximum safe water level that will not overflow the riverbank or cause any damage within the reach of the river.[5]

Base flood elevation (BFE)   The elevation associated with the flood having a 1 percent chance annually of being equaled or exceeded in any given year; typically as shown on the Flood Insurance Rate Map.[3]

Base flow   Sustained, low flow in a stream maintained by groundwater discharge, rather than storm runoff.[1,7]

Base map   Map of the watershed that depicts cultural features (roads, railroad, bridges, dams, culverts, etc.), drainage features, and the watershed boundary.[3]

Base station   A computer that accepts radio signals from IFLOWS or ALERT gaging sites, decodes the data, places the data in the database, and makes the data available to other users.[5]

Basic fixed sites   Sites on streams at which stream flow is measured and samples are collected for temperature, salinity, suspended sediment, major ions and metals, nutrients, and organic carbon to assess the broad-scale spatial and temporal character and transport of inorganic constituents of stream water in relation to hydrologic conditions and environmental settings.[7]

Basin   An area having a common outlet for its surface runoff.[5]

Bedload   Sediment that moves on or near the streambed and is in almost continuous contact with the bed.[7]

Bedrock   General term for consolidated (solid) rock that underlies soils or other unconsolidated material.[7]

Bed sediment   The material that is temporarily stationary at the bottom of a stream or other watercourse.[7]

Bed sediment and tissue studies   Assessments of concentrations and distributions of trace elements and hydrophobic organic contaminants in streambed sediment and tissues of aquatic organisms to identify potential sources and to assess spatial distribution.[7]

Beneficial use   Cardinal principle of the Appropriation Doctrine, which has two components: the nature or purpose of the use and the efficient or nonwasteful use of water.[6]

Benthic   Refers to plants or animals that live on the bottom of lakes, streams, or oceans.[7]

Benthic invertebrates   Insects, mollusks, crustaceans, worms, and other organisms without a backbone that live in, on, or near the bottom of lakes, streams, or oceans.[7]

Best management practice (BMP)   A method, activity, maintenance procedure, or other management practice for reducing the amount of pollution entering a water body.[1]

Bioaccumulation   The biological sequestering of a substance at a higher concentration than that at which it occurs in the surrounding environment or medium.[7]

Bioavailability   The capacity of a chemical constituent to be taken up by living organisms either through physical contact or by ingestion.[7]

Biochemical   Refers to chemical processes that occur inside or are mediated by living organisms.[7]

Biochemical oxygen demand (BOD)   The amount of oxygen (milligrams per liter) that is removed from aquatic environments by the life processes of microorganisms.[7]

Biodegradation   Transformation of a substance into new compounds through biochemical reactions or the actions of microorganisms such as bacteria.[7]

Biomass   The amount of living matter, in the form of organisms, present in a particular habitat, usually expressed as weight per unit area.[7]

Biota   Living organisms.[7]

Blue-baby syndrome   A condition that can be caused by ingestion of high amounts of nitrate, resulting in the blood losing its ability to effectively carry oxygen. It is most common in young infants and certain elderly people.[7]

Breach   The failed or human-induced opening in a dam.[5]

Breakdown   A natural process that may result in a more toxic or a less toxic compound and a more persistent or less persistent compound.[7]

Buffer zone   An area of specified distance (radius) around a map item or items, such as a stream.[3]

Canopy angle   Generally, a measure of the openness of a stream to sunlight. Specifically, the angle formed by an imaginary line from the highest structure (for example, tree, shrub, or bluff) on one bank to eye level at mid-channel to the highest structure on the other bank.[7]

Carbonate rocks   Rocks (such as limestone or dolostone) that are composed primarily of minerals (such as calcite and dolomite) containing the carbonate ion (CO_3^{2-}).[7]

CFS/Day   The volume of water represented by a flow of 1 cubic foot per second for 24 hours.[6]

Channel routing   The process of determining progressively the timing and shape of the flood wave at successive points along a river.[5]

Channelization   Modification of a stream, typically by straightening the channel, to provide more uniform flow; often done for flood control or for improved agricultural drainage or irrigation.[7]

Classic watershed   A land and water area that has all the surface drainage within its boundary converging to a single point.[4]

Clastic   Rock or sediment composed principally of broken fragments that are derived from preexisting rocks that have been transported from their place of origin, as in sandstone.[7]

Climate   The sum total of the meteorological elements that characterize the average and extreme conditions of the atmosphere over a long period of time at any one place or region of the earth's surface.[7]

Cold-deciduous broadleaf   Woody angiosperms with wide, flat leaves (e.g., paper birch) that are shed by plants during the dormant season (that portion of the year when frosts occur).[2]

Colluvium   A general term applied to any loose, heterogeneous, and incoherent mass of soil material and/or rock fragments deposited by rainwash, sheetwash, or slow continuous downslope creep, usually collecting at the base of gentle slopes or hillsides (Bates and Jackson, 1980).[2]

Combined sewer overflow   A discharge of untreated sewage and stormwater to a stream when the capacity of a combined storm/sanitary sewer system is exceeded by storm runoff.[7]

Community   In ecology, the species that interact in a common area.[7]

Community identification (CID)   A unique six-digit number assigned to each community by the Federal Emergency Management Agency (FEMA) and used for identity in computer databases; it is shown on the FIS and FIRM and in the Q3 Flood Data files.[3]

Community Rating System (CRS)   A program created by FEMA to provide new incentive for activities that reduce flood losses and support the sale of flood insurance.[3]

Compact disk read-only memory (CD-ROM)   A digital medium for the storage of data in a standardized format.[3]

Compilation   Preparation of a new or revised map from existing maps, field surveys, and other sources.[4]

Composite hydrologic unit   A land and water area that receives surface flow from an upstream watershed(s) and drains to one outlet.[4]

Comprehensive plan   Regional, state, or local documents that describe community visions for future growth.[1]

Concentration   The amount or mass of a substance present in a given volume or mass of sample, usually expressed as microgram per liter (water sample) or micrograms per kilogram (sediment or tissue sample).[7]

Conceptual model   A general idea or understanding of an existing stream/lake/aquifer system from which it is possible to mathematically simulate that system.[1]

Confined aquifer   An aquifer, overlain by a confining bed, in which groundwater is under pressure that is significantly greater than atmospheric pressure.

Confining layer   A layer of low-permeability material with low hydrologic conductivity that bounds an aquifer.[1]

Confluence   The flowing together of two or more streams; the place where a tributary joins the main stream.[7]

Conglomerate   A rock consisting of pebbles embedded in a finer (usually sandstone) matrix.[1]

Constituent A chemical or biological substance in water, sediment, or biota that can be measured by an analytical method.[7]

Consumptive use The quantity of water that is not available for immediate reuse because it has been evaporated, transpired, or incorporated into products, plant tissue, or animal tissue.[6]

Contamination Degradation of water quality due to human activity, as compared to original or natural conditions.[7]

Contents An expression of the volume of water in a reservoir or lake in acre-feet.

Contiguous boundaries Hydrologic unit boundaries shared in whole or in part by different hydrologic units.[4]

Contour An imaginary line of equal elevation on the ground representing the land surface.[4]

Contraction Reduced area of flow in a weir due to the weir's crest and sides being sufficiently removed from the bottom and sides of the weir box or channel in which it is set.[6]

Contributing area The area in a drainage basin that contributes water to stream flow or recharge to an aquifer.[7]

Control point Any station in a horizontal or vertical control network that is identified in a dataset or photograph and used for correlating the data shown in the dataset or photograph.[4]

Coordinate system A particular kind of reference frame or system, such as plane rectangular coordinates or spherical coordinates, that uses linear or angular values to designate the position of points within that particular reference frame The three major coordinate systems used in the United States are the Geographic Coordinate system, State Plane Coordinate system, and Universal Transverse Mercator Grid system.[4]

Coordinates Linear and/or angular quantities designating the position of a point in relation to a given reference frame. In a two-dimensional system, x and y coordinates are used to designate locations.[4]

Crest 1. The top of a dam, dike, weir, or spillway, which water must reach before passing over the structure. 2. The highest elevation reached by floodwaters.[5,6]

Criterion A standard rule or test on which a judgment or decision can be based.[7]

Crystalline rocks Rocks (igneous or metamorphic) consisting wholly of crystals or fragments of crystals.[7]

Cubic feet per second (ft³/s or cfs) The flow rate or discharge equal to 1 cubic foot of water per second.[7]

Cuesta A hill or ridge with a gentle slope on one side and a steep slope on the other; formed by uplifted rock outcrop consisting of strata having different resistances to erosion.[2]

Current The horizontal movement of water.[6]

Current meter A device used to measure the speed of flowing water.[6]

Database A collection of information related by a common fact or purpose.[3]

Data capture Series of operations required to encode data in a computer-readable form (digitizing).[3]

Data layer Data with similar characteristics contained in the same plane or overlay (e.g., roads, rivers).[3]

Dataset or data file A named collection of logically related data records arranged in a prescribed manner.[3]

Datum A basis and reference for a geodetic survey. The vertical datum is usually mean sea level or mean low water, and is usually referred to as National Geodetic Vertical Datum, or NGVD.[4]

Degradation products Compounds resulting from transformation of an organic substance through chemical, photochemical, and/or biochemical reactions.[7]

Delaware River Basin Commission (DRBC) Formed in 1961 by the signatory parties to the Delaware River Basin Compact (Delaware, New Jersey, New York, Pennsylvania, and the United States government). The DRBC is responsible for managing the water resources of the basin.[1]

Delineation The boundaries of a hydrologic unit with a line.[4]

Denitrification A process by which oxidized forms of nitrogen such as nitrate (NO_3^-) are reduced to form nitrites, nitrogen oxides, ammonia, or free nitrogen. Denitrification is commonly brought about by the action of denitrifying bacteria and usually results in the escape of nitrogen to the air.[7]

Deposition The act or process of settling solid material from a fluid suspension.[1]

Detection limit The concentration below which a particular analytical method cannot determine, with a high degree of certainty, a concentration.[7]

Dewatering Withdrawal of groundwater in order to proceed with activities such as mining or construction.[1]

Diabase A dark-colored, finely crystalline, hard igneous rock.[1]

Diatoms Single-celled, colonial, or filamentous algae with siliceous cell walls constructed of two overlapping parts.[7]

Dichloro-diphenyl-trichloroethane (DDT) An organochlorine insecticide. DDT is no longer registered for use in the United States, since it was recognized as the cause of the near extinction of the bald eagle.[7]

Dieldrin An organochlorine insecticide no longer registered for use in the United States. Also a degradation product of the insecticide aldrin.[7]

Digital data Data displayed, recorded, or stored in binary notation.[3]

Digital elevation model (DEM) Georeferenced digital files consisting of terrain elevations for exposed or submerged ground positions at regularly spaced horizontal intervals, generally 10 to 30 meters.[4]

Digital Flood Insurance Rate Map (DFIRM) Comprised of all digital data required to create the hard copy Flood Insurance Rate Map (FIRM).[3]

Digital Line Graph (DLG) A computer file format for mapping data that provides a topological structure to describe points, lines, and polygons.[3]

Digital raster graphic (DRG) A scanned georeferenced image of a USGS standard series topographic map or NOAA nautical chart.[4]

Digitizing The process of converting an analog image or map into a digital format usable by a computer.[3]

Discharge 1. Rate of fluid flow passing a given point at a given moment in time, expressed as volume per unit of time. 2. Water that leaves an aquifer or the groundwater body.[6,7]

Discharge area An area in which there are upward components of hydrologic head in an aquifer.[1]

Dissolved constituent Operationally defined as a constituent that passes through a 0.45-micrometer filter.[7]

Dissolved solids Amount of minerals, such as salt, that are dissolved in water.[7]

Diversion A turning aside or alteration of the natural course of a flow of water.[7]

Division An ecological unit in the ecoregion planning and analysis scale of the National Hierarchical Framework corresponding to subdivisions of a domain that have the same regional climate.[4]

Dolomite A rock consisting largely of calcium magnesium carbonate.[1]

Domain An ecological unit in the ecoregion planning and analysis scale of the National Hierarchical Framework corresponding to subcontinental divisions of broad climatic similarity that are affected by latitude and global atmospheric conditions.[4]

Drainage area An area having common outlet for its surface runoff.[5]

Drainage basin The portion of the surface of the earth that contributes water to a watercourse, including tributaries and impoundments.[3,5]

Drainage divide The rim of a drainage basin. The drainage divide, or divide, is used to denote the boundary between one drainage basin and another.[1]

Drawdown The lowering of the water level caused by pumping. Also, for flowing wells, the reduction of the pressure head as a result of the discharge of water.[6,7]

Drinking-water standard or guideline A threshold concentration in a public drinking-water supply, designed to protect human health.[7]

Drip irrigation An irrigation system in which water is applied directly to the root zone of plants by means of applicators (orifices, emitters, porous tubing, perforated pipe, and so forth) operated under low pressure.[7]

Drought Commonly defined as being a time of less-than-normal or less-than-expected precipitation.[7]

Drumlin   An elongated hill or ridge of glacial drift.[2]

Ecological studies   Studies of biological communities and habitat characteristics to evaluate the effects of physical and chemical characteristics of water and hydrologic conditions on aquatic biota and to determine how biological and habitat characteristics differ.[7]

Ecoregion   An area of similar climate, landform, soil, potential natural vegetation, hydrology, or other ecologically relevant variables. Ecoregions include Domain, Division, and Province ecological units.[2,7]

Ecosystem   The interacting populations of plants, animals, and microorganisms occupying an area, plus their physical environment.[1,7]

Edge matching   The comparison and graphic adjustment of features to obtain agreement along the edges of adjoining map sheets.[3]

Effluent   Outflow from a particular source, such as a stream that flows from a lake or liquid waste that flows from a factory or sewage treatment plant.[7]

Elevation Derivatives for National Applications (EDNA)   The development of a hydrologically correct version of the National Elevation Dataset (NED) and systematic derivation of standard hydrologic derivatives (previously known as NED-H).[4]

Elevation reference mark (ERM)   A point of vertical ground elevation reference shown on the FIRM for comparison to the BFE.[3]

Endocrine system   The collection of ductless glands in animals that secrete hormones, which influence growth, gender, and sexual maturity.[7]

Environmental framework   Natural and human-related features of the land and hydrologic system, such as geology, land use, and habitat, that provide a unifying framework for making comparative assessments of the factors that govern water quality conditions.[7]

Environmental sample   A water sample collected from an aquifer or stream for the purpose of chemical, physical, or biological characterization of the sampled resource.[7]

Environmental setting   Land area characterized by a unique combination of natural and human-related factors, such as meadows or hydric soils.[7]

Ephemeral stream   A stream or part of a stream that flows only in direct response to precipitation or snowmelt. Its channel is above the water table at all times.[7]

EPT Richness Index   An index based on the sum of the number of taxa in three insect orders, *Ephemeroptera* (mayflies), *Plecoptera* (stoneflies), and *Trichoptera* (caddisflies), that are composed primarily of species considered to be relatively intolerant to environmental alterations.[7]

Equal-width increment (EWI) sample   A composite sample across a section of stream, with equal spacing between verticals and equal transit rates within each vertical, that yields a representative sample of stream conditions.[7]

Erosion The process whereby materials of the earth's crust are loosened, dissolved, or worn away and simultaneously moved from one place to another.[7]

Estuary The wide lower course of a water passage where its current is met by the tides.[4]

Eutrophication The process by which water becomes enriched with plant nutrients, most commonly phosphorus and nitrogen.[7]

Evaporite minerals (deposits) Minerals or deposits of minerals formed by evaporation of water containing salts. These deposits are common in arid climates.[7]

Evapotranspiration The combination of evaporation from free water surface and the transpiration of water from plant surfaces to the atmosphere.[5,7]

Fecal bacteria Microscopic single-celled organisms (primarily fecal coliforms and fecal streptococci) found in the wastes of warm-blooded animals. Their presence in water is used to assess the sanitary quality of water for body-contact recreation or for consumption. Also, their presence indicates contamination by the wastes of warm-blooded animals and the possible presence of pathogenic (disease-producing) organisms.[7]

Fecal coliform *See* Fecal bacteria.

Federal Emergency Management Agency (FEMA) The agency reporting directly to the president and responsible for identifying and mitigating natural and man-made hazards.[3]

Federal Geographic Data Committee (FGDC) An interagency committee, established by the Office of Management and Budget, that promotes the coordinated development, use, sharing, and dissemination of geospatial data on a national basis.[3]

Federal Information Processing Standards (FIPS) The official source within the federal government for information processing standards, which are developed and published by the Institute for Computer Sciences and Technology at the National Institute of Standards and Technology (NIST).[3]

Federal Insurance Administration (FIA) An administration within FEMA that provides flood and crime insurance.[3]

Fertilizer Any of a large number of natural or synthetic materials, including manure and nitrogen, phosphorus, and potassium compounds, spread on or worked into soil to increase its fertility.[7]

Fish community *See* Community.

Fixed sites USGS's National Water Quality Assessment Program's (NAWQA's) most comprehensive monitoring sites.[7]

Flash flood A flood that follows within a few hours (usually less than six) of heavy or excessive rainfall, dam or levee failure, or the sudden release of water impounded by an ice jam.[5]

Flash flood warning A warning by the National Weather Service (NWS) issued to warn of flash flooding that is forecasted, imminent, or occurring.[5]

Flash flood watch   An announcement by the National Weather Service that alerts the public and need-to-know persons to the possibility of flash flooding in specific areas.

Flood   An overflow or inundation from a water source.[6]

Flood crest   The maximum height of a flood wave at a particular location.[5]

Flood Insurance Rate Map (FIRM)   A map on which the 100- and 500-year floodplains, base flood elevations (BFEs), and risk premium zones are delineated, to enable insurance agents to issue accurate flood insurance policies to homeowners in communities participating in the National Flood Insurance Program (NFIP).[3]

Flood Insurance Rate Map Digital Line Graph (FIRM-DLG)   A product developed by digitizing and/or scanning the existing hard-copy FIRM to create a thematic overlay of flood risks.[3]

Flood Insurance Study (FIS)   An examination, evaluation, and determination of the flood hazards and, if appropriate, the corresponding water-surface elevations.[3]

Flood irrigation   The application of irrigation water where the entire surface of the soil is covered by ponded water.[7]

Flood of record   The highest observed river stage or discharge at a given location during a period of record-keeping, but not necessarily the highest known stage.[5]

Floodplain   The relatively level area of land bordering a stream channel and inundated during moderate to severe floods.[7]

Flood potential outlook   An NWS outlook issued to alert the public of potentially heavy rainfall that could send area rivers and streams into flood or aggravate an existing flood.[5]

Flood routing   Process of determining progressively the timing, shape, and amplitude of a flood wave as it moves downstream to successive points along the river.[5]

Flood stage   A height usually higher than or equal to bankfull at which a watercourse overtops its banks and begins to cause damage to any portion of the defined reach.[5]

Flood warning   A release by the NWS to warn the public that the stream or river is forecasted or observed to equal or exceed flood stage. A flood warning will usually contain river stage (level) forecasts.[5]

Flood wave   A rise in stream flow to a crest and its subsequent recession, caused by precipitation, snowmelt, dam failure, or reservoir releases.[5]

Flow path   One of the paths followed by groundwater as it flows through the saturated zone from the area of recharge to discharge, or by surface water.[1]

Flow path study   Network of clustered wells located along a flow path extending from a recharge zone to a discharge zone, preferably a shallow stream.[7]

Flume A specially shaped open-channel flow section installed in a canal, stream, or ditch to measure the rate of flow.[6]

Fluvial deposit A sedimentary deposit consisting of material transported by suspension or laid down by a river or stream.[7]

Free flow When used to describe the operation of a flume, free flow is when discharge depends solely on the width of the throat and the depth of water in the converging section.[6]

Freshwater chronic criteria The highest concentration of a contaminant to which freshwater aquatic organisms can be exposed for an extended period of time (four days) without adverse effects.[7]

Frontal hydrologic unit A land and water area where surface flow originates entirely within the hydrologic unit and drains to multiple points along a large water body such as the ocean or large lake.[4]

Gage (or gauge) A device for indicating the elevation of a water surface, the velocity of flowing water, the pressure of water, the amount or intensity of precipitation, and the depth of snowfall.[1]

Gage datum The datum elevation from which all stage measurements are made.[5]

Gage height The water surface elevation referred to an arbitrary or predetermined gage datum. Also known as *stage*.[6]

Gage zero The elevation of zero stage. Same as *gage datum*.[5]

Gaging (or gauging) The determination of the quantity of water flowing per unit of time in a stream channel or conduit at a given point by means of current meters, rod float, weirs, or other measuring devices or measures.[1]

Gaging station A site on a watercourse where systematic observations of stage and/or flow are measured.[5,6]

Gaining stream A stream or reach of a stream whose flow is being increased by inflow of groundwater.[1]

Geocoding Associating either geographic coordinates or grid cell identifiers to data, points, lines, and shapes.[3]

Geographic coordinates Coordinate system in which horizontal and vertical distances on a planimetric map are represented in units of latitude and longitude, rather than feet or meters.[3]

Geographic Information System (GIS) A computer system for capturing, storing, checking, integrating, manipulating, analyzing, and displaying data related to positions on the earth's surface.[1,3,4]

Geophysical log A record of certain physical properties, measured by instruments and plotted against depth, for a well.[1]

Georeference system An X,Y or X,Y,Z coordinate system that locates points on the surface of the earth as a reference to points on a map.[6]

Georelational A geometry of spatial data. Housed separately from its attributes.

Geospatial   The type of data in a GIS representing the shape, location, or appearance of geographic objects.[4]

Geothermal   Relating to the earth's internal heat; commonly applied to springs or vents discharging hot water or steam.[7]

Gneiss   A coarse-grained metamorphic rock made up of minerals in bands and streaks.[1]

Granite   A coarse-grained igneous or metamorphic rock composed chiefly of orthoclase feldspar and quartz, with lesser amounts of dark minerals.[1]

Granitic rock   A coarse-grained igneous rock.[7]

Groundwater   Water in the zone of saturation where all openings in rocks and soil are filled, the upper surface of which forms the water table.[7]

Groundwater discharge   Discharge of water from the saturated zone by (1) natural processes such as groundwater runoff and groundwater evapotranspiration, and (2) discharge through wells and other man-made structures.[1]

Groundwater divide   A ridge in a water table from which the water table slopes downward on both sides.[1]

Groundwater flow   Stream flow that results from precipitation that infiltrates into the soil and eventually moves through the soil into the stream channel. Also referred to as *base flow* or *dry weather flow.*

Groundwater mining   The practice of withdrawing groundwater in a specific area at rates in excess of the natural recharge.

Groundwater protected area (GWPA)   An area within which individual groundwater withdrawals of greater than so many gallons per day (gpd) require a permit.[1]

Groundwater recharge   Water that is added to the saturated zone.[1]

Growth management   Government programs that control timing, location, and character of land use and development.[1]

Habitat   The part of the physical environment where plants and animals live.[7]

Head, static   The height above or below a standard datum of the surface of a column of water (or other liquid) that can be supported by the static pressure at a given point.[1,6]

Headwaters   The source and upper part of a stream.[7]

Headwaters basin   A basin at the headwaters of a river.[5]

Health advisory   Nonregulatory levels of contaminants in drinking water that may be used as guidance in the absence of regulatory limits.[7]

Herbicide   A chemical or other agent applied for the purpose of killing undesirable plants.[7]

Holistic   Relating to or concerned with wholes or complete systems.[1]

Horizontal control   Network of stations of known geographic or grid positions referred to a common horizontal datum, which controls the horizontal positions of mapped features with respect to parallels and meridians or northing and easting grid lines shown on the map.

Human health advisory Guidance provided by the U.S. Environmental Protection Agency (EPA), state agencies, or scientific organizations, in the absence of regulatory limits, to describe acceptable contaminant levels in drinking water or edible fish.[7]

Hydraulic conductivity The volume of water at the existing kinematic viscosity that will move in unit time under unit hydraulic gradient through a unit area measured at right angles to the direction of flow.[1]

Hydraulic gradient The change in head per unit of distance in a given direction.[1]

Hydrogeology The branch of hydrology dealing with the occurrence and distribution of groundwater, particularly as it relates to the properties of the rock materials in which it occurs. Also called *groundwater hydrology.*[1]

Hydrograph A graph showing, for a given point on a stream or conduit, the discharge, stage, velocity, available power, or other property of water with respect to time.[1,5]

Hydrography The scientific description, study, and analysis of the physical conditions, boundaries, measurement of flow, investigation and control of flow, and related characteristics of surface water such as rivers, lakes, and oceans.[4]

Hydrologic budget An accounting of the inflow to, outflow from, and storage in a hydrologic unit such as a drainage basin, aquifer, soil zone, lake, reservoir, or irrigation project.[1]

Hydrologic cycle The circuit of water movement from the atmosphere to the earth and return to the atmosphere through various stages or processes such as precipitation, interception, runoff, infiltration, percolation, storage, evaporation, and transpiration. Also called *water cycle.*[1,5]

Hydrologic unit (HU) A drainage area delineated to nest in a multilevel, hierarchical drainage system.[4]

Hydrologic unit code (HUC) The numerical identifier of a specific hydrologic unit consisting of a two-digit sequence for each specific level within the delineation hierarchy.[4]

Hydrologic unit name A name assigned to hydrologic units to better enable identifying and understanding of the geographic location of the hydrologic unit.[4]

Hydrology The science dealing with the properties, distribution, and circulation of water on the surface of the land, in the soil and underlying rocks, and in the atmosphere.[4]

Hypsography A branch of geography that deals with the mapping of the topography of earth above sea level.[4]

Ice jam A stationary accumulation of ice in a river that blocks or restricts stream flow.[5]

Igneous Rocks that were formed from the molten state.[1]

Index of Biotic Integrity (IBI) An aggregated number, or index, based on several attributes or metrics of a fish community that provides an assessment of biological conditions.[7]

Indicator sites Stream sampling sites located at outlets of drainage basins with relatively homogeneous land use and physiographic conditions; most indicator site basins have drainage areas ranging from 20 to 200 square miles.[7]

Induced infiltration The process by which water moves into an aquifer from an adjacent surface-water body as a result of reversal of the hydraulic gradient in response to withdrawal.[1]

Induced recharge The amount of water entering an aquifer from an adjacent surface-water body by the process of induced infiltration.[1]

Infiltration Movement of water, typically downward, into soil or porous rock.[7]

Insecticide A substance or mixture of substances intended to destroy or repel insects.[7]

Instantaneous discharge The volume of water that passes a point at a particular moment in time.[7]

Instream flow requirement The amount of water flowing through a natural stream course that is needed to sustain the instream values at an acceptable level.[6]

Instream use Water use taking place within the stream channel for such purposes as hydroelectric power generation, navigation, water quality improvement, fish propagation, and recreation. Sometimes called *nonwithdrawal use* or *in-channel use.*[7]

Integrator or mixed-use site Stream sampling site located at an outlet of a drainage basin that contains multiple environmental settings.[7]

Intensive fixed sites Basic fixed sites with increased sampling frequency during selected seasonal periods and analysis of dissolved pesticides for one year.[7]

Interception The process by which precipitation is caught and held by foliage, twigs, and branches of trees, shrubs, and other vegetation, and lost by evaporation, never reaching the surface of the ground.[1]

Interflow The lateral movement of water in the unsaturated zone during and immediately following a precipitation event. The water moving as interflow discharges directly into a stream or lake.[1]

Intermittent stream A stream that flows only when it receives water from rainfall runoff or springs or from some surface source such as melting snow.[7]

Intolerant organisms Organisms that are not adaptable to human alterations to the environment and thus decline in numbers where human alterations occur.[7]

Intrusive (geologic) Term applied to igneous rock that was forced into pre-existing rock while in a molten state, as opposed to *extrusive*.[1]

Inundation map A map delineating the area that would be submerged at various levels of flooding.[5]

Invertebrate An animal having no backbone or spinal column.[7]

Irrigation return flow The part of irrigation applied to the surface that is not consumed by evapotranspiration or uptake by plants and that migrates to an aquifer or surface-water body.[7]

Isohyet A line that connects points of equal rainfall.[5]

Joint A fracture plane in rock, frequently in intersecting sets.[1]

Karst A type of topography that is formed over carbonate rocks such as limestone and dolomite and characterized by closed depressions or sink-holes, caves, and underground drainage.[4]

Kill Dutch term for stream or creek.[7]

Lag The time it takes a flood wave to move downstream.[5]

Lake hydrologic unit A large lake, reservoir, or playa that may have flow from adjacent frontal, composite, and classic hydrologic units.[4]

Layer In GIS, refers to the various *overlays* of data, each of which normally deals with one thematic topic.[3]

Leaching The removal of materials in solution from soil or rock to ground-water; refers to movement of pesticides or nutrients from land surface to groundwater.[7]

Length The distance, in the direction of flow, between two specific points along a river, stream, or channel.[5]

Letter of Map Amendment (LOMA) An official determination that a specific structure or property is not within the 1 percent annual chance flood-plain. A LOMA amends the effective FIRM.

Letter of Map Change (LOMC) A letter that revises FIRMs, both LOMAs and LOMRs.

Letter of Map Revision (LOMR) A letter that revises BFEs, flood hazard zones, floodplain boundaries, or floodways that are shown on an effective FIRM or FBFM.

Levee (dike) A long, narrow embankment usually built to protect land from flooding.[5]

Limestone A sedimentary rock consisting largely of calcium carbonate.[1]

Lithification The process of changing unconsolidated sediment into solid rock.[1]

Lithologic log Description of the geologic material collected during the drilling of wells.[1]

Load General term for a material or constituent in solution, in suspension, or in transport; usually expressed in terms of mass or volume.[7]

Local flooding Flooding conditions over a relatively limited (localized) area.[5]

Loess Homogeneous, fine-grained sediment made up primarily of silt and clay, and deposited over a wide area (probably by wind).[7]

Long-term monitoring Data collection over a period of years or decades to assess changes in selected hydrologic conditions.[7]

Low-flow measurement A stream-discharge measurement taken when flow is near minimum; consists primarily of base flow.

Lowland flooding Inundation of low areas near the river, often rural, but may also occur in urban areas.[5]

Macro A series of instructions combined to be executed with a single command.

Main stem The principal course or reach of a river or a stream formed by the tributaries that flow into it.[5,7]

Major flooding A general term indicating extensive inundation and property damage.

Major ions Constituents commonly present in concentrations exceeding 1.0 milligram per liter. Dissolved cations generally are calcium, magnesium, sodium, and potassium; the major anions are sulfate, chloride, fluoride, nitrate, and those contributing to alkalinity, most generally assumed to be bicarbonate and carbonate.[7]

MapInfo A desktop mapping system, from MapInfo Corporation, Troy, New York, that combines map graphics and a relational database for cartographic display and spatial analysis.[3]

Maximum contaminant level (MCL) Maximum permissible level of a contaminant in water that is delivered to any user of a public water system. MCLs are enforceable standards established by the U.S. Environmental Protection Agency.[7]

Mean The average of a set of observations, unless otherwise specified.[7]

Mean discharge (MEAN) The arithmetic mean of individual daily mean discharges during a specific period, usually daily, monthly, or annually.[7]

Median The middle or central value in a distribution of data ranked in order of magnitude. The median is also known as the fiftieth percentile.[7]

Menu A list of options on a screen display or pallet from which an operator can select the next operations by indicating one or more choices with a pointing device.[3]

Merge In a GIS, to combine items from two or more similarly ordered sets into one set that is arranged in the same order.[3]

Mesozoic A geologic era that is popularly referred to as the "Age of Dinosaurs," extending from approximately 240 to 66 million years before the present.[1]

Metabolite A substance produced in or by biological processes.[7]

Metadata   Descriptive information about the contents of a digital geospatial data file; often defined as "data about data."[4]

Metamorphic rock   Rock that has formed in the solid state in response to pronounced changes of temperature, pressure, and chemical environment.[1,7]

Method detection limit   The minimum concentration of a substance that can be accurately identified and measured with present laboratory technologies.[7]

Micrograms per liter (μg/L)   A unit expressing the concentration of constituents in solution as weight (micrograms) of solute per unit volume (liter) of water; equivalent to one part per billion in most stream water and groundwater.[7]

MicroStation   A software package from Bentley Systems, Inc., Exton, Pennsylvania, that provides a menu and key-in operator interface with commands for generating and editing graphics and data.

Midge   A small fly in the family *Chironomidae*. The larval (juvenile) life stages are aquatic.[7]

Milligram (mg)   A mass equal to 10^{-3} grams.[7]

Milligrams per liter (mg/L)   A unit expressing the concentration of chemical constituents in solution as weight (milligrams) of solute per unit volume (liter) of water.[7]

Minimum reporting level (MRL)   The smallest measured concentration of a constituent that may be reliably reported using a given analytical method. In many cases, the MRL is used when documentation for the method detection limit is not available.[7]

Minor flooding   A general term used to describe minimal or no property damage but possibly some public inconvenience.[5]

Moderate flooding   Flooding level marked by the inundation of secondary roads, making transfer to higher elevations necessary to save property; some evacuation may be required as well.[5]

Monitoring   Repeated observation or sampling at a site, on a scheduled or event basis, for a particular purpose.[7]

Monitoring well   A well used for repeated water sampling or for measurements of groundwater levels.[1,7]

Monocyclic aromatic hydrocarbons   Constituents of lead-free gasoline; also used in the manufacture of monomers and plasticizers in polymers.[7]

Moraine   A mound, ridge, or other distinct accumulation of unsorted, unstratified glacial drift, predominantly till, deposited chiefly by direct action of glacier ice, in a variety of topographic landforms that are independent of control by the surface on which the drift lies.[2]

Mouth   The place where a stream discharges to a larger stream, a lake, or the sea.[7]

National Academy of Sciences/National Academy of Engineering (NAS/NAE)   Numerical guidelines recommended by two joint NAS/NAE committees for the protection of freshwater and marine aquatic life, respectively.[7]

National Elevation Dataset (NED)   Dataset developed by merging the highest-resolution, best-quality elevation data available across the United States into a seamless raster format.[4]

National Flood Insurance Program (NFIP)   The federal regulatory program under which flood-prone areas are identified and flood insurance is made available to residents of participating communities.[3]

National Hydrography Dataset (NHD)   Combination of the US-EPA River Reach files and USGS Digital Line Graph (DLG) hydrography geodata files.[4]

National Map Accuracy Standards (NMAS)   Specifications governing the accuracy of topographic, base, orthophoto, and other maps produced by federal agencies.[4]

National Pollution Discharge Elimination System (NPDES)   The national program for issuing, modifying, revoking and revising, terminating, monitoring and enforcing permits, and imposing and enforcing pretreatment requirements, under Sections 307, 318, 402, and 495 of the Clean Water Act.[1]

Neatline   The lines on a map serving as the outside border or limits of the map sheet.[4]

Needle-leaved evergreen   Woody gymnosperms with green, needle-shaped, or scalelike leaves (e.g., black spruce) that are retained by plants throughout the year.[2]

Nitrate   An ion consisting of nitrogen and oxygen (NO_3^-). Nitrate is a plant nutrient and is very mobile in soils.[7]

Node   In the GIS, a point at which two or more lines meet; called an *edge* or *vertex* in graph theory.

Noncontact water recreation   Recreational activities, such as fishing or boating, that do not include direct contact with the water.[7]

Noncontributing area   An area within a hydrologic unit that normally does not contribute directly to the surface runoff to the river or stream at the outlet of the hydrologic unit.[4]

Nonpoint source   A pollution source that is distributed over an area rather than limited to an identifiable point.[1,7]

Nonpoint-source water pollution   Water contamination that originates from a broad area and enters the water resource diffusely over a large area.[7]

Nonselective herbicide   Kills or significantly retards growth of most higher plant species.[7]

Nutrient   Element or compound essential for animal and plant growth, such as nitrogen, phosphorus, and potassium.[7]

Observation well   A well used for repeated water-level measurements.[1]

Occurrence and distribution assessment   Characterization of the broad-scale spatial and temporal distributions of water quality conditions in relation to major contaminant sources and background conditions for surface water and groundwater.[7]

Open-water flow channel   The deeper portion of an estuary or sound that is large enough to be defined as a separate hydrologic unit (HU).[4]

Open-water hydrologic unit   A water-based hydrologic unit (HU) existing within a large open body of water.[4]

Open-water mixing basin   A relatively broad portion of an estuary or sound confined by shoals and/or coastal land hydrologic units (HUs) that has a long retention time for inflows.[4]

Operating system (OS)   The master control program that governs the operation of a computer system, running job entry, input/output services, data management, and supervision or housekeeping.

Ordinance   A law or rule made by a government or authority.[1]

Organic detritus   Any loose organic material in streams, such as leaves, bark, or twigs removed and transported by mechanical means, such as disintegration or abrasion.[7]

Organochlorine compounds   Synthetic organic compounds containing chlorine.[7]

Organochlorine insecticide   A class of organic insecticides containing a high percentage of chlorine, which includes dichlorodiphenylethanes (such as DDT), that tends to bioaccumulate in organism tissue.[7]

Organonitrogen herbicides   A group of herbicides consisting of a nitrogen ring with associated functional groups; includes such classes as triazines and acetanilides.[7]

Organophosphorus insecticides   Insecticides derived from phosphoric acid. Generally, these are the most toxic of all pesticides to vertebrate animals.[7]

Outwash   Soil material washed down a hillside by rainwater and deposited upon more gently sloping land.[7]

Overland flow   The part of surface runoff flowing over land surfaces toward stream channels.[7]

Paleozoic   The era of geologic time preceding the Age of Dinosaurs, extending from about 570 to 240 million years before the present.[1]

Part per million (ppm)   Unit of concentration equal to 1 milligram per kilogram or 1 milligram per liter.[7]

Peak discharge   The highest rate of flow passing a given location, usually in cubic feet per second (CFS).[5]

Perennial stream   A stream that normally has water in its channel at all times.[7]

Periphyton   Organisms that grow on underwater surfaces, including algae, bacteria, fungi, protozoa, and other organisms.[7]

Permeability   The capability of a rock or sediment to permit the flow of fluids through its pore spaces.[1]

Pesticide   A chemical applied to crops, rights-of-way, lawns, or residences to control weeds, insects, fungi, nematodes, rodents, or other "pests."

pH   The logarithm of the reciprocal of the hydrogen ion concentration (activity) of a solution; a measure of the acidity (pH less than 7) or alkalinity (pH greater than 7) of a solution; a pH of 7 is neutral.[7]

Phenols   A class of organic compounds containing phenol (C_6H_5OH) and its derivatives used to make resins and weed killers, and as a solvent, disinfectant, and chemical intermediate.[7]

Phosphorus   A nutrient essential for growth and that can play a key role in stimulating aquatic growth in lakes and streams.[7]

Photosynthesis   Synthesis of chemical compounds by organisms with the aid of light. Carbon dioxide is used as raw material for photosynthesis, and oxygen is a product.[7]

Physiographic province   A region delineated on the basis of its typical landforms that are usually the result of having underlying rock types, geologic history, and climate in common.

Physiography   The natural surface features of the earth, its landforms, and its drainage features, or the study of them.[1,7]

Phytoplankton   Floating or weakly swimming plants that are at the mercy of the waves and currents.[7]

Pixel   In the GIS, the smallest discrete element that makes up an image. Short for "picture element."[6]

Planimetric map   Map representing only horizontal positions from features represented.[3]

Plankton   Floating or weakly swimming organisms that are at the mercy of the waves and currents.[7]

Plant association   A potential natural plant community of definite floristic composition and uniform appearance.[2]

Plant community   A group of one or more populations of plants in a common spatial arrangement.[2]

Playa   A term used in the southwestern United States for a dry, vegetation-free, flat area at the lowest part of an undrained desert basin, underlain by stratified clay, silt, or sand, and commonly by soluble salts.[2]

Point   In the GIS, a level of spatial measurement referring to an object that has no dimension.[3]

Point data   In a vector structure, data consisting of a single, distinct X,Y coordinate pair. In a raster structure, point data are represented by single cells.[3]

Point source   A source at a discrete location such as a discharge pipe, drainage ditch, tunnel, well, concentrated livestock operation, or floating craft.[7]

Point-source contaminant Any substance that degrades water quality and originates from discrete locations such as discharge pipes, drainage ditches, wells, concentrated livestock operations, or floating craft.[7]

Pollutant Any substance that, when present in a hydrologic system at sufficient concentration, degrades water quality in ways that are or could become harmful to human and/or ecological health or that impair the use of water for recreation, agriculture, industry, commerce, or domestic purposes.[7]

Polychlorinated biphenyls (PCBs) A mixture of chlorinated derivatives of biphenyl, marketed under the trade name Aroclor, with a number designating the chlorine content (such as Aroclor 1260). Sale of PCBs for new use was banned by law in 1979.[7]

Polygon A closed loop of X,Y coordinate pairs defining the boundary of an area.[3,4]

Pool A small part of the stream reach with little velocity, commonly with water deeper than surrounding areas.[7]

Porosity In rock or rock material, the ratio of pore space to the volume of the entire mass.[1]

Potability The quality of water making it fit for drinking.[1]

Potential natural community The biotic community that would be established if all successional sequences of its ecosystem were completed without additional human-caused disturbance under present environmental conditions.[2]

Potential natural vegetation The vegetation that would exist today if humans were removed from the scene and if the plant succession after their removal were telescoped into a single moment.[2]

Potentiometric gradient The rate of change in hydraulic head in a given direction.[1]

Potentiometric surface An imaginary surface representing the static head of groundwater in tightly cased wells that tap a water-bearing rock unit (aquifer); or, in the case of unconfined aquifers, the water table.[1]

Pothole A shallow depression, generally less than 10 acres in area, occurring between dunes on a prairie, often containing an intermittent pond or marsh and serving as a nesting place for waterfowl.[2]

Precipitation Any or all forms of water particles that fall from the atmosphere, such as rain, snow, hail, and sleet.[7]

Preemergence herbicide Herbicide applied to bare ground after planting the crop but prior to the crop sprouting aboveground to kill or significantly retard the growth of weed seedlings.[7]

Pressure The force over a unit area.[6]

Priority Determines the order of rank of the rights to use water in a system under the Appropriation Doctrine.[6]

Projection The mathematical transformation of three-dimensional space into a two-dimensional (flat) surface.[3]

Province An ecological unit in the ecoregion planning and analysis scale of the National Hierarchical Framework corresponding to subdivisions of a division that conform to climatic subzones controlled mainly by continental weather patterns.[4]

Public-supply withdrawals Water withdrawn by public and private water suppliers for use within a general community.[7]

Q3 Flood Data A digital FIRM product developed and distributed by FEMA.[3]

Quad (also USGS Quad) A U.S. Geological Survey (USGS) topographic map or "quadrangle."[3]

Quantitative precipitation forecast (QPF) A spatial and temporal forecast that predicts the amount of future precipitation for a specific time period and area or region.

Radon A naturally occurring, colorless, odorless, radioactive gas formed by the disintegration of the element radium; damaging to human lungs when inhaled.[7]

Raster data The representation of spatial data as regular grid cells.[3]

Rating curve A graph showing the relationship between the river stage, usually plotted vertically (Y-axis), and the discharge, plotted horizontally (X-axis).[5]

Rating table A table of stage values and the corresponding discharge for a river gaging site.[5]

Reach A portion of a river between two points to which the forecast applies.[5]

Read-only memory (ROM) A microcircuit containing programs or data that cannot be erased.[3]

Recharge Water that infiltrates the ground and reaches the saturated zone.[1,7]

Recharge area An area in which there are downward components of hydraulic head (pressure) in an aquifer.[1]

Recorder, stream data A mechanical or electrical apparatus that records a continuous record of water level.[6]

Reference gage Usually, a nonrecording gage; for example, a staff gage used for checking and resetting the gage height indicated on the stream data recorder.[6]

Reference site A sampling site selected for its relatively undisturbed conditions.[7]

Region Regions are the largest drainage basins, containing either the drainage area of a major river or the combined drainage areas of several rivers.[4]

Regionalization A mapping procedure in which a set of criteria are used to subdivide the earth's surface into smaller, more homogeneous units that display spatial patterns related to ecosystem structure, composition, and function.[2]

Relative abundance The number of organisms of a particular kind present in a sample relative to the total number of organisms in the sample.[7]

Remnant areas Areas typically formed as residual areas after delineation of classic watersheds.[4]

Reservoir A man-made facility for the storage, regulation, and controlled release of water.[5]

Response time The amount of time it will take a watershed to react to a given rainfall event.[5]

Retrospective analysis Review and analysis of existing data in order to address the National Water Quality Assessment program (NAWQA) objectives, to the extent possible, and to aid in the design of NAWQA studies.[7]

Riffle A shallow reach in a stream that produces a stretch with increased velocity and ruffled or choppy water.[6,7]

Riparian Describes areas adjacent to rivers and streams with a high density, diversity, and productivity of plant and animal species relative to nearby uplands.[7]

Riparian Doctrine The doctrine granting all riparian owners the right to make reasonable use of the water as long as the water use does not interfere with the reasonable use of water by other riparian users. Used in most eastern states, some midwestern and southern states, and in California (which also uses the Appropriation Doctrine).[6]

Riparian zone Pertaining to or located on the bank of a body of water, especially a stream.[7]

River basin Drainage area of a river and its tributaries.[5]

River forecast Stage and/or flow forecasts for specific locations based on existing and forecasted hydrometeorologic conditions.[5]

River forecast center (RFC) A center that serves groups of National Weather Service forecast offices in providing hydrologic guidance; a first-echelon office for the preparation of river and flood forecasts and warnings.

River gage A device for measuring the river stage or level.[5]

River system Comprises all of the streams, creeks, and channels draining a river basin.[5]

Routing Method of predicting the attenuation of a flood wave as it moves down the course of a river.[5]

Runoff (total) That part of the precipitation that flows toward the streams on the surface of the ground or within the ground. Runoff is composed of base flow and surface runoff.[1,5]

Sandstone A sedimentary rock consisting of sand, usually quartz, cemented by various substances such as calcium carbonate, clay, iron compounds, or silica.[1]

Saturated zone That part of a water-bearing material in which all voids, large and small, are ideally filled with water under pressure greater than atmospheric.[1]

Scale A representative fraction of a paper map distance to ground distance.[3]

Scanner Any device that systematically decomposes a sensed image or scene into pixels and then records some attribute of each pixel.[3]

Scanning Process of using an electronic input device to convert analog information such as maps, photographs, overlays, and so on into a digital format usable by a computer.[3]

Secondary Maximum Contaminant Level (SMCL) The maximum contamination level in public water systems that, in the judgment of the U.S. Environmental Protection Agency (U.S. EPA), is required to protect the public welfare.[7]

Section An ecological unit in the subregion planning and analysis scale of the National Hierarchical Framework corresponding to subdivisions of a province and having broad areas of similar geomorphic process, stratigraphy, geologic origin, drainage networks, topography, and regional climate.[2]

Sediment Particles, derived from rocks or biological materials, that have been transported by a fluid or other natural process, suspended or settled in water.[7]

Sediment guideline Threshold concentration above which there is a high probability of adverse effects on aquatic life from sediment contamination, as determined using modified U.S. EPA (1996) procedures.[7]

Sedimentary Term applied to rocks that were formed by the deposition of sediment.[1]

Selective herbicide A herbicide that kills or significantly retards growth of an unwanted plant species without significantly damaging desired plant species.[7]

Semivolatile organic compound (SVOC) Operationally defined as a group of synthetic organic compounds that are solvent-extractable; includes phenols, phthalates, and polycyclic aromatic hydrocarbons (PAHs).[7]

Sewage The spent water of a community. Synonym for *wastewater*.[1]

Shale Rock of laminated structure formed from the consolidation of clay or other very fine material.[1]

Shoal An underwater permanent geomorphic feature, usually a sandbank or sandbar, that makes the water become shallow and impedes or directs flow.[4]

Sideslope gradient The representative change in elevation in a given horizontal distance perpendicular to a stream.[7]

Siltstone A sedimentary rock composed of particles finer than sand and coarser than clay, cemented together.[1]

Sinuosity   The ratio of the channel length between two points on a channel to the straight-line distance between the same two points; a measure of meandering.[7]

Small-stream flooding   Flooding of small creeks, streams, or runs.[5]

Snow cover   The accumulated snow and ice on the surface of the ground at any time.[1]

Snow storage   Storage of water in the hydrologic cycle in the form of snow and ice.

Snow water equivalent (SWE)   The water content of melted snow. On average, 1 inch of rain would be equivalent to 10 inches of snow.

Snowmelt flooding   Flooding caused primarily by the melting of snow.

Soil moisture   Water contained in the upper levels of the soil above the water table.[5]

Sole-source aquifer   A groundwater system that supplies at least 50 percent of the drinking water to a particular human population.[7]

Solution   Formed when a solid, gas, or another liquid in contact with a liquid becomes dispersed homogeneously throughout the liquid.[7]

Sorption   General term for the interaction (binding or association) of a solute ion or molecule with a solid.[7]

Source rocks   The rocks from which fragments and other detached pieces have been derived to form a different rock.[7]

Source water protection   The protection of water in its natural state, prior to any treatment for drinking.[1]

Spatial Data Transfer Standards (SDTS)   An FIPS standard (FIPS 173) developed as a mechanism for the transfer of digital spatial data between different computer systems.[3]

Special flood hazard area (SFHA)   Any area inundated by the base (1 percent annual chance) flood. These areas are identified on the FIRM as Zones A, AE, AH, AO, AR, A1-30, A99, V, VE, and V1-30.[3]

Species   Populations of organisms that may interbreed and produce fertile offspring having similar structure, habits, and functions.[7]

Species diversity   An ecological concept that incorporates both the number of species in a particular sampling area and the evenness with which individuals are distributed among the various species.[7]

Species (taxa) richness   The number of species (taxa) present in a defined area or sampling unit.[7]

Specific conductance   A measure of the capability of a liquid to conduct an electrical current.[7]

Spillway   A structure over or through which excess or flood flows are discharged.[5]

Spillway crest   The elevation of the highest point of a spillway.[5]

Split sample A sample prepared by dividing it into two or more equal volumes, where each volume is considered a separate sample but representative of the entire sample.[7]

Spring A concentrated discharge of groundwater coming out at the surface as flowing water.[6]

Staff gage A vertical staff graduated in appropriate units that is placed so that a portion of the gage is in the water at all times.[5]

Stage The height of the water surface above an established datum plane at a given location.[3,6,7]

Statistics A branch of mathematics dealing with the collection, analysis, interpretation, and presentation of masses of numerical data.[7]

Steady flow The flow that occurs when, at any point in a flow system, the magnitude and direction of the specific discharge are constant in time.[1]

Steady state Equilibrium water levels or heads; aquifer storage and water levels do not vary with time.[1]

Stilling well A container that houses a float that is coupled to a stream data recorder in order to obtain water stage data that is not affected by surges or choppy water that may occur in the channel.[6]

Stormwater (or storm water) Stormwater runoff, snowmelt runoff, and surface runoff and drainage.[1]

Stream-aquifer interactions Relations of water flow and chemistry between streams and aquifers that are hydraulically connected.[7]

Stream gaging A process of determining the rate of flow (i.e., the discharge) of streams.[6]

Stream mile A distance of 1 mile along a line connecting the midpoints of the channel of a stream.[7]

Stream order A ranking of the relative sizes of streams within a watershed based on the number of their tributaries.[7]

Stream reach A continuous part of a stream between two specified points.[7]

Streambed conductance The property of a reach of stream (or river) that describes the capability to transmit or receive water from underlying sediments.[1]

Stream flow That part of surface runoff in a stream. Often used interchangeably with *discharge*.[5]

Stream gaging The process of measuring the flow of a stream.[1]

Study area The region under investigation.

Study unit A major hydrologic system of the United States in which National Water Quality Assessment (NAWQA) studies are focused.[7]

Study-unit survey Broad assessment of the water quality conditions of the major aquifer systems of each study unit.[7]

Subbasin *See* Subwatershed.

Subenvelope   The general altitude of the drainage network that portrays differences in stream gradient from one geomorphic unit to another.[2]

Submerged flow   When the resistance to flow due to increased discharge becomes sufficient to reduce the velocity, increase the flow depth, and cause a backwater effect.[6]

Subregion   A scale of planning and analysis in the National Hierarchical Framework that has applicability for strategic, multiforest, statewide, and multiagency analysis and assessment. Subregions include section and subsection ecological units.[2]

Subsection   An ecological unit in the subregion planning and analysis scale of the National Hierarchical Framework corresponding to subdivisions of a section into areas with similar surficial geology, lithology, geomorphic process, soil groups, subregional climate, and potential natural communities.[2]

Subsidence   Compression of soft aquifer materials in a confined aquifer due to pumping of water from the aquifer.[7]

Substrate size   The diameter of streambed particles such as clay, silt, sand, gravel, cobble, and boulders.[7]

Subsurface drain   A shallow drain installed in an irrigated field to intercept the rising groundwater level and maintain the water table at an acceptable depth below the land surface.[7]

Subwatershed   A drainage area or small basin located within a larger watershed.[1]

Superfund site   A contamination site designated for cleanup by the EPA under provisions of a special fund established by Congress.[1]

Surface runoff   The runoff that travels overland to the stream channel.[5]

Surface water   Water above the ground surface that may infiltrate into the ground, run off into watercourses, or be stored on the surface.[1,7]

Survey   Sampling of any number of sites during a given hydrologic condition.[7]

Suspended   The amount (concentration) of undissolved material in a water-sediment mixture.[7]

Suspended sediment   Particles of rock, sand, soil, and organic detritus carried in suspension in the water column, in contrast to sediment that moves on or near the streambed.[7]

Suspended-sediment concentration   The velocity-weighted concentration of suspended sediment in the sampled zone.[7]

Suspended solids   Different from *suspended sediment* only in the way that the sample is collected and analyzed.[7]

Sustainable development   Development with the goal of preserving environmental quality, natural resources, and livability for present and future generations.[1]

Synoptic sites Sites sampled during a short-term investigation of specific water quality conditions during selected seasonal or hydrologic conditions to provide improved spatial resolution for critical water quality conditions.[7]

Tagged Image File Format (TIFF) A standard exchange format for raster or image files.

Taxon (plural taxa) Any identifiable group of taxonomically related organisms.[7]

Tertiary-treated sewage The third phase of treating sewage that removes nitrogen and phosphorus before it is discharged.[7]

Thematic layer A data layer containing selected information relating to a specific theme, such as soils, vegetation, land use, and so on.[3]

Tier 1 sediment guideline Threshold concentration above which there is a high probability of adverse effects on aquatic life from sediment contamination, determined using modified U.S. EPA (1996) procedures.[7]

Tile drain A buried perforated pipe designed to remove excess water from soils.[7]

Tissue study The assessment of concentrations and distributions of trace elements and certain organic contaminants in tissues of aquatic organisms.[7]

Toe of the shore face The depth to which seasonal storms and prevailing winds and resultant waves and currents move shallow sediments to and from the shore.

Tolerant species Species adaptable to (tolerant of) human alterations to the environment and that often increase in number when human alterations occur.[7]

Topographic quadrangle A USGS lithographic printed map that represents the horizontal and vertical positions of cultural, political, and land features, usually at scales of 1:100,000 or 1:24,000.[4]

Topologically Integrated Geographic Encoding and Referencing File (TIGER) The nationwide digital database of planimetric base map features developed by the U.S. Bureau of the Census for the 1990 census.[3]

Topology The spatial relationships between connecting and adjacent features.[3,4]

Total concentration The concentration of a constituent regardless of its form (dissolved or bound) in a sample.[7]

Total DDT The sum of DDT and its metabolites (breakdown products), including DDD and DDE.[7]

Trace element An element found in only minor amounts (concentrations less than 1.0 milligram per liter) in water or sediment; includes arsenic, cadmium, chromium, copper, lead, mercury, nickel, and zinc.[7]

Tracer A stable, easily detected substance or a radioisotope added to a material to follow the location of the substance in the environment or to detect any physical or chemical changes it undergoes.[7]

Transducer A device that converts electrical energy into pressure pulses, or the converse.[6]

Transmissivity The rate at which water is transmitted through a unit width of aquifer under unit hydraulic gradient. Transmissivity is equal to the product of hydraulic conductivity and saturated thickness of the aquifer, expressed in units of feet squared per day.[1]

Transpiration Water discharged into the atmosphere from plant surfaces.[5]

Tributary A river or stream flowing into a larger river, stream, or lake.[7]

Turbidity Reduced clarity of surface water because of suspended particles, usually sediment.[7]

Unconfined aquifer An aquifer whose upper surface is a water table; an aquifer containing unconfined groundwater.[7]

Unconsolidated deposit Deposit of loosely bound sediment that typically fills topographically low areas.[7]

Un-ionized The neutral form of an ionizable compound (such as an acid or a base).[7]

Un-ionized ammonia The neutral form of ammonia-nitrogen in water, usually occurring as NH_4OH.[7]

Unit hydrograph (or unitgraph) The discharge hydrograph from 1 inch of surface runoff distributed uniformly over the entire basin for a given time.[5]

Universal Transverse Mercator (UTM) grid A system of plane coordinates based on 60 north–south trending zones, each 6 degrees of longitude wide, that circle the globe.

Unsaturated zone The zone between the land surface and the water table.[1]

Up-gradient Of or pertaining to the place(s) from which groundwater originated or traveled through before reaching a given point in an aquifer.[7]

Upland Elevated land above low areas along a stream or between hills; elevated region from which rivers gather drainage.[7]

Urban flooding Flooding of streets, underpasses, low-lying areas, or storm drains.[5]

Urban site A site that has greater than 50 percent urbanized and less than 25 percent agricultural area.[7]

Vector data A coordinate-based data structure used to represent geographic features in line, point, and polygon.[3,4]

Vector digitizing The process of using a digitizer to encode the locations of geographic features by converting their map positions to a series of X,Y coordinates.[4]

Vested right An appropriative right established by actual use of water prior to enactment of a state water right permit system.

Volatile organic compounds (VOCs) Organic chemicals that have a high vapor pressure relative to their water solubility, including components of gasoline, fuel oils, and lubricants, as well as organic solvents, fumigants, some inert ingredients in pesticides, and some by-products of chlorine disinfection.[7]

Wastewater The spent water of a community. Synonym for *sewage*.[1]

Wasteway A waterway used to drain excess irrigation water dumped from the irrigation delivery system.[7]

Water balance A quantitative statement of the amounts of water circulating through various paths of the hydrologic cycle.[1]

Water budget An accounting of the inflow, outflow, and storage changes of water in a hydrologic unit.[1,7]

Water column studies Investigations of physical and chemical characteristics of surface water, which include suspended sediment, dissolved solids, major ions, and metals, nutrients, organic carbon, and dissolved pesticides, in relation to hydrologic conditions, sources, and transport.[7]

Water quality criteria Specific levels of water quality which, if reached, are expected to render a body of water unsuitable for its designated use.[7]

Water quality guidelines Specific levels of water quality which, if reached, may adversely affect human health or aquatic life.[7]

Water quality standards State-adopted and U.S. Environmental Protection Agency-approved ambient standards for water bodies where the water quality criteria must be met to protect the designated use or uses.[7]

Water table The surface of a groundwater body at which the water pressure equals atmosphere pressure and is very nearly equivalent to the upper surface of the saturated zone.[1]

Water table map A specific type of potentiometric-surface map for an unconfined aquifer that shows lines of equal elevation of the water table.[1]

Water year The time period from October 1 through September 30.[5]

Watershed The region or land area that contributes to the drainage or catchment area above a specific point on a stream or river.[1,5]

Watershed Boundary Dataset National geospatial database containing the hydrologic unit boundaries for the first- through sixth-level units.[4]

Weathering The processes by which various natural agents, such as wind, ice, and water, act on rock at or near the land surface to cause it to disintegrate.[1]

Weir A structure placed in a canal, stream, or ditch to measure the rate of flow of water.[6]

Wellhead protection area (WHPA) The surface and subsurface area surrounding a water well or well field, supplying a public water system, through which contaminants are reasonably likely to move toward and reach such water well or well field.[1]

Well log A record of lithologic data with respect to depth in a well.[1]

Wetlands Ecosystems whose soil is saturated for long periods seasonally or continuously and supports hydrophytic vegetation.[7]

Withdrawal   The act or process of removing; such as removing water from a stream for irrigation or public water supply.[7]

Yield   The mass of material or constituent transported by a river in a specified period of time divided by the drainage area of the river basin.[7]

Zero datum   A reference "zero" elevation for a stream or river gage.[5]

Zoning   Classification of land in a community into different areas and districts to promote sound land use practices.[1]

Zooplankton   Floating or weakly swimming animals that are at the mercy of the waves and currents.[7]

1. Bucks County Planning Commission (BCPC). "Pennridge Water Resources Plan, BCPC" (Bucks County, PA: BCPC), July 2002.
2. ECOMAP. "National Hierarchical Framework of Ecological Units," unpublished administrative paper (Washington, DC: U.S. Department of Agriculture, Forest Service), 1993.
3. Federal Emergency Management Agency (FEMA). Map Modernization Guidelines and Specifications for Hazard Mapping Partners. Washington, DC, http://www.fema.gov/pdf/fhm/frm-gsgl.pdf, accessed April 2003.
4. Federal Geographic Data Committee (FGDC). FGDC Proposal Version 1.0, Standards for Delineating Hydrologic Unit Boundaries, Washington, DC, ftp://ftp-fc.sc.egov.usda.gov/NCGC/products/watershed/hu-standards.doc, accessed March 1, 2002.
5. National Weather Service, NOAA. IFLOWS—Glossary of Hydrologic Related Terms, Washington, DC, http://docs.afwsinet/supportsite/iflows/manual/Glossary.htm, accessed June 13, 2003.
6. U.S. Fish and Wildlife Service (U.S./WS). Water Rights Definitions, Denver, CO, http://mountain-prairie.fws.gov/wtr/water_rights_def.htm, accessed June 10, 2003.
7. U.S. Geological Survey (USGS). National Water Quality Assessment Glossary, Washington, DC, http://usgs.gov/nawga/glos.html.

ADDITIONAL REFERENCES

Alert News. Quarterly Newsletter of the Alert Users Group, Connecticut Department of Environmental Protection, vol. IV, no. 4, Fall 1996.

Bailey, R.G. "Description of the Ecoregions of the United States," Miscellaneous Publication 1391 (Washington, DC: U.S. Department of Agriculture), 1980.

———. "Description of the Ecoregions of the United States," 2d ed. Miscellaneous Publication 1391, Washington, DC: U.S. Department of Agriculture Forest Service, in process.

Bailey, Robert G., Peter E. Avers, Thomas King, and W. Henry McNab (eds.). "Ecoregions and Subregions of the United States" (Washington, DC: U.S. Geological

Survey), 1994. Map: Scale 1:7,500,000; colored. Accompanied by a supplementary table of map unit descriptions compiled and edited by W. Henry McNab and Robert G. Bailey. Prepared for the U.S. Department of Agriculture, Forest Service.

Bates, Robert L., and Julia A. Jackson (eds.). *Glossary of Geology* (Falls Church, VA: American Geological Institute), 1980.

Blair, Thomas A. *Climatology, General and Regional* (New York: Prentice-Hall, Inc.), 1942.

Cowardin, L.M., Virginia Carter, F.C. Golet, and E.T. LaRoe. "Classification of wetlands and deepwater habitats of the United States: U.S. Fish and Wildlife Service Report," FWS/OBS-79/31, 131 p. 1979.

Ecoregions of the United States, map (revised edition); scale 1:7,500,000; colored; Robert G. Bailey, cartographer (Washington, DC: U.S. Department of Agriculture, Forest Service), 1994.

Linsley, Ray K. Jr., Max A. Kohler, and Joseph L.H. Paulhaus. *Applied Hydrology* (New York: McGraw-Hill Book Company, Inc.), 1949.

National Weather Service (NWS) Central Region, Hydrologic Services Division. "Helpful Information on Hydrologic Operations," Flood-Frequency Analysis, *Manual of Hydrology:* Part 3, "Flood-Flow Techniques" (Kansas City, MO: Geological Survey Water-Supply Paper 1543-A), October 1996.

Ostrowski, Joe. Flood Hydrographs for the Saddle River at Lodi, New Jersey, Middle Atlantic River Forecast Center, State College, PA, February 1997.

Pennsylvania Emergency Management Agency. "Flash Flood Handbook," Harrisburg, PA, August 1992.

U.S. Environmental Protection Agency (EPA). "The National Sediment Quality Survey: A Report to Congress on the Extent and Severity of Sediment Contamination in Surface Waters of the United States," U.S. Environmental Protection Agency, Office of Science and Technology, Draft Report EPA 823-D-96-002, 1996.

INDEX